Women in Artificial Intelligence (AI)

Women in Artificial Intelligence (AI)

Editors

Aida Valls
Karina Gibert

MDPI • Basel • Beijing • Wuhan • Barcelona • Belgrade • Manchester • Tokyo • Cluj • Tianjin

Editors
Aida Valls
Universitat Rovira i
Virgili Tarragona
Spain

Karina Gibert
Universitat Politècnica de
Catalunya-BarcelonaTech (UPC)
Spain

Editorial Office
MDPI
St. Alban-Anlage 66
4052 Basel, Switzerland

This is a reprint of articles from the Special Issue published online in the open access journal *Applied Sciences* (ISSN 2076-3417) (available at: https://www.mdpi.com/journal/applsci/special_issues/Women_in_AI).

For citation purposes, cite each article independently as indicated on the article page online and as indicated below:

LastName, A.A.; LastName, B.B.; LastName, C.C. Article Title. *Journal Name* **Year**, *Volume Number*, Page Range.

ISBN 978-3-0365-5531-7 (Hbk)
ISBN 978-3-0365-5532-4 (PDF)

Cover image courtesy of Aida Valls

Contents

About the Editors

Aida Valls

Aida Valls has been an Associate Professor at the Department of Computer Science and Mathematics in Universitat Rovira i Virgili (URV) since 1998. She received a PhD in Artificial Intelligence from the Polytechnical University of Catalonia in 2002. She is the head of the ITAKA research group and is expert in the management of uncertainty, decision support systems and data mining. Her work is mainly focused on linguistic and semantic data. She has participated in several Spanish and EU research projects, with applications in Tourism, Environment Management and Health Care. She is the author of more than 150 papers in international journals and conferences. She was vice-president of the Catalan Association for Artificial Intelligence (2014–2018). She has been Head of the PhD Program in Computer Science and Mathematics of Security at URV (2012–2021) and Coordinator of Mobility of Computer Science at URV (2017–2022). She is now the secretary of the Management Team at the School of Engineering, URV.

Karina Gibert

Karina Gibert is a Full Professor at Universitat Politècnica de Catalunya-BarcelonaTech (UPC), with a teaching position at UPC from 1990. ORCID: 0000-0002-8542-3509, H-index: 29 (www.eio.upc.edu/en/homepages/karina, karina.gibert@upc.edu, Twitter: @karinagibertk). Bachelor and PhD in Informatics Engineering. Director and cofounder of IDEAI-UPC, (Jan 2018–). Vice-dean for Equity & Ethics of COEINF (June 2020–). Expert and co-writer of the Catalan Strategy for AI of the Catalan government, Catalonia.ai (from Oct 2018, Generalitat de Catalunya, (Feb 2018, 28th–), she participates in its current deployment.

She is interested in extracting strategic knowledge from data and ethical and explainable intelligent systems. Founded up to five gender working groups to bridge the gender gap in STEAM. WiDS (Standford) ambassador for Barcelona (2021–). Editor of JRC journal Environmental Modeling and Software, Elsevier, (Jan 2018–).

Awards: Ada Byron 2022, (ANoite). Honorific Mention of Creu Casas 2021 and 2022 (IEC), donaTIC 2018 (Gencat). Finalist at AMETIC 2021 and ESSA Awards 2021. First HackingBullipedia contest (nov 2013). Elected Fellow from iEMSs (Jul 2007).

Preface to "Women in Artificial Intelligence (AI)"

Artificial Intelligence (AI) research has expanded quickly in recent years due to the increase in data and resources, and companies' engagement in proposing many challenging applications. However, there is a lack of women working and conducting research in this field, as in other technological disciplines. This male bias influences the way that intelligent systems are conceived, designed, and developed. This may have a significant impact on the future world, where digital transformation seems to be intrinsically connected to the new Artificial Intelligence developments.

Even if they represent a small proportion of the sector, women conduct interesting research in the AI field. However, sometimes remain invisible for different reasons, thus increasing the lack of female referents for new generations and contributing to the imbalance. This Special Issue aims to contribute to the dissemination and promotion of the research done by women in AI. It contains an attractive compendium of 17 papers that represent multiple AI fields and applications. In all papers, the first author is a woman, which was a requirement for the submission of works in this journal. From these papers, the reader will discover women working on cutting-edge topics in AI, both from a theoretical and applied perspective.

This Special Issue was conceived during the 23rd International Conference of the Catalan Association of Artificial Intelligence (CCIA-2021), organized by the Catalan Association for Artificial Intelligence (ACIA). We want to thank to the ACIA board for its support in the dissemination of the call for papers, and the ACIA members for their contributions.

We would like to take this opportunity to express our gratitude to the MDPI Book staff, and the editorial team of the *Applied Sciences* journal, especially Ms. Christine Zhang, the assistant editor of this Special Issue, for her continuous and excellent support. We also want to thank all the leading female researchers and their research teams, who wanted to contribute to this Special Issue and sent us their high-quality works. Finally, we also want to thank the 52 international reviewers who helped us to select and improve the best papers.

We planned to issue this book on the on the Ada Lovelace Day (11/10/2022), an international day dedicated to the first computer programmer, a woman who had to fight the gender difficulties of her times, in the XIX century. We also thank the publisher for making this possible, thus allowing for this book to become a part of the international activities dedicated to celebrating the value of women in ICT worldwide. With this book, we want to pay homage to all the women that contributed over the years to the field of AI.

Aida Valls and Karina Gibert
Editors

applied sciences

MDPI

Editorial

Women in Artificial Intelligence

Aida Valls [1,2,*] and Karina Gibert [2,3,*]

1 Research Group on Intelligent Technologies for Advanced Knowledge Acquisition, Department of Computer Science and Mathematics, Universitat Rovira i Virgili, 43005 Tarragona, Spain
2 donesIAcat Working Group on Gender of the Catalan Association for Artificial Intelligence, 08193 Bellaterra, Spain
3 Research Group on Knowledge Engineering and Machine Learning at Intelligent Data Science and Artificial Intelligence Research Center, Universitat Politècnica de Catalunya, 08034 Barcelona, Spain
* Correspondence: aida.valls@urv.cat (A.V.); karina.gibert@upc.edu (K.G.)

Citation: Valls, A.; Gibert, K. Women in Artificial Intelligence. *Appl. Sci.* **2022**, *12*, 9639. https://doi.org/10.3390/app12199639

Received: 19 September 2022
Accepted: 19 September 2022
Published: 26 September 2022

Artificial Intelligence (AI) research has expanded very quickly in recent years due to the increase in data and resources, along with the engagement of companies in proposing many challenging applications. AI is a field at the intersection of Computer Science and Mathematics, with a high Engineering component. It belongs to the area of STEM: Science, Technology, Engineering, and Mathematics.

Several studies have demonstrated the gender bias existing in STEM, as well as in AI [1]. The number of male researchers working in STEM is much larger than that of women, and this is a constant in most countries all over the world. In the field of Artificial Intelligence, this male bias influences the way intelligent systems are conceived, designed, and developed. This may have a significant impact on the future world, where digital transformation seems to be intrinsically connected to new Artificial Intelligence developments.

Even though they represent a small proportion of the sector, women produce interesting research in the AI field. However, sometimes they remain invisible for different reasons, thus increasing the lack of female referents for new generations and contributing to perpetuation of the imbalance. For this reason, it is relevant to lead actions that help to make visible the valuable work of female researchers in the different STEM sectors, especially in AI, which is now having an increasing influence in the construction of the new digital society.

This Special Issue aims to contribute to this task by providing an attractive compendium of AI research led by women over a wide range of fields and applications. The volume contains 17 papers for which the first author is a woman. From those papers, the reader will discover women working on cutting-edge topics in AI, from both theoretical and applied points of view. Papers have been classified according to two dimensions.

The first classification regards the eight domains of AI defined by AIWatch 2021 [2]. The distribution can be seen in Figure 1. It is worth nothing that the Special Issue includes some papers from each one of these eight main areas.

The 17 works constituting the Special Issue also show wide representation of different application fields. We have taken the taxonomy of economic sectors proposed in the AIWatch 2021 report [2] (see Figure 2). The taxonomy covers 16 sectors, one of them being "M. Other technical and/or scientific sectors". We considered "Gender" as part of this category.

We find the majority of papers in two domains: Ethics and Philosophy (five papers) and Learning (five papers). Let us begin by commenting about **Ethics and Philosophy**; this is particularly interesting because three papers analyse different aspects related to gender bias, which is an uncommon topic in scientific publications. Such a group of papers providing different perspectives in this field is a relevant contribution of this Special Issue.

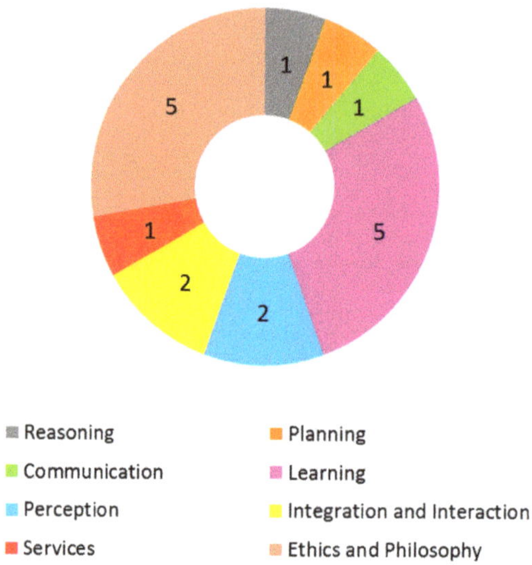

Figure 1. Distribution of the papers in the Special Issue by AI domain of research.

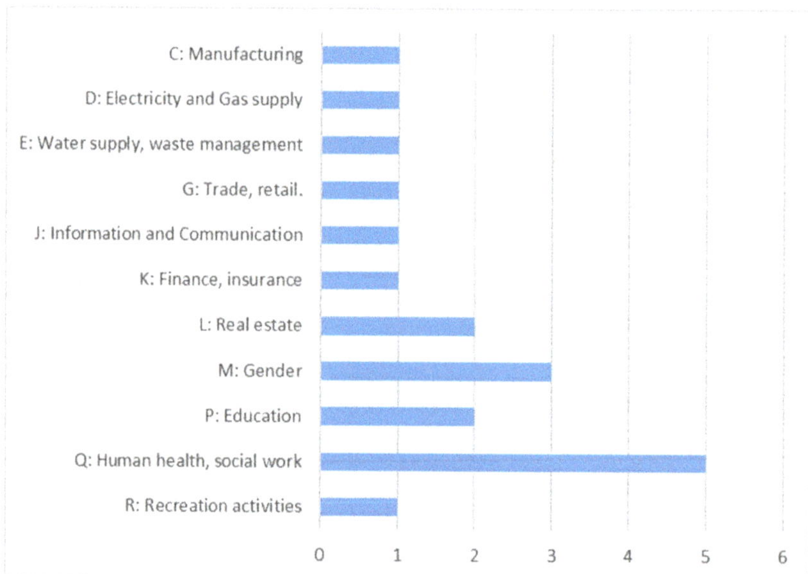

Figure 2. Distribution of the papers of the Special Issue by application sector.

Among those papers we have one work of the two Guest Editors (Dr. Aïda Valls and Pr. Karina Gibert) entitled "Building a Territorial Working Group to Reduce Gender Gap in the Field of Artificial Intelligence" [3], which revises the gender imbalance in AI during several stages throughout the life of a person (primary scholar, graduate, and professional) and proposes a network-based organisation general model for constructing AI-centred working groups that engage young females from the beginning until they become role models. The proposal was put into place with great success in the creation of several gender working groups in Catalonia and Spain; the specific example of donesIAcat, the gender working group of the Catalan Association of Artificial Intelligence, created in 2019, is analysed as a real example of the proposed methodology.

In the same line, the paper [4] led by Dr. Lenka Lhotska from Czech Republic illustrates, through examples of women researchers and selected AI projects in medicine, the wide spectrum of applications developed by female researchers during the last fifteen years in the Czech Republic and, in particular, at the Czech Technical University in Prague. Women researchers have played an important and irreplaceable role in the construction of AI research in the Czech Republic, inspiring many young female students to join the community and start their research career in this area.

The paper by Dr. Genoveva Vargas [5] discusses the problem of the lack of datasets containing the relevant information required to analyse the role of women in Computer Science and Artificial Intelligence in depth. The author highlights that current datasets miss the information relevant to identify invisible patterns when studying the gender gap in different STEM disciplines. Therefore, the first step to understanding gender imbalance is building women's history by "completing" existing datasets. The lack of relevant data to analyse the needs and problems of women is not specific to the AI field but is a serious and general problem that affects all aspects, from economy to working conditions, including health, education, and all other fields. Traditionally, systemic datasets have been designed by male researchers and often lack the relevant indicators that would allow the elicitation of women's specific needs and problems; this paper is a small example within the ICT field of this critical phenomenon that requires urgent solutions in general.

There are other two papers about ethical issues in AI. In the paper "Ethical Issues in AI-Enabled Disease Surveillance: Perspectives from Global Health" [6], headed by Dr. Ann Borda, the authors present a study of qualitative perspectives for a responsible AI framework, to explore its potential application to disease surveillance in a global health context. AI-based disease surveillance helps to collect and analyse vast amounts of unstructured and real-time data to inform epidemiological and public health emergency responses, especially in poor countries.

The paper "The digital revolution in the urban water cycle and its ethical–political implications: a critical perspective", led by Lucia Popartan [7], provides a critical overview and interesting socio-political and ethical concerns related to water digitalisation and the role of AI techniques in this process. The study concludes that a hydro-social approach to digital water management is timely and necessary to guarantee the human right to water. The authors indicate that AI methods may be a relevant factor, but they need to have a non-discriminatory design, with democratic and participatory access and with an interdisciplinary view.

The second larger group of papers refers to **Learning** and contains a total of five papers, two of them designing and using novel techniques related to deep learning (a field currently undergoing great expansion), and three using other types of machine learning methods (clustering and classification).

The paper led by Nuria Valls [8] applies to the energy sector (D. Electricity and gas) and proposes a deep neural network built from a pipeline of simple neural networks to reproduce the steps of a benchmark algorithm for calorimetry reconstruction for the Large Hadron Collider beauty Experiment (LHCb) developed at CERN. The proposed model achieves to efficiently solve the problem in nearly constant time by reducing the computational complexity of the classic algorithms.

The third paper on deep learning is entitled "Recognition of mental workload of pilots in the cockpit using EEG signals" [9] and led by Aura Hernández-Sabaté. The authors propose a convolutional neural network to classify EEG (electroencephalography, sector Q. Health, social work) features across different mental workloads in a continuous performance task test. They present two different approaches to the fusion of EEG sensor signals with DL models trained and validated on self-designed games (one serious game and one flight simulator with specific scenarios).

The Special Issue also represents other areas of the **Machine Learning** field. The paper led by Shikha Suman [10] refers to the application field of unsupervised learning in Industry 4.0 (sector C. Manufacturing), in particular, additive manufacturing and large 3D printers. It presents a modification of the classical CURE strategy that scales up hierarchical clustering to large datasets by introducing bootstrap techniques into the basic algorithm and an automatic criterion to detect the number of clusters in the dendrogram. The proposal, named bootstrap-CURE, is applied to identify operation modes of 3D printers by analysing multivariate sensor data. The proposed methodology is scalable and significantly reduces computational costs, and it is being currently used at a leading real 3D printer manufacturer.

The work by Jessie C. Martín et al. [11] proposes a new predictive support tool to optimise the number of copies to print when a new book is published (sector G. Trade and retail), using data from the book, from the authors' social networks, and from the author's web mentions. The tool introduces the Combined model of Artificial Intelligence techniques (CAIT) that combines a classifier with a predictive model. First, it applies an XGBoost algorithm to classify the book into one of the possible book market segments. Next, a regressor also based in XGBoost is used to predict the appropriate number of copies to print.

The paper [12] led by Najlaa Maaroof is centred on the explainability of the output of complex automatic classifiers, a key step to allow users to make a good and informed decision, particularly in medical applications. The paper by Maaroof et al. proposes a method to generate explanations of classifiers in the form of a minimal set of short rules, using both numerical and qualitative variables. Two different machine learning methods for creating classification rules are analysed and compared in the case of RETIPROGRAM, an intelligent clinical decision support system that computes the personalised risk of developing diabetic retinopathy. Short explanations are obtained consisting of one representative rule and some counter-examples.

The issue includes two works on **Perception**: one on computer vision and the other on voice recognition (audio processing).

The paper [13] led by Syeda Furruka Banu contributes to early diagnosis of lung cancer. It proposes an accurate lung nodule detection and segmentation in computed tomography (CT) images. The proposed system combines nodule detection, based on fine-tuned Faster R-CNN to localise the nodules in CT images, with nodule segmentation, to enhance the ability to discriminate between nodule and non-nodule feature representations. The work shows promising experimental results.

In the paper "An Approach for Pronunciation Classification of Classical Arabic Phonemes using Deep Learning" [14], a group of researchers from Saudi Arabia, led by Amna Asif, face the problem of recognition of precise pronunciation of the large number of short vowels in Arabic alphabets, which cannot be dealt with using traditional audio processing techniques, contributing to the sector of P. Education. They present a new classification architecture based on convolutional neural networks. Identifying the vowels correctly is crucial in the Arabic language since a mistake in a short vowel can change the meaning of a complete sentence.

The **Integration and Interaction** area has two papers devoted to research on multiagent systems.

The paper led by Maite Lopez-Sanchez [15] focuses on a sensible topic in social networks: hate messages (sector L. Real estate). Although hate propagators are less than 1% of participants, they create a high amount of hate content (racial, gender, religion, etc.). The goal is to detect that hate content as soon as possible to avoid its propagation. In this paper, the authors propose an agent-based model to reproduce how the hate speech phenomenon spreads within social networks. Three countermeasures are modelled, simulated and evaluated: education, deferring hateful content, and cyber activism. Their effectiveness in containing the spread of that kind of messages is studied.

The second paper in this block is entitled "Probabilistic Models for Competence Assessment in Education" [16] and is led by Alejandra López De Aberasturi-Gómez. Current trends for competence assessment propose hybrid solutions combining the benefits of automation with human judgment, such as using peer assessments to help the teacher in the evaluation of students in large classrooms (sector P. Education). The authors of this paper start with a probabilistic model based on Bayes inference and compare it with a model based on multiagent systems, called PAAS, where each actor relies on the judgment of others as long as their opinions coincide. To reconcile the benefits of Bayesian inference with the concept of trust posed in PAAS, the paper proposes a third peer evaluation model that considers the correlations between any pair of peers who have evaluated someone in common. An empirical study is done on synthetic data to determine the drawbacks and advantages of these three solutions.

The issue contains a single paper that simultaneously represents two areas: **Planning** (in its specific field of optimisation) and **Communication** (natural language processing). The research led by Tere Alsinet [17] deals with online discussions in social networks and a combination of sentiment analysis, a graph representation of the interactions among participants in the discussion, and a greedy local search optimisation algorithm to develop a quantitative model for measuring the polarisation degree in an online debate, such that this behaviour can be monitored to generate a warning signal when the debate polarisation reaches some threshold value (sector L. Real estate).

Two more papers complete the issue, which then covers all branches of AI, according to the AI Watch classification: one about **Reasoning** and the other on **AI services**.

The paper led by Zuzana Janková [18] studies several models of type-2 fuzzy sets based on different definitions of membership functions to handle uncertainty in international market financial data, which are inaccurate and incomplete (sector K: Finance, insurance). The results of this research show that type-2 fuzzy sets with dual membership functions are the most suitable for making predictions on the highly chaotic and unstable international stock markets.

Finally, a paper led by Immaculada Rodríguez [19] belongs to the AI services branch of AI and proposes a dynamic adaptive gamification method which takes into account initial players' profiles and also considers how these profiles may slightly change over time based on their interactions and opinions (sector R. Recreation activities). Then, the users are provided with a personalised experience through the use of game elements that correspond to their dynamic playing profile.

As we have seen, the issue has a wide representation of AI areas of specialisation and different application fields. In Figure 3, the global thematic coverage is visualised.

	AI branches							
AI application sectors	Reasoning	Planning	Learning	Communication	Perception	Integration and Interaction	AI Services	Ethics and Philosophy
A. Agriculture, forestry and fishing								
B. Oil and gas								
C. Manufacturing			[10]					
D. Electricity and Gas supply			[8]					
E. Water supply, Waste management								[7]
F. Construction								
G. Trade, retail			[11]					
H. Transportation and storage								
I. Accommodation and Food								
J. Information and Communication								[5]
K. Finance, insurance	[18]							
L. Real estate		[17]		[17]		[15]		
M. Other: Gender								[3][4][5]
P. Education					[14]	[16]		
Q. Human health, Social work			[9][12]		[13]			[4][6]
R. Recreation activities							[19]	

Figure 3. Classification of the papers regarding both AI branches and AI applications.

Analysing the configuration of authors of the papers, Figure 4 shows that the papers in this Special Issue have between 1 and 6 authors, but never more than 4 female authors. From a specific call for women-led papers, still, 35.3% of the papers have a minority of female authors; however, the first author is always a woman, as required in the call for papers.

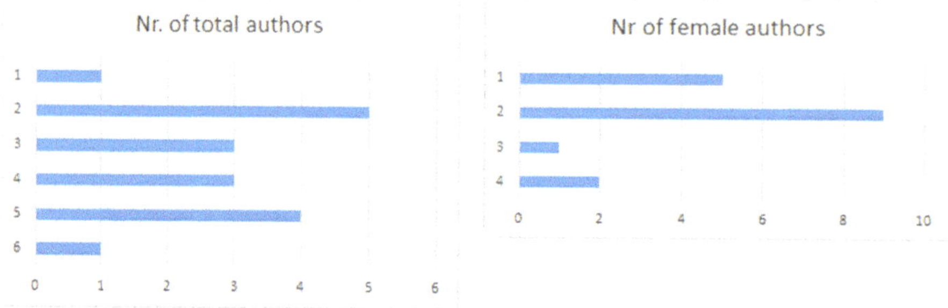

Figure 4. Distribution of the papers according to the number of authors (**left**); Distribution of the papers according to the number of female authors (**right**).

In Figure 5, it can be seen that the number of female authors slightly increases with the total number of authors, but the most frequent situation is to have 2 female authors (with 10 papers), and often they work alone (4 papers with a total of 2 authors, all female).

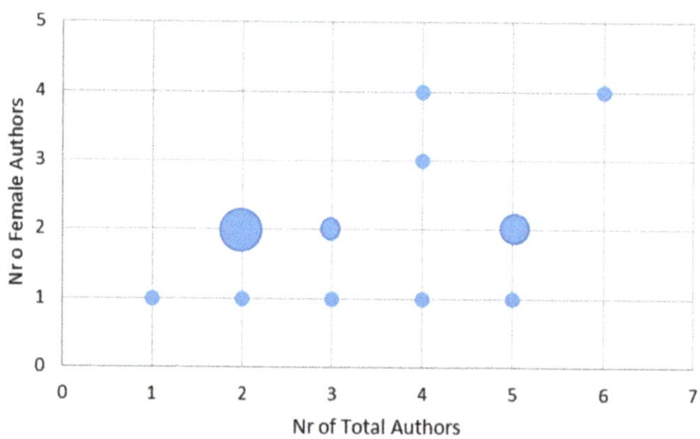

Figure 5. Relationship between total number of authors of the paper and the number of female authors. Size of the circle represents the number of papers at each point.

Regarding the origin of the working teams represented in this issue, this Special Issue was conceived during the 23rd International Conference of the Catalan Association of Artificial Intelligence, and although there was an international open call, it was intensively promoted during the conference. This explains why most of the papers reflect research conducted in Catalonia. However, the issue also contains five international papers with research done in France, Czech Republic, the U.K., and Saudi Arabia, also involving authors from other countries like Mexico and Australia. From the 12 remaining papers, international working teams are also represented in 7 of them, involving authors from other countries. In five of them, the first author is an international female under mobility who is developing research in Catalonia. They come from Bangladesh, Cuba, India, Yemen, and Romania. In fact, from the 58 authors integrating the issue, 26 (44.8%) of them are international, involving more than 12 countries, providing quite a wide international perspective of the female talent in AI.

The papers were evaluated by 52 international researchers from all over the world. Even though the usual proportion of women in AI is about 10% [3], we had 25% female researchers involved as reviewers. As guest editors, we thank all of them for their effort and great reviews, which made possible the publication of this issue.

In conclusion, apart from its intrinsic scientific value as a Special Issue itself combining interesting research works, the effort made in building this Special Issue intends to help enhance the visibility of where women in AI are, what they do, and how they contribute to Artificial Intelligence developments from different places, positions, research branches, and application fields in AI.

Funding: This research received no external funding.

Conflicts of Interest: The authors declare no conflict of interest.

References

1. European Commission. Women in the Digital Age. 2018. Available online: https://op.europa.eu/en/publication-detail/-/publication/84bd6dea-2351-11e8-ac73-01aa75ed71a1 (accessed on 1 September 2022).
2. Samoili, S.; López Cobo, M.; Delipetrev, B.; Martínez-Plumed, F.; Gómez, E.; De Prato, G. *AI Watch. Defining Artificial Intelligence 2.0. Towards an Operational Definition and Taxonomy for the AI Landscape*; EUR 30873 EN; Publications Office of the European Union: Luxembourg, 2021; ISBN 978-92-76-42648-6. [CrossRef]
3. Gibert, K.; Valls, A. Building a Territorial Working Group to Reduce Gender Gap in the Field of Artificial Intelligence. *Appl. Sci.* **2022**, *12*, 3129. [CrossRef]
4. Lhotska, L.; Stepankova, O. Artificial Intelligence and Women Researchers in the Czech Republic. *Appl. Sci.* **2022**, *12*, 1465. [CrossRef]
5. Vargas-Solar, G. Intersectional Study of the Gender Gap in STEM through the Identification of Missing Datasets about Women: A Multisided Problem. *Appl. Sci.* **2022**, *12*, 5813. [CrossRef]
6. Borda, A.; Molnar, A.; Neesham, C.; Kostkova, P. Ethical Issues in AI-Enabled Disease Surveillance: Perspectives from Global Health. *Appl. Sci.* **2022**, *12*, 3890. [CrossRef]
7. Popartan, L.A.; Cortés, À.; Garrido-Baserba, M.; Verdaguer, M.; Poch, M.; Gibert, K. The Digital Revolution in the Urban Water Cycle and Its Ethical–Political Implications: A Critical Perspective. *Appl. Sci.* **2022**, *12*, 2511. [CrossRef]
8. Valls Canudas, N.; Calvo Gómez, M.; Golobardes Ribé, E.; Vilasis-Cardona, X. Use of Deep Learning to Improve the Computational Complexity of Reconstruction Algorithms in High Energy Physics. *Appl. Sci.* **2021**, *11*, 11467. [CrossRef]
9. Hernández-Sabaté, A.; Yauri, J.; Folch, P.; Piera, M.A.; Gil, D. Recognition of the Mental Workloads of Pilots in the Cockpit Using EEG Signals. *Appl. Sci.* **2022**, *12*, 2298. [CrossRef]
10. Suman, S.; Karna, A.; Gibert, K. Bootstrap–CURE: A Novel Clustering Approach for Sensor Data—An Application to 3D Printing Industry. *Appl. Sci.* **2022**, *12*, 2191. [CrossRef]
11. Martín Sujo, J.C.; Golobardes i Ribé, E.; Vilasís Cardona, X. CAIT: A Predictive Tool for Supporting the Book Market Operation Using Social Networks. *Appl. Sci.* **2022**, *12*, 366. [CrossRef]
12. Maaroof, N.; Moreno, A.; Valls, A.; Jabreel, M.; Szeląg, M. A Comparative Study of Two Rule-Based Explanation Methods for Diabetic Retinopathy Risk Assessment. *Appl. Sci.* **2022**, *12*, 3358. [CrossRef]
13. Banu, S.F.; Sarker, M.M.K.; Abdel-Nasser, M.; Puig, D.; Raswan, H.A. AWEU-Net: An Attention-Aware Weight Excitation U-Net for Lung Nodule Segmentation. *Appl. Sci.* **2021**, *11*, 10132. [CrossRef]
14. Asif, A.; Mukhtar, H.; Alqadheeb, F.; Ahmad, H.F.; Alhumam, A. An Approach for Pronunciation Classification of Classical Arabic Phonemes Using Deep Learning. *Appl. Sci.* **2022**, *12*, 238. [CrossRef]
15. Lopez-Sanchez, M.; Müller, A. On Simulating the Propagation and Countermeasures of Hate Speech in Social Networks. *Appl. Sci.* **2021**, *11*, 12003. [CrossRef]
16. López de Aberasturi Gómez, A.; Sabater-Mir, J.; Sierra, C. Probabilistic Models for Competence Assessment in Education. *Appl. Sci.* **2022**, *12*, 2368. [CrossRef]
17. Alsinet, T.; Argelich, J.; Béjar, R.; Martínez, S. Measuring Polarization in Online Debates. *Appl. Sci.* **2021**, *11*, 11879. [CrossRef]
18. Janková, Z.; Rakovská, E. Comparison Uncertainty of Different Types of Membership Functions in T2FLS: Case of International Financial Market. *Appl. Sci.* **2022**, *12*, 918. [CrossRef]
19. Rodríguez, I.; Puig, A.; Rodríguez, À. Towards Adaptive Gamification: A Method Using Dynamic Player Profile and a Case Study. *Appl. Sci.* **2022**, *12*, 486. [CrossRef]

applied sciences

MDPI

Article

Building a Territorial Working Group to Reduce Gender Gap in the Field of Artificial Intelligence

Karina Gibert [1,2] and Aida Valls [2,3,*]

1 Intelligent Data Science and Artificial Intelligence Research Center, Universitat Politècnica de Catalunya (IDEAI-UPC), Catalonia, 08034 Barcelona, Spain; karina.gibert@upc.edu
2 donesIAcat, Gender Working Group of the Catalan Association of Artificial Intelligence, 08034 Catalonia, Spain
3 Intelligent Technologies for Advanced Knowledge Acquisition Research Group, Departament d'Enginyeria Informàtica i Matemàtiques, Universitat Rovira i Virgili, 43005 Tarragona, Spain
* Correspondence: aida.valls@urv.cat

Abstract: The gender gap (both at vocational and professional sides) in Artificial Intelligence (AI), and scientific and technological fields in general, is one of the most critical challenges that the current digital society must solve. This paper describes the proposal of the gender commission donesIAcat to create a gender working group formed by Catalan AI scientists and professionals who work in a network for bridging this gap. The main objectives for letting girls know that they can study and work in the AI field are presented in this paper. A general methodological framework is proposed, following the internal organization of the Catalan group donesIAcat. Several key actions are explained and classified into six blocks. A relevant contribution of the paper is the definition of the guidelines required to build a territorial network-based structure capable of launching several AI-related activities targeting people at different stages of their life. The activities done at donesIAcat illustrate the possible outcomes of the proposed methodology and show successful initiatives to engage girls in technology and AI. The paper shows the validity of this model for small homogeneous territories where activities can be suitable for the different cities in the region. Proximity is one of the advantages of such a model and one of the reasons for its success.

Keywords: Artificial Intelligence; gender gap; equity

Citation: Gibert, K.; Valls, A. Building a Territorial Working Group to Reduce Gender Gap in the Field of Artificial Intelligence. *Appl. Sci.* **2022**, *12*, 3129. https://doi.org/10.3390/app12063129

Academic Editor: Federico Divina

Received: 2 February 2022
Accepted: 16 March 2022
Published: 18 March 2022

1. Introduction

Jobs in the area of STEM (Science, Technology, Engineering, and Mathematics) constitute a large proportion of the available professions nowadays. Among them, those related to Computer Science and Information and Communication Technologies (ICT) are highly demanded in Europe and worldwide. As it is said in the 2018 report of the European Center of Development for vocational training [1] entitled, Skills forecast: trends and challenges to 2030, the demand for ICT professionals is expected to increase 12.8% by 2030. The US census bureau published Figure 1, where we can see that women are nearly half of the workforce, but only 27% are working in STEM. We can observe that, since 2000, the number of women employed in Engineering is quite stable at around 10%, but Computer Science is the only field showing a decreasing trend, with only about 20% of women in this area in 2019 [2]. Social sciences, mathematics, physics, and life sciences have a percentage of women above 40% and below 60%, showing a good balance between genders. The numbers shown for computer workers are alarming, considering that we are working towards a digital society in the near future where many jobs will be in the STEM fields and women should be there together with men.

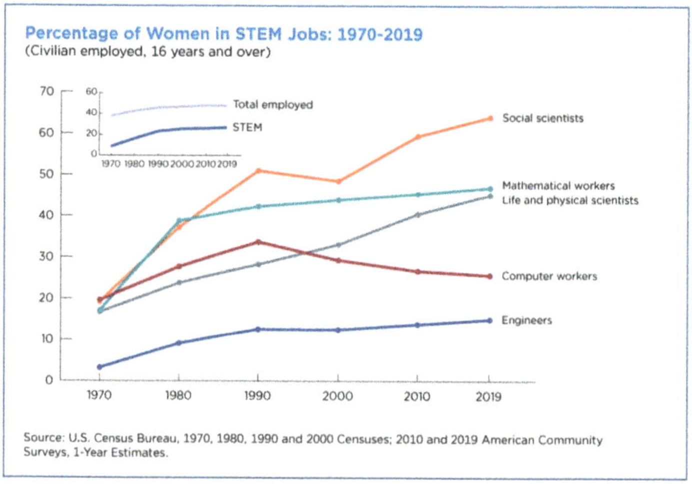

Figure 1. Employment of women in the US from 1970 to 2019.

In the region of study of this paper, Catalonia (Spain), the Catalan Talent Observatory regularly publishes job demands. Figure 2 shows the five most demanded occupancy profiles in 2021; computer programming and consultancy being the largest with 67% of the share, followed by the other three ICT-related jobs and total 78% of the job offers in Catalonia.

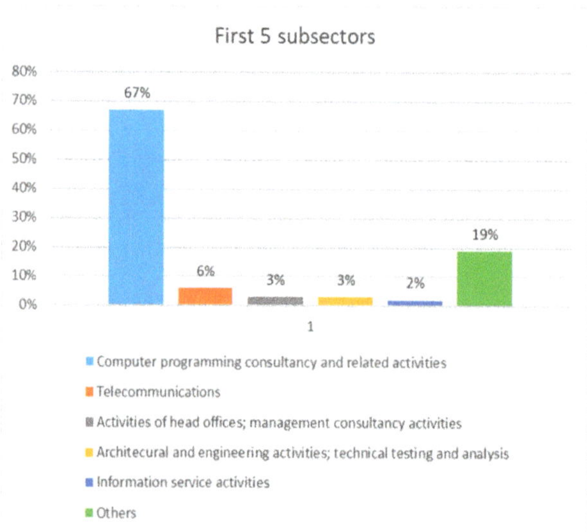

Figure 2. Occupational demand in Catalonia in the last 12 months (published December 2021), from The Catalan Talent Observatory (aqu.cat).

The presence of women in ICT educational courses and jobs is known to be smaller than men. Traditionally, those professions have been associated with males, leading to gender inequality in the digital area. The European Commission in its 2018 report [3] highlights this bias, observing that there are four times more men taking ICT-related courses than women. The same report also detects that there are 313% more men than women working in digital professions. These differences lead to the problem of leaving women out of some of the most relevant professions in the near future, such as the ones related to Data Science and Artificial Intelligence. Since the presence of women in STEM and AI is too low in the professional sector, it is difficult to claim there are balanced panels in round tables, professional teams, or women in directive positions in the field. Thus, in the particular case of Informatics or AI, a previous challenge is now a priority: to inspire girls to choose a degree in the field so they can get the proper training to become professionals.

Artificial Intelligence (AI) is a discipline highly connected with Computer Science as it emerged from the idea of building intelligent software running in a standard computer or embedded in a robot. Therefore, AI methods are studied in bachelor's degrees in Informatics although recently some universities have started to offer specific AI degrees. The field of AI devoted to data mining or data science is also strongly connected with degrees in Mathematics and Statistics. AI-focused masters and doctorates also permit students to specialize in this field. These courses are given in Engineering and Polytechnic universities. The majority of students in these branches are male, which raises an alarm about the low influence that women will have in AI and, consequently, in the design of the near-future technology. Figure 3 shows the gender distribution progression in Spain in the last five years, taking three university years (2016–2017, 2018–2019, and 2020–2021) [4]. The percentage of women is increasing slightly in Engineering and Informatics, but has decreased a bit in Mathematics and Statistics (also related to ICT jobs). The total sum of these three fields of the study indicate a very small increase in women's share. It is clearly seen that informatics has only 12–14% of women, which highlights that the presence of women in this sector is significantly lower than other STEM degrees.

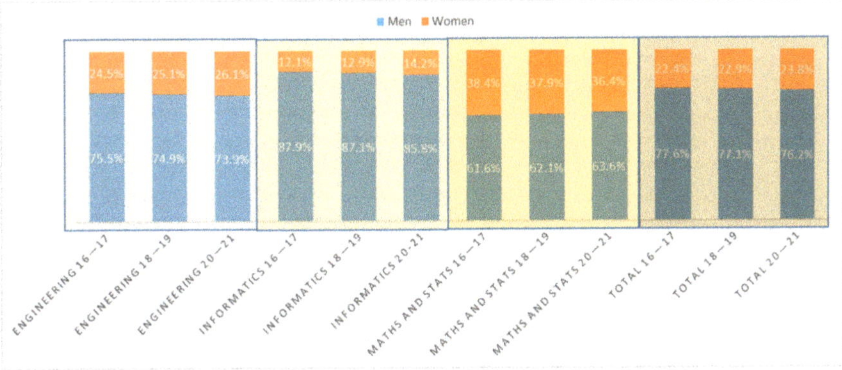

Figure 3. Gender distribution of university students by field in Spain 2016–2021.

The loss of interest in technology occurs at very early ages [5], near the moment when children identify their gender. Family and culture seem to be determinants [6] in the association of technology roles with boys, while girls are pushed more towards roles linked to care and humanistic activities. This push is supported by the messages unconsciously transmitted by the media. In [7], authors analyzed how media contributes to gender stereotypes in children and youth, observing that girls are underrepresented and their characters usually have limited personality traits and show stereotypes about beauty and sex appeal. Among other biases, the work in [7] also includes research papers that show how youth's expectations and preferences concerning academic degrees and jobs are influenced by the media. In particular, they highlight that the lack of women professionals in the media diminishes the aspirations of girls. Among this lack of female role models, the case in Engineering and Informatics is even higher, leading to the gender-driven division of interests in youth. Engagement in technology or, more precisely, in robotics and AI-related fields, has been seen to be different by gender. As an example, the US statistics company Statista published a graphic illustrating the familiarity with some virtual digital assistants by gender in 2016 (see Figure 4). It can be seen that men duplicate or triplicate the familiarity in the majority of voice assistants with respect to women (except for Siri).

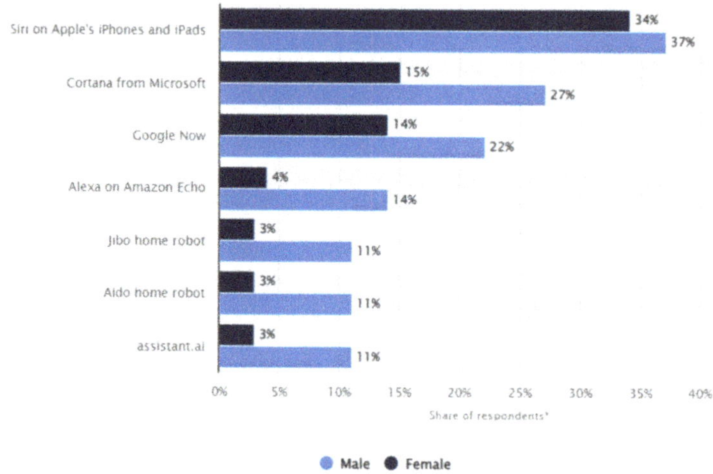

Figure 4. Familiarity with virtual digital assistants in the United States (from Statista 2021). * 492 respondents, 18 years and older.

Regarding women's interests, the Pisa Report 2015 states that the majority of girls enrolled in scientific careers declared expectations of working in the health sector [8]. Researchers have found that women search for societally meaningful jobs, that is, jobs to improve the lives of people [9]. The tacit pressure of the entire society to push girls and women to play a social and humanistic role is so heavy that, unless girls are exposed to explicit messages about other possibilities, they tend to align their vital decisions with those stereotyped systemic expectations, choosing those humanistic and social components in their training programs. In Catalonia, the students formalize an application to the Education Department from the Catalan government indicating their preferences about higher education around March so that they follow the admission processes in the different High Schools. Data from 2021 indicates that applications to pure ICT bachelor's degrees (like informatics engineering or telecommunications engineering) show 16% of females. However, when double training programs combine an ICT bachelor with some more "social" degree (like biotechnology or business administration), the applications of females raise to 30% (data extracted from the Department of Education of the Catalan government). This increase indicates that girls tend to feel more comfortable applying for training programs including this social and humanistic component, even if they are interested in technology.

However, the lack of association of Informatics with improving the lives of people is wrong. The reality is that developments in computers have led to huge benefits in the quality of life of many people. In the same line, Artificial Intelligence is also strongly connected to meaningful jobs as shown in [9], where different applications of Machine Learning and Data Science positively influence wealth in areas such as health care, business, education, or environmental protection. The possibilities of AI for society are enormous, and women can certainly find their place in it if they are able to discover the possibilities of the tools in data analysis, machine learning, robotics, and computer vision among others.

In this paper, we will use the term *gender gap* to refer to all inequalities related to gender observed in the STEM sector. The main inequality is the presence in the sector. Moreover, other expressions of the gender gap are the difficulties of the STEM women to promote their careers, for example. Therefore, in this paper this term will have a wide interpretation and, depending on the context, it will refer to the lack of female professionals in the sector, to inequalities in salaries, or promotion in career, etc.

The need to engage women and bridge the gender gap in STEM and AI has been recognized by several local and international institutions, and some actions to mitigate it have started in academic and professional sectors [10,11]. Some teams have been created to promote STEM in girls [12]. Other initiatives try to gather professionals in these fields to make them visible, such as *Women in AI*, an international non-profit organization to create an AI female community for inspiring and empowering women to become AI and data experts [13]. They organize large events such as AI camps, master classes, or hackathons. Similarly, *Humans for AI* is a non-profit organization focused on democratizing AI to attract minorities, not only to diversify gender but also to include different socioeconomic statuses or races [14]. Regarding research, there are many conferences for women to facilitate networking, such as Women in Data Science (WiDS), Women in Robotics, or Women in Machine Learning (WiML). However, to spread the AI interest to all girls at different ages, having local territorial structures may be more appropriate since we can find the girls instead of waiting for them to discover and join a specialized international association.

In this paper, we present a set of objectives, a conceptual model, and a work methodology for addressing the gender gap in AI in the different stages of women's lives. This methodology is based on building a working group of AI professionals with a territorial structure which facilitates the efficient de-centralization of cooperative actions for promoting the AI field from girls' childhood. This project is aligned with the Sustainable Development Goals of the United Nations for 2030, in particular, with the goal SDG5: achieve gender equality and empower all women and girls [15]. The proposal has been built from the experience of the donesIAcat group, which is a team of AI women who belong to the scientific Catalan Association for Artificial Intelligence.

The main contribution of this paper is a conceptual framework for reducing the gender gap in Artificial Intelligence at a national or regional level. This model has two main distinctive characteristics: first, members are organized in a territorial network, which facilitates the rapid dissemination of information and activities in the region of influence, with rapid and solid connections in each remote corner of the territory; second, the model identifies six types of actions to pursue goals addressed to the different target population (students and professionals) and at different ages. Section 2 presents the Catalan gender working group, donesIAcat, which will be used to illustrate the proposed methodology. Section 3 explains the general conceptual model and its agents. Sections 4–7 explain each of the actions and provide examples from the experience of donesIAcat in the Catalan-speaking regions. Finally, Section 8 presents some conclusions and future work.

2. The Gender Working Group: donesIAcat

The gender working group donesIAcat belongs to the Catalan Association of Artificial Intelligence (ACIA, www.acia.cat (accessed on 17 March 2022)). First, we want to make a short presentation of the framework of the scientific association ACIA, and then we will describe the gender group and its mission, structure, and activities.

2.1. The Catalan Association of Artificial Intelligence (ACIA)

The Catalan Association of Artificial Intelligence (ACIA) was created in 1994 as a non-profit association to gather the scientists, professionals, and students that work in Artificial Intelligence in Catalonia. ACIA was born with the aim of being a meeting point for all the researchers who were pioneers in the study of AI in Catalonia. The main goals of ACIA include (1) the facilitation of the communication between professionals and organizations that work in AI, (2) the promotion of AI courses, techniques, and applications within the Catalan society, and (3) the organization of social and scientific events (e.g., conferences, workshops, and meetups) to disseminate the knowledge about AI. Among its many activities, we highlight the periodic publication of the association's magazine (ACIA's Bulletin, later called *Nodes*), and the organization of an annual International Congress about AI (called CCIA) since 1998, whose proceedings are published by IOS Press. ACIA also provides some benefits for its members, especially youth, such as discounts in sponsored conferences or the annual award prizes to the best AI doctoral thesis and the best master thesis.

Today, ACIA gathers most of the Catalan scientific community as well as alumni and professionals in the sector. Currently, the association has more than 220 individuals and institutional partners. Despite being the association of a small territory, ACIA has been a member of the European Association for Artificial Intelligence since 1995 (first denoted ECAI and now named EurAI). As an example of its international scientific leadership, we can mention that 10 members of ACIA have been distinguished as EurAI fellows, a program that started in 1999. This distinction is given to less than 3% of the EurAI associates. Only France, Germany, the United Kingdom, and Italy associations have more awarded scientists than ACIA.

These numbers show that ACIA is the largest and oldest inter-institutional AI association in Catalonia. It has served as a connection point for the scientific community and as a link with companies and society in general.

2.2. *donesIAcat (WomenAIcat), the Gender Commission of ACIA*

On 8 March 2019, International Women's Day, donesIAcat was founded by Prof. Karina Gibert as a working group inside ACIA. The main goal of this working group was to increase the presence of women in the Artificial Intelligence sector in Catalonia, thus reducing the gender differences. Currently, the association only has 17.5% of female members. The first board of donesIAcat involved five members from the different areas where ACIA had associates which soon organized in a territorial structure as will be explained later in this paper.

In line with other recent Catalan actions about gender, some of the women in ACIA decided to create the working group donesIAcat to make visible the work we were already doing to reduce the gender gap in AI and achieve a more balanced AI community. Being born from a scientific association, donesAIcat is mainly constituted of members with a scientific profile who work at universities or research centers in Catalonia. Due to this profile, some of the members of ACIA have been working to stimulate technological vocations in girls for years, but the lack of coordination of these individual initiatives made the work much harder. The formalization of this group permits them to have a supporting structure for coordinating and disseminating their actions, as well as to jointly identify new lines of work. The existing previous contacts between those women facilitated the decision of creating donesIAcat as well as the engagement in defining, sharing, and participating in gender-based activities for the dissemination of AI.

The mission of donesIAcat is to contribute to bridging the gender gap in the Artificial Intelligence field with a focus on the territorial scope of ACIA, which is the area known as Catalan Countries (70,000 km^2) that includes all territories where the Catalan language is spoken.

The working group donesIAcat is connected to other gender organizations in Catalonia and Spain. For example, it has close relations with donesCOEINF (the gender commission of the Official Professional Chamber of Informatics Engineering in Catalonia) and the two collaborate in some initiatives with the aim of being more effective together.

3. Conceptual Working Framework for Gender-Bias Reduction

The discovery of the interest in Artificial Intelligence (or in STEM in general) may happen at different moments in life. Moreover, it can be triggered in different ways and by different activities and goals, depending on the target audience. Establishing a well-defined framework is crucial to structure the actions done by women working groups in order to materialize effective lifelong support of women.

In this paper, we propose a conceptual framework based on different kinds of goals for the different target populations and different ages. This proposal is the result of synthesizing the expertise accumulated over the years from developing gender-oriented activities and collecting personal evidence of what works, for better or worse. The joint experiences of all members of the working group allowed us to find the essential actions that have positive impacts and has permitted the authors to elaborate on this conceptual framework. Figure 5 shows a graphical representation of this circular framework which is organized around a central node (the gender working group). The outer circles represent the female population, separated into students from primary school to postgraduate, and working people (professionals and scientists). The actions of the working group towards this population are grouped into six lines:

- Inspiring new vocations in children and undergraduates.
- Talent generation and training at university age.
- Talent Up-skilling of students and professionals.
- Talent Re-skilling of students and professionals.
- Networking of AI professionals and scientists.
- Making female AI professionals visible to create role models.

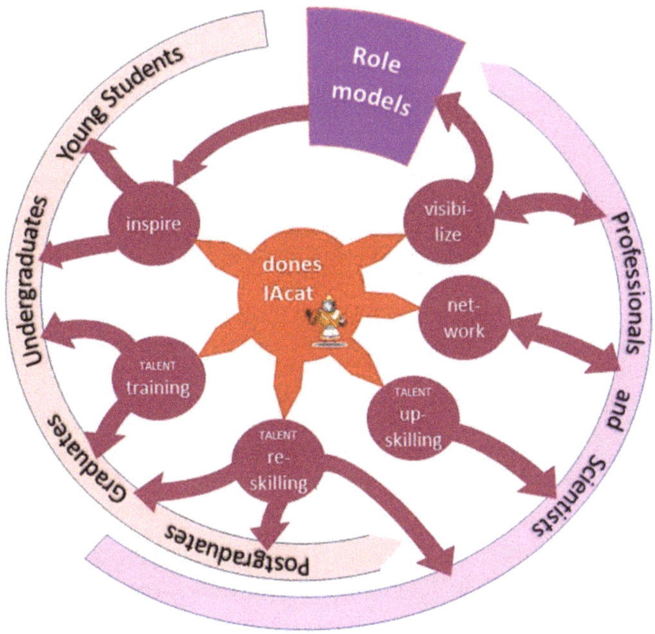

Figure 5. Conceptual working framework for the gender-bias reduction in AI.

A crucial step in this conceptual model is the feedback that arrives to the central women association by means of visibility and networking actions. AI professionals and scientists who received inspiration and talent formation can later become part of the working group, becoming new role models who inspire the next generation of girls. Maintaining and growing a territorial network structure is the core of this organizational model.

4. Inspiring New Vocations in Children and Teens

The first line of action is targeted at K-12 students, from elementary school to university undergraduates. Activities must consider the age of the participants to appropriately present the field of Artificial Intelligence and its potential impact on society. The inspirational message must be given by AI women, that will implicitly act as a role models for these young ages, as the experience of current AI females indicates [16]. The presentation of relevant women in AI and Informatics is a good starting point (e.g., Ada Lovelace, Grace Hooper, Margaret Hamilton, etc.). However, explaining AI from a personal perspective, such as presenting personal motivation and ongoing AI projects, usually has a greater impact on girls. Additionally, presenting current professionals that can be perceived as closer to the girls is important. We propose two different types of activities for this period: inside or outside school.

- Inspirational activities in the school: These kinds of activities are done by a female role model in the student's classroom and during school time. They can be done just with the girls or with both genders together, but it is worth noting that students do not choose to participate in the activity, as it is mandatory for them. This permits us to arrive at all possible kinds of children's profiles, even if one of them has never thought of technology or AI. Thus, they can learn about the field and maybe discover an uncovered interest not manifested before. These activities usually consist of a presentation of the role model and some short participatory activities (question-answer, logic puzzles, etc.) for 60 or 90 min. Some examples of activities made by donesIAcat in this line are talks at primary school that might be part of a frame project or not. Among those framed in bigger projects, members of donesIAcat participate in the program Aquí-STEAM (https://aquisteam.upc.edu/ca (accessed on 17 March 2022)) for children between 9 and 14 (from University Politècnica de Catalunya in the province of Barcelona) and INSPIRA-STEAM (https://inspirasteam.net/ (accessed on 17 March 2022)), which is a Spanish project addressed to children aged 11–12 lead by the University of Deusto and the University Rovira i Virgili in the province of Tarragona; some members of donesIAcat also visit specific schools on their own with similar sessions or in the framework of Catalan events for female empowerment (like Girls Day (https://www.urv.cat/en/campus-life/corporate-responsibility/equality-observatory/girlsday/ (accessed on 17 March 2022)) or 100tífiques (https://100tifiques.cat/en/home/ (accessed on 17 March 2022)). We also mentor the research works of students at secondary school (16–18 years) by participating in their interviews with AI experts, providing them suitable references to AI, or supervising some practical experiments with AI tools.
- Open activities outside school: These inspirational actions may include invited conferences to some technology-related forums (e.g., giving a talk at the closure of the Technovation Girls (https://technovationchallenge.org/ (accessed on 17 March 2022)) event for developing apps, at the province of Lleida), being members of the evaluation jury in tech-related activities for juniors (such as the First Lego League (https://www.etse.urv.cat/flltarragona/ (accessed on 17 March 2022)) in Tarragona-Reus), or participating in special workshops for schools on programming and robotics organized during the weekends in social innovation labs (Citilab, Barcelona). The previous actions belong to projects led by other entities with synergic aims within the territory. However, the gender working group itself has also initiated and led some activities open to teenagers. In the framework of the annual Mobile World Congress in Barcelona, donesIAcat organizes an AI stand in a co-located event called YOMO (Youth Mobile Festival) which is usually visited by 25,000 students from many secondary schools in Catalonia. The teens participating in these kinds of activities are usually interested in STEM in advance and they choose to attend those events. Therefore, the role models should focus on giving more detailed knowledge about Artificial Intelligence and its multiple applications and sub-fields, as well as encouraging them to choose the university degrees that give the formation needed to become an AI expert.

5. Talent Generation

The previous inspirational actions were aimed at enabling young people to discover the Artificial Intelligence field and its diverse applications. Female role models try also to convince youth that this is not a male field, but a nice job for any person, encouraging girls to break stereotypes. Making the fundamental role of AI in the current health or environmental fields visible becomes crucial to stimulate the interest of girls in the field. The second main line of work is devoted to the education of women in Artificial Intelligence.

This goal should be addressed differently depending on the life stage of the participants. We propose three different groups of actions:

- Training: Regular courses about AI in bachelor's and master's degrees for the students mainly in Informatics or AI degrees.
- Re-skilling: Short courses for giving insights in particular areas of AI, addressed to university students, postgraduates, or professionals with a STEM background who want to work in AI-related projects, such as students of Mathematics, Statistics, Business, or other Engineering specializations.
- Up-skilling: Short courses or training camps for professionals of other disciplines who want to get some general concepts to begin to understand AI.

5.1. Talent Training

Having scientists and researchers in a women association is a great value for taking certain actions. These AI researchers may promote and participate in teaching actions within universities, as some of its members are involved in the coordination of courses or degree programs. It is important to try to be influential in the curricula definition with regards to AI, giving a female perspective. Being involved as teachers is also a way of being role models for the next generation at bachelor and master levels. It is indicated in Figure 5 as talent training actions.

The experience of donesIAcat in this line is quite large, since many of their members are researchers at universities with more than 25 years of teaching experience in regular training programs, both at public and private centers. Some of our members created the first AI subjects in the Informatics degrees at different universities of Catalonia since 1995. Currently, donesIAcat has members in 12 different universities spread throughout the Catalan Countries. Recently, some of our members have to lead the definition of two Artificial Intelligence bachelor's degree programs, started in 2021 at the Universitat Politècnica de Catalunya (official degree recognized by the Spanish Education Ministry) and the Universitat Autònoma de Barcelona (specific title offered by UAB, currently in process of officialization). In both cases, the participation of donesIAcat in the definition of competencies and elaboration of the curricula of these degrees has been acknowledged. We have also been involved in master courses and the coordination of AI master studies in several universities. It is worth mentioning the inter-university Masters on Artificial Intelligence, given jointly by three Catalan universities, where our members give eight different specialized courses, such as multi-agent systems, intelligent decision support systems, machine learning (including artificial neural networks and deep learning, among others) or computational intelligence.

5.2. Talent Re-Skilling and Up-Skilling

Artificial Intelligence is a field of specialization related to many other disciplines (Statistics, Business Intelligence, Telecommunications, etc.) and with many different fields of application. For this reason, having some knowledge about AI may be of interest to students and professionals that have not studied AI at university.

Re-skilling and up-skilling courses could be promoted from an AI female association. These training programs can be addressed only to women or they can be open to any attendant. The support and orientation from the AI women association are important to define appropriate training program contents that include the discussion about gender bias in AI and how the gender perspective must be included in the design of AI apps, databases, machine learning algorithms, and reports. Additionally, if women from the association participate in teaching, this implicitly generates role models and inspires the female attendants to become AI professionals too.

Some members of the group donesIAcat participate and promote a recent re-skilling training program on AI for girls that are at the end of their STEM bachelor's degrees or have recently graduated. The program is called Top Secret Rosies (https://topsecretrosies.soko.tech/ (accessed on 17 March 2022)) and the women of the research center IDEAI-UPC (with 13 women out of the 70 AI researchers) have the academic and scientific direction of the program and have designed and elaborated the training contents and the methodology. In its first edition, it has filled the 21 seats available (from 53 application forms received) with women from many different countries and backgrounds. Some of them have won an internship to work in AI at international companies.

The women association members can also take part in professionalizing masters, giving courses on high specialization where gender perspective can be applied as well, and their professional experience can be communicated to students. From donesIAcat, we participate in many initiatives. The members of donesIAcat from UPC take part in many advanced courses from UPC Foundation, the permanent training program at UPC, participating as lecturers and advisors in programs like Master on Industry4.0, Master on eHealth, Master on digital transformation, or Master for CIOs in innovation. Specialized professionals enroll in these masters, and can apply what they learn directly to their jobs, so the presence of donesIAcat brings to those professionals the gender perspective and new ways to address their daily work.

6. Visibilization of AI Female Talent

Women do not appear in media or other public technological dissemination events as frequently as men do. According to the last study in 2020 of UN Women [17] on women's representation in society, only 24% of persons read, heard, or seen in the news are women. In the science and health fields, it raises to 35%, but it is still much lower than men's visibility. It is known, as said in Section 3, that this biased view induces gender stereotypes in the young. Therefore, it should be a priority goal of AI women associations to make visible the female talent by means of different actions. We propose two basic directions:

- To explain and make visible what female professionals (from our association or not) do in the AI field.
- To identify role models among STEM women and engage them in activities that increase their visibility (as keynote speakers in international conferences, panelists in round tables, etc.).

In the first line of work, donesIAcat has started to make some dissemination publications. We have addressed scientific publications, such as a position paper in a specialized STEM conference [10] and a paper in a scientific journal [12], which discuss the gender bias in technology in order to make this fact visible. Other kinds of publications include articles in magazines, such as Forbes [18], with a text explaining the challenge of being a woman in technology, or an article about why girls are not engaged in STEM at The Conversation (in Spanish) [6].

Promoting the publication of scientific papers can be a strategic action to explain what women make in AI. For this reason, editing special issues in relevant journals focused on women authors is important for visibilization. The current special issue is edited by two members of donesIAcat, who spread the call for papers not only in the Catalan Countries but all over the world by means of several AI mailing lists. In addition, donesIAcat presents the work done over the year at the yearly conference of ACIA in a public session of the international CCIA conference.

Regarding the second line of work, the first set of actions of women associations is devoted to monitoring that the organizers of AI-related events avoid gender bias. This bias can happen on the invited speakers, on the selection of round table members, or the configuration of panels of experts. In this sense, donesIAcat achieved the inclusion of these ideas for gender gap reduction in the guide for organizing the CCIA conference (International Conference of the Catalan Association for Artificial Intelligence).

Another set of actions is focused on promoting and participating in specific events in technology and AI. For example, some members of donesIAcat became ambassadors of the Women in Data Science (WiDS (https://sites.google.com/isglobal.org/widsbarcelona (accessed on 17 March 2022))) for Barcelona and were the leading organizers of the WiDS international event 2021 (they are currently preparing the one for 2022). Others participated in the organization of a round table with the title "Women in Tech; an inclusive approach", as an event scheduled in 4YFN 2020 (part of the Mobile World Congress), and still others participated in the panels at the Catalan forum Cicle Hipàtia (https://www.50a50.org/es/iniciamos-el-ciclo-hipatia/ (accessed on 17 March 2022)) (50 × 50 shared leading association), or on the set of talks about "Sex and Gender Bias in Artificial Intelligence and Health: Building a Future for Equality", held in Barcelona during 3 months at Caixaforum.

This leadership in some events, as well as the promotion of role models made from the association, brings new opportunities to participate in other activities organized at the different locations of the territory. Our associates are now included in the list of AI women who work in the inspirational activities done with students, presented in Section 2. Moreover, our members are called to participate in initiatives of other entities for gender visibilization like the PreInf (https://enginyeriainformatica.cat/preinf/ (accessed on 17 March 2022)), the portal for promotion of Informatics in women (from the Illustrious Official Chamber of Informatics Engineering of Catalonia), TEDx events (e.g., a motivational talk about Data Science by Dr. Karina Gibert at TEdxIgualada in 2021), or The Women in Computer Science and STEM workshop at the AICCSA 2021 conference.

These two lines of promotion of women in AI facilitate the consolidation of the AI working group, increasing the opportunities of being enrolled in future AI activities. It is worth noting that, although many actions are focused on making AI female experts visible to society, we have also presented specific initiatives for professionals and AI scientists.

7. Constructing a Network

This paper addresses the case of creating a women's AI working group for serving a certain territory (i.e., a country or a subset such as the Catalan Countries). This locality may be beneficial for many people of all ages by means of the activities previously presented, which range from inspirational sessions at elementary schools to training courses for professionals.

Most gender groups and associations are flat structures, with a board and a set of members. However, as the main goal is to arrive at different locations of the territory, we propose building the association with a territorial organization network structure. The area of influence is divided into sub-areas according to some administrative structures (such as provinces, regions, or others) and a node is created for each of these parts. There is one representative of the commission at each territorial node who knows the local context regarding AI well. She will be in charge of implementing the gender actions in her area with the support of other members who live in the same area. This node representative works in line with the board of the commission which is integrated by the president of the commission and all the other territorial delegates.

This network structure brings several advantages:

- A better understanding of the reality of AI expertise and knowledge in each part of the territory. This local knowledge is crucial to adapt the actions to the particularities of specific areas.
- Ease in finding local professionals with the availability for deploying the initiatives in their local city, reducing effort and travel time, and facilitating the scheduling.
- Providing a capacity to react very quickly to interventions in all territorial areas, not just in big cities as it usually happens.
- Providing territorial proximity that facilitates the success of certain actions.

These advantages are aligned with significantly reducing the reaction time on the appearance of new activities, which increases the potential of the network.

As this network structure depends on each territory, in the following subsections we will explain the network organization of donesIAcat.

7.1. Organizing Based on a Territorial Structure for donesIAcat

Catalan Countries include the Spanish regions of Catalonia, the Balearic Islands, Valencia, and parts of Aragon (La Franja) and Murcia (Carche), as well as the Principality of Andorra, the department of Pyrénées-Orientales (including Cerdagne, Roussillon, and Vallespir) in the south of France, and the city of Alghero in Sardinia Island (Italy); all of the territories where Catalan is spoken. Based on these administrative regions and taking into account the population of the areas and the number of schools, donesIAcat has defined seven district nodes, each one with a local representative as shown in Figure 6.

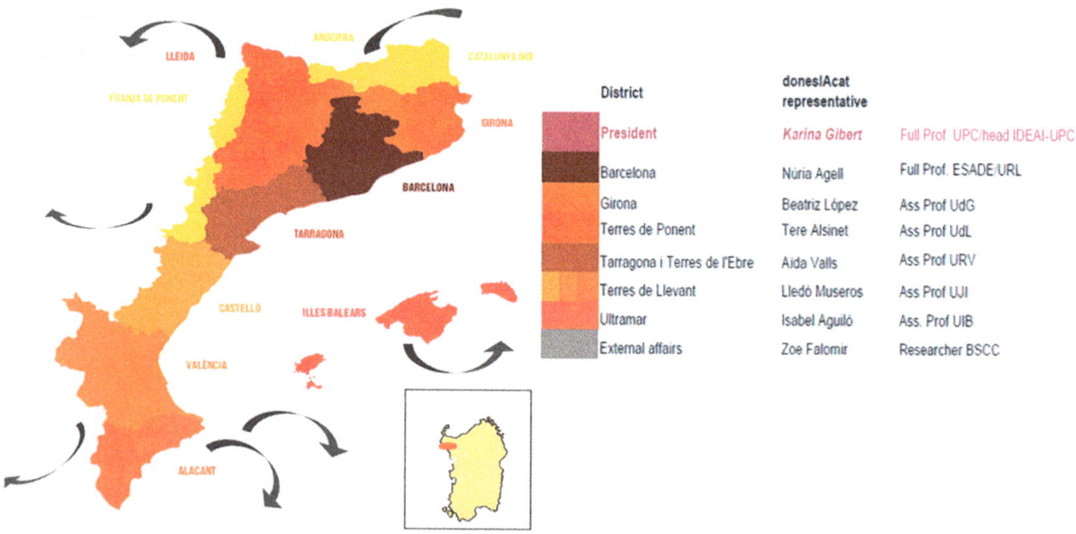

Figure 6. Territorial network organization of donesIAcat.

We have four nodes in Catalonia, one for each of its provinces and neighboring areas: Barcelona, Girona (including south of France), Terres Ponent (for Lleida province and Aragon area), and Tarragona i Terres de l'Ebre. Another node in the south corresponds to the Terres de Llevant region which gathers Castelló, Valencia, Alacant, and Carche. Finally, a node for the territories in islands is denoted Ultramar (which includes the Balearic Islands and Alghero).

In addition to these six local nodes, we propose to establish at least one international node with the aim of gathering national women that are now working abroad. A good connection with the professionals and academics that work in other countries is extremely for international role models, to make our talent and experts visible outside the territory, and to facilitate connections to the new AI professionals that participate in our talent programs.

7.2. Corporative Image

Defining a brand image is relevant for providing a unified view of the actions that come from the gender commission and facilitating the identification of the entity. In this line, a logo or icon image can help the user easily associate an activity with the commission. Logos are the symbol of the organization and should inspire an immediate recognition of the working group. They can be composed of an emblem or symbol and a wordmark. Meanwhile, an icon is an artistic picture that represents what is being offered by the organization. Icons help viewers to understand immediately what the organization is about.

In the gender commission donesIAcat, we have both a logo and an icon. As donesIAcat is a workgroup inside the ACIA association (Catalan Association for AI), the logo was designed using the same brand image from ACIA but adding some new elements that represent women. In Figure 7 (left) you can see the ACIA logo and (right) the logo of donesIAcat. This new logo adds the flower to represent women and uses a purple color which is associated with female actions. The original typography, background color form, and style were maintained to have a unified image with ACIA and let viewers identify the link between both.

Figure 7. ACIA and donesIAcat logos.

In addition to this abstract logo, donesIAcat also has an icon image, displayed in Figure 8. It represents a friendly female robot dressed in traditional Valencian clothes, mixing the idea of intelligence in an artificial body, feminization, and Catalan Countries. This icon was designed by a professional illustrator in 2015 as an image for the CCIA conference held in Valencia. She was called Rita, a typical Valencian female name. One of the organizers of CCIA 2015, who is also a member of the donesIAcat, gently transferred the use of Rita as a donesIAcat icon with a previous modification to add the donesIAcat brand in the dress of the robot. Rita officially became the icon of donesIAcat in 2020.

Figure 8. The icon image of donesIAcat called Rita.

The design of the logo and the icon should be carefully studied. Colors and symbols may have different meanings depending on the culture. In the case of donesIAcat, the members of the working group did a selection among different options. For example, the purple color is used in Spain to represent actions that are pro women's rights. However, it is important to take into account who is the target audience of the actions and then define a logo and icon that do not reinforce the stereotypes we are seeking to eliminate.

7.3. Communication Channels

The communication strategy, both with internal and external people, must be also appropriately defined. First, it is important to establish communication protocols with the members of the working group that enable agile decision-making. In our case, WhatsApp lists among members of donesIAcat permit quick dissemination of information and communication of new events or interesting issues. We have a second list for the board members that includes all the territorial representatives. Inside each territory, a specific list helps to spread the information in that location. This channel guarantees that information is spread quickly to all the territorial structures. Whatsapp, Slack, or Telegram might be good possibilities to fulfill this purpose. The advantage of the laters is that they are more robust and have more elaborate privacy protocols. Nevertheless, it is also important to have the possibility of doing at least one live meeting every year. In this regard, the working group donesIAcat has a yearly meeting of the majority of its members during the annual conference organized by ACIA to share their experiences, discuss improvement actions, and strengthen the relations between the members. Additionally, when convenient, online meetings are arranged throughout the year for specific issues that become difficult to manage through the Whatsapp or telegram lists. We must note that the kind of interactions in online meetings are not the same as presential yearly meetings. However, with the territorial scope of donesIAcat so wide, it is difficult to organize more than one presential meeting per year.

For communication outside the group, we created a Twitter channel (@donesIAcat) and have a section on the webpage of the ACIA association. (https://www.acia.cat/catala-dones/ (accessed on 17 March 2022)).

7.4. Institutional Relationships

Gender committees should be connected with the relevant institutions in their countries not only to be known but also to establish relationships and participate in joint activities. The working group donesIAcat is in close contact with *donesCOEINF*, the gender commission of the Official Professional Chamber of Informatics Engineering of Catalonia, and with Mujeres en IA (from AEPIA, the Spanish Association on Artificial Intelligence). At the same time, they are linked to other gender structures in Spain, as well as at the international level. Establishing synergies may help us to be more effective in our goals. In Catalonia, there are some other institutions that work to reduce gender bias in a more general framework. Some members of donesIAcat are representatives in the associations "50 × 50 lideratge compartit" (https://www.50a50.org/es/ (accessed on 17 March 2022)) (association created by the Commerce Chamber in Barcelona) and promoters and leaders of the "Red de Mujeres del Sector Digital" (https://sheleader.eu/web/es/redes/mujeres-sector-digital (accessed on 17 March 2022)) (an ICT network created from donesIAcat, SheLeader, donesCOEINF, and Telecos.cat). As gender issues are often related to ethics as well, it is relevant to mention that some donesIAcat members are members of the Ethics Advisory boards of entities like the Catalan Observatory on Ethics in AI (OEIAC) or the Centre of Innovation and Development of Artificial Intelligence in Catalonia (CIDAI). DonesIAcat also has two women that are part of the experts' group of the Artificial Intelligence strategy of the Catalan Government (catalonia.ai) which is now deploying. Fortunately, the Catalan Parliament has a special interest in gender equality and equity and has started several initiatives to achieve them. Gender working groups should be aware of these governmental activities and participate in them. In our case, every year since 2019 one representative from donesIAcat participates in the panel sessions of the Catalan Parliament entitled "Effective women and men equality: a country challenge" and follows meetings led by the Catalan Parliament President around different topics in the field.

8. Discussion and Conclusions

The implications of Artificial Intelligence in current and future societies are enormous. Traditionally, we developed machines and software programs for solving specific and concrete problems. In contrast, today we are developing technologies that can learn and adapt dynamically, being more autonomous. Knowing, understanding, and participating in the design and implementation of these new AI systems will be important for future generations. Women must be working in the field with equal capabilities and opportunities with respect to men.

This paper has started by showing that the number of women studying Informatics is around 14%, which significantly reduces the presence of women with knowledge about AI and consequently, there are few women working in the field and influencing the design of new AI technology. Having recognized that reducing the gender gap in this sector is a challenge, we have proposed a conceptual model for creating national or regional working groups to work effectively and efficiently to bridge this gap. This model is organized into six axes around three main lines from which we can extract some conclusions:

- Inspiration and talent training: Activities to stimulate AI interest in girls and to give AI knowledge by means of educational programs designed for different stages of life and serving different purposes (regular university degrees, up-skilling courses, or re-skilling training). This paper has presented different activities conducted inside and outside educational centers. As female AI experts are the teachers and mentors of these activities, they become role models for the youths. This is a key point in the proposed conceptual framework since it is known that gender stereotypes can be reduced by making female talent visible as role models.
- Visibilization actions: Actions presented to make AI experts visible were targeted mainly to impact the general society by means of more appearances in public events to help women have equal opportunities in job promotions, in the market, and in leadership positions. These women can then become part of the gender commission or participate in inspirational activities for girls, becoming new role models. In addition, it is also important to carry out some activities for professionals or AI scientists (e.g., in fairs, congresses, and journals), as women in these fields are currently in a minority. Only a global view of visibilization will be able to effectively make a change.
- Networking: We have seen some useful tools for the internal organization of the network and for having a unified image for communication. Our model is based on the creation of territorial structures that contribute to decentralizing the activities and facilitating the dissemination to every city in an easy way. Once the internal aspects are solved, it is important to find collaborations with other entities, creating synergies with existing initiatives already established in the territory.

This network-based organizational model is suitable for relatively small territories or countries where it is feasible that women in the working group know each other and contacts between them are easy to establish. For larger areas, a different kind of organization, such as a hierarchy, would be probably better. This model is also appropriate for territories sharing a common history, language, and way of living, so that the activities proposed in the group are suitable for the different cities in the region. Proximity is one of the advantages of such a model and one of the reasons for the success of the inspiration and visibilization actions.

This conceptual model is supported by the experience of the gender working group donesIAcat of the ACIA association which holds its activities in the Catalan Countries. Since its foundation at the beginning of 2019, the group has been able to consolidate a large number of activities following the proposed methodology.

One of the limitations encountered is the lack of funded projects oriented to the development of activities to reduce gender bias. Although there is a general agreement about the importance of increasing female presence in AI, until now, most of the activities done in this commission are currently made voluntarily, without any return, and using the non-working time of the members of the commission to develop and prepare the activities. This is of course limiting the kind of actions that can be activated since we cannot, at the moment, think of working plans that require too much personal effort from the members of the working group or too many material resources. Some concrete actions have accounted for specific sponsors from companies, but they are not easy to find. Universities have also provided some funds for some actions organized with them, especially inspirational activities and talent training courses. We must highlight that recently, the Catalan government approved the strategic plan on AI from Catalonia (https://politiquesdigitals.gencat.cat/ca/tic/catalonia-ai (accessed on 17 March 2022)) which recognizes the gender gap in AI and identifies gender gap as a strategic priority in Catalonia [19]. We want to highlight this pioneering policy in favor of promoting the societal transformation required to balance the presence of women in STEM. This policy may also open the possibility of new financial resources for the actions done in gender groups.

The activities of donesIAcat started just three years ago and some of them will show results in the mid-term, so they cannot be properly measured yet. However, we can provide some insights about the kind of impact related to the actions developed from donesIAcat. The first inspirational and dissemination actions were with children in schools, as well as in the YOMO congress. After 3 years, some of the girls that participated have arrived at the university and even though we have not made a formal study, we can observe an increase in the number of female students in the Informatics degrees. For example, at UPC (in Barcelona) there has been an increase in the presence of women from 7% in 2019 to 14% in 2021; similarly, in the Universitat Rovira i Virgili (in Tarragona) the increase was from 8.7% in 2019 to 11.5% in 2021.

Moreover, building a network of female professionals creates the atmosphere to share glass-ceiling experiences and can provide resources to help the affected woman better manage the situation and minimize the impact on her career. After several years, the awareness and capacity of all members of the working group to identify gender inequalities in their working contexts has increased and they have acquired expertise in gender issues so that they can also help other women in their companies. This training helped, for example, to build a new type of promotional speech to promote the new official degree on Artificial Intelligence in the Barcelona School of Informatics that started in September 2020 and has attracted 45% of female students. Based on this evidence, the school is now redesigning the promotion programs of the bachelor of Informatics Engineering using the same technique of enhancing and making explicit the social role of technology and how these professions are useful in contributing to health, wellness, or sustainability from a professional position different from that of doctors, psychologists, or environmental scientists, for example.

We certainly need a more temporal perspective to collect rigorous evidence associated with our actions, but we can already observe some promising results that encourage us to continue working in the proposed directions. However, after analyzing the work done by donesIAcat until now, we can say that we are proud of being quite active along the territory. We must continue to engage women to increase our presence in some parts of the Catalan Countries.

We would also like to get men involved in this gender balance goal. As said before, ACIA only has 17.5% of female associates, and being a minority, it is possible that the board of ACIA, composed of 77% of men in 2019, did not accept the creation of the donesIAcat working group. Fortunately, the situation was contrary and the initiative to create donesIAcat was highly welcomed by the ACIA board. This guaranteed that the working group was officially integrated into the association structure and approved by the government of the association. In addition, our experience is that 82.5% of male associates in ACIA support the activities of the donesIAcat and encourage us to continue working to bridge the gender gap in the AI sector. However, we would like to design specific activities to also involve men in the dissemination of our message and the development of activities where they can participate to contribute in a more effective way to bridge the gender gap in AI.

Some other tasks are also in our plans, such as the creation of an online procedure for subscription to the working group or to receive news from the group. Another line of work is the development of a communication plan that facilitates the presence in the media to disseminate the activities done and to consolidate activities to reach more girls and to support more professionals in their professional development. We should have a community manager and then create specific content on social networks like Instagram or Linkedin. We strongly believe that our young working group will be able to reach these goals and give girls the opportunity to know what Artificial Intelligence is and that they can study and work in the AI field if they want to.

Author Contributions: All authors have equally participated in the conceptualization, methodology, writing, review and edition of this paper. All authors have read and agreed to the published version of the manuscript.

Funding: This research received no external funding.

Institutional Review Board Statement: Not applicable.

Informed Consent Statement: Not applicable.

Acknowledgments: The authors want to acknowledge all other members of donesIAcat working group for their free dedication to bridging the gender gap in AI: Núria Agell, Beatriz López, Lledó Museros, Isabel Aguiló, Zoe Falomir, Teresa Alsinet, Emilia López-Iñesta, Àngela Nebot, Atia Cortés, Maite López, Pilar Dellunde, Maria Salamó, Mònica Sànchez, Eva Armengol, and Elisabet Golobardes. Additionally, our acknowledgment to the ACIA association for the support in allowing the creation of donesIAcat.

Conflicts of Interest: The authors declare no conflict of interest.

Abbreviations

The following abbreviations are used in this manuscript:

ACIA	Catalan Association of Artificial Intelligence
AEPIA	Spanish Association on Artificial Intelligence
AI	Artificial Intelligence
AICCSA	ACS/IEEE International Conference on Computer Systems and Applications
CCIA	International Conference of the Catalan Association for Artificial Intelligence
CIDAI	Centre of Innovation and Development of Artificial Intelligence in Catalonia
EurAI	European Association for Artificial Intelligence
ICT	Information and Communication Technologies
IDEAI	Intelligent Data Science and Artificial Intelligence Research Center
OEIAC	Catalan Observatory on Ethics in Artificial Intelligence
STEM	Science, Technology, Engineering, and Mathematics
UPC	Universitat Politècnica de Catalunya—BarcelonaTech
WiDS	Women in Data Science
WiML	Women in Robotics or Women in Machine Learning
YOMO	Youth Mobile Festival

References

1. CEDEFOP. *Skills Forecast: Trends and Challenges to 2030*; Publications Office CEDEFOP Reference Series; European Center for the Developement of Vocational Training: Luxembourg, 2018. Available online: http://data.europa.eu/doi/10.2801/4492 (accessed on 12 December 2021).
2. Martinez, A.; Christnacht, C. Women Are Nearly Half of U.S. Workforce but Only 27% of STEM Workers; US Census Bureau Report, Online. 26 January 2021. Available online: https://www.census.gov/library/stories/2021/01/women-making-gains-in-stem-occupations-but-still-underrepresented.html (accessed on 12 December 2021).
3. European Commission. Women in the Digital Age. 2018. Available online: https://op.europa.eu/en/publication-detail/-/publication/84bd6dea-2351-11e8-ac73-01aa75ed71a1 (accessed on 12 December 2021).
4. Ministerio de Universidades, Gobierno de España. Datos y Cifras del Sistema Universitario Español 2020–2021. Available online: https://public.tableau.com/views/Academica20_EEU/InfografiaEEU?%3AshowVizHome=no&%3Aembed=true#7 (accessed on 12 December 2021).
5. Bian, L.; Leslie, S.J.; Cimpian, A. Gender stereotypes about intellectual ability emerge early and influence children's interest. *Science* **2017**, *27*, 389–391. [CrossRef] [PubMed]
6. López-Iñesta, E.; Forte, A.; Botella-Mascarell, C.; Marzal, P.; Rueda, S. Niñas y Disciplinas STEM: Si no Están, Será Porque no les Gusta. Available online: https://theconversation.com/ninas-y-disciplinas-stem-si-no-estan-sera-porque-no-les-gusta-155339 (accessed on 12 December 2021).
7. Ward, L.M.; Grower, P. Media and the development of gender role stereotypes. *Annu. Rev. Dev. Psychol.* **2020**, *2*, 77–99. [CrossRef]
8. OECD. *Pisa 2015 Results (Vol 1: Excellence an Equity in Education)*; OECD: Paris, France, 2016.
9. Hoffman, S.F.; Friedman, H.H. Machine Learning and Meaningful Careers: Increasing the Number of Women in STEM. *J. Res. Gend. Stud.* **2018**, *8*, 11–27. [CrossRef]
10. Gibert, K.; Pérez, C.; Castell, N. Deployment of territorial structures to reduce gender gap in technology and some real cases in Catalonia. In Proceedings of the 2019 24th IEEE International Conference on Emerging Technologies and Factory Automation (ETFA), Zaragoza, Spain, 10–13 September 2019; pp. 1823–1830. [CrossRef]
11. Samuel, Y.; George, J.; Samuel, J. Beyond STEM, How Can Women Engage Big Data, Analytics, Robotics and Artificial Intelligence? An Exploratory Analysis of Confidence and Educational Factors in the Emerging Technology Waves Influencing the Role of, and Impact Upon, Women. 2018. Available online: https://doi.org/10.2139/ssrn.3735279 (accessed on 12 December 2021).

12. Benavent, X.; De Ves, E.; Forte, A.; Botella-Mascarell, C.; López-Iñesta, E.; Rueda, S.; Roger, S.; Perez, J.; Portalés, C.; Dura, E.; et al. Girls4STEM: Gender Diversity in STEM for a Sustainable Future. *Sustainability* **2020**, *20*, 6051. [CrossRef]
13. Roopaei, M.; Horst, J.; Klass, E.; Foster, G.; Salmon-Stephens, T.; Grunow, J. Women in AI: Barriers and solutions. In Proceedings of the IEEE World AI IoT Congress, Seattle, WA, USA, 10–13 May 2021. [CrossRef]
14. Prives, L. AI for all: Drawing women into the artificial intelligence field. *IEEE Women Eng. Mag.* **2018**, *12*, 30–32. [CrossRef]
15. United Nations. Transforming Our World: The 2030 Agenda for Sustainable Development. 2015. Available online: https://sustainabledevelopment.un.org/post2015/transformingourworld (accessed on 12 December 2021).
16. Callaghan, S.; Darbyshire, T.; Rafajnia, S. Ada Lovelace day and celebrating women in STEM, Editorial. *Patterns* **2020**, *1*, 3. [CrossRef] [PubMed]
17. UN Women. Visualizing the Data: Women's Representation in Society. UN Digital Library. February 2020. Available online: https://www.unwomen.org/en/digital-library/multimedia/2020/2/infographic-visualizing-the-data-womens-representation (accessed on 12 December 2021).
18. Agell, N. Women in Technology, a Major Challenge for the Future. Available online: https://www.forbes.com/sites/esade/2021/03/08/women-and-technology-a-major-challenge-for-the-future-of-work/?sh=444ee95aeeeb (accessed on 12 December 2021).
19. GENCAT. *Catalonia.AI: L'estratègia d'Intel·Ligència Artificial a Catalunya*; Department of Digital Policies: Barcelona, Spain, 2019. Available online: https://politiquesdigitals.gencat.cat/ca/tic/catalonia-ai (accessed on 12 December 2021).

applied sciences

MDPI

Article

Artificial Intelligence and Women Researchers in the Czech Republic

Lenka Lhotska [1,2,*] and Olga Stepankova [2]

1 Faculty of Biomedical Engineering, Czech Technical University in Prague, 272 01 Kladno, Czech Republic
2 Czech Institute of Informatics, Robotics and Cybernetics, Czech Technical University in Prague, 160 00 Praha, Czech Republic; olga.stepankova@cvut.cz
* Correspondence: lhotska@cvut.cz

Abstract: Artificial intelligence as a research area has been continuously growing for several decades. Many applications were developed in various domains. Medicine and health care have attracted more intensive attention thanks to rapid technological development that has accelerated generation of large volumes of data requiring intelligent analysis and evaluation. This article illustrates, through examples of women researchers and selected AI projects in medicine, the wide spectrum of applications developed during the last fifteen years in the Czech Republic, and in particular at the Czech Technical University in Prague. Women researchers played an important and irreplaceable role since the advent of AI research in the Czech Republic. By their example, they motivated many young female students to join the community and start their research career in the AI area. They frequently participated in research projects led by the senior women researchers. The presented overview of projects illustrates the diversity of the medical area and the potential of AI methods that can be used for solving data- and knowledge-intensive problems. We briefly touch on the AI study programs. In conclusion, we point out the future challenges in AI and its applications in medicine and health care.

Keywords: artificial intelligence; machine learning; women; research

Citation: Lhotska, L.; Stepankova, O. Artificial Intelligence and Women Researchers in the Czech Republic. *Appl. Sci.* **2022**, *12*, 1465. https://doi.org/10.3390/app12031465

Academic Editor: Federico Divina

Received: 19 December 2021
Accepted: 24 January 2022
Published: 29 January 2022

1. Introduction

Artificial intelligence (AI) has been with us for several decades. We can observe both intensive theoretical research and the development of AI applications in various areas. The research and development of AI are growing continuously as technologies generate larger volumes of data that require intelligent analysis and evaluation. One of the application areas that has been developing fast is health care, where rapid technological development during the last two decades of the 20th century and both decades of the 21st century changed the character of the health-care services and many types of medical examinations. Advanced technology was introduced to diagnostics and therapy. As a consequence, thanks to direct connection of devices to computers, the volume of collected data has been continuously and tremendously growing. Better communication channels allow speeding up the exchange and use of data, information and knowledge, and are eliminating geographical and temporal barriers. All these developments increased the importance of medical informatics as a discipline that builds on the results of computer science, artificial intelligence, image processing, etc. The crucial issue is the suitable design of health information systems, in particular electronic health/patient records that should contain all the important data and information about the patient. As has been discussed many times, the critical point is represented by semantic interoperability and satisfaction of standards. Without that, it is difficult to share the data and information properly. The requirements for sharing data are gradually rising as population mobility is increasing and patient data is being accumulated in different places. However, the data needs to be accessible in an organized manner; that means the systems must satisfy defined standards and interoperability requirements.

This approach has been adopted within *the integrated care cycle* (ICC). Its idea is based on the vision of such an environment where the patient-centered approach is applied, where care is personalized, where prevention and maintaining health is central to health-care solutions, where hospitals provide the least-invasive interventions with the shortest length-of-stay for any condition, and where recovery occurs outside the hospital. ICC can enhance both patient centricity and care cycle efficiency through embedded technology and through a holistic approach to integrating health data, health workflows and health systems and devices. ICC may strengthen the health-care ecosystem of hospitals, care providers, health insurers, research institutes and industry in healthcare-embedded systems by bringing people-centric innovations to the market faster. The ICC is built on the following steps: prevention, screening for early detection and diagnosis, discovery to treatment, minimally invasive interventions, management and monitoring, and chronic disease management. Obviously, the activities must be coordinated, and allow for adequate data and information exchange. This requires well-designed data and information models and formats satisfying the interoperability requirements. It is necessary to stress that both data and information need to be accompanied with contextual information. AI methods and systems can contribute to this effort since many appropriate tools and methods have been developed, in particular in machine learning, knowledge representation, image processing, natural language processing and related areas.

The next sections focus on the development of AI in the Czech Republic, the active role of female researchers and examples of research projects that were successfully solved in the health-care area.

2. Women in Artificial Intelligence Research in the Czech Republic

Development of AI in the Czech Republic is similar to that in other European countries. In the 1970s, more intensive activities appeared, including first workshops and conferences. Since the beginning, several women researchers actively participated and successfully developed their research areas and research groups. During the decades they participated in education of new generations of researchers. Through the positive examples they set, they succeeded to attract more women to AI research. Many of them are still active researchers both in the theoretical foundations of AI and applications of AI to various areas. Starting in the early 1990s, a series of textbooks on artificial intelligence In Czech [1–6] was written with the aim to bring the topics to a broader audience. Many women researchers in AI were invited as (co-)authors of chapters in the individual volumes that covered not only theoretical issues but also development of numerous AI applications, including those designed for medical purposes. In the next paragraphs, we present several female personalities from the Czech AI scene. We are well aware of the fact that we cannot accomplish a full list of all AI researchers and research groups in the country.

Eva Hajičová (* 1935) is a professor of linguistics at the Faculty of Mathematics and Physics, Charles University, Prague. She graduated in English and Czech at the Faculty of Philosophy of Charles University and got her PhD degree as well as the highest academic degree of DrSc in general and computational linguistics at Charles University. Her interests cover theoretical linguistics as well as computational applications [7]; she has concentrated on the semantic structure of the sentences, on the discourse phenomena and on different topics in computational linguistics [8].

She is/was a member of a number of editorial boards of international journals (*Journal of Pragmatics, Computers and Artificial Intelligence, Linguistica Pragensia* and *Kybernetika*) and she was the editor-in-chief of *Prague Bulletin of Mathematical Linguistics*. She was the first president of the European Chapter of the Association of Computational Linguistics (1982–1987) and the president of the international Association for Computational Linguistics in 1998; she was the President of the Societas Linguistica Europaea, 2006–2007, and since 1978 she has been a member of the International Committee of Computational Linguistics.

She was the chairperson of the Prague Linguistic Circle (1997–2006), is a Fellow of the Association of Computational Linguistics and na honorary member of Societas Linguistica Europaea, as well as a member of several Czech scientific societies.

She was awarded the international Alexander von Humboldt Research Prize in 1995, the Medal of the Minister of Education of Czech Republic in recognition of the pedagogical and scientific work in computational linguistics in 2003 and the ACL Life Achievement Award in 2006. In 2017, she received the Josef Hlávka Medal awarded to Czech scientists in recognition of their life-time achievements in science and arts, and in 2018 she received the international Antonio Zampolli Prize for outstanding contributions to the advancement of language resources and language technology evaluation within human language technologies.

Jana Zvárová (1943–2017) graduated in Mathematics in 1965 at the Faculty of Mathematics and Physics of Charles University in Prague. She collaborated in several disciplines offered at Charles University (Medicine, and Mathematics and Physics). She founded the Medical Informatics section of the Czech Society of Biomedical Engineering and Medical Informatics in 1978. She was nominated in 1999 as full professor at Charles University and received in the same year the highest Czech scientific degree of Doctor of Sciences at the Academy of Sciences of the Czech Republic.

She systematically applied new theoretical knowledge in biomedicine, particularly in epidemiology and public health [9]. Since 1994, she chaired the *European Center of Medical Informatics, Statistics, and Epidemiology* (EuroMISE) of Charles University and the Academy of Sciences. Between 2006 and 2011, she was the director of the *Center of Biomedical Informatics*. She was the representative of the Czech Republic in IMIA (the International Medical Informatics Association) and in EFMI (the European Federation for Medical Informatics). She was a member of editorial boards of several national and international journals. The results of her research are documented in 10 monographs and more than 300 articles in peer-reviewed journals.

In the framework of European projects, she started new lines of research and education concerning electronic health records, knowledge representation in clinical guidelines, decision support systems, and methods for evaluation of knowledge. She organized several IMIA and EFMI international conferences and workshops in Prague. Jana Zvárová also initiated the foundation of the EuroMISE Mentor Association focusing on international cooperation in student mentoring activities.

Věra Kůrková (* 1948) graduated from the Faculty of Mathematics and Physics of Charles University, specialization in Topology (1972) and later obtained the Ph.D. degree in the same field from the Academy of Sciences of the Czech Republic (CAS). Between the years 1975 and 1989, she worked as a researcher at the Research Institute of Mathematical Machines. In 1990 she joined the Institute of Computer Science of the CAS, where she remains until now. This position gave her an opportunity to participate in shaping the field of artificial neural networks at the turbulent moment when development of hardware enabled the implementation of algorithms for efficient learning of networks composed of biologically inspired perceptron units that were complemented soon by computational units chosen purely for their suitable mathematical properties. As a mathematician, she focused on theoretical questions inspired by experimental research and worked on building the theory of nonlinear approximation, which involves the approximation of functions of many variables by neural networks of different types, the study of the dependence of network complexity on increasing data dimensionality and the connection between inverse problem theory (which was developed for solving physical problems) and generalization in machine learning.

From 2002 to 2008 she held the position of the head of the Department of Theoretical Computer Science and currently she is a senior researcher in the Department of Machine Learning. Her research interests include the mathematical theory of artificial neural networks and learning, in particular the analysis of the capabilities and limitations of shallow and deep neural networks, the dependence of network complexity on increasing input dimension, optimization of networks computing randomly chosen classifiers [10], the relationship between inverse problem theory and generalization in machine learning and a new branch of function approximation theory involving neural networks. She is a member of the editorial boards of *Neural Networks* (Elsevier) and *Neural Processing Letters* (Springer), and has been a chair or a vice-chair of the program committees of several European conferences. Since 2008, she has been a member of the Council of the European Neural Network Society, of which she is currently President. She was awarded the Bolzano Medal by the Czech Academy of Sciences in 2010.

Iveta Mrazová (* 1966) finished her studies at the Friedrich Schiller University in Jena, Germany, in 1989. She worked on her thesis dedicated to neural nets at the Institute of Informatics of the CAS. In 1996, she was awarded the Annual Prize of the Bolzano Foundation for the collection of original publications on neural networks and a year later she gained a Ph.D. degree. In 2002 she won the Fulbright Commission scholarship, which gave her an opportunity to gain international experience during her stay at the Missouri University of Science and Technology, USA. Since 1992 she has been teaching at the Faculty of Mathematics, Physics and Informatics of the Charles University and in 2007 she became associate professor at the Department of Theoretical Computer Science and Mathematical Logic—she is currently the head of this department. She has received numerous awards for her work in science, including the annual Bolzano Foundation Award and the Prof. Babuska Award, given by the Union of Czech Mathematicians and Physicists and the Czech Society for Mechanics. She has long-term interest in the utilization of neural nets in the context of data-mining. The paper "Enforced Knowledge Extraction with BP-Networks" [11], co-authored by her and Zuzana Reitermanová, obtained the Best Paper Award of the conference Artificial Neural Networks In Engineering (ANNIE) in 2007. Since 2014 she has been cooperating on analysis of data gathered in the Czech Insolvency Proceedings [12].

Vlasta Radová (* 1969) graduated in 1992 from the Faculty of Applied Sciences of West Bohemian University, majoring in technical cybernetics. Here she continued her doctoral studies and subsequently habilitated in 2005. She remained faithful to the Faculty throughout her professional life. She works at the Department of Cybernetics and at the university research center dedicated to new technologies for information society, NTIS. Her research focuses on artificial intelligence, speech technologies, speech processing [13] and speech recognition [14]. She has completed internships abroad in the USA and Hungary, and is a member of several professional societies and committees. In 2009–2014, she represented her alma mater on the Board of the Council of Universities; she is also a long-time member and former chair of the Academic Senate of the Faculty of Arts and Sciences and the Academic Senate of the University of West Bohemia.

Hana Rudová studied mathematical informatics at the Faculty of Informatics, Masaryk University, in Brno, where she obtained a master's degree in 1995 and entered for her doctoral studies, too. In 2001, she gained a Ph.D. degree for her thesis *Constraint Satisfaction with Preferences*. Since then she has been working on various problems broadly related to scheduling and routing, such as educational timetabling, scheduling for distributed environments or transport planning [15]. Her work is inspired by real-life problems coming from practice, and she concentrates on approaches that can contribute to solving practical problems, namely, meta-heuristics, constraint programming or mixed integer programming. She applies her knowledge in problems arising, e.g., during course timetabling in the UniTime system, computer job scheduling in CERIT national infrastructure or vehicle routing [16]. She generously shares her practical experience with students during the courses on Scheduling, Constraint Programming, and Artificial Intelligence. In 2010, she gained the title associate professor in Informatics at Masaryk University, Brno. From 2011 to 2015, she served as the vice-dean of bachelor's and master's studies at her alma mater.

Hana Rudová is a co-author of more than 120 research papers. She is an associate editor of *Journal of Scheduling* and member of the PATAT steering committee. She co-chaired the Novel application track at the ICAPS conference in 2017 and 2018 and co-chaired as well as co-organized the PATAT 2006 conference in her home town. She regularly serves in program committees of conferences such as ICAPS, AAAI, IJCAI, PATAT or MISTA, and co-organizes the ongoing International Timetabling Competition (ITC 2019) with more than 300 registered users from about 60 countries. She spent a half year both at Carnegie Mellon University in 2016 and Purdue University in the USA in 2001.

Jiřina Vejnarová (* 1962) obtained the RNDr. degree (corresponding to M.Sc.) in Probability and Statistics from the Faculty of Mathematics and Physics, Charles University, Prague, in 1986, and the CSc. degree (corresponding to Ph.D.) in Theoretical Cybernetics from the Czechoslovak Academy of Sciences in 1991. Since 1986, she has been with the Institute of Information Theory and Automation at the Czech Academy of Sciences. In 2009, she became the deputy director for research of this institute and in May 2017 its director. Her research interest is in the field of structured multidimensional models in the framework of imprecise probabilities, particularly in possibility and evidence theories [17]. She has authored (and co-authored) more than 50 research publications in these areas. She has been involved in several research projects (both Czech and international ones). Since 1996 she has been teaching at the Faculty of Informatics and Statistics, University of Economics, Prague, and since 2008 at the Faculty of Nuclear Science and Physical Engineering, Czech Technical University in Prague. At present she is responsible for the undergraduate courses on Theoretical Computer Science, Information and Inference Theory and Probabilistic Models of Artificial Intelligence. In 2011, Jiřina Vejnarová was appointed associate professor of the Czech Technical University in Prague. She has been a member of numerous program and organizing committees at international conferences and personally organized the 5th International Symposium on Imprecise Probabilities: Theory and Applications ISIPTA'07 in Prague. She also co-chaired the program committee of this conference. From 2003 to 2010 she was a member of the Board of Czech Society for Cybernetics and Informatics; she is also a member of the Editorial Board of the journal *Kybernetika*.

Barbara Zitová received her Ph.D. degree in software systems from the Charles University, Prague, Czech Republic, in 2000. Her research interests cover all aspects of digital image processing and pattern recognition; particularly, object recognition by invariants [18], degraded image recognition [19], geometric invariants, theory of moments, remote sensing and medical imaging applications and cultural heritage applications. She is a head of Department of Image Processing at the Institute of Information Theory and Automation, Czech Academy of Sciences, Prague. She teaches advanced courses on Digital Image Processing and Wavelets in Image Processing for students of the Faculty of Mathematics, Physics and Informatics at Charles University and of the Faculty of Nuclear Science and Physical Engineering, Czech Technical University. She has authored/co-authored more than 70 research publications in these areas, including two internationally recognized monographs on pattern recognition [20] and image analysis [21]. Barbara Zitová has many editorial and organizational activities. Among others, she has been an Associate Editor of the journal *Pattern Recognition*.

Olga Štěpánková (* 1949) graduated in Theoretical Cybernetics at the Faculty of Mathematics and Physics, Charles University, Prague, in 1972, and received her Ph.D. in Mathematical Logic there later. In 1972, she joined as a researcher The Computer Science Institute of the Czech Technical University in Prague. Since 1988 she has been working at the AI group at the Faculty of Electrical Engineering of the Czech Technical University in Prague (CTU). In 1998 she became professor of technical cybernetics at CTU. During the period 2004–2012 she worked as deputy head of the Department of Cybernetics at CTU. After establishment of the Czech Institute of Informatics, Robotics and Cybernetics at CTU she joined the institute in 2016 and since then she is head of the department of Biomedical Engineering and Assistive Technologies.

When starting her research carrier, she wanted to test the strength of mathematical logic as a tool for accomplishing some of the AI tasks [22]; e.g., goal-oriented action planning for an autonomous agent. That topic led her later to investigation of the properties and limits of the logic programming paradigm and of the reasoning necessary for building distributed systems [23]. Both these research streams share an interest in the efficient performance of the suggested approaches. The sources of inefficiency in a specific solution can be identified through careful analysis of the data produced by the system during its problem-solving activity. She started to search for machine learning tools that could do the job and she became fascinated by their ability to find structure in extensive complex data sets [24–26]. She saw promising potential machine learning offers for data interpretation in the domain of biomedical engineering and since then her team has been cooperating with numerous experts from the medical or biochemical domains, not only to design and implement complex SW solutions for medical data collection and interpretation but also to study how AI and robotics can contribute to better care through the design of novel assistive technologies. She was always aware of the responsibility researchers have towards society and as an active member of the Czech Society Cybernetics and Informatics (CSKI) she promoted utilization of standardized and continuously innovated ICDL/ECDL international certification of digital literacy in the Czech labor market. She supports this concept even now as the CSKI chairwoman.

Lenka Lhotská (* 1961) graduated as Master of Science in Electrical Engineering at the CTU. In 1989 she got her PhD degree in Cybernetics from CTU. In 1984 she joined the Department of Control Engineering, CTU. In 1997 she became associated professor of Cybernetics and head of the Biocybernetic Lab, AI Division, of the Department of Control Engineering. In 2016 she joined the Czech Institute of Informatics, Robotics and Cybernetics, CTU. Currently, she is head of the COGSYS Department (Cognitive Systems and Neurosciences) at the Czech Institute of Informatics, Robotics and Cybernetics and deputy head of the Department of Natural Sciences of the Faculty of Biomedical Engineering, CTU.

Lenka Lhotská joined the AI area in mid 1980s and since that time she has been focusing on decision support systems, knowledge representation, machine learning and multi-agent systems. More than 25 years ago, she started more intensive research and development of the medical applications of AI methods. Currently, her research focuses on following areas: knowledge-based systems, data and knowledge representation, application of artificial intelligence methods to medicine, digital signal processing, machine learning, feature extraction and feature selection, semantic interoperability, mobile technologies in healthcare and electronic health records. She is chair of the Working Group Personal Portable Devices of European Federation for Medical Informatics, member of the Council of the Czech Society for Biomedical Engineering and Medical Informatics, national representative in the International Society for Telemedicine and eHealth (IsfTeH), national representative in International Federation for Medical and Biological Engineering (IFMBE) and Member of the Engineering Academy of the Czech Republic.

She has supervised 20 PhD students (5 of them were women), who all successfully defended their theses. Currently, she is supervising 7 Ph.D. students. She has authored/co-authored more than 100 research publications in the abovementioned research areas, including several book chapters. She was co-chair of the program committee of the 2018 IUPESM World Congress on Biological and Medical Engineering and Medical Physics. She has many editorial and organizational activities. Among others, she has been an Associate Editor of the journal *Health & Technology*.

All these personalities contributed and still contribute to the development of AI methods and their applications. Jana Zvárová († 2017) was among the first who initiated research in medical informatics and decision-support systems. She envisaged the importance of the relation between data, information and knowledge, in particular in medical applications [27,28].

All of these women serve till now as positive role models for students and young researchers. So, there is no wonder that they found their followers, among which many talented female researchers appeared and became members of their research teams or finally organized their own teams.

3. AI Project Applications in Medicine

Modern health care is highly specialized. Complex examination of a single patient involves many expert consultations and laboratory tests. Medical knowledge, examinations and treatment are distributed functionally, geographically and also temporally. There is a need for reliable and consistent information flow among all participating subjects, with the aim to satisfy the global goal—improved health of a patient. Of course, the necessary information flow is not predictable in its extent or structure, but develops and changes in time due to new knowledge and reactions. To satisfy these requirements and provide adequate decision support, the use of flexible intelligent software support is becoming increasingly desirable. Many of these systems utilize artificial intelligence methods as a suitable approach to big heterogeneous multidimensional and multimodal data analysis.

When we look at the topics of our research projects during the past decades we can find several common lines, namely, large multimodal and multidimensional data, continuous data (e.g., signals), machine learning, feature extraction, feature selection using optimization methods and semi-automated classification methods [29–32].

The first application areas in medicine were those that are data intensive. In our case, the collaboration started in cardiology, electrophysiology, neurology and diabetology. Later, rehabilitation was added. With the advancement of the Internet of Things and wearables, new areas opened in home care and telemedicine [33].

In the following, we present the most important relevant projects we have been involved in and their results. We selected as examples those projects in which the principal investigator or even the co-investigator was a woman. This information is shown at each project.

Knowledge-based support of diagnostics and prediction in cardiology (PI female—Lenka Lhotska)

In this project, a set of theoretically well-developed methods and algorithms for knowledge-based support of diagnostics and prediction in cardiology was designed and developed [34,35]. Developed methods were implemented. The implemented system allows to analyze data from multi-electrode records of surface ECG [36], solve inverse problem, integrate results of further methods and suggest diagnosis on their basis [37–39]. A database of interpreted data was developed. The project supported especially interconnection of basic research in the area of artificial intelligence with application of developed methods to medicine. A real-world task provided a platform for verification of both the functionality of the proposed methods and the medical application itself. The project results are used in electrophysiology and constituted a good theoretical platform for a successive project in cardiology, as described in the next paragraph.

Features of electromechanical dyssynchrony that predict effect of cardiac resynchronization therapy (PI female—Lucie Riedlbauchova; co-investigator—Lenka Lhotska)

Impaired coordination between the ventricles or uncoordinated contraction of individual ventricular walls against each other (dyssynchrony) contribute to further heart failure progression in some patients. Reduction or elimination of dyssynchrony through cardiac resynchronization therapy (CRT) is able to slow or even stop pathologic remodeling, improve heart failure symptoms and induce reverse remodeling, leading to a decreased hospitalization rate and mortality. However, our abilities to identify the presence and type of dyssynchrony are still limited and although the primary goal of CRT is to restore the normal ventricular contraction sequence (i.e., to eliminate mechanical dyssynchrony), our current CRT indication criteria are based solely on an electrical dyssynchrony, specifically on assumption that it is present based on a 12-lead ECG. Therefore, the aim of this study was to assess the relationship between electrical and mechanical dyssynchrony, their contribution to ventricular contraction inefficiency and to identify the relevant markers of the presence and extent of mechanical dyssynchrony [40,41]. The main tasks were development of methods for intelligent analysis of body surface potential mapping (128-channel ECG) [42–44] and MRI analysis of heart mechanical activity.

Intelligent methods for evaluation of long-term EEG recordings (PI female—Lenka Lhotska)

In this project, we designed a methodology for automated processing and evaluation of long-term EEG records, based on a set of theoretically well-developed methods and algorithms. Developed methods were implemented and tested on real biomedical signals recorded at a neurology department. The implemented system allows automated detection and classification of specific graphoelements (parts of signals with a characteristic shape and defined diagnostic value) in long-term EEG records [45–48]. A database of interpreted EEG segments was developed. The recorded signals provided a platform for verification of the functionality of the proposed methods as well [49]. The project results were directly applied at several neurology departments of Czech hospitals. Later, we applied the developed methods to processing of a similar type of data from various areas of medicine and technology.

When finishing the project, we found that in some cases it is more suitable to approach the data in a personalized way, namely, in experimental medicine, there are usually complex data of small number of patients, frequently with high interpersonal variability. In such cases, many methods fail or do not give satisfactory results. This was a motivation for us to design a new project Temporal context in analysis of long-term non-stationary multidimensional signal.

Temporal context in the analysis of long-term non-stationary multidimensional signals (co-investigator female—Lenka Lhotska)

The project focused on temporal context-aware approaches in active learning. We hypothesized that the utilization of the contextual information within the query selection or generation can make the active learning more efficient and available for all temporal context-aware methods. This approach is novel, as it transfers a large portion of temporal context handling directly to active learning approaches and thus makes the active learning more efficient and available for any temporal context-aware supervised machine learning method. The proposed research aimed to answer three main questions: (1) How to utilize a given information about temporal context within the active learning approaches? (2) How to utilize a learned temporal context within the active learning approaches? (3) How to extend the application area of the developed approaches?

We consider as the main results the proposal of the methodology of a semi-automatic approach to long-term signal classification utilizing active learning. The designed and implemented methods were tested and validated on real data—polysomnographic recordings.

The proposed semi-supervised method enables fast and objective evaluation of PSG data compared to the gold standard manual scoring done by a certified sleep expert. The undisputed advantage of the proposed method is its independence in the configuration and amount of training data. The method can work with any available PSG recording. The algorithm does not require a specific location of EEG electrodes or montages and has greater resistance to unexpected artifacts in the data; thus, it is robust to noise. We proposed to use the hidden Markov model for the detection of the transitional instances, and proved that it supports the active learning better than the label-based method.

Publications [50–52] present the important results achieved in the area of classification of complex long-term data, tested on real clinical data. They show two different approaches—fully automated unsupervised and semi-automatic supervised. In the first case, active learning can be added as the next step. The other two present advantages of active learning combined with hierarchical clustering and hidden Markov models, respectively. Automated behavioral state classification in intracranial EEG (iEEG) recordings is beneficial for iEEG interpretation and quantifying sleep patterns to enable behavioral state-dependent neuromodulation therapy in next-generation implantable brain stimulation devices. The suggested cluster-based approach can be linked with currently available methods for evaluation of long-term EEG/PSG. It can create a revolutionary and comprehensive tool for the field of experimental and clinical medicine. The active learning outperforms the random sampling in semi-automatic EEG-based sleep scoring in terms of mean class error. Its main conclusion is a solution of the problem of detection of potentially ambiguous data instances that should be not queried for labeling. The method based on the most probable state sequence of HMM can find data instances whose deletion from the training set can statistically significantly improve both the random sampling and the active learning procedure.

Evaluation of cardiotocography using artificial intelligence methods (co-investigator female—Lenka Lhotska)

The aim of the project was a comprehensive computer analysis of intrapartum fetal monitoring. Using artificial intelligence methods and the synergy of technical approach and clinical experience, a system was developed to support the diagnosis of intrapartum fetal hypoxia [53]. A dataset of approximately 10,000 pregnant women was be created containing intrapartum monitors and other data assessing the incidence of hypoxia in the peripartum period. The data were evaluated by independent experts. The final annotated database was verified also internationally and then published on Physionet. The database serves as a platform for data mining and a tool for validation of the methods being developed. Hypotheses were tested as to whether artificial intelligence and digital signal processing algorithms are able to reliably discriminate abnormal CTGs and whether they outperform the results and/or variability of physician decision making [54–56]. Furthermore, the importance of using additional information from medical history and the qualitative impact of individual clinical symptoms on the accuracy of the automatic classification and individual algorithms were tested.

Individual dynamics of glycaemia excursions identification in diabetic patients to improve self-managing procedures influencing insulin dosage (PI female—Katerina Stechova; co-investigator female—Lenka Lhotska)

The aim of the project was interdisciplinary research focused on decision support in diabetic patient treatment. The core is the design of a software prototype (SW)—an application for cell phones—that offers to a diabetic patient treated by an insulin pump an advanced advice on insulin dose based on previous individual experience (self-learning algorithm) [57–59]. Inputs for the SW are (1) complex information on food content (including glycaemia index and reflecting the previous patient's glycaemia reaction to the similar type of food); (2) the amount of active insulin (information available from the insulin pump); and (3) additional information on physical activity and stress (psychic, illness, etc.). The SW uses information about food to be consumed, logbook data analysis and online connection to the food database [60,61]. Improving fitting the insulin dose to real needs improves diabetes stabilization further, with important individual as well as socio-economic impacts. Several project outputs are planned to be applied in other patients requiring precise diet management.

COGAIN project (Computer Gaze Interaction) (co-investigator female—Olga Stepankova)

The COGAIN project was a network of excellence supported by the European Commission's IST 6th framework program from 2004 to 2009, which aimed to study the possibilities of alternative gaze-based tools for interaction with a computer, namely, for persons who could not utilize the classical interfaces for diverse reasons [62]. Various implementations were developed—special attention was given to eye movement recognition and various approaches that can improve the quality of the acquired data. Infra-red illumination is one of these approaches. The project tried to answer questions related to the advantages and dangers related to long-term use of gaze-based solutions. The project started investigating the optical safety of extensive exposure of a human eye to infrared eye trackers and the results of its activity contributed to accepting the corresponding standard by Commission Internationale de L'Eclairage in 2021 [63]. The project founded the COGAIN Association that aims to promote research and development in the field of gaze-based interaction in computer-aided communication and control.

MAS Nanoelectronics for Mobile Ambient Assisted Living (AAL) Systems, EU ENIAC project, No. 120228 (2010–2013) (co-investigator female—Olga Stepankova)

The objective of the MAS project was to develop a common communication platform and nanoelectronics circuits for health and wellness applications to support the development of flexible, robust, safe and inexpensive mobile AAL systems, to improve the quality of human life and improve the well-being of people. In this context, reference architectures were defined in order to enable system development from devices to complete mobile AAL systems, and to enable cooperative clusters of such systems for specific environments and applications. Various approaches to visualization of the data and results were developed and tested on real data [64].

MAS focused on the development of an integrated approach for the areas of health monitoring and therapy support at home, and mobile health, wellness and fitness. The systems were intended for remote patient supervision using multi-parameter biosensors and secure communication networks [65], and health and wellness monitoring in the home environment. Mixed healthcare and consumer markets were targeted with the MAS-platform-based devices, with five application demos: (1) Health and Activity Monitor; (2) Point of Care Terminal and Gateway; (3) Cardiovascular Monitor; (4) Diabetes Monitor; and (5) Mobile Cardiotocography [66].

The key developments addressed mobile, unobtrusive sensor systems with standardized interfaces linked by secure wireless communication to a managing controller. User-friendly interfaces, multiple heterogeneous sensors networks, low power and power management formed key elements of the platform. Seamless connectivity, interoperability and cooperation across mobile AAL systems, health service providers and patients were field tested in different environments.

SPES (Support Patients through E-services Solutions, 2011–2014, 3CE286P2, Operational Program CENTRAL EUROPE) (co-investigator female—Olga Stepankova)

The SPES project demonstrated that technological solutions can improve the quality of life of patients with chronic and serious diseases and of their families. Most of these patients are aged people who have reduced movement capabilities and some difficulties in going to periodic medical examinations to the hospital or medical centers. The technological solutions in the field of telemedicine, such as the patients' home health care, social inclusion tools not only addressed to young people or monitoring systems for non-self-sufficient people, are examples that can be easily evaluated both by decision makers and patients. The SPES project implemented tele-health and entertainment platform in four cities—Ferrara, Vienna, Brno and Kosice—focusing on dementia, handicapped people, respiratory problems and social exclusion. Patients had the opportunity to exploit an easy-to-use telemedicine solution, lowering their displacement costs and the time necessary for going to the care service providers (hospitals, grand physicians and medical centers).

DISTINCT project (Dementia: Intersectorial strategy for training and innovation network for current technology, 2019–2022, EU Horizon 2020 Marie Skłodowska-Curie grant, No. 813196) (co-investigator female—Olga Stepankova)

The rapid growth of the technological landscape and related new services have the potential to improve the cost-effectiveness of health and social services and facilitate social participation and engagement in activities. It is generally believed that well designed and reliable technology has the potential to simplify our daily lives, compensate for disability and promote social inclusion. However, most products available on the market are only ready to support and solve the problems as perceived by an average adult. They do not work for people with cognitive impairment. Even worse, high penetration of technological gadgets throughout all places in society places at high risk the exclusion of all who fail to upgrade or maintain their competencies to manage technology. This risk pertains to daily life at work, in public space as well as at home as the complexity of both realms is continuously increasing. The users' ability to manage products and services has been largely neglected or taken for granted.

The DISTINCT project first identified more than a dozen hot research topics to be resolved within the frame of a Ph.D. study that can contribute to one of the three domains of social health the project is organized around: Technology to enable people to fulfil their potential and obligations; Technology to enable people to manage their life with some degree of independence; and Technology to enable people to participate in social and meaningful activities. Second, the project partners recruited for their topics enthusiastic early stage researchers (ESR) who decided to enter Ph.D. studies at the partner institutions and make from this topic the subject of their dissertation. The main task of the project is to ensure the recruited students have an inspiring, truly international environment; e.g., by offering them a possibility to cooperate for several months with two additional co-supervisors at two different partner institutions. Further, a significant part of the ESR's education is provided in a block format during regular project gatherings (DISTINCT schools) where all the ESRs can share their experience and discuss their recent achievements. The project pays special attention to the top quality content of its schools organized twice a year.

The ESR at our institution is developing a low-cost pet-bot well informed about the activities of its owner through a data stream provided by smart home sensors and their customized interpretation [67]. We intent to use such an interpretation to trigger timely unobtrusive reminders to be provided by a care-bot for patients with dementia.

4. Education in AI

High quality research is very important for successive development of practically applicable solutions. However, research cannot exist without the education of a young generation of researchers. The AI topics penetrated first as topics into existing courses in the early 1990s and gradually developed into separate courses at several Czech universities in study programs of Cybernetics, Computer Science or Information Technology. During the last two decades, artificial intelligence became an independent study program at many universities. At CTU in Prague, there are two study programs explicitly mentioning in their name or in the name of their specializations Artificial intelligence, namely, a bachelor study program in Artificial Intelligence at the Faculty of Informatics and the study program Open Informatics of the Faculty of Electrical Engineering with its bachelor specialization in Artificial Intelligence and Computer Science and master specialization in Artificial Intelligence.

The subject AI is included as a compulsory one also in more study programs. One of these is the study program Biomedical and Clinical Informatics, where the AI-related courses constitute a substantial part of the curricula. This study program stresses the importance of the close connection of the theoretical methods with the large application area of medicine. The students get insight into the terminology, data acquisition, medical procedures and other related topics. They gain knowledge for the design, development and security of biomedical applications. In frame of their projects and finally master's theses, they work on topics coming from the health-care area. After graduation, they are able to implement algorithms for processing, interpreting and evaluating biomedical data, apply their knowledge in the development of mobile and telemedicine applications, and implement personalized solutions for care in medical facilities, etc. The program graduates are also able to analyze biomedical data, design procedures and algorithms for their evaluation and implement them in clinical practice. The acquired knowledge enables the graduates to understand the genesis of biomedical data and to acquire the relevant terminology that facilitates professional cooperation with doctors and researchers.

The next one is the program Assistive Technologies, leading to master's degree and offering education at the doctoral level, too. This study program is accredited at the Faculty of Biomedical Engineering and its first graduates passed their final exams in the year 2021. Design of its curriculum has been extensively inspired by rich experience in the multidisciplinary research to which all the faculty members are highly dedicated. The projects mentioned earlier offer a representative sample of topics that are now further developed by current master's and doctoral students. These students have a solid background not only in signal processing, programming, AI and data mining, but their curriculum covers even user-centered design as well as electronics, health information systems, Healthcare Interoperability Standards and safety issues.

We observe in all the above-mentioned study programs a slow, but gradual, increase in female students, which is a positive sign of change in their thinking about their future career. We are convinced that the positive examples set by successful female Ph.D. graduates also contribute to this trend. The CTU supports the activities of female students. There exists a female student club, wITches, that organizes events for children with the aim to increase their interest in engineering and informatics. We are well aware of the fact that it is a long-distance race. However, we see the first results as more female students have applied for the study of AI at CTU in recent years.

5. Conclusions

The task to describe the artificial intelligence research and most important women in this area in the Czech Republic was not easy. Although there are many personalities that reached excellent results, they are not so known outside the AI community. In addition to the research and educational activities, we see their role in mentoring and supervising young researchers, in particular female researchers, for which they might serve as models of their future career. We tried to select those researchers who were/are active in AI research aiming at medical applications. In the project description we selected a broader variety of our projects to show how colorful and variable the topics in medicine and health care are and how AI methods can be applied.

Health care as an application area has great potential for the implementation of many AI systems to various phases of the care cycle, population and epidemiological studies, and other connected areas. In all cases, data, information and knowledge represent the core of the decision-making process. Decision-support systems were originally intended to be applications for a small group of specialists; in recent years, we can observe a transition to more widely used applications. All applications process data and information. The final result, having the form of a recommendation or even decision, is highly dependent on the quality of the data and information and not only on the quantity of data. That means that the quality must be understood properly and the data and information verified accordingly. The main properties that should be checked are completeness, consistency, validity, precision, redundancy, readability, accessibility, confidence, credibility and usefulness. There are still many open issues that will require our attention in future research as new demands on the functionalities of applications appear. Just to mention a few of them:

- Exchange of data, information and knowledge is desirable; however, it imposes requirements on high-level communication protocols and data structures so that the sender and receiver understand each other. In this context, we speak about interoperability at the data, information, knowledge and process levels.
- Current Health Information systems and electronic patient records place great demands on the time their user has to dedicate to input of all the requested data [68]. It is important to bring into focus a user-centered design of these systems and various possibilities of how this process could benefit from the application of recent AI achievements.
- Last but not least, it is necessary to pay due attention to ensuring data privacy, ethics and the corresponding legal regulations.

Some challenges are directly open in the AI domain; the most significant ones are explainable and trustworthy AI that is fully in line with the requirements already appearing in the medical domain, where an explanation of the proposed solution (diagnosis or therapy) is of utmost importance.

All mentioned issues open new fields for new research projects, topics for Ph.D. theses and finally applications that can find their place in routine practice.

Author Contributions: The article was jointly written by both authors. All authors have read and agreed to the published version of the manuscript.

Funding: Research of L. Lhotska and O. Stepankova was supported by institutional resources of Czech Technical University in Prague.

Institutional Review Board Statement: Not applicable.

Informed Consent Statement: Not applicable.

Conflicts of Interest: The authors declare no conflict of interest.

References

1. Marik, V.; Stepankova, O.; Lazansky, J. *Artificial Intelligence 1*; Academia: Prague, Czech Republic, 1993; ISBN 80-200-0496-3. (In Czech)
2. Marik, V.; Stepankova, O.; Lazansky, J. *Artificial Intelligence 2*; Academia: Prague, Czech Republic, 1997; ISBN 80-200-0504-8. (In Czech)
3. Marik, V.; Stepankova, O.; Lazansky, J. *Artificial Intelligence 3*; Academia: Prague, Czech Republic, 2001; ISBN 80-200-0472-6. (In Czech)
4. Marik, V.; Stepankova, O.; Lazansky, J. *Artificial Intelligence 4*; Academia: Prague, Czech Republic, 2003; ISBN 80-200-1044-0. (In Czech)
5. Marik, V.; Stepankova, O.; Lazansky, J. *Artificial Intelligence 5*; Academia: Prague, Czech Republic, 2007; ISBN 80-200-0502-1. (In Czech)
6. Marik, V.; Stepankova, O.; Lazansky, J. *Artificial Intelligence 6*; Academia: Prague, Czech Republic, 2013; ISBN 80-200-0502-1. (In Czech)
7. Hajič, J.; Bejček, E.; Bémová, A.; Buráňová, E.; Hajičová, E.; Havelka, J.; Homola, P.; Kárník, J.; Kettnerová, V.; Klyueva, N.; et al. Prague Dependency Treebank 3.5; Institute of Formal and Applied Linguistics, LINDAT/CLARIN, Charles University, LINDAT/CLARIN. 2018. Available online: http://hdl.handle.net/11234/1-2621 (accessed on 15 January 2022).
8. Rysová, K.; Rysová, M.; Novák, M.; Mírovský, J.; Hajičová, E. EVALD—A Pioneer Application for Automated Essay Scoring In Czech. *Prague Bull. Math. Linguist.* **2019**, *113*, 9–30. [CrossRef]
9. Kalina, J.; Zvárová, J. Decision support systems in the process of improving patient safety. In *Bioinformatics: Concepts, Methodologies, Tools, and Applications*; IGI Global: Hershey, PA, USA, 2013; pp. 1113–1125.
10. Kůrková, V.; Sanguineti, M. Correlations of random classifiers on large data sets. *Soft Comput.* **2021**, *25*, 12641–12648. [CrossRef]
11. Mrázová, I.; Reitermanová, Z. Enforced knowledge extraction with BP-networks. *Intell. Eng. Syst. Through Artif. Neural Netw.* **2007**, *17*, 285–290.
12. Mrázová, I.; Zvirinský, P. Subjects Involved In Czech Insolvency Proceedings: An Assessment of Their Future Impact. *Procedia Comput. Sci.* **2021**, *185*, 63–72. [CrossRef]
13. Kunešová, M.; Zajíc, Z.; Radová, V. Experiments with segmentation in an online speaker diarization system. In *International Conference on Text, Speech, and Dialogue*; Springer: Cham, Switzerland, 2017; pp. 429–437.
14. Pražák, A.; Loose, Z.; Psutka, J.V.; Radová, V.; Psutka, J. Live TV subtitling through respeaking with remote cutting-edge technology. *Multimed. Tools Appl.* **2020**, *79*, 1203–1220. [CrossRef]
15. Isukapati, I.K.; Rudová, H.; Barlow, G.J.; Smith, S.F. Analysis of trends in data on transit bus dwell times. *Transp. Res. Rec.* **2017**, *2619*, 64–74. [CrossRef]
16. Dang, Q.V.; Nguyen, C.T.; Rudová, H. Scheduling of mobile robots for transportation and manufacturing tasks. *J. Heuristics* **2019**, *25*, 175–213. [CrossRef]
17. Vejnarová, J. Compositional models for credal sets. *Int. J. Approx. Reason.* **2017**, *90*, 359–373. [CrossRef]
18. Flusser, J.; Suk, T.; Zitová, B. Complete and Incomplete Sets of Invariants. *J. Math. Imaging Vis.* **2021**, *63*, 917–922. [CrossRef]
19. Kamenický, J.; Šroubek, F.; Zitová, B.; Hannuksela, J.; Turtinen, M. Image restoration in portable devices: Algorithms and optimization. *J. Signal Process. Syst.* **2019**, *91*, 9–20. [CrossRef]
20. Flusser, J.; Zitova, B.; Suk, T. *Moment Invariants in Pattern Recognition*; John Wiley & Sons: Hoboken, NJ, USA, 2009; ISBN 978-0-470-69987-4.
21. Flusser, J.; Suk, T.; Zitova, B. *2D and 3D Image Analysis by Moments*; John Wiley & Sons: Hoboken, NJ, USA, 2016; ISBN 978-1-119-03935-8.
22. Štěpánková, O.; Havel, I.M. A logical theory of robot problem solving. *Artif. Intell.* **1976**, *7*, 129–161.
23. Oliveira, E.; Fischer, K.; Štěpánková, O. Multi-agent systems: Which research for which applications. *Robot. Auton. Syst.* **1999**, *27*, 91–106, ISSN 0921-8890. [CrossRef]
24. Kléma, J.; Nováková, L.; Karel, F.; Štěpánková, O.; Železný, F. Sequential data mining: A comparative case study in development of atherosclerosis risk factors. *IEEE Trans. Syst. Man Cybern. Part C* **2008**, *38*, 3–15, ISSN 1094-6977. [CrossRef]
25. Anýž, J.; Vysloužilová, L.; Vaculovic, T.; Tvrdonova, M.; Kanicky, V.; Haase, H.; Horak, V.; Stepankova, O.; Heger, Z.; Vojtech, A. Spatial mapping of metals in tissue-sections using combination of mass-spectrometry and histology through image registration. *Sci. Rep.* **2017**, *7*, 40169, ISSN 2045-2322. [CrossRef] [PubMed]
26. Němý, M.; Cedres, N.; Grothe, M.J.; Muehlboeck, J.-S.; Lindberg, O.; Nedelska, Z.; Štěpánková, O.; Vysloužilová, L.; Eriksdotter, M.; Barroso, J.; et al. Cholinergic white matter pathways make a stronger contribution to attention and memory in normal aging than cerebrovascular health and nucleus basalis of Meynert. *NeuroImage* **2020**, *211*, 116607, ISSN 1053-8119. [CrossRef]

27. Zvara, K.; Tomeckova, M.; Peleska, J.; Svatek, V.; Zvarova, J. Tool-supported Interactive Correction and Semantic Annotation of Narrative Clinical Reports. *Methods Inf. Med.* **2017**, *56*, 217–229. [CrossRef]
28. Nagy, M.; Seidl, L.; Zvarova, J. Evaluation of Possibilities in Demographic Data Exchange Support In Czech Healthcare. In *E-Health across Borders without Boundaries*; Book Series: Studies in Health Technology and Informatics; Stoicu Tivader, L., Blobel, B., Marcun, T., Orel, A., Eds.; IOS Press: Amsterdam, The Netherlands, 2011; Volume 165, pp. 143–148. [CrossRef]
29. Burša, M.; Lhotská, L. The Use of Convolutional Neural Networks in Biomedical Data Processing. In *Information Technology in Bio- and Medical Informatics*; Springer International Publishing: Cham, Switzerland, 2017; Volume 10443, pp. 100–119. ISBN 978-3-319-64265-9.
30. Bursa, M.; Lhotska, L. Evaluation of various classifiers performance on biomedical datasets. In Proceedings of the 5th IEEE International Conference on E-Health and Bioengineering, Iasi, Romania, 19–21 November 2015; ISBN 978-1-4673-7545-0.
31. Burša, M.; Lhotská, L. Ant-Inspired Algorithms in Health Information System Data Mining, Classification and Visualization. In *Proceedings of the XIV Mediterranean Conference on Medical and Biological Engineering and Computing 2016 (MEDICON 2016), Paphos, Cyprus, 31 March–2 April 2016*; IFMBE Series; Efthyvoulos, K., Stelios, C., Pattichis, C.S., Eds.; Springer: Berlin/Heidelberg, Germany, 2016; pp. 868–873. ISBN 978-3-319-32703-7.
32. Macaš, M.; Lhotská, L.; Gabrys, B.; Ruta, D. Particle Swarm Optimization of Multiple Classifier Systems. In *Computational and Ambient Intelligence*; Springer: Berlin/Heidelberg, Germany, 2007; pp. 333–340.
33. Lhotska, L.; Stechova, K.; Pharow, P. Personal Portable Devices in the Light of Internet of Things. In *pHealth 2017*; Blobel, B., Goossen, W., Eds.; IOS Press: Amsterdam, The Netherlands, 2017; pp. 34–36.
34. Chudáček, V.; Georgoulas, G.; Lhotská, L.; Stylios, C.; Petrík, M.; Cepek, M. Examining Cross-Database Global Training to Evaluate Five Different Methods for Ventricular Beat Classification. *Physiol. Meas.* **2009**, *30*, 661–678, ISSN 0967-3334. [CrossRef]
35. Lhotská, L.; Chudáček, V.; Huptych, M. ECG Processing. In *Data Mining and Medical Knowledge Management: Cases and Applications*; IGI Publishing: Hershey, PA, USA, 2009; pp. 137–160. ISBN 978-1-60566-218-3.
36. Palova, S.; Szabo, K.; Charvat, J.; Slavicek, J.; Medova, E.; Mlcek, M.; Kittnar, O. ECG body surface mapping changes in type 1 diabetic patients with and without autonomic neuropathy. *Physiol. Res.* **2009**, *59*, 203–209. [CrossRef] [PubMed]
37. Huptych, M.; Lhotská, L. ECG Beat Classification Using Feature Extraction from Wavelet Packets of R Wave Window. In *Proceedings of the World Congress on Medical Physics and Biomedical Engineering, Munich, Germany, 7–12 September 2009*; Springer Science+Business Media: Berlin/Heidelberg, Germany, 2009; ISBN 978-3-642-03897-6.
38. Chudáček, V.; Lhotská, L.; Georgoulas, G.; Stylios, C. Is it Possible to Distinguish Different Types of ECG-Holter Beats Based Solely on Features Obtained from Windowed QRS Complex? In *Proceedings of the World Congress on Medical Physics and Biomedical Engineering, Munich, Germany, 7–12 September 2009*; Springer Science+Business Media: Berlin/Heidelberg, Germany, 2009; pp. 918–921. ISBN 978-3-642-03897-6.
39. Křemen, V.; Lhotská, L.; Macaš, M.; Čihák, R.; Kautzner, J.; Wichterle, D. A New Approach to Automated Assessment of Fractionation of Endocardial Electrograms During Atrial Fibrillation. *Physiol. Meas.* **2008**, *29*, 1371–1381. [CrossRef]
40. Kittnar, O.; Riedlbauchová, L.; Tomis, J.; Ložek, M.; Valeriánová, A.; Hrachovina, M.; Mlček, M.; Huptych, M.; Janoušek, J.; Lhotská, L. Electrocardiographic Outcome of Resynchronization Therapy. *Physiol. Res.* **2017**, *66* (Suppl. S4), S523–S528. [CrossRef]
41. Kittnar, O.; Riedlbauchová, L.; Adla, T.; Suchánek, V.; Tomis, J.; Ložek, M.; Valeriánová, A.; Hrachovina, M.; Popková, M.; Veselka, J.; et al. Outcome of resynchronization therapy on superficial and endocardial electrophysiological findings. *Physiol. Res.* **2018**, *67* (Suppl. S4), S601–S610. [CrossRef] [PubMed]
42. Hrachovina, M.; Lhotská, L.; Huptych, M. Preprocessing and filtration techniques of BSPM signals in a small-scale study. Precision Medicine Powered by pHealth and Connected Health. In *Proceedings of the International Conference on Biomedical and Health Informatics 2017, Thessaloniki, Greece, 18–21 November 2017*; Springer Nature: Singapore, 2018; Volume 66, pp. 127–132. ISSN 1680-0737. ISBN 978-981-10-7418-9. Available online: https://link.springer.com/chapter/10.1007/978-981-10-7419-6_24 (accessed on 10 December 2021).
43. Huptych, M.; Hrachovina, M.; Lhotská, L. Preprocessing of the BSPM Signals with Untraditionally Strong Baseline Wandering. In *Proceedings of the World Congress on Medical Physics and Biomedical Engineering, Prague, Czech Republic, 3–8 June 2018*; Part of the IFMBE Proceedings Book Series; Lhotska, L., Sukupova, L., Lackovic, I., Ibbott, G.S., Eds.; Springer: Berlin/Heidelberg, Germany, 2018; Volume 2, pp. 463–467. Available online: https://link.springer.com/book/10.1007/978-981-10-9038-7 (accessed on 15 January 2022).
44. Huptych, M.; Hrachovina, M.; Lhotska, L. Software for Preprocessing Experimental BSPM Signals for a CRT Study. *Proceedings* **2019**, *31*, 69. Available online: https://www.mdpi.com/2504-3900/31/1/69 (accessed on 15 January 2022). [CrossRef]
45. Djordjevic, V.; Reljin, N.; Gerla, V.; Lhotská, L.; Krajča, V. Feature Extraction and Classification of EEG Sleep Recordings in Newborns. In Proceedings of the 9th International Conference on Information Technology and Applications in Biomedicine, Larnaka, Cyprus, 4–9 November 2009; ISBN 978-1-4244-5379-5.
46. Gerla, V.; Djordjevic, V.; Lhotská, L.; Krajča, V. Visualization Methods Used for Evaluation of Neonatal Polysomnographic Data. In Proceedings of the 9th International Conference on Information Technology and Applications in Biomedicine, Larnaka, Cyprus, 4–9 November 2009; ISBN 978-1-4244-5379-5.
47. Gerla, V.; Burša, M.; Lhotská, L.; Paul, K.; Krajča, V. Newborn Sleep Stage Classification Using Hybrid Evolutionary Approach. *Int. J. Bioelectromagn.* **2007**, *9*, 28–29, ISBN 978-4-9903873-0-3.

48. Gerla, V.; Paul, K.; Lhotská, L.; Krajča, V. Multivariate Analysis of Full-Term Neonatal Polysomnographic Data. *IEEE Trans. Inf. Technol. Biomed.* **2009**, *13*, 104–110. [CrossRef] [PubMed]
49. Rieger, J.; Lhotská, L.; Krajča, V.; Matoušek, M. Development of the Long-Term EEG Processing Software. In *Analysis of Biomedical Signals and Images—Proceedings of the Biosignal 2006*; VUTIUM Press: Brno, Slovenia, 2006; pp. 149–151. ISBN 80-214-3152-0.
50. Křemen, V.; Brinkmann, B.H.; Van Gompel, J.J.; Stead, M.; St Louis, E.K.; Worrell, G.A. Automated unsupervised behavioral state classification using intracranial electrophysiology. *J. Neural Eng.* **2019**, *16*, 026004. [CrossRef] [PubMed]
51. Macaš, M.; Grimová, N.; Gerla, V.; Lhotská, L. Semi-Automated Sleep EEG Scoring with Active Learning and HMM-Based Deletion of Ambiguous Instances. *Proceedings* **2019**, *31*, 46. [CrossRef]
52. Gerla, V.; Kremen, V.; Macas, M.; Dudysova, D.; Mladek, A.; Sos, P.; Lhotska, L. Iterative expert-in-the-loop classification of sleep PSG recordings using a hierarchical clustering. *J. Neurosci. Methods* **2019**, *317*, 61–70. [CrossRef]
53. Burša, M.; Lhotská, L. Applying Ant-Inspired Methods in Childbirth Asphyxia Prediction. In *Information Technology in Bio- and Medical Informatics. ITBAM 2016*; Renda, M.E., Bursa, M., Holzinger, A., Khuri, S., Eds.; Springer: Berlin/Heidelberg, Germany, 2016; pp. 192–207.
54. Spilka, J.; Chudáček, V.; Koucký, M.; Lhotská, L.; Huptych, M.; Janku, P.; Georgulas, G.; Stylios, C. Using Nonlinear Features for Fetal Heart Rate Classification. *Biomed. Signal Process. Control.* **2012**, *4*, 350–357, ISSN 1746-8094. [CrossRef]
55. Chudáček, V.; Spilka, J.; Janků, P.; Koucký, M.; Lhotská, L.; Huptych, M. Automatic Evaluation of Intrapartum Fetal Heart Rate Recordings: A Comprehensive Analysis of Useful Features. *Physiol. Meas.* **2011**, *32*, 1347–1360. [CrossRef]
56. Spilka, J.; Chudacek, V.; Janku, P.; Hruban, L.; Bursa, M.; Huptych, M.; Zach, L.; Lhotska, L. Analysis of obstetricians' decision making on CTG recordings. *J. Biomed. Inform.* **2014**, *51*, 72–79. [CrossRef]
57. Saiti, K.; Macaš, M.; Štechová, K.; Pithová, P.; Lhotska, L. A Review of Model Prediction in Diabetes and of Designing Glucose Regulators Based on Model Predictive Control for the Artificial Pancreas. In *Information Technology in Bio- and Medical Informatics, ITBAM 2017*; Lecture Notes in Computer Science; Bursa, M., Holzinger, A., Renda, M., Khuri, S., Eds.; Springer: Cham, Switzerland, 2017; Volume 10443.
58. Macas, M.; Lhotska, L.; Stechova, K.; Pithova, P.; Saiti, K. Particle Swarm Optimization Based Adaptable Predictor of Glycemia Values. In Proceedings of the 2017 3rd IEEE International Conference on Cybernetics (CYBCONF), Exeter, UK, 21–23 June 2017; pp. 1–6.
59. Saiti, K.; Macaš, M.; Štechová, K.; Pithová, P.; Lhotska, L. A Combined-Predictor Approach to Glycaemia Prediction for Type 1 Diabetes. In *World Congress on Medical Physics and Biomedical Engineering*; Springer: Singapore, 2018; Volume 2, pp. 753–756. ISSN 1680-0737.
60. Lhotska, L.; Štechová, K.; Hlúbik, J. Improving prediction of glycaemia course after different meals—New individualized approach. In *World Congress on Medical Physics and Biomedical Engineering 2018*; Springer: Singapore, 2018; Volume 2, pp. 757–762.
61. Stechova, K.; Hlubik, J.; Pithova, P.; Cikl, P.; Lhotska, L. Comprehensive Analysis of the Real Lifestyles of T1D Patients for the Purpose of Designing a Personalized Counselor for Prandial Insulin Dosing. *Nutrients* **2019**, *23*, 1148. [CrossRef]
62. Fejtová, M.; Figueiredo, L.; Novák, P.; Stepankova, O.; Gomes, A. Hands-free interaction with a computer and other technologies. *Univ. Access Inf. Soc.* **2009**, *8*, 277. [CrossRef]
63. Sliney, D.H.; Mulvey, F.; Charlier, J.; Cleveland, D.; Daunys, G.; Donegan, M.; Droege, D.; Joos, M.; Liggins, E.; Schulmeister, K.; et al. *CIE 245:2021 Optical Safety of Infrared Eye Trackers Applied for Extended Duration, Standard by Commission Internationale de L'Eclairage, 10/01/2021*; International Commission on Illumination: Vienna, Austria, 2021; ISBN 978-3-902842-14-5. [CrossRef]
64. Nováková, L.; Štěpánková, O. Visualization of trends using RadViz. *J. Intell. Inf. Syst.* **2011**, *37*, 355–369, ISSN 0925-9902. [CrossRef]
65. Havlik, J.; Dvorak, J.; Parak, J.; Lhotska, L. Monitoring of Physiological Signs using Telemonitoring System. In *Information Technology in Bio- and Medical Informatics*; Springer: Berlin/Heidelberg, Germany, 2011; pp. 66–67. ISBN 978-3-642-23207-7.
66. Zach, L.; Chudacek, V.; Kuzılek, J.; Spilka, J.; Huptych, M.; Bursa, M.; Lhotska, L. Mobile CTG—Fetal Heart Rate Assessment Using Android Platform. In Proceedings of the 2011 Computing in Cardiology, Hangzhou, China, 18–21 September 2011; p. 66, ISBN 978-1-4577-0612-7.
67. Ozdemir, D.; Cibulka, J.; Štěpánková, O.; Holmerová, I. Design and implementation framework of social assistive robotics for people with dementia—A scoping review. *Health Technol.* **2021**, *11*, 367–378. [CrossRef]
68. Shiells, K.; Holmerova, I.; Steffl, M.; Stepankova, O. Electronic patient records as a tool to facilitate care provision in nursing homes: An integrative review. *Inform. Health Soc. Care* **2019**, *44*, 262–277. [CrossRef] [PubMed]

applied sciences

MDPI

Article

Intersectional Study of the Gender Gap in STEM through the Identification of Missing Datasets about Women: A Multisided Problem

Genoveva Vargas-Solar

CNRS, University Lyon, INSA Lyon, UCBL, Ecole Centrale Lyon, University Lyon 2, LIRIS, UMR5205, 69622 Villeurbanne, France; genoveva.vargas-solar@cnrs.fr

Abstract: This paper discusses the problem of missing datasets for analysing and exhibiting the role of women in STEM with a particular focus on computer science (CS), artificial intelligence (AI) and data science (DS). It discusses the problem in a concrete case of a global south country (i.e., Mexico). Our study aims to point out missing datasets to identify invisible information regarding women and the implications when studying the gender gap in different STEM disciplines. Missing datasets about women in STEM show that the first step to understanding gender imbalance in STEM is building women's history by "completing" existing datasets.

Keywords: data; missing datasets; gender gap; data-driven studies; women in artificial intelligence; women in data science; women in STEM

Citation: Vargas-Solar, G. Intersectional Study of the Gender Gap in STEM through the Identification of Missing Datasets about Women: A Multisided Problem. *Appl. Sci.* **2022**, *12*, 5813. https://doi.org/10.3390/app12125813

Academic Editors: Aida Valls and Karina Gibert

Received: 1 April 2022
Accepted: 2 June 2022
Published: 8 June 2022

1. Introduction

Many science and engineering occupations are predicted to grow faster than the average rate for all professions. Workforce projections in 2018 by the U.S. Department of Labour showed that nine of the ten fastest-growing occupations that require at least a bachelor's degree would need significant scientific or mathematical training [1]. This prediction has been confirmed in recent years, and computer science fields with great potential like artificial intelligence (AI) and data science (DS) can be certainly included. More recent sources anticipate that 75% of future jobs will be related to these fields (https://lac.unwomen.org/en/digiteca/publicaciones/2020/09/mujeres-en-ciencia-tecnologia-ingenieria-y-matematicas-en-america-latina-y-el-caribe, accessed on 1 June 2022), given that 7.1 million jobs were expected to be displaced by 2020, and half of existing jobs will disappear by 2050. Some of the most significant increases will be in engineering (and computer-related) fields in which women currently hold one-quarter or fewer positions [2,3]. Indeed, only 22% of all professionals working in AI around the world are women (World Economic Forum. Global Gender Gap Report 2018, https://www.weforum.org/reports/the-global-gender-gap-report-2018/, accessed on 1 June 2022). The ubiquitous male dominance in countries in different regions results in a feedback loop shaping gender bias in AI and machine learning systems used in DS experiments [4–6]. These fields are particularly fast-moving [7] both in industry and academia, so it is essential to map how gender gaps (Davenport, T.H. and Patil, D.J. (2012) Data Scientist: The Sexiest Job of the 21st Century. Harvard Business Review. Retrieved from: https://hbr.org/2012/10/data-scientist-the-sexiest-job-of-the-21st-century, accessed on 1 June 2022) [8] are manifest in data-driven solutions comprehensively. If not addressed soon, the gender gap in STEM will widen during the Fourth Industrial Revolution. Science and engineering address key challenges of our time:

- Finding cures for diseases like cancer and malaria, tackling global warming, providing people with clean drinking water, developing renewable energy sources, and understanding the origins of the universe;

- Engineers design many of the things we use daily, including buildings, bridges, computers, cars, wheelchairs, and X-ray machines.

The presence of women in teams seem to increase collective intelligence, and if those women have decision making positions, success probabilities increase in startups. When women are not involved in designing products and addressing social and political problems, then needs and desires unique to women may be overlooked. For example, a predominantly male group of engineers tailored the first generation of automotive airbags to adult male bodies, resulting in avoidable deaths for women and children [9]. With a more diverse workforce, scientific and technological products, services, and solutions are likely to be better designed and more likely to represent all users.

The consolidation of AI and DS (World Economic Forum. Global Gender Gap Report 2018. Retrieved from: https://www.weforum.org/reports/the-global-gender-gap-report-2018, accessed on 1 June 2022, World Economic Forum. Global Gender Gap Report 2020. Retrieved from: https://www.weforum.org/reports/gender-gap-2020-report-100-years-pay-equality, accessed on 1 June 2022, World Economic Forum. The Future of Gender Parity: A Labour Market Shift. Retrieved from: https://reports.weforum.org/global-gender-gap-report-2020/the-future-of-gender-parity/?doing_wp_cron=1612906910.1824119091033935546875, accessed on 1 June 2022, as backbone tools to promote data-driven solutions opens the risk of overlooking women in the loop. Women are underrepresented in observation and data collection tasks. Their workforce in AI and DS is small, and they are underrepresented in algorithm design and testing teams. Thus, their characteristics are not represented because of how data are engineered. Their needs are not visible for designing units, even if more than the 50% of the planet's population are women (The National Centre for Women & Information Technology (NCWIT https://ncwit.org, accessed on 1 June 2022.) argues about the importance and interest for organizations to search for gender and race equality. Besides the economic welfare, teams respecting gender equality seem to be more creative and effective in performing tasks and sharing knowledge).

When science is combined with economy, politics and social impact, it is possible to classify disciplines and reason about the capital they produce and who produces that capital. To activate the economy, every asset that is potentially productive must participate in the active STEM market. Women are a massive part of the assets that remain excluded from the equation, weakening the economy. Countries with dynamic economies like Germany lack human force in these disciplines and import brains to fulfil the requirements of their economy. According to the economist François Lenglet in his TV show "La guerre des âges", two strategies are possible for these economies: either hiring an immigrant brain force or integrating women into the production market. Surprisingly, Germany is developing a very aggressive strategy for attracting immigrant experts (!) (Is the old German saying that associates women to the three K's (i.e., Kirche, Küche, Kinder), still valid? The German gender issues are out of the scope of this paper. Thus, we leave this topic open and we invite the interested reader willing to seek for more information to visit [10,11]). Why are women excluded from the equation? Are they recognised as a potential labour force, or are they hidden Cinderellas invisible to statistics? Are gender bias and discrimination playing a role in women exclusion in STEM?

We believe that it is critical to study the phenomenon of data and women's absence in STEM from many different perspectives. Since women come in many different sizes, colours, and formats, the study must consider several communities from the global north and south instead of general studies.

This paper studies missing data under a qualitative approach [12] that combines grounded theory and use case strategies for answering the following research questions:

RQ1 Do missing data concerning women's STEM research participation prevent a comprehensive view of their contribution across history?

RQ2 Is it possible to design viable data science experiments that can estimate the female workforce in STEM and answer the question, "where are women in STEM"?

RQ3 Do missing intersectional perspectives prevent performing representative quantitative analysis about women's activity and contribution in STEM?

Our work adopts a qualitative methodology that does not rely upon mathematical, data mining or artificial intelligence models. Still, this "alternative" analytics choice can be seen as a preparation protocol for data science experiments that use AI models because of the critical role of data in ensuring "representative" results. These models are adapted in cases with datasets that are representative observation samples of a phenomenon, even if raw. Yet, the hypothesis that drives this study is that missing datasets giving insight into female contribution are missing. We question the possibility of producing statistical and induced or deduced knowledge from incomplete observations.

1.1. Context and Methodology

For performing a study on missing data and female absence as two perspectives of a mirror, it is essential to consider the conditions in which data were absent before and after the pandemic years. In the global north, there are programs like UN Women (https://lac.unwomen.org/en, accessed on 1 June 2022), She Figures (https://ec.europa.eu/assets/rtd/shefigures2021/index.html, accessed on 1 June 2022), and Davos reports (https://www.eversheds-sutherland.com/global/en/what/articles/index.page?ArticleID=en/Data-Protection/World_Economic_Forum_Report_2022, accessed on 1 June 2022) that have collected data about women in the economy, particularly in STEM. They perform serious data collection, and they often rely on governmental data. Other works like [13] perform analytics processes to measure the gender gap index in schools (before university and possible activities that can generate scientific contribution) by combining qualitative and quantitative methods. However, data are not intersectional (Intersectionality is an analytical framework for understanding how aspects of a person's social and political identities converge to create different modes of discrimination and privilege, https://en.wikipedia.org/wiki/Intersectionality, accessed on 1 June 2022); companies and institutions do not share them, so they are not included in studies. In consequence, despite these efforts, data about women are still missing. Thus, it is capital to understand the forces that play in the persistence of this discriminatory situation and the distribution of opportunities [14]. We need datasets as complete as possible, and as fine-grained and representative as possible, to perform clear data-driven studies about gender balance. Yet, these studies' fuel (i.e., datasets) is partial, incomplete or absent. It is not a matter of the analytics processes (data preparation, cleaning, engineering) nor the analytics models applied to extract knowledge. The point is that without intersectional data, which is often missing in available datasets, the gender issue cannot be modelled and understood in its complexity. These studies can lead to conclusions but do not wholly understand the gender gap problems. Yet, to efficiently address gender gap problems, we believe that we need to identify which datasets are missing. Our work focuses on tracking data that provide evidence about women's participation in STEM, particularly in CS, AI and DS.

Therefore, our work focusses on missing datasets instead of analysing the gender gap with available ones. These studies have been already done at the national level in different countries (e.g., in the UK [15], or specific datasets [16]) by international bodies like UN Women, the European Commission with the "She Figures" report, and by editorial bodies. These studies remain partial because they concentrate on quantitative approaches with datasets of one type (i.e., publications, student numbers); they lack intersectional approaches for collecting data [17,18].

Our first strategy is to identify women's influential role and contribution to the global north (The concept of Global North and Global South is used to describe a grouping of countries along socio-economic and political characteristics. The Global South identifies the regions of Latin America, Asia, Africa, and Oceania. It denotes regions outside Europe and North America, mostly (though not all) low-income and often politically or culturally marginalised countries on one side of the so-called divide, the other side being the countries of the Global North (often equated with developed countries). The term does not inherently refer to a geographical south) in CS represented in the annals of relevant awards in the field. We have performed an exploratory study of the sources where data lies that is explicit or hidden about women in STEM.

1.2. Contributions

As a result of the grounded theory and use case based approach:

1. We enumerate and characterise datasets regarding women in DS and IA, which are timidly emerging;
2. We report on a data analytics use case in Mexico with information about the National System of Researchers, the Mexican Academy of Sciences, the Mexican National Award of Sciences and elements of the history of Mexican women in STEM.

The choice of Mexico as a case study is because our vision is to perform regional studies about women, mainly located in the global south.

Accordingly, the remainder of the paper is organised as follows. Section 2 introduces existing projects and initiatives devoted to building the history of women in CS with a particular interest in works addressing their role in AI and DS. It gives a timeline of women that have contributed to pin milestones in the history of CS. Section 3 describes a use case about Mexico highlighting possible sources where data refer to female scientific and technological contributions. Section 4 concludes the paper and discusses open issues and future work.

2. Making Women's Contribution in Computer Science Visible

Studying and understanding the gender gap in STEM is a complex problem because each discipline and scientific community has different characteristics and histories. To analyse missing datasets about the gender gap in STEM and identify the type of datasets missing, we decided to focus on CS and two related sub-disciplines, AI and DS. However, at some point, we refer to other disciplines, which shows to which extent data are missing that we need to look at the datasets that we have at hand and derive guesses about related fields.

Figure 1 shows the workflow of our qualitative approach and how it is related to the steps of a data science workflow. Our approach opens the data collection box and exhibits three main phases, namely, (1) selecting and choosing data providers and (2) building datasets both systematically interacting with a (3) qualitative assessment phase. In our approach, the first phase applies the grounded theory methodology to identify providers and the type and characteristics of provided data. It drives conclusions about them, launching the qualitative assessment phase. Once data providers have been selected, datasets are built by applying fusion, integration and correlation to the data stemming from different providers. The objective of this phase is to "complete" data representative for answering a research question that drives a data science experiment. Recall that data science uses quantitative methodology implemented by applying data mining, machine learning and other artificial intelligence techniques.

Qualitative phases of data science and data-driven studies are mainly done manually or with an active presence of a human in the loop. This characteristic explains why this study does not report the application of (semi)-automatic techniques for performing analytics. Still, we argue that qualitative methodologies preliminary to the automatic phases of a data science workflow are critical because they show the interaction between DM, ML and AI and human tasks.

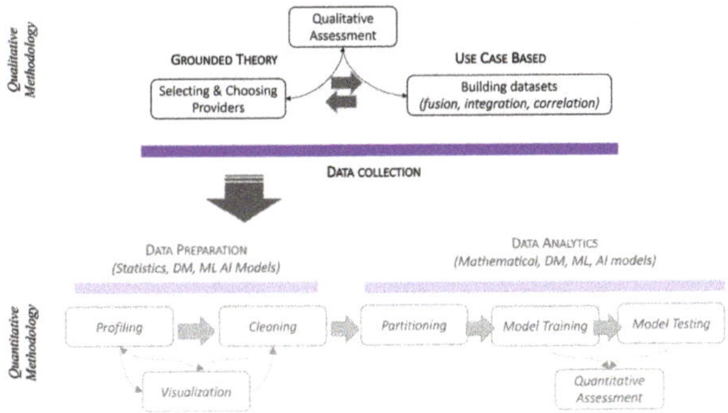

Figure 1. Qualitative methodology for identifying missing datasets.

We instantiated this approach in our study as follows. For selecting and choosing data providers (phase 1), we adopted the hypothesis that female contribution (from the global north and south) in STEM should be visible through the data representing their work and contribution. Data should refer to actions intended to create women collectivities in the area, the female workforce in promising topics like AI and DS, and prestigious awards. Research questions (RQ1, RQ2, RQ3) guide our grounded theory strategy, and it leads to conclusions about the qualitative characteristics of available datasets presented in the following sections. Similarly, the building datasets phase is driven by a use case strategy. Therefore, we report on observations and datasets available in Mexico that can compose a representative dataset about female contribution in STEM, particularly in computer science (see Section 3).

2.1. Women in STEM Studies and Inclusion Initiatives

The exclusion of the contribution of women scientists in STEM history is a concern in specific academic and industrial sectors around the world. Some studies suspect that the lack of critical mass of visible female scientists and professionals is one of the reasons for the increasing desertification of women in the field over the last three to four decades.

Different academic organisations in the global north like IEEE (women in engineering), ACM (women in ACM) in the US, the European Institute of Gender Equality in Europe, the CNRS through the "Mission pour la place des Femmes" and the "Comité parité-égalité", the Institute of Gender in France, and major technology companies (e.g., Microsoft, Google, Facebook) have recognised the importance of understanding and organising actions that can promote a more gender-balanced, diverse and inclusive STEM (science, technology, engineering and mathematics) community. The opportunities and economic share, and return of investment in different disciplines in STEM are not homogeneous.

Some international actions in industry and academia have been organised to make women's contributions to computing visible. One of the most established actions is the Anita Borg Institute, founded by computer science PhDs Anita Borg and Telle Whitney to recruit, retain and advance women in technology. Other international organisations are working to promote the visibility of women in computing:

- the Association for Computing Machinery (ACM) Committee on Women;
- the Association for Women in Computing;
- the Center for Women in Technology;
- Girl Develop It;
- Girls Who Code;
- In Latin America and the Carribean, there is Meninas na Computaçao and Meninas Digitais of the Brazilian Computer Society; Mexicanas en Computació, of the Mexican Academy of Computing, the University of Chile organises the Latinity conference "Latinas in Computing"; the conference CLEI organises a special call for women in STEM in Latin America;

- Similar organisations in Africa, the Middle East, Asia-Pacific and Oceania have emerged.

This list is not representative of the diversity of groups and forums working for the inclusion and visibility of the contribution of women in computing. There are too many, and there is no integrated map enumerating them. This diversity demonstrates both the interest in the issue, the lack of coordinated actions, and the lack of scientific gender studies applied to women in computing and women in STEM considering fine-grained analysis, for example, by field or geographic region. Some governments and universities have created gender studies institutes with dedicated chairs for science, technology, engineering and mathematics. Yet, despite exceptions, the topic remains considered of second class and not an absolute priority.

Missing datasets 1: Who is studying and making women's work and contribution in STEM visible? In the datification era, the gender imbalance issue, women's contribution in raw datasets and representative samples for statistics are missing. For example, who is measuring the role of women in professional social networks, in DS forums like Kaggle, in technical discussions on Stack Overflow? The "absence" of (integrated) datasets and sources about women in STEM studies is a big issue. Without intersectional datasets, it is difficult to respond to the initial question: Where are women in STEM? What are the topics and problems they are addressing? How are they contributing to the economy of STEM? Who is collecting evidence and promoting studies to answer these questions?

2.2. Women in Artificial Intelligence and Data Science

The persistent absence of women employed in the AI and DS fields is troubling. According to the report of the World Economic Forum in 2018 (https://www.weforum.org/reports/the-global-gender-gap-report-2018, accessed on 1 June 2022), over three-quarters of professionals in these fields globally are male (78%); less than a quarter are women (22%). What about other underrepresented communities [19]? How are they represented in the DS and AI workforces, and which are the part of opportunities offered by these promising areas taken by these communities? Of course, to acquire a complete understanding of this phenomenon, it is necessary to treat the female community [1] and other underrepresented communities [19,20] as multifaceted and heterogeneous groups, with a plurality of experiences, and where gender intersects with multiple aspects of difference and disadvantage [21]. Discrimination at work must indeed be studied with an intersectional approach to acquire better and fine-grained understanding of the problem [22].

In the last decade, the role of data and scientific and technical skills used to exploit it have created promising career and economic spaces. AI and DS have emerged as promising areas for developing careers with critical financial benefit perspectives. Nevertheless, professional career perspectives in these disciplines are not equal depending on gender [23,24] and other criteria [25], including race/ethnicity, socio-economic level, the institution's reputation where people did their studies, etc. For example, in the United Kingdom, the House of Lords Select Committee on Artificial Intelligence in 2018 advocated increasing gender and ethnic diversity amongst AI developers. In France, several companies like Renault and Engie, through the "Laboratoire de l'Egalité", signed a call for widespread awareness of the discriminatory effects of AI and a commitment by its supporters to correct them. It is addressed to leaders in the public and private sectors, research and training organisations, companies that produce digital technology, companies that use digital technology and AI consultants. In 2020, the European Commission (European Commission (2019). 'Women in Digital Scoreboard'. Retrieved from: https://digital-strategy.ec.europa.eu/en/library/women-digital-scoreboard-2020, accessed on 1 June 2022. European Commission (2020a). Opinion on Artificial Intelligence—opportunities and challenges for gender equality. Advisory Committee on Equal Opportunities for Women and Men. (18 March). European Commission (2020b). Gendered Innovations 2: How Inclusive Analysis) noted that it is time to reflect on the interplay between AI and gender equality. French, European, and international organisations and agencies [8,26,27] perform studies for observing workforce shares in industry and sometimes in academia from a global perspective. Few fine-grained studies have studied underrepresented communities' workforce evolution and gaps from the gender perspective in disciplines such as AI and DS [6,28].

A thorough understanding of the way the workforce accesses the AI and DS opportunities in industry [29] and academia [30,31] is essential for building fair and inclusive societies [32]. This understanding can also be crucial for ensuring that countries obtain the full benefits of developing these areas to achieve better economic and social conditions and leading positions in the international arena through technology self-sufficiency.

Despite the economic and symbolic capital investment seeking a fair distribution of AI and DS opportunities for women and underrepresented communities, the crystal ceiling must still be broken [33]. Part of the explanation resides in data (!) [34]. Indeed, many studies agree to consider that quality, disaggregated, intersectional data are still missing. These data are essential to interrogate and tackle inequities in the AI and data science labour force [6]. As stated in the Alan Turing Institute study, "Where are women?" [6], the Royal Society has noted that a significant barrier to diversity is the lack of access to data on diversity statistics. The AI Roadmap recognises diversity and inclusion as a priority to make data-driven decisions to determine where to invest and ensure that underrepresented groups are given equal opportunity.

Missing datasets 2: Intersectional data about AI and DS female labour force. The existing evidence base about gender diversity in the AI and DS workforce is minimal. The available data is fragmented, incomplete and inadequate for investigating the career trajectories of women and men in the fields. Public data sets often rely upon data produced through proprietary analyses and methodologies. Governmental statistics lack detailed information about job titles and pay levels within ICT, computing, and technology. This partial vision of the workforce status is a significant barrier to examining the emerging hierarchy between data science, AI, and other subdomains. Furthermore, available data about the global AI & DS workforce is often aggregated and rarely broken down by age, race, geography, (dis)ability, sexual orientation, socioeconomic status, and gender. As stated by [6,35,36], "this is particularly concerning since it is those at the intersections of multiple marginalised groups who are at the greatest risk of being discriminated against at work and by resulting AI bias".

Integrating intersectional datasets allows understanding discrimination from different perspectives and mainly exhibits the various aspects that contribute to such discrimination. The absence of datasets about the female labour force in AI and DS is a form of discrimination. We believe it is necessary to promote actions applying mathematical and machine learning techniques to integrate complete and high quality, privacy-preserving data collections. Studies can use these data collections to drive conclusions about the gender gap in these disciplines.

2.3. *Awarded Women in Computer Science*

The history of computing seems to have an equivalent gender balance issue [9], as it has acknowledged with difficulty and marginally the contributions of women or at least their participation in the advances of this young science. Few documents outline the history of computer science, including women, and documents systematically include men: Alan Turing, Charles Babbage, Herman Hollerith, etc. The documents that list female computer scientists include Ada Lovelace, Hedy Lamarr, the ENIAC programmers (although their names and faces are unknown), Grace Hooper, Mary Allen Wikes, Lois Haibt or Radia Perlman. However, even members of the computing community are probably unaware of who these women scientists were and what their contributions were besides Ada Lovelace and Grace Hooper (the Appendix A lists a non-exhaustive but a more extended set of contributions authored by female computer scientists).

The history of women's contribution to computing is spread across blogs, websites and news articles. Some films acknowledge women's role as "calculators" [9,37–39], and female labour is mentioned as a curiosity in the history of science, insisting on "pencil-dragging" tasks rather than on how they developed as programmers and became part of the core of digital computing advances [40]. The objective in referring to the term "calculator" is intended to show that contribution of women has been considered paper dragging even if they played a relevant role in the projects they participated in. Despite the importance of their contribution, they were not regarded as leaders of projects, and they remained invisible for a long time for history. The films have contributed to give visibility to these women. Still their stories and their contributions must be studied with methodological approaches and then included in books and in study programs, etc.

Histories of computing are not abundant, but they do exist; for example [41,42]. Reconstructing history is a complicated undertaking, to the extent that there are special series on the subject in well-known publishers such as Springer and IEEE. In Springer, the series is entitled "History of Computing" and in IEEE, it is entitled Annals of the History of Computing. These papers name the works of the illustrious scientists who have become the pillars on which computing is founded. Not surprisingly, the most prestigious prize in the field, the Turing Award, has only been awarded to three female scientists since it was first awarded in 1966. This event first occurred in 2006 (40 years later!). Can you name three or five names of people who have received the award? Are there any women's names on your list? The names of the three female Turing Award laureates are:

- Frances E. Allen, pioneer in compiler optimisation, IBM Emeritus (2006);
- Barbara Liskov, programming languages, operating systems and innovations that have led to data abstraction, modularity, fault tolerance, persistence and distributed computing, Massachusetts Institute of Technology (MIT, 2008);
- Shafi Goldwasser, complexity theory, cryptography and number theory, MIT and Weizmann Institute of Science (2012).

The cases of other awards are similar. For example, the ACM SIGMOD Contributions Award in the area of databases, initiated in 1992, has recognised the work of Maria Zmankova (1992), Laura Haas (2000 with Michael Carey), Marianne Wenslett (2012) and Meral Özsoyoğlu (2018), Juliana Freire, Ioana Manolescu with four other male colleagues in (2020). The ACM Edgar F. Codd Innovations Award, also in databases initiated in 1992, has recognised Patricia Selinger (2002), Jennifer Widom (2007), Laura Haas (2015), and Anastasia Ailamaki (2019)—four in thirty years by 2022. The Internet recognises approximately thirty-six women who contributed to advances in computing; Wikipedia lists about sixty-four women's contributions between 1842 and 2022 (in 180 years). Since the beginning of this science, outstanding contributions have been made by women in programming languages and programming (Fortran, Smalltalk C, C# Ruby). Many famous women programmers are clustered in video games, operating systems, software engineering and software evaluation. There is also an ambition to disseminate knowledge and active participation in computer education. These names do not include, for example, the recipients of the ACM SIGMOD Award Contributions or the ACM Edgar F. Codd Innovations Award. This situation demonstrates the dispersion when it comes to reconstructing history and accessing the memory of forgotten science when it comes to remembering the names of women scientists [43].

Missing datasets 3: Female contributions in CS organised by sub-discipline and geographic region. The discipline has different areas, including AI, DS, and much more, yet there is no database collecting significant contributions per discipline. Datasets are missing about the test of time and best paper awards concerning papers that have been highly cited during a ten-year interval. Indeed, major database conferences VLDB, SIGMOD/PODS, EDBT/ICDT, ICDE, and major data mining conferences like KDD have adopted this practice. Few databases collect information about papers with female partial or complete authorship, including, for example, the exhibition Women in Computing of the Science Museum in the UK (https://www.sciencemuseum.org.uk/objects-and-stories/women-computing, accessed on 1 June 2022). What happens at national conferences? What is their role in the local generation of significant knowledge? What is the part of female contribution to this knowledge production?

Missing datasets 4: Contributions in CS of female scientists of the global south. We remark that these lists do not include women scientists working in the global south (i.e., Latin America and the Caribbean, Asia Pacific, Africa, and the Middle East). Who are the Latin American, Caribean, Asian, and African women who have contributed to advances in computing science in different fields? What have been their contributions? How have their contributions improved their regions' development and knowledge? The first action by the United Nations recognised female scientists in Latin America and the Caribbean today. There is considerable work to collect data about female CS and STEM contributions.

2.4. Discussion

As a result of the identified datasets and observations done, we can derive answers to our research questions:

> *[RQ1] Do missing data concerning women's STEM research participation prevent a comprehensive view of their contribution across history?*

Through the conclusions discussed in the three sources that we have analysed, we observed that women's history in STEM is partial and mainly located in the global north. We also observed that data are not organised according to disciplines and subdisciplines. This situation is particularly true in "young" sciences such as computer science and its subdisciplines. Regarding data science, if contributions to statistics and numerical methods should also be considered, it is true that very few and sparse data concern women, while women have been there contributing to mathematics and other sciences for ages. Data is sparse in that there are few or almost no datasets devoted to the topic, collected and built according to scientific methodologies. Any theory about women's contribution and absence in science history is anecdotal with these missing datasets.

> *[RQ2] Is it possible to design viable data science experiments that can estimate the female workforce in STEM and answer the question, "where are women in STEM"?*

Data about the female workforce in STEM is provided in big grain, and the presence of women in different disciplines indeed changes a lot. The study presented in [44] shows, for instance, that the female authors in different computer science disciplines are not evenly distributed; areas like software engineering report more female publications co-authorships than in human performance architectures, for instance. With current datasets, even combining different datasets, such as awards and census of women scientists in universities, it is impossible to provide realistic cartography of women in the various STEM disciplines and subdisciplines. Besides, the location aspect remains since fewer datasets are available regarding the female workforce in the global south, even if they have widely contributed to STEM evolution. It is not realistic to design data science workflows for modelling women in STEM. Using, for example, clustering methods nor correlating variables that determine the choice of women choosing STEM careers and the evolution of their careers can find behavioural patterns that can explain their presence and absence in STEM. We should devote efforts to building datasets about women in STEM that are (i) intersectional, (ii) geographically representative, (iii) organised by discipline and subdiscipline, and (iv) show their experience in academia and industry.

3. The Mexican Case

Our vision is to address the study of missing datasets for studying the gender gap in STEM focuses on specific countries, considering that the context is essential to understanding the issue. Thus, we chose the Mexican case as a case study. The methodology adopted was to look for data about visible scientists in CS. The case study includes scientists recognised by the National System of Researchers (SNI), the Mexican Academy of Sciences, female scientists awarded by the National Award of Sciences, the history of Mexican women in STEM, and female professional situation and perspectives in STEM. We use these references because they provide an integrated national view of the Mexican scientific community. A complete list of the acronyms used in this section is given in Appendix B.

3.1. Mexican National System of Researchers

According to their scientific production, the SNI classifies Mexican scientists with PhDs into five levels (candidate being the lowest one, and emeritus being the highest one). The evaluation focuses mainly on publications appearing in the ISI Thompson List and the impact factor of these publications. For the candidate level, applicants (less than 40 years old) must have published three journal papers recognized in the JCR or the list of journals of the CONACyT. To apply to level I, the scientist must have published 5 JCR papers, three of them in the last three years before the application. For level II, scientists must have published 15 JCR papers, 5 or 6 in the previous 5 or 6 years before the application. The scientist must also have advised graduate students. Finally, to apply to level 3, scientists must have published 15 to 30 JCR journal papers, and 8–9 papers must have been published during the last three years before the application. The awards can be renewed, and for SNI 3, emeritus scientists can have lifetime fellowship. Awarded scientists at SNI are granted fellowships that range between 375 USD– and 1750 USD.

Beyond the research are domains that cover a large spectrum of disciplines, organised in seven areas: I. mathematics, physics and earth sciences; II. biology and chemistry; III. medicine and health sciences; IV. humanities and behavioural sciences; V. social sciences; VI. biotechnology and agricultural sciences; and VII. engineering. What attracts attention is the distribution of the population across the SNI classification levels between men and women. The numbers show that 34% of Mexican scientists are women (!). What happened to all those women doing graduate studies, more significant in number than men? From this 34%, the majority of the female scientists are classified as grade I. This grade corresponds to people with at least 3 or 4 years of experience. Next, 15% of females are classified as grade II and 5% as grade III. These statistics show that it seems complicated for women to consolidate their careers and access recognised senior positions independently of the discipline. To gender inequality, we could add the age dimension, as Paloma Alcalá stated (Paloma Alcalá, seminario "El sexo de la Ciencia" held in the Faculty of Philosophy and Education Sciences of the Universidad del País Vasco, San Sebastián, 1 y 2 March 2000) that when women achieve the highest positions in the organisation, they have invested 16 or 20 years more than men. STEM, as other profession possibilities for women, are fields where women make less money and advance through the ranks more slowly [45]. Women's absence at the top provides arguments to understand women's underrepresentation in STEM.

During the new presidential six-year term (2018–2024), the SNI has made an effort in including women in evaluation committees. Recent raw data has been exported recently on the official governmental site; our study can be eventually completed with this new release. The gender gap is still wide, particularly in the highest grades of the SNI.

In 2014, the Mexican Council of Science and Technology (CONACyT) published the distribution of scientists doing research in the Mexican system recognised by the SNI. Only 15% of researchers in Mexico were women, with the criterion of having obtained the distinction of emeritus granted by the SNI. Later, according to official report of the CONACyT in 2020, there were 33,165 researchers recognised by the SNI, of which 8727 were candidates (26.31%), 17,091 level I (51.53%), 4793 level II (14.36%) and 2584 level III and emeritus (7.79%). In 2021, the percentage of women in the SNI achieved 38.2%. In 2022, the SNI granted a group of senior scientists with the emeritous grade: 183 scientists in all disciplines; of this number, 38 correspond to female researchers, the highest number of women who have been awarded this distinction. In 2022, there are 102 female researchers with emeritus status in the SNI out of a total of 462.

Let us analyse people doing research in computer science recognised in the SNI. We find eighty-two women researchers identified out of four hundred and thirty-seven recognised between candidate and three consecutive grades. There is only one female scientist in grade three, Elba Patricia Melin Olmeda of the Instituto Tecnológico de Tijuana, a specialist in artificial intelligence, compared to thirteen male researchers. There are two female researchers in grade II, against thirty-eight male researchers with the same grade. The rest are distributed among forty female researchers at grade one against two hundred and seventeen male researchers. There are thirty-eight female researchers with grade "candidate" against ninety-seven male researchers. Regardless of observing an evident and pronounced gender disparity, sociological and economic studies are still needed to explain this phenomenon in the Mexican context.

The CONACyT and the Mexican government implemented inclusion policies seeking gender balance, promoting it in evaluation committees with the controversial idea that with more women in committees, they would encourage gender balance. In 2016, the SNI decision-making commissions, composed of 14 members each, were published on the CONACyT Web Site (http://conacyt.gob.mx/index.php/convocatorias-conacyt/informacion-importante-sni/miembros-de-comisiones-dictaminadoras/12183-integrantes-de-las-comisiones-dictaminadoras-2016/file (accessed on 20 August 2016)). The gender proportion was as follows:

- Physics, mathematics and earth sciences: 3 women:11 men;
- Biology, chemistry and life sciences: 5 women:9 men;
- Medicine and health sciences: 7 women:7 men;
- Humanities and behaviour sciences: 7 women:7 men;
- Social sciences: 3 women:11 men;
- Biotechnology and agricultural sciences: 3 women:11 men;
- Engineering: 1 woman:13 men;
- Technology: 2 women:12 men (female president).

In 2021, the distributions were as follows. Note that the classification of disciplines evolved, more women have the presidency of the commission, and their participation is higher in areas where they are underrepresented:

- Physics, mathematics and earth sciences: 3 women:11 men (male president);
- Biology, chemistry and life sciences: 8 women:6 men (male president);
- Medicine and health sciences: 11 women:3 men (female president);
- Behaviour sciences and education: 9 women:5 men (female president);
- Humanities: 9 women:5 men (female president);
- Social sciences: 10 women:4 men (female president);
- Agriculture, agriculture, forestry and ecosystem sciences : 5 women:9 men (female president);
- Engineering and technological development: 5 women:9 men (male president);
- Interdisciplinary: 5 women:9 men (male president).

Figure 2 shows a comparison between the number of female and male members of the evaluation commissions. The difference is still significant even if it lowered between 2016 and 2022. Note that the CONACyT merged engineering and technology and created a multidisciplinary area. This reorganisation of disciplines might be why the number of women in engineering and technology, for example, increased.

Missing datasets about the female scientific force in STEM at SNI: The Mexican government has adopted an open data initiative that includes the SNI. This initiative gives access to the SNI community, and it is possible to perform simple statistics about women in different disciplines. However, no contextual data can make it possible to perform more profound studies about the professional career vs. the evolution of SNI awarded grades. Are women working in universities located in urban areas? Are women working in institutions located close to the capital more likely to be awarded by the SNI? What kind of contact and working network do SNI-awarded women have? Does this network play a role in their evolution in the SNI grades? There are also missing datasets about the situations in which scientists, particularly women, lose the SNI grant for some time and then are granted again when their scientific production takes off. Additionally, datasets about the metrics used for evaluation are missing. They could be helpful to observe how their values evolve according to gender and SNI level of scientists. This data could provide a more representative understanding of the "productivity" of scientists and its evolution correlated with the development of their career (e.g., administrative positions, classification in their institutions, projects coordination, student advising).

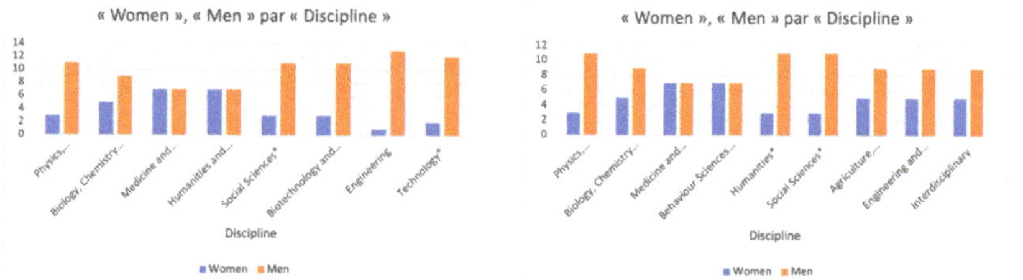

Figure 2. Comparison between the composition of commission members by gender and discipline between 2016 and 2020. Commissions with a * have female presidents. The complete names of the items with "..." are given in the text.

3.2. *Mexican Academy of Sciences*

Another reference showing the unbalanced number of men and women in STEM is the current membership numbers in the Mexican Academy of Sciences (AMC). The AMC is a non-profit organization comprised of distinguished Mexican scientists attached to various institutions in the country and several eminent foreign colleagues, including different Nobel Prize winners. According to the Web site of the AMC, by March 2022, there were 1376 members in "Exact Sciences" (astronomy, physics, engineering, chemistry and geosciences)—1158 men and 218 women. The discipline with fewer women was engineering, with 43 women and 318 men (see Figure 3).

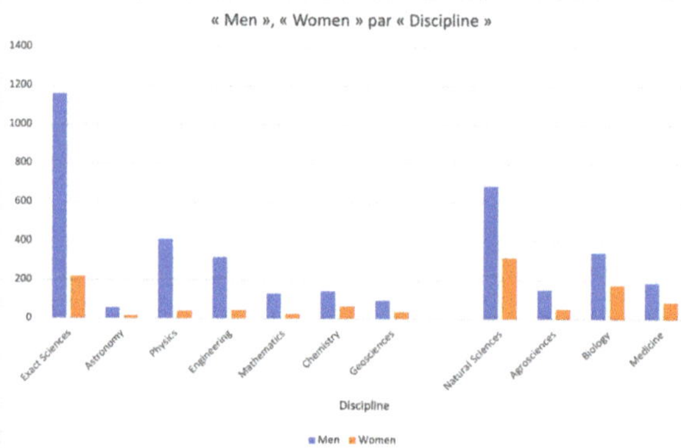

Figure 3. Members of the AMC by gender and discipline.

These numbers show that the AMC is even progressive considering that women's access to university was granted recently. For example, in Europe, between 1860–1900, Cambridge accepted women without restrictions until 1947, and the National Academies of Sciences opened spaces for women later. Marjory Stephenson and Kathleen Londsdale were the first scientists to join the Royal Society in 1945, an institution with 300 years of tradition. Yvonne Choquet-Bruhat was the first scientist accepted in the Academie de Sciences in France in 1979. This institution was founded in 1666. The first Spanish women received in the Royal Academy of Pharmacy and the Royal Academy of Sciences, Physics, and Natural Sciences were María Cascales (1987) and Margarita Salas (1988).

The AMC has already had its first female president, Rosaura Ruíz, who actively promoted the Science L'Oréal-UNESCO-AMC Award. She also encouraged the creation of fellowship programs for women studying Humanities and Social Sciences. She was awarded the National Award of Sciences and Arts in Technology and Design in 2009. In 2007, Esther Orozco was elected director of the Institute of Science and Technology of Mexico City, and she was awarded the L'Oréal UNESCO award in 2006. In 2009 Yoloxóchitl Bustamante was named the first female president of the Mexican Institute of Technology (Instituto Politécnico Nacional), 73 years after being founded. Some Mexican female scientists are influential and have occupied important positions in national and international institutions, including, for example, Ana María Cetto in the International Agency of Atomic Energy. She is a senior scientist at the Institute of Physics of the National Autonomous University of Mexico (UNAM). She has been head of the Faculty of Sciences and president of the Pugwash Conferences, a committee member that received the Nobel Peace Prize in 1995. In 1991, Ana María Cetto was president of the Organization for Women in Science for the Developing World (OWSD) for the Caribbean and Latin American Region. Mayra de la Torre was elected vice-president of the same institution. She collaborates in Science and Technology and the renewal of the program for Biotechnology/Biosecurity for the Americas. She has been awarded the National Award for Sciences and Arts, the Award of Sciences in the Third World, the Manuel Noriega Morales Award of the OEA and the CIBA GEIGY award in Innovation, Technology and Ecology.

Missing datasets about females in the AMC. Detailed datasets are missing regarding the contributions of women in the AMC. The open question is similar to data about women at the SNI: their connection networks, possible mentors, and mentoring roles. Which Mexican female scientists did they support to be accepted as members of the AMC?

3.3. Mexican National Award of Sciences

Let us take the Mexican National Award of Sciences as another possible data provider that shares little data about awardees. This award started being granted in 1945, and it recognises individuals or groups working in the following areas:

- Linguistics and literature beaux arts;
- History, social sciences and philosophy;
- Physics and mathematics, and natural sciences;
- Technology and design;
- Art and popular traditions.

Female scientists started to be awarded in 1979 with Guillermina Bravo, the first woman awarded for her work in Beaux-Arts. In 1996, María Luisa Ortega Delgado was awarded the National Award for Technology and Design with Adolfo Guzmán Arenas. In 2008, María de los Ángeles Valdés Ramírez was awarded (biology), and in 2009, Blanca Elena Jiménez Cisneros was awarded (environmental engineering) with José Luis Leyva Montiel. Thus, only one woman in CS has been awarded in the whole history of the awards program. Gender inequality is evident, and importance is given to gender issues in the country, considering that women acquired the right to vote in 1945 and that sexual liberation happened in the 60s.

The situation is even more precarious when making decisions and defining plans, policies and programs in committees. Indeed, few women participated in the structure of the Scientific and Technological Consultative Forum that coordinated the development of the National Science and Technology 2006–2012 (María Valdés, La mujer mexicana en la ciencia, https://www.cronica.com.mx/notas-la_mujer_mexicana_en_la_ciencia-946726-2016.html, accessed on 1 June 2022).

Missing datasets about female contributions in STEM and its implications. We have highlighted the lack of data about female contributions in STEM. The lack of data leads to the underrepresentation of women in visible databases that can help choose them to be granted awards. Their role is not highlighted in scientific teams and labs, project coordination, keynote speeches, or innovation authorships. Therefore, they disappear from the scientific committees' memory, and thus, they are rarely granted prestigious awards.

3.4. History of Mexican Women in STEM

The history of Mexican women in science is still to be written and disseminated as part of the education of young Mexicans. Women have contributed to scientific production since ancient times. An example of one of the first Mexican women to attend University and obtain a diploma is Matilde Montoya (1859–1939). President Porfirio Díaz granted her permission to perform her professional examination, becoming the first Mexican Surgeon. Later, Helia Bravo (1901–2001) was the first Biologist in Mexico to develop scientific work on cactus. She published more than 160 papers and 3 books about these plants and the arid Mexican ecosystems. She described 57 species, 2 genders and 8 species that bear her name. Paris Pishmish (1911–1999) was an astronomer. She proposed a theory for explaining the origin and the spiral structure of galaxies. She discovered three open stellar cumuli, now named by her name. She also found that stellar galactic associations move away from the galaxy's centre. Then, Luz María del Castillo Fregoso, chemist and pioneer in biotechnology and zymology in Mexico, was internationally recognized for her work on physical chemistry, particularly on enzymatic reactions. She was the first woman to obtain the Sciences Award of the AMC, and she was a member of the SNI grade III from her first application.

Missing datasets about Mexican women in Science. The work of reconstructing the history of computing in Mexico from a gender perspective remains to be done. Naming all-female forgotten science contributors to STEM is important because this can let emerge forgotten activities and contributions that would be erased from the collective memory. It can make it possible to understand the conditions in which STEM fields produce results and capital, the real opportunities for professional careers, what people (men and women) should expect and how they can become agents of change.

3.5. Female Professional Situation and Perspectives in STEM

Women's representation varies by discipline and professorship status in the academic labour force. The majority of full-time female faculty in STEM disciplines are significantly low in computer and information sciences, maths, physical sciences, and engineering. In the life sciences, an area in which many people assume that women have achieved parity, women made up only one-third of faculties. The professional expectations do not seem to be very promising for women if we analyse the numbers reflecting female professionals accessing decision-making positions, including, for example, the number of female deans in engineering schools in universities and the number of university presidents in technology institutes globally. Of course, these numbers change among countries depending on culture, race, political trend and prominence of the institutions.

Studies show how women leaders in the US have a more significant impact on a company's bottom line (https://www.fastcompany.com/90733328/the-secret-to-womens-leadership-that-can-drive-such-a-positive-impact, accessed on 1 June 2022). However, only an estimated 15% of C-suite executives and 51% of managers are women. Additionally, women make up only 30% of full professors and 26% of college deans in US universities (https://www.catalyst.org/research/women-in-academia/, accessed on 1 June 2022). However, in STEM, gender inequality is about leadership and inequality in terms of critical mass. It is impossible to choose female leaders where few women have the professional and/or scientific consolidation to access such positions. Referring to our use case, in Mexico, according to the Asociación Nacional de Universidades e Instituciones de Educación Superior (ANUIES (http://www.anuies.mx), accessed on 1 June 2022) between 2014–2015, 1 million 842 thousand 978 women and 1 million 876 thousand 17 men did studies in STEM fields. Thus, more or less, the numbers are balanced. Moreover, more women (167,967 women) than men do graduate studies in STEM (146,030 men). Therefore, what happens in professional life?

According to "Attitudes and experiences of engineering alumni" by [46–48] workplace environment, bias, and family responsibilities all play a role. Many women appear to encounter a series of challenges at mid-career that contribute to their leaving careers in STEM industries. Women cited feelings in studies of isolation, an unsupportive work environment, extreme work schedules, and unclear rules about advancement and success as significant factors in their decision to leave. Departmental culture includes the expectations, assumptions, and values that guide the actions of professors, staff, and students. Individuals may or may not be aware of the influence of departmental culture as they design and teach classes, advise students, organise activities, and take classes. For example, people tend to view women in "masculine" fields, such as most STEM fields, as either competent or likeable but not both, according to Madeline Heilman, an organizational psychologist at New York University [49,50]. Although being both competent and well-liked are essential for advancement in the workplace, this balance may be more difficult for women than men to achieve in science and engineering fields. STEM fields are perceived as male, including fields like chemistry and maths, where almost one-half of degrees awarded now go to women. Heilman's research shows how, in the absence of clear performance information, individuals view women in male-type occupations as less competent than men [49,50]. There is a need to understand how people are evaluated in professional settings and understand which features in organisations give men more success opportunities. The first false assumption is that superior intelligence is required to address STEM fields. More than intelligence that can be numerically measured, the prejudice is that STEM calls for talent and that there is only one single talent that ensures success in such fields. The discrimination further argues about the possibility of numerically measuring such aptitude. The worst is the idea that such talent is a fixed gift, that talent is not malleable. Only those people with the highest standardised test scores and the most confidence will step into STEM fields and develop a career. Many believe that all these prejudices reduce the number of people deciding to make a career in STEM fields and even break the glass ceiling. The good news is that this talent is not a unitary thing. It is multidimensional and difficult to quantify and measure; many different skills are critical to step into STEM fields; talent can be developed and enhanced by education, encouragement, self-confidence, and hard work [45].

Few institutions promote reflection about policies for addressing gender issues in their organizations. In Mexico, there are three representative centres: the Centro de Investigaciones y Estudios de Género (CIEG) and the *Unidad Politécnica de Gestión con perspectiva de Género* (Polytechnic Unit for Management with a Gender Perspective) (http://www.genero.ipn.mx/, accessed on 1 June 2022) at the Universidad Autónoma de México and the Instituto Politécnico Nacional; and the Project on Gender Issues of the Humanities and Social Sciences School of the Monterrey Institute of Technology (Tecnológico de Monterrey). The CIEG has a unique chair on Gender in Science, Technology and Innovation, in addition to the Mexican Network of Science, Technology and Gender.

Missing datasets about working conditions of women in STEM. Women's labour conditions when they develop a professional career in STEM are taboo both in academia and industry. Few or no data are collected by institutions, industry and independent bodies about the labour conditions: working hours, positions, evolution, environment, stress, entrepreneurship, glass ceiling, etc. How can the gender gap be understood and reduced with little input to drive studies and design policies? Data about these issues are disseminated in online forums (e.g., Quora, Medium, Glassdoor), blogs and social networks. They must be collected and integrated into datasets that can drive seminal studies.

3.6. Discussion

After applying a grounded study strategy for approaching datasets about women in STEM in general and then focusing on computer science, we observed that datasets are missing. The history of female contribution in STEM is often silent about women. We concluded that currently available data make it questionable to apply DM, ML and AI methods for picturing women's history in STEM and discovering patterns to understand their apparent absence. With these observations, we approximated conjectural answers for research questions RQ1 and RQ2. This first study could not provide elements to answer RQ3 because the studied datasets do not provide variables or perspectives about the conditions in which reported contributions were produced, only scientists' year, institution, discipline, and administrative nationality (nationality of birth and age). RQ3 questions about intersectionality, and therefore we chose a concrete case, that of Mexico, for which we observed the contribution statistics that show representative scientific production and excellence in the country. We enumerated the criteria used to evaluate scientific contribution in the National System of Researchers (SNI). The question was whether these criteria could provide elements to build an intersectional perspective and a quantitative study to approach an intersectional gender gap index.

> *[RQ3] Do missing intersectional perspectives prevent performing representative quantitative analysis about women's activity and contribution in STEM?*

Existing datasets in Mexico providing data about women in STEM are sparse, partial and reduced to a tiny group of scientists (both female and male). For example, scientists working in private universities were excluded from the possibility of obtaining a fellowship related to SNI recognition. This exclusion hides an essential group of scientists from statistics. Universities, polytechnic centres, and other institutions that develop science provide different facilities to scientists to do their research. This difference depends on their location, type, and the disciplines they support. Female scientists and their conditions almost disappear in this complex context. Building intersectional datasets combining all these elements is a project to run. In a country such as Mexico, these socio-economic and political factors produce clusters of scientific production with different characteristics and conditions. Being a female scientist at the CINVESTAV (the most important research centre in Mexico) in Mexico City is entirely different than being a scientist in Huajuapan de Leon in Oaxaca (the poorest state in Mexico). The studies about women must address these aspects that call for thorough data collection strategies (by region and by the institution) and then running analytics at a small scale before scaling to the whole country. To answer RQ3, it is true that without intersectional datasets built with methodological data collection strategies, women will continue to be hidden and difficult to locate in the scientific cartography.

4. Conclusions

Gender gap issues in STEM are a matter of money and power and, more precisely, about capital in the sense of the theory of the French sociologist, Pierre Bourdieu (who defines three types of capital: education, material and cultural. Pierre Bourdieu proposes a complex machinery of interrelationships of these three types of capital that impose power and influence on each other and organise activities and social markets). Indeed, gender gap issues are all about an unbalanced problem among the different capitals activating STEM, particularly in promising disciplines like the AI and DS markets. These issues have been animated by bias associated with the roles and capacities related to men and women. Society believes and promotes that men are more apt to address STEM disciplines than women. However, the gender divide in STEM is more complex because of socio-cultural and racial factors. The official social order has viciously controlled access to specific knowledge disciplines.

The absence of women in STEM denotes the possibility of significant losses of brains with potential. If the gender perspective is not relevant or a priority from a human point of view, perhaps it can acquire value when it translates into millions of dollars lost by not valuing and encouraging the scientific contribution of a part of its population. In any case, this lack of balance in the STEM professional market attracts the attention of different actors willing to activate the global economy. For example, AI and DS are widely regarded as critical to the national economies in many "developed" countries. Concern about America's ability to be competitive in the global economy has led to several calls to action to strengthen the pipeline into these fields [51–53]. Thus, this unfair situation in countries of the global south like Mexico creates an economic loss. Countries where inequality is more critical compete with less capital and thus with fewer possibilities to obtain benefits.

There is still a lot to be done, mainly because the capital conditions in which women as minorities enter the STEM professional and scientific market are very different from those in which men do. This issue is also related to education in STEM, the way young students perceive it and how they decide to step into STEM studies and careers depending on whether they are men or women. Many departments and universities worldwide develop studies about the aspects that move female students to inscribe in STEM majors and the conditions of their permanence from bachelor to graduate studies eventually. This collected data must be integrated and shared, but it has the merit of existing. Postcolonialism and decolonialism have already explained that minorities are never the same and cannot be studied and analysed as a whole. Racial aspects must be considered as well. In Mexico, indigenous women living in small villages cannot aspire to the same opportunities as women with more European racial heritage living in a big city. Nevertheless, the participation of women in STEM is still considered a special event that should have special treatment.

Data-driven studies are crucial to understanding the gender gap from an intersectional point of view, considering the particular conditions of the global north and south countries. This paper has shown that data are the fuel that can provide perspectives about the conditions in which women participate in the STEM market. The paper has also demonstrated that datasets are cruelly missing and that the challenge of collecting representative data is significant. Data-driven studies also need algorithms that can exploit them. These algorithms must be designed with care so that they do not bias observations and conclusions about the gender gap and do not further marginalise and discriminate against women in STEM. We need to open the door and allow and encourage women to contribute to the development of STEM, and this perspective depends on data and fair algorithms. Here is the view that should drive gender studies. Our current work addressed these issues in the context of the JOWDISAI and SINFONIA projects that are willing to create intersectional data collections about the women labour force in AI and DS in academia and industry in France. Other initiatives are seeking to reason about datasets, their content and collection conditions from decolonisation and feminist perspective, for example, the movement Tierra Común (https://www.tierracomun.net, accessed on 1 June 2022), and the A+ Alliance (https://feministai.pubpub.org, accessed on 1 June 2022). Our future work will include data collection strategies and algorithms to propose an intersectional analysis of the gender gap in STEM.

Funding: This research was partially funded by the projects JOWDISAI program AAP of the CNRS, France and SINFONIA of the interlaboratory program of the Institute of Gender, France.

Institutional Review Board Statement: Not applicable.

Informed Consent Statement: Not applicable.

Data Availability Statement: Not applicable.

Conflicts of Interest: The author declares no conflict of interest.

Appendix A

The following list shows contributions in CS authored by leading women. This list is complementary to the list of women granted with prestigious awards. The list shows the controversy about the relation between awards and contributions to knowledge with a gender perspective.

1. In 1843, Ada Lovelace (1815–1852) became the first programmer by designing the first algorithm and explaining how it would work in Babbage's non-existent analytical engine.
2. In 1942, Hedy Lamarr invented the "frequency-hopping" technology that enabled the invention of wireless signals such as Wi-Fi and Bluetooth.

3. In 1945–46, Jean Bartik (1924–2011) and five women developed and coded the fundamentals of software programming working on the ENIAC. In spite of their decisive work, these women programmers were not invited to the dinner celebrating the construction of the ENIAC. These are their names: Frances Spence (1922–2012), Marlyn Meltzer (1922–2008), Kathleen Antonelli (1921–2006), Betty Holberton (1917–2001), Ruth Teitelbaum (1924–1986), and Gloria Gordon Bolotsky (1921–2009).
4. In 1951, Admiral Grace Hooper (1906–1992) created the first compiler for the programming language she proposed, COBOL.
5. Mary Allen Wikes (1937–), computer programmer and designer. Her best known work is on the LINC computer, now recognised as the first personal computer.
6. Lois Haibt (1934–) is known as one of the ten people who formed the IBM group that developed FORTRAN, the first successful high-level programming language.
7. Radia Perlman (1951–) is famous for inventing the spanning-tree protocol fundamental to network operation while working at Digital Equipment Corporation.
8. Elsie Shutt founded Computation Incorporated in 1957.
9. Danese Cooper (1959–) is a programmer, computer scientist, and activist in favour of open software.
10. Rebecca Heineman (1963–) is a video game programmer, veteran of the video game industry, founding member of Interplay Productions, Logicware, and Contraband Entertainment.
11. Jamie Fenton (1964–) is a programmer of the famous 1981 arcade hit game Gorf. In 1978, she created an early example of glitch art entitled Digital TV Dinner.
12. Corrinne Yu (1979–) is game programmer.
13. Audrey Tang (1981–) is a Taiwanese free software programmer described as one of the top ten Taiwanese computer programmers.
14. Lorinda Cherry, contributor to the Plan 9 Operating System, co-wrote with Mike Leks "Typing Documents on the UNIX System: Using the -ms and -mcs Macros with Troff" for the tenth edition of the Unix Handbook. She coded a non-dictionary-based spelling checker proposed by Bob Morris.
15. Xiaoyuan Tu is co-founder and lead scientist at AiLive Inc. (formerly iKuni Inc.) , a Silicon Valley startup working on artificial intelligence effects in computer entertainment. Dr. Tu has received awards for her technical contributions in the fields of computer animation, behavioural modelling and artificial life. In 1996, she received the prestigious ACM Doctoral Dissertation Award for her doctoral dissertation entitled "Artificial Animals for Computer Animation: Biomechanics, Locomotion, Perception, and Behavior". Although other women have received awards for their doctoral dissertations, Dr. Tu is the first and only woman so far to receive first place.
16. Carla Meninsky was a video game designer and programmer during her early career at Atari VCS. Along with Carol Shaw, Meninsky was one of two engineers at Atari, Inc. who developed video game cartridges.
17. Leah Culver (1982–) co-founded the micro-blogging site Pownce which was bought by Six Apart in December 2008.
18. Amanda Wixted (1982–) is a game programmer and founder of Meteor Grove Software.
19. Dona Bailey is an American game programmer and educator who together with Ed Logg in 1981 created the game Centipede. Jade Raymond (1975–) is a Canadian video game executive and founder of Electronic Arts' Motive Studios and Ubisoft Toronto.
20. Dina St. Johnston (1930–2007) was an English programmer credited with founding the first software house in the UK in 1959.
21. Linda Liukas (1986–) is a Finnish programmer, programming instructor and children's book writer. In 2014, her children's coding book "Hello Ruby" became the most financially supported book on Kickstarter.
22. Edith Windsor (1929–2017) was a chief technology officer at IBM.
23. Elaine Wayuker (1945–) is an ACM Fellow, IEEE Fellow, and AT&T Fellow at Bell Labs and a member of the National Academy of Engineering for her research in software metrics and evaluation.

24. Zoë Quinn (1987–) is a video game developer, programmer, writer and artist. She developed the interactive fiction Depression Quest.
25. Amy Briggs (1962–) is a video game programmer best known for creating "Plundered Hearts", an interactive video game published by Infocom in 1987.
26. Molly Holzschlag (1963–) is an author, teacher and Open Web advocate. She has written and co-authored 35 books on Web design and open standards including "The Zen of CSS Design: Visual Enlightenment for the Web".
27. Sandi Metz is a software engineer and author of "Practical Object-Oriented Design in Ruby".
28. Ellen Ullman is a programmer and writer. She has written novels and articles for various publications including Harper's Magazine, Wired, New York Times and Salon.
29. Coraline Ada Ehmke is a speaker, writer, open software advocate, technologist with over 20 years experience in Web application development. She is the creator of Covenant Contributor, the most popular open source code with 35,000 adoptions including JRuby, Swift, F#, GitLab, and Rails.
30. Brianna Wu (1980–) is a video game developer and programmer. She co-founded Giant Spacekat, an independent video game development studio with Amanda Warner in Boston, Massachusetts.

Appendix B

This section provides a list of acronyms with their definition.

- ACM—Association for Computing Machinery
- AMC—Mexican Academy of Sciences (Academia Mexicana de Ciencias)
- ANUIES—National Association of Universities and Institutions of Higher Education (Asociación Nacional de Universidades e Instituciones de Educación Superior)
- CIEG—Centro de Investigaciones y Estudios de Género
- CONACyT—National Council of Science and Technology (Consejo Nacional de Ciencia y Tecnología)
- EDBT—Extending Database Technology
- ICDE—International Conference on Data Engineering
- ICDT—International Conference on Database Theory
- IEEE—Institute of Electrical and Electronics Engineers
- ITESM—Monterrey Institute of Technology (Instituto Tecnológico de Monterrey)
- JCR—Journal Citation Reports
- KDD—Knowledge Discovery and Data Mining
- NCWIT—National Centre for Women & Information Technology
- OEA—Organisation of American States (Organización de los Estados Americanos)
- OWSD—Women in Science for the Developing World
- PODS—Principles of Database Systems
- SIGMOD—Special Interest Group on Management of Data
- SNI—National System of Researchers (Sistema Nacional de Investigadores)
- STEM—Science, Technology, Engineering and Mathematics
- UNAM—National Autonomous University of Mexico (Universidad Nacional Autónoma de México)
- VLDB—Very Large Data Base

References

1. Hill, C.; Corbett, C.; St Rose, A. *Why So Few? Women in Science, Technology, Engineering, and Mathematics*; American Association of University Women: Washington, DC, USA, 2010.
2. Keller, E.F.; Scharff-Goldhaber, G. *Reflections on Gender and Science*; American Association of Physics Teachers: College Park, MD, USA, 1987.
3. Lehming, R.F.; Alt, M.N.; Chen, X.; Hall, L.; Burton, L.; Burrelli, J.S.; Kannankutty, N.; Proudfoot, S.; Regets, M.C.; Boroush, M.; et al. *Science and Engineering Indicators 2010*; NSB 10-01; National Science Foundation: Arlington, VA, USA, 2010.

4. Eubanks, V. *Automating Inequality: How High-Tech Tools Profile, Police, and Punish the Poor*; St. Martin's Press: New York, NY, USA, 2018.
5. Faulkner, W. Doing gender in engineering workplace cultures. II. Gender in/authenticity and the in/visibility paradox. *Eng. Stud.* **2009**, *1*, 169–189. [CrossRef]
6. Young, E.; Wajcman, J.; Sprejer, L. *Where Are the Women? Mapping the Gender Job Gap in AI*; Policy Briefing: Full Report; The Alan Turing Institute: London, UK, 2021.
7. Berman, F.D.; Bourne, P.E. Let's make gender diversity in data science a priority right from the start. *PLoS Biol.* **2015**, *13*, e1002206. [CrossRef]
8. Alfrey, L.; Twine, F.W. Gender-fluid geek girls: Negotiating inequality regimes in the tech industry. *Gend. Soc.* **2017**, *31*, 28–50. [CrossRef]
9. Margolis, J.; Fisher, A. *Unlocking the Clubhouse: Women in Computing*; MIT Press: Cambridge, MA, USA, 2002.
10. Bulmahn, E. Women in science in Germany. *Science* **1999**, *286*, 2081. [CrossRef]
11. Costas, I. Women in science in Germany. *Sci. Context* **2002**, *15*, 557–576. [CrossRef]
12. Mello, P.A. *Qualitative Comparative Analysis: An Introduction to Research Design and Application*; Georgetown University Press: Washington, DC, USA, 2022.
13. Makarova, E.; Aeschlimann, B.; Herzog, W. The gender gap in STEM fields: The impact of the gender stereotype of math and science on secondary students' career aspirations. *Front. Educ. Front.* **2019**, *4*, 60. [CrossRef]
14. Fayyad, U.; Hamutcu, H. Toward Foundations for Data Science and Analytics: A Knowledge Framework for Professional Standards. 2020. Available online: https://hdsr.mitpress.mit.edu/pub/6wx0qmkl/release/4?readingCollection=70ac5c46 (accessed on 1 June 2022).
15. Kemp, P.E.; Wong, B.; Berry, M.G. Female performance and participation in computer science: A national picture. *ACM Trans. Comput. Educ. (TOCE)* **2019**, *20*, 1–28. [CrossRef]
16. Maltese, A.V.; Cooper, C.S. STEM pathways: Do men and women differ in why they enter and exit? *AERA Open* **2017**, *3*, 2332858417727276. [CrossRef]
17. D'ignazio, C.; Klein, L.F. *Data Feminism*; MIT Press: Cambridge, MA, USA, 2020.
18. Hearn, J.; Louvrier, J. Theories of difference, diversity, and intersectionality. In *The Oxford Handbook of Diversity in Organizations*; Oxford University Press: Oxford, UK, 2015; p. 62.
19. Benjamin, R. Race after technology: Abolitionist tools for the new jim code. *Soc. Forces* **2019**, *98*, 1–3. [CrossRef]
20. Herring, C. Does diversity pay? Race, gender, and the business case for diversity. *Am. Sociol. Rev.* **2009**, *74*, 208–224. [CrossRef]
21. Collins, P.H. It's all in the family: Intersections of gender, race, and nation. *Hypatia* **1998**, *13*, 62–82. [CrossRef]
22. Bobbitt-Zeher, D. Gender discrimination at work: Connecting gender stereotypes, institutional policies, and gender composition of workplace. *Gend. Soc.* **2011**, *25*, 764–786. [CrossRef]
23. Wajcman, J.; Young, E.; Fitzmaurice, A. *The Digital Revolution: Implications for Gender Equality and Women's Rights 25 Years after Beijing*; United Nations: New York, NY, USA, 2020.
24. Myers West, S. *Discriminating Systems: Gender, Race and Power in Artificial Intelligence*; Georgia Institute of Technology: Atlanta, GA, USA, 2020.
25. Freire, A.; Porcaro, L.; Gómez, E. Measuring diversity of artificial intelligence conferences. In Proceedings of the Artificial Intelligence Diversity, Belonging, Equity, and Inclusion, Online, 21–31 January 2021; pp. 39–50.
26. Abbate, J. *Recoding Gender: Women's Changing Participation in Computing*; MIT Press: Cambridge, MA, USA, 2012.
27. Alegria, S. Escalator or step stool? Gendered labor and token processes in tech work. *Gend. Soc.* **2019**, *33*, 722–745. [CrossRef]
28. UNESCO. *Artificial Intelligence and Gender Equality: Key Findings of UNESCO'S Global Dialogue*; UNESCO: Paris, France, 2020.
29. Cardador, M.T.; Hill, P.L. Career paths in engineering firms: Gendered patterns and implications. *J. Career Assess.* **2018**, *26*, 95–110. [CrossRef]
30. Ensmenger, N.L. *The Computer Boys Take Over: Computers, Programmers, and the Politics of Technical Expertise*; MIT Press: Cambridge, MA, USA, 2012.
31. Foulds, J.R.; Islam, R.; Keya, K.N.; Pan, S. An intersectional definition of fairness. In Proceedings of the 2020 IEEE 36th International Conference on Data Engineering (ICDE), Dallas, TX, USA, 20–24 April 2020; IEEE: Piscataway, NJ, USA, 2020; pp. 1918–1921.
32. Simonite, T. AI Is the Future—But Where Are the Women. 2018. Available online: https://www.wired.com/story/artificial-intelligence-researchers-gender-imbalance (accessed on 1 June 2022).
33. Dobbin, F.; Kalev, A. Why diversity programs fail. *Harv. Bus. Rev.* **2016**, *94*, 14.
34. Buolamwini, J.; Gebru, T. Gender shades: Intersectional accuracy disparities in commercial gender classification. In Proceedings of the Conference on Fairness, Accountability and Transparency, New York, NY, USA, 23–24 February 2018; pp. 77–91.
35. Wajcman, J. *Feminism Confronts Technology*; Penn State Press: University Park, PA, USA, 1991.
36. Wajcman, J. Feminist theories of technology. *Camb. J. Econ.* **2010**, *34*, 143–152. [CrossRef]
37. Allen, K.J. Hidden Figures: The American Dream and the Untold Story of the Black Women Mathematicians Who Helped Win the Space Race by Margot Lee Shetterly. *IEEE Ann. Hist. Comput.* **2017**, *39*, 70–71. [CrossRef]
38. Marino, K. The Glass Universe: How the Ladies of the Harvard Observatory Took the Measure of the Stars. *Hist. J. Mass.* **2018**, *46*, 172–177.

39. Light, J. Rise of the Rocket Girls: The Women Who Propelled Us, from Missiles to the Moon to Mars. *Nature* **2016**, *532*, 34–35. [CrossRef]
40. Light, J.S. When computers were women. *Technol. Cult.* **1999**, *40*, 455–483. [CrossRef]
41. Coello, C.A.C. *Breve Historia de la Computación y Sus Pioneros*; Fondo de Cultura Económica: México City, Mexico, 2003.
42. Copeland, B.J. The Modern History of Computing. 2000. Available online: https://plato.stanford.edu/entries/computing-history/ (accessed on 1 June 2022).
43. Galpin, V. Women in computing around the world. *ACM SIGCSE Bull.* **2002**, *34*, 94–100. [CrossRef]
44. Vela, B.; Cavero, J.M.; Vargas-Solar, G.; Espinosa-Oviedo, J.A.; Cáceres, P. A Geo-Gender Study of Indexed Computer Science Research Publications. *arXiv* **2021**, arXiv:2105.00972.
45. Valian, V. *Why So Slow? The Advancement of Women*; MIT Press: Cambridge, MA, USA, 1999.
46. Hewlett, S.A.; Luce, C.B.; Servon, L.J.; Sherbin, L.; Shiller, P.; Sosnovich, E.; Sumberg, K. The Athena factor: Reversing the brain drain in science, engineering, and technology. *Harv. Bus. Rev. Res. Rep.* **2008**, *10094*, 1–100.
47. Frehill, L.; Javurek-Humig, A.; Jeser-Cannavale, C. Women in Engineering: A review of the 2005 literature. *Mag. Soc. Women Eng.* **2006**, *52*, 34–63.
48. Interactive, H. Attitudes and Experiences of Engineering Alumni, Prepared for the Society of Women Engineers. *unpublished*.
49. Heilman, M.E.; Wallen, A.S.; Fuchs, D.; Tamkins, M.M. Penalties for success: Reactions to women who succeed at male gender-typed tasks. *J. Appl. Psychol.* **2004**, *89*, 416. [CrossRef]
50. Heilman, M.E.; Okimoto, T.G. Why are women penalized for success at male tasks? The implied communality deficit. *J. Appl. Psychol.* **2007**, *92*, 81. [CrossRef]
51. McClure, C.R. *A Test of Leadership: Charting the Future of US Higher Education*; U.S. Department of Education Contract: Jessup, MD, USA, 2007.
52. Committee on Prospering in the Global Economy of the 21st Century. *Rising above the Gathering Storm: Energizing and Employing America for a Brighter Economic Future*; National Academies Press: Washington, DC, USA, 2007.
53. Ashby, C. Science, technology, engineering, and mathematics: Trends and the role of federal programs. In *Testimony before the Committee on Education and the Workforce, House of Representatives*; National Center on Educational Outcomes (NCEO): Minneapolis, MN, USA, 2006.

applied sciences

MDPI

Article

Ethical Issues in AI-Enabled Disease Surveillance: Perspectives from Global Health

Ann Borda [1,2,*], Andreea Molnar [3], Cristina Neesham [4] and Patty Kostkova [5]

1 Centre for Digital Transformation of Health, University of Melbourne, Parkville, VIC 3010, Australia
2 Department of Information Studies, University College London, London WC1E 6BT, UK
3 Department of Computing Technologies, Swinburne University of Technology, Hawthorn, VIC 3122, Australia; amolnar@swin.edu.au
4 Newcastle University Business School, Newcastle University, Newcastle upon Tyne NE1 7RU, UK; cristina.neesham@newcastle.ac.uk
5 Institute for Risk and Disaster Reduction, University College London, London WC1E 6BT, UK; p.kostkova@ucl.ac.uk
* Correspondence: aborda@unimelb.edu.au or a.borda@ucl.ac.uk

Abstract: Infectious diseases, as COVID-19 is proving, pose a global health threat in an interconnected world. In the last 20 years, resistant infectious diseases such as severe acute respiratory syndrome (SARS), Middle East respiratory syndrome (MERS), H1N1 influenza (swine flu), Ebola virus, Zika virus, and now COVID-19 have been impacting global health defences, and aggressively flourishing with the rise of global travel, urbanization, climate change, and ecological degradation. In parallel, this extraordinary episode in global human health highlights the potential for artificial intelligence (AI)-enabled disease surveillance to collect and analyse vast amounts of unstructured and real-time data to inform epidemiological and public health emergency responses. The uses of AI in these dynamic environments are increasingly complex, challenging the potential for human autonomous decisions. In this context, our study of qualitative perspectives will consider a responsible AI framework to explore its potential application to disease surveillance in a global health context. Thus far, there is a gap in the literature in considering these multiple and interconnected levels of disease surveillance and emergency health management through the lens of a responsible AI framework.

Keywords: AI; disease surveillance; pandemics; global public health; ethics

Citation: Borda, A.; Molnar, A.; Neesham, C.; Kostkova, P. Ethical Issues in AI-Enabled Disease Surveillance: Perspectives from Global Health. *Appl. Sci.* **2022**, *12*, 3890. https://doi.org/10.3390/app12083890

Academic Editors: Karina Gibert and Aida Valls

Received: 15 February 2022
Accepted: 8 April 2022
Published: 12 April 2022

Publisher's Note: MDPI stays neutral with regard to jurisdictional claims in published maps and institutional affiliations.

1. Introduction

Infectious diseases such as COVID-19 pose a global health threat in an interconnected world. In the last 20 years, resistant infectious diseases such as severe acute respiratory syndrome (SARS), Middle East respiratory syndrome (MERS), H1N1 influenza (swine flu), Ebola virus, Zika virus, and now COVID-19 are impacting global health defences, and aggressively flourishing due to the rise of global travel, urbanization, climate change, and ecological degradation.

In parallel, this episode in global human health highlights the potential for artificial intelligence (AI)-enabled disease surveillance to collect and analyse vast amounts of unstructured and real-time data to inform epidemiological responses. The exploration of the requirements of a global data-driven and epidemic intelligence surveillance system finds precedence in the literature [1], but the implementation has still a long way to go. A 2020 Lancet commentary provides insights into what might be possible in a collective approach to address the unpredictable and evolving COVID-19 transmission rates using AI [2]. However, this promise is also hampered by the challenges of AI, including the opaqueness of the outputs derived from informal sources, and the commensurate need for transparency at multi-stakeholder levels [3,4].

AI as a component of global health surveillance systems is relatively new and largely bounded by both specific applications and geographies [5,6]. Global health itself is built on national public health actions and institutions, primarily concerning population-wide interventions. However, global health is ideally concerned with all strategies for health improvement, whether population-wide or individual-based health care actions. In other words, it goes across geographical administrative boundaries. There is no formalized definition for *global health*, but Koplan et al. [7] is widely cited for defining it as: "an area for study, research, and practice that places a priority on improving health and achieving health equity for all people worldwide". Taking on board the parameters of this approach, we are faced with both opportunities and ethical stumbling blocks of an AI-enabled disease surveillance for global health. For example, how might we take into account the interconnected levels of individual and population disease contexts and stakeholders, multiple data sources, and geographical boundaries, and the implementation of equitably applied interventions?

2. Disease Surveillance

Surveillance is already a core function of both public health and global health practice. It is defined by the World Health Organization (WHO) as the "continuing scrutiny of all aspects of the occurrence and spread of disease that are pertinent to effective control". It is characterised by "methods distinguished by their practicability, uniformity, and frequently by their rapidity, rather than complete accuracy" [8].

The enabling components of surveillance systems may include laboratory diagnostics to detect or confirm health conditions; information technologies to support the surveillance processes of data collection, analysis, and dissemination; clinician consultation and reporting; public health education and training; and legislation, regulations, and policies that support the conduct of surveillance and response. Communicable disease surveillance can operate at international, national, state, and local levels. The primary responsibility for public health action often lies with state and national health departments or public health agencies (e.g., Public Health England in the UK). For the purposes of this paper, WHO [9] has defined a disease epidemic as "the occurrence of cases of disease in excess of what would normally be expected in a defined community, geographical area, or season". In practice, infectious disease data collection and analysis is complicated and encompasses a multi-stage process with several stakeholders across various organizational boundaries [10]. The main objective of infectious disease surveillance is to identify changes in incidence, either in the form of an acute outbreak (e.g., COVID-19) or a change in long-term trends [11].

Disease surveillance intelligence includes activities related to the prompt identification of potential health hazards and their verification, assessment, and investigation to enable public health control recommendations [12]. The 2005 International Health Regulations (IHR) are designed to ensure the timely recognition of outbreaks of infectious disease with the potential to spread widely [13]. The IHR 2005 also serves as a foundation for the Global Health Security (GHS) agenda [14]. The GHS agenda is "an effort by nations, international organizations, and civil society to accelerate progress toward a world safe and secure from infectious disease threats; to promote global health security as an international priority; and to spur progress toward full implementation of the IHR". They require WHO member nations to report outbreaks of international concern to the WHO within 24 h of discovery [15].

For case-based surveillance, member states can report using the WHO outbreak toolkit [16] to develop an initial case report. For example, consistent with the IHR, during the initial months of the H1N1 pandemic in 2009, the WHO requested that countries report the initial cases and, thereafter, the number of confirmed cases and deaths throughout the H1N1 pandemic. The resulting database represented one of the most comprehensive and timely outbreak reporting databases available to the public on the Internet at the time [13]. By contrast, the overwhelming human resource demands to effectively undertake COVID-19 data collection and analysis over the initial months of the pandemic swiftly required a shift from case-based reporting to aggregate and accelerated forms of reporting [17].

A critical component of disease surveillance which is growing in importance due to the availability of big data is 'epidemic intelligence' [18,19]. Epidemic intelligence incorporates two components: an indicator-based component and an event-based component [9,12,20,21]. The goal of indicator-based surveillance is to find increased numbers or clusters at a specific time, period, and/or geolocation that may indicate a threat [22]. The indicator-based component refers to structured or formal data collected through routine surveillance systems, such as the number or rates of cases based on standard case definitions, and the computation of indicators upon which abnormal disease patterns to investigate are detected [12]. Traditional indicator-based surveillance systems are based on the obligatory reporting of certain diagnosed diseases to a central health agency.

Event-based disease surveillance systems use information on events impacting human health from Internet-based sources, including news aggregators and social media channels [21–23]. Event-based surveillance complements traditional approaches to public health surveillance and can provide early warning of emerging events, whereas there may be a lag in other forms of data aggregation due to delays in sample collection, laboratory confirmation, and country reporting, for example [24,25]. The Medi + Board is an example of a dashboard framework which integrates public health surveillance data streams with zoonotic surveillance data [26], illustrating multiple geo-referenced and time series data.

There are several established and active disease surveillance systems utilized to monitor disease trends, mainly using online news sources, across the globe. Among these are: the Global Public Health Intelligence Network (GPHIN), HealthMap, and ProMED [20,21,25,27,28]. A short description of these systems is outlined in Table 1.

Table 1. Disease Surveillance Systems.

Global Public Health Intelligence Network (GPHIN)

- GPHIN was established by the Public Health Agency of Canada in the late 1990s.
- GPHIN is an automated surveillance tool, which collects non-structured, event-based, digital data from news feed aggregators and then reports to national and international health agencies, such as the WHO Global Outbreak Alert and Response System (GOARN).
- In 2004, GPHIN detected severe acute respiratory syndrome (SARS) more than two months before the first publications by the WHO [29].
- URL: https://gphin.canada.ca/cepr/aboutgphin-rmispenbref.jsp?language=en_CA (accessed on 1 November 2021)

HealthMap

- HealthMap was established in 2006.
- HealthMap uses online informal sources, such as ProMed and official-validated outbreak RSS feeds, for disease outbreak monitoring and real-time surveillance of emerging public health threats.
- HealthMap also brings together disparate data sources, including online news aggregators, eyewitness reports, expert-curated discussions, and validated official reports.
- The freely available website 'healthmap.org' and mobile app 'Outbreaks Near Me' deliver real-time intelligence on a broad range of emerging infectious diseases.
- URL: https://healthmap.org/en/ (accessed on 1 November 2021)

Program for Monitoring Emerging Diseases (ProMED)

- ProMED is a program of the International Society for Infectious Diseases (ISID).
- ProMED was launched in 1994 as an internet service to identify unusual health events related to emerging and re-emerging infectious diseases and toxins affecting humans, animals, and plants.
- ProMED is one of the largest publicly accessible surveillance systems conducting global reporting of infectious disease outbreaks.
- URL: https://promedmail.org/ (accessed on 1 November 2021)

The IHR has emphasized the importance of both indicator-based and event-based components of epidemic intelligence for the early detection of events [30]. Informal information sources are important in this context, with the WHO reporting that more than 60% of initial disease epidemic reports come from unofficial sources and this percentage continues to increase with the availability of internet-based and social media sources [25]. During the COVID-19 crisis, the frequently changing government platforms used to publicly disseminate data included Facebook and Twitter, as well as government-authorized websites [31].

Common informal sources found on the internet include search queries, news feeds (e.g., Google and Baidu news), blogs, and social media channels [25,32]. Social media platforms, such as Twitter, are perceived as an increasingly useful form of real-time data with a geolocation resolution that can be systematically mined, aggregated, and analysed by algorithmic applications to inform public health agencies [30,33,34]. The increasing number of smartphones worldwide and lower policy barriers in many lower-middle income countries (LMICs) also provide an opportunity for the usage of AI-enabled mobile applications in event-based surveillance [18,35,36].

There are also emergent forms of epidemic intelligence [12] which are arising from newly identified microbes through advances in DNA sequencing technology. This primarily relates to genomic epidemiology in which lab-based molecular diagnostics and genotyping methods are being enhanced or replaced by rapid genomics- and AI-based methods to aid epidemiologic investigations of communicable and outbreak diseases [37]. From DNA to environmental exposures, climate change is similarly modifying the public and global health environment such that disease spread can take on new patterns, and subsequently new types of surveillance data are required, e.g., meteorological data [38].

3. AI and Public Health Disease Surveillance

Rapid epidemic detection and real-time monitoring are critical objectives of event-based surveillance to minimize the morbidity and mortality caused by infectious diseases and to enable the preparation of rapid response public health services and personnel. Automated event-based surveillance systems have the potential capacity to aggregate and sort vast amounts of heterogeneous data for patterns and abnormalities using forms of AI [39].

AI machine learning classifiers play an important role in real-time analysis and trend predictions [40]. AI applications used with social media sources can be effective in providing trends for climatic and socio-economic contexts [38,41]. Social networks can further provide an efficient method for risk communication to prevent disease outbreaks and serve as real-time dissemination channels for declaration of pandemic events [42].

Machine learning models analysing internet and open health data sources have precedence across a number of disease surveillance projects to improve infectious disease surveillance and prediction, e.g., malaria [43], dengue fever [44], and cholera [45]. In the example of influenza, algorithmic and machine learning models have been applied to influenza tracking for nearly two decades. Bernardo et al. [32] indicate that the use of internet-based sources for disease surveillance is traceable to 2006, with early work focusing on influenza. Influenza can be highly contagious and easily spreads as people move about and travel, for instance, making tracking and forecasting flu activity a challenge. To address these challenges, researchers have proposed methods utilizing internet-based sources, mobile phones, and other event-based surveillance methods. Research in the use of internet data has demonstrated a positive predictive value of the incidence of infectious diseases, e.g., seasonal influenza [46], but it has shown in some cases to have a low predictive value [47].

To monitor the volume of influenza cases, there is an increasing use of "flu tracker" apps, whereby flu symptoms are reported using crowdsourced platforms to improve the global surveillance of influenza [35,48]. The U.S. Centres for Disease Control and Prevention (CDC) developed the FluView app, which draws on the U.S. influenza-like illness surveillance network (ILINet) that records the percentage of patients reporting to outpatient clinics with symptoms of influenza-like illness (ILI), such as fever, a sore throat, or cough, over the total number of patient visits, for example. The data are integrated in the weekly joint WHO and European Centre for Disease Prevention and Control (ECDC) COVID-19 and influenza (www.flunewseurope.org (accessed on 15 January 2022)) bulletins [49].

These symptom measurements are an established indicator of historical ILI activity, but an important drawback is that they require several days to collect from individual health-care providers, causing a delay and reducing the opportunity for real-time situational awareness and data analysis. In an evaluation of surveillance systems, it was found that the systems issued alerts for human cases between 1.9 and 6.1 days, on average, before WHO reports [30].

As monitoring is improved over time, the heuristics improve as their outputs are confirmed against a set threshold for the epidemiological attributes of extracted health events [22]. Prior to returning the extracted information, the surveillance system aggregates the extracted events into outbreaks, across several documents and sources based on this 'learning'.

The Computational Health Informatics Program (CHIP) at Boston Children's Hospital is among numerous research groups developing forecasting techniques combined with machine learning to provide more predictive indicators and estimates of local flu activity. An application called ARGONet (autoregression with general online information) leverages information from electronic health records, flu-related Google searches, and historical flu activity in a given location. In a research study, ARGO was shown to outperform Google Flu Trends (GFT) [50], a flu forecasting system developed by Google that operated from 2008 to 2015, but notably failed to accurately predict peaks in the flu season [25,48], as described by Daughton et al. 2020. To improve its accuracy, ARGONet also draws on the spatial-temporal patterns of flu spread in neighbouring areas. "It exploits the fact that the presence of flu in nearby locations may increase the risk of experiencing a disease outbreak at a given location" [50].

In disease outbreaks before COVID-19, the application of AI was limited because of a shortage of the data needed to provide rapid updates. The millions of feeds about coronavirus on social media and news sites are allowing algorithms to generate near-real-time information for the public health services tracking its spread. For instance, the disease surveillance system HealthMap based at Boston Children's Hospital uses natural language processing and machine leaning to analyse data from government reports, social media, news sites in 15 languages, and other informal sources to constantly track the spread of outbreaks [51]. The future developments at the time of this study include building a symptom-checker to assess the symptoms of coronavirus that distinguish it from seasonal flu.

In the emergence of COVID-19, risk analysis companies using AI-enabled tools were able to provide early detection of the coronavirus, before the WHO was informed. An AI epidemiologist in the Canadian-based company BlueDot (https://bluedot.global/ (accessed on 1 November 2021)) is widely reported as the first source to reveal news of the COVID-19 outbreak in late December 2019 [2]. BlueDot analysed data from news reports, airline ticketing, and animal disease outbreaks to predict areas that would be prone to the outbreak, expanding from regions in China [51]. Data sources and AI played an important role in various countries' responses to the pandemic [18].

Another company using AI capabilities is Metabiota (https://www.metabiota.com/ (accessed on 1 November 2021)) based in San Francisco, which offers an epidemic tracker and a near-term forecasting model of disease spread. Metabiota, which also used analytics to track flight data accurately, anticipated that countries such as Japan, Thailand, Taiwan, and South Korea would be at risk of a coronavirus outbreak days before any case was reported in any of those countries [51,52].

Interest in syndromic surveillance based on social media data has greatly increased in recent years, leading to what might be termed "nowcasting" [50,53,54], that is providing predictions of disease levels which incorporate data from current social media activity but can include rolling data sources such as ever-hospitalized cases and persons tested [53]. The early detection and analysis of epidemics based on mining Twitter messages is a common approach in nowcasting [53,54]. The VAC Medi + Board dashboard provides visualizations of vaccination threads extracted from Twitter, for example, illustrating the diffusion of information, the most influential users, and impact groups [55,56].

Over the course of the swine flu (H1N1) pandemic in 2009, novel big data streams from social media channels, in correlation with traditional surveillance reporting, demonstrated a potential for early-warning systems for infection disease outbreaks detecting a peak in an outbreak of swine flu up to two weeks before the official public health authorities [57]. Broniatowski et al. [34] reports that including Twitter data on influenza or influenza symptoms can improve nowcasting performance to a greater degree than was possible with search surveillance using Google Trends. During the 2012–2013 influenza season, a Twitter-based AI surveillance application also consistently predicted and mirrored CDC data with 85% accuracy [34].

COVID-19 big data applications are drawing on increasing aggregations of both formal and informal sources, including Twitter and major published open data platforms, e.g., Johns Hopkins University's COVID-19 global map dashboard. Pilot initiatives for COVID-19 forecasting, modelling, and machine learning rose exponentially in 2020, demonstrating the capability of AI for disease surveillance applications [58]. The Global Partnership on Artificial Intelligence (GPAI) identified 84 initiatives supporting some form of AI tool, application, and/or platform, from across the global north and south, related to the pandemic [59].

4. Ethical Considerations

There is a noted asymmetry in the literature on the ethics of AI in healthcare, with less attention granted to global health, particularly disease surveillance [4,19,60,61]. Much of the AI-driven intervention research in global health has limited descriptions of the ethical, regulatory, or practical considerations specifically required for cases of widespread use or deployment at scale [59,62]. Not least, there is the interconnectedness of individual and population levels of disease surveillance, diverse data types, and sources which are required for reporting and decision-support [17–19].

Making AI systems transparent, fair, secure, and inclusive is essential in this context, but how these systems are interpreted and operationalized can vary [6]. The OECD has recognized that AI systems deployed in specific contexts require different approaches to policy-making and governance, and have produced an evaluation toolkit to address variations using scenario-based examples [63]. The OECD toolkit shares principles with increasingly pervasive responsible AI frameworks [59,61,64,65], although this still remains a relatively unexplored area when applied to disease surveillance systems.

The recent literature focusing on the COVID pandemic has only just touched on the potentially important application of responsible AI [59,64,65]. The Future Society in collaboration with the GPAI launched a responsible AI assessment framework aimed at applying a number of normative criteria to AI-related initiatives arising in the COVID-19 pandemic. This assessment is based on human rights, inclusion, diversity, innovation, and economic growth [59].

In an earlier and wider-ranging study, a meta-analysis of ethical AI frameworks by Floridi et al. [66] resulted in the development of the AI4People framework for a "good AI society". The five principles underlying the AI4People framework are [66]:

- Beneficence: promoting well-being, preserving dignity, and sustaining the planet
- Non-maleficence: privacy, security and "capability caution"
- Autonomy: the power to decide (whether to decide)
- Justice: promoting prosperity and preserving solidarity
- Explicability: enabling the other principles through intelligibility and accountability.

Of particular relevance to this study is the grounding of the AI4People framework in bioethics principles, which more closely aligns it with an ecosystem of agents, patients, systems, and environments [60–62,66]. In applying the AI4People principles in the context of global public health disease surveillance, we reflect on how such principles could inform both ethical and practical considerations for future disease surveillance system practices and applications.

4.1. Beneficence: Promoting Well-Being, Preserving Dignity, and Sustaining the Planet

This principle establishes the requirements for doing good, in the sense of pursuing the good pro-actively, as a constructive outcome, adding value to existing benefits. In this context, the role of a public health disease surveillance system in itself depends on a wider global health network, through which we can promote collaborative actions for the protection of people and the environment [38,63]. Those systems of IHR member states already have some form of embedded transparency in protecting and promoting the value of human dignity in the time of pandemics and in the process of disease surveillance. However, the implementation of public health responses, particularly as post-surveillance events, as well as whole-of government approaches to public health disease outbreaks, need to also support and ensure a continuity of beneficence [4,61,63,66].

4.2. Non-Maleficence: Privacy, Security and "Capability Caution"

This principle promotes the idea of avoiding harm, in particular to people, human society, and the natural environment. In the context of this paper, key aspects of this "do no harm" norm are data sharing, data tracking, data quality, and algorithm benchmarking.

4.2.1. Data Sharing

The quality and completeness of the data on which the WHO relies depends on the capacity of its member states for surveillance and their willingness to share surveillance data. As the COVID-19 pandemic progressed, member states were asked by the WHO to submit detailed case data via line lists on more than 80 variables, the vast majority of which were incompletely reported [17]. This recent pandemic, in particular, has highlighted significant challenges in sharing, storing, and linking data from public, private, and quasi-public sectors, as well as between, across, and outside jurisdictions. Depending on the nature of the data, there may also be restrictions on linking or sharing outside of the geographical source of that data [19]. In the context of new technology and innovation, it is becoming increasingly important for people to share their data, so that disease prevention advances can be more impactful. For example, the sharing of SARS-CoV-2 genetic sequences has allowed scientists to track bio evolution and variants [37,67].

4.2.2. Data Tracking

Here, there are concerns related to the privacy of the people whose data is being used, especially when the data is linked from different sources such as self-reporting mobile apps or trackers, or CCTV video, making it easier to identify the people involved. Many countries relied on an extrapolation of infection-control and public-health measures to contain the COVID-19 pandemic. These have ranged from extreme quarantine measures in China to detailed contact tracing in South Korea to enforced stay-at-home policies and travel restrictions [51,68]. Such measures have been variously implemented as COVID-19 contact tracing has revealed ethical trade-offs between public health and privacy. For instance, in South Korea, there were human rights concerns that the excessive disclosure of private information could cause people with symptoms to avoid testing [68].

Bluetooth privacy-presenting, anonymous, voluntary opt-in apps have been championed by human rights and data protection groups and supported by the WHO [69]. The decentralized privacy-preserving proximity tracing (DP-3T) protocol developed by several European academic institutions was among those developed in response to the COVID-19 pandemic to safely facilitate digital contact tracing of infected individuals at the population level [70]. These were intended to offer a comprehensive set of technical and legal safeguards against the potential misuse of personal data. However, a majority of individuals have not downloaded national digital contact tracing apps [70]. According to the Future Society report [59], public hesitancy querying the purpose of COVID-19 contact tracing apps has also emerged, caused by mistrust in public authorities and by fears around the establishment of continuing forms of government surveillance. Additional means of tracking and screening such as thermal image scanning are among the potential privacy breaches and creep of mass surveillance [71].

What should the ethically acceptable trade-offs be between stopping the spread of disease (a public good) and protecting a person's privacy from potential breaches (a private good)? This is a sensitive issue.

4.2.3. Data Quality

Real-world disease data tends to be noisy and incomplete. Although reporting of most notifiable diseases through various public health agencies is required by law, for the most part hospitals, laboratories, and clinicians participate voluntarily. This reinforces the need for careful evaluation of data sources and collection procedures. Moreover, event-based surveillance systems can generate a sizable amount of information on any given outbreak topic, sometimes overwhelming the systems themselves [21]. For example, the Argus system generated approximately 22,000 reports on the pandemic (H1N1) 2009 from April 2009 to March 2010 [20]. In stark contrast, the COVD-19 case reports captured early in the pandemic tallied 49,659 daily cases and 2739 deaths for 27 March 2020 only [17]. Given the serious consequences of using poor quality data to manage global health related decisions, improving data quality becomes a crucial ethical concern [63].

4.2.4. Algorithm Benchmarking

Each syndromic surveillance system also implements a unique set of outbreak detection algorithms. This requires a better understanding of the strengths and limitations of various detection techniques and their applicability, which may require capability caution in the process [71]. For instance, algorithms can perform differently in the field than in the lab, and this may create a major challenge in ensuring high performance, accurate benchmarking against any current standard care, and continuously assessing performance after deployment [63,72]. Given the wide and powerful impacts of the decisions informed by such benchmarking, again, considerations of technical accuracy become ethical issues.

4.3. *Autonomy: The Power to Decide (Whether to Decide)*

The use of AI-based technology for global health surveillance also raises ethical questions around who should have the power to decide on these matters, and who should decide whether (and when) to decide at all. Concerns have been expressed about significant shifts of the locus of control on such decisions from humans to machines. In the context of global health surveillance, we note specificity, misinformation, consent, and data governance as the main aspects that would require further ethical analysis (and, possibly, regulation).

4.3.1. Specificity

In event-based surveillance, reports may be generated by automated machine-based processes, e.g., HealthMap, or by human analysts or subject matter experts, e.g., GPHIN, ProMED [25,27,28]. For the automated systems, manual report examination for relevancy typically occurs post-dissemination. ProMed utilizes local observers on the ground for some of its outbreak reporting; otherwise event-based biosurveillance systems often disseminate reports that are not observer or laboratory verified [27,37]. Thus, although the reports provide near real-time alerts to users, the data they provide may not be specific enough. One consequence of this on human autonomy may be that individual human agents end up being constrained by limited one-size-fits-all (general) data and may not be able to see where (and how) they can act in different circumstances [63].

4.3.2. Misinformation

Syndromic surveillance systems often generate false detections because it is sometimes difficult to distinguish natural data variations resulting from outlier and dynamic system changes, e.g., with time or space variables, from real outbreaks [73]. Human reviews and follow-up investigations are necessary for verification, which can be costly in time and labour. The prompt identification of novel microbes is often not matched with equally prompt epidemiological assessments [37]. The scarcity of such data, and the lack of transparency in algorithms could compound challenges of misinformation or false alarms that could potentially occur with AI-enabled platforms, especially if they rely on social media or non-authoritative data sources, for example. False detections can erode public trust in these surveillance systems, and such misinformation could weaken an agency's or individual's power to effectively make decisions [74–76]. The role of social media in trending public attitudes can also be put to effective use to disseminate risk communication in real time [26,42], but could be a double-edged sword increasingly prone to misinformation and the spread of fake news [74–76].

4.3.3. Consent

In an article by Blasimme and Vayena [60], the authors raised ethical questions around consent when employing AI-driven social media analysis for digital epidemiology. For example, ubiquitous surveillance for use by AI systems through personal devices, such as mobile phones, introduces the concern that granular data can be re-identified, and personal health information can be hacked [35,71]. The main concern here is that this personalized data can be used by third parties without the subject's knowledge or approval. In the case of contact tracing apps, for example, there are ethical considerations to gather and act on information without consent [69]. Such digital surveillance on individual-level data can lead to unique complications [74], and relates to the trade-offs between the public good and individual rights [75].

4.3.4. Data Governance

The COVID–19 pandemic has exposed long-standing data governance issues such as intellectual property rights, data sharing, reuse, and ownership [59,63,76,77]. Globally, persistent data gaps and fragmented approaches to governing disease surveillance data need to be addressed through global cooperation and clear, unified direction. The Global Pandemic Data Alliance (https://gpdahub.org/, (accessed on 1 November 2021)) is a collaboration of data-focused organizations across the G7 countries meeting the challenge of establishing health data as a global public good. This need for globally integrated data governance for the public good is highlighted in the Future Society report [59], the OECD Framework [63], and in the WHO Health Data Governance Summit [78].

4.4. Justice: Promoting Prosperity and Preserving Solidarity

It is increasingly accepted that principles of justice should apply when AI is used to support event-based surveillance efforts in global health policy. Justice is not just a lofty ethical ideal, and an end in itself. It is also a means to distribute prosperity widely within society and to foster social cohesion and solidarity across different social groups. Geographic scope, human rights, and equity in predictive decisions are the three main aspects of ethical concern to be discussed here.

4.4.1. Geographic Scope

The geographic scope of an event-based disease surveillance system may vary [21,27]. It could cover a region, a country, a continent, or the entire globe. The following jurisdictions have established, hosted, and maintained such systems: United States, European Union, Canada, Japan, Australia, Brazil, Singapore, Mexico, and Thailand [11,15]. While there is immense potential opportunity for AI to support event-based surveillance of aggressive disease outbreaks such as COVID-19, the issue of missing data is especially present in low- and middle- income countries (LIC and LMIC), which may also lack the infrastructure and human capital required to maintain these systems [17,36,74,75,79]. Geographic inequalities in event-based system maintenance may constitute significant barriers to achieving justice in global health outcomes, hence improvements in geographic coverage should become a priority.

4.4.2. Human Rights

The protection of human rights is paramount and, as stated in the UN Declaration of 1948, should be universally guaranteed. However, individuals from vulnerable groups and stateless communities may discover that current disease surveillance practices are still a long way from safeguarding them from discrimination and oppression [79]. The IHR [15] has introduced important safeguards to protect the rights of travellers and other persons in relation to the treatment of personal data, informed consent, and non-discrimination in the application of health measures under the regulations. The effectiveness of these measures on the ground remains to be determined.

4.4.3. Predictive Decisions

With respect to ensuring equity in predictive decisions, there is some debate as to whether responsible AI frameworks can address the explicit and implicit biases embedded within these systems [62]. To verify that an AI system is not using data in ways that result in bias or discriminatory outcomes, some level of transparency is necessary to explicitly addresses diversity and inclusion [59,60].

4.5. Explicability: Enabling the Other Principles through Intelligibility and Accountability

It is now recognized that the 'black-box' nature of first-generation AI systems can be a source of injustice, dominance, and harm [4,72]. The need for explainability (or explicability), in the sense of having algorithmic processes and outcomes explained in clear terms to the human intellect, has become evident. None of these processes and outcomes can be submitted to human judgment and scrutiny without certain explainability requirements being met [72]. The main aspects we discuss here are: data noise; meeting regulatory standards or policy requirements; assessing risks, robustness, and vulnerability; and understanding and verifying the outputs from a system.

4.5.1. Data Noise

Noisy data are a result of data corruption which can carry a large amount of additional or meaningless information. High noise data are a weakness of real-time data streams analysis which can introduce small errors that in turn can have an outsized effect on large-scale predictive models [80]. Some systems, such as HealthMap, relieve noise by integrating data from an assortment of online sources that have been moderated already [73].

4.5.2. Meeting Regulatory Standards or Policy Requirements

Transparency is essential to legal compliance. Without it, enforcing individuals' legal rights in relation to the uses and applications of a technological system, establishing that a service meets regulatory standards, and determining liability may prove impossible [41,63,75]. While most nations have in place the procedures needed to foster compliance with existing data protection and privacy regulations, many such procedures are time consuming and create significant barriers to explicability [17,41,74].

4.5.3. Assessing Risk, Robustness and Vulnerability

Understanding how a system works can be important in assessing risk [41]. This can be particularly important if a system is deployed in a new environment, where the user cannot be sure of its effectiveness [63,80]. Interpretability can also help developers understand how a system might be vulnerable to so-called adversarial attacks, in which actors seeking to disrupt a system identify a small number of carefully chosen data points that they could alter in order to prompt an inaccurate output from the system [72,81,82].

4.5.4. Understanding and Verifying the Outputs from a System

Interpretability can be useful in verifying the outputs from a system, by tracing how modelling choices, combined with the data used, affect the results. In some applications, this can help developers understand cause-and-effect relationships in their analysis. Furthermore, people are more likely to trust an algorithm when they can easily understand its implications and can modify it [72,81,82].

5. Key Issues of Normative Ethics

We preface this discussion by noting that our conceptual approach inevitably presents limitations in the application of responsible AI principles in general. Challenges that may not be covered by such principles, for example, point to finer grained considerations relating to the complexity of disease surveillance systems, from a system alert to reporting to different levels of health authority with varying governance capabilities and a reliance on automated data generation and analysis in a global health emergency scenario.

What is summarily highlighted in the process is that AI is both a technical problem and a human problem. AI models are increasingly used in decision-making contexts in global health disease surveillance. At the same time, they "are more complex and less interpretable than ever" [83]. Scientific and practical work on the application of AI in supporting disease surveillance must consider that it is ultimately humans who need to understand the technological parameters. From a normative-ethical perspective, we see two issues bearing on AI-based decision-making in disease surveillance systems: (1) it is 'ethically blind' unless curated for value alignment (i.e., values shared by the human community are embedded in the algorithm); and (2) it complicates the issue of who is responsible for the consequences/outcomes of this decision-making process and their impact on various stakeholders.

The application of responsible AI has considerable potential for supporting transparency which, in itself, promotes respect for the human dignity of stakeholders in the decision-making process. It also makes value alignment and judicious responsibility attribution possible in the first place. Without responsible AI principles, meeting the two conditions systematically would be impossible.

6. Conclusions

Introducing *responsible AI* applications could support a more ethical and equitable approach [66] to combatting pandemics and maintaining global public health [59]. However, the layered complexity and scale of global disease surveillance systems raise challenges for the integration of *responsible AI* at different levels. These go beyond data, algorithm, prediction, and privacy concerns. The long-term goal relies on responsible health actors limiting the power of private corporate interests, ensuring transparency and protection of individual rights, and not least fostering trusted networks of partnerships and shared commitment through responsible global governance [63,66,84]. Further research should be engaged in developing effective approaches for value alignment in AI-assisted global health surveillance systems, as well as in normative-ethical analyses of responsibility attributions in the context of global health decision-making—the latter particularly in reference to the management of volatile pandemics, such as COVID-19.

Author Contributions: Conceptualization, A.B., A.M., C.N. and P.K.; methodology, A.B., A.M., C.N. and P.K.; investigation, A.B., A.M., C.N. and P.K.; writing—original draft preparation, A.B., A.M., C.N. and P.K.; writing—review and editing, A.B., A.M., C.N. and P.K. All authors have read and agreed to the published version of the manuscript.

Funding: This research received no external funding.

Institutional Review Board Statement: Not applicable.

Informed Consent Statement: Not applicable.

Data Availability Statement: Not applicable.

Conflicts of Interest: The authors declare no conflict of interest.

References

1. Kostkova, P. A roadmap to integrated digital public health surveillance: The vision and the challenges. In Proceedings of the WWW'13: 22nd International World Wide Web Conference, Rio de Janeiro, Brazil, 13–17 May 2013; pp. 687–694. [CrossRef]
2. McCall, B. COVID-19 and artificial intelligence: Protecting health-care workers and curbing the spread. *Lancet Digit. Health* **2020**, *2*, e166–e167. [CrossRef]
3. Gordon, D.; Stavrakakis, I.; Gibson, J.P.; Tierney, B.; Becevel, A.; Curley, A.; Collins, M.; O'mahony, W.; O'sullivan, D. Perspectives on computing ethics: A multi-stakeholder analysis. *J. Inf. Commun. Ethics Soc.* **2022**, *20*, 72–90. [CrossRef]
4. Morley, J.; Machado, C.C.; Burr, C.; Cowls, J.; Joshi, I.; Taddeo, M.; Floridi, L. The ethics of AI in health care: A mapping review. *Soc. Sci. Med.* **2020**, *260*, 113172. [CrossRef] [PubMed]
5. Khemasuwan, D.; Colt, H.G. Applications and challenges of AI-based algorithms in the COVID-19 pandemic. *BMJ Innov.* **2021**, *7*, 387–398. [CrossRef]
6. Piccialli, F.; di Cola, V.S.; Giampaolo, F.; Cuomo, S. The Role of Artificial Intelligence in Fighting the COVID-19 Pandemic. *Inf. Syst. Front. A J. Res. Innov.* **2021**, *23*, 1467–1497. [CrossRef]
7. Koplan, J.P.; Bond, T.C.; Merson, M.H.; Reddy, K.S.; Rodriguez, M.H.; Sewankambo, N.K.; Wasserheit, J.N. Towards a common definition of global health. *Lancet.* **2009**, *373*, 1993–1995. [CrossRef]
8. Last, J.M. *A Dictionary of Epidemiology*; Oxford University Press: New York, NY, USA, 1988.
9. World Health Organization. Disease Outbreaks. 2016. Available online: http://www.emro.who.int/health-topics/disease-outbreaks/index.html (accessed on 12 July 2016).
10. Zeng, D.; Cao, Z.; Neill, D.B. Artificial intelligence–enabled public health surveillance—from local detection to global epidemic monitoring and control. *Artif. Intell. Med.* **2020**, 437–453. [CrossRef]
11. World Health Organization. Communicable Disease Surveillance and Response Systems: Guide to Monitoring and Evaluating. 2006. Available online: https://www.who.int/csr/resources/publications/surveillance/WHO_CDS_EPR_LYO_2006_2.pdf (accessed on 8 November 2021).
12. Paquet, C.; Coulombier, D.; Kaiser, R.; Ciotti, M. Epidemic intelligence: A new framework for strengthening disease surveillance in Europe. *Eurosurveillance* **2006**, *11*, 212–214. [CrossRef]
13. Baker, M.G.; Forsyth, A.M. The new International Health Regulations: A revolutionary change in global health security. *N. Z. Med. J.* **2007**, *120*, 1267.
14. Global Security Agenda. Available online: https://ghsagenda.org (accessed on 14 February 2021).
15. World Health Organization. *International Health Regulations*, 2nd ed.; WHO Press: Geneva, Switzerland, 2008; Available online: http://whqlibdoc.who.int/publications/2008/9789241580410_eng.pdf (accessed on 21 October 2021).
16. World Health Organization. Outbreak Tookit. Available online: https://www.who.int/emergencies/outbreak-toolkit (accessed on 14 November 2021).
17. World Health Organization. Rapid Review of WHO COVID-19 Surveillance: External Review. 2021. Available online: https://www.who.int/publications/m/item/rapid-review-of-who-covid-19-surveillance-external-review-27-october-2021 (accessed on 27 October 2021).
18. Kostkova, P.; Saigí-Rubió, F.; Eguia, H.; Borbolla, D.; Verschuuren, M.; Hamilton, C.; Azzopardi-Muscat, N.; Novillo-Ortiz, D. Data and Digital Solutions to Support Surveillance Strategies in the Context of the COVID-19 Pandemic. *Front. Digit. Health* **2021**, *3*, 707902. [CrossRef]
19. Kostkova, P. Disease surveillance data sharing for public health: The next ethical frontiers. *Life Sci. Soc. Policy* **2018**, *14*, 16. [CrossRef] [PubMed]
20. Nelson, N.P.; Yang, L.; Reilly, A.R.; Hardin, J.E.; Hartley, D.M. Event-based internet biosurveillance: Relation to epidemiological observation. *Emerg. Themes Epidemiol.* **2012**, *9*, 4. [CrossRef] [PubMed]
21. O'Shea, J. Digital disease detection: A systematic review of event-based internet biosurveillance systems. *Int. J. Med. Inform.* **2017**, *101*, 15–22. [CrossRef] [PubMed]
22. Velasco, E.; Agheneza, T.; Denecke, K.; Kirchner, G.; Eckmanns, T. Social Media and Internet-Based Data in Global Systems for Public Health Surveillance: A Systematic Review. *Milbank Q.* **2014**, *92*, 7–33. [CrossRef]
23. Gajewski, K.N.; Peterson, A.E.; Chitale, R.A.; Pavlin, J.A.; Russell, K.L.; Chretien, J.-A. A review of evaluations of electronic event-based biosurveillance systems. *PLoS ONE* **2014**, *9*, e111222. [CrossRef]
24. Chunara, R.; Freifeld, C.; Brownstein, J. New technologies for reporting real-time emergent infections. *Parasitology* **2012**, *139*, 1843–1851. [CrossRef]
25. Yan, S.J.; Chughtai, A.A.; Macintyre, C.R. Utility and potential of rapid epidemic intelligence from internet-based sources. *Int. J. Infect. Dis.* **2017**, *63*, 77–87. [CrossRef]
26. Kostkova, P.; Szomszor, M.; St. Louis, C. #swineflu: The use of twitter as an early warning and risk communication tool in the 2009 swine flu pandemic. *ACM Trans. Manag. Inf. Syst.* **2014**, *5*, 1–25. [CrossRef]
27. Hartley, D.M.; Nelson, N.P.; Walters, R.; Arthur, R.; Yangarber, R.; Madoff, L.; Lightfoot, N. The landscape of international event-based biosurveillance. *Emerg. Health Threat. J.* **2010**, *3*, 7096. [CrossRef]

28. Walters, R.; Harlan, P.; Nelson, N.P.; Hartley, D.M. Data sources for biosurveillance. In *Wiley Handbook of Science and Technology for Homeland Security*; Voeller, J., Ed.; John Wiley & Sons: New York, NY, USA, 2010; pp. 1–17.
29. Milinovich, G.J.; Williams, G.M.; Clements, A.C.A.; Hu, W. Internet-based surveillance systems for monitoring emerging infectious diseases. *Lancet Infect. Dis.* **2014**, *14*, 160–168. [CrossRef]
30. Barboza, P.; Vaillant, L.; Mawudeku, A.; Nelson, N.P.; Hartley, D.; Madoff, L.C.; Linge, J.P.; Collier, N.; Brownstein, J.S.; Yangarber, R.; et al. Evaluation of Epidemic Intelligence Systems Integrated in the Early Alerting and Reporting Project for the Detection of A/H5N1 Influenza Events. *PLoS ONE* **2013**, *8*, e57252. [CrossRef] [PubMed]
31. Tsao, S.F.; Chen, H.; Tisseverasinghe, T.; Yang, Y.; Li, L.; Butt, Z.A. What social media told us in the time of COVID-19: A scoping review. *Lancet Digit. Health* **2021**, *3*, e175–e194. [CrossRef]
32. Bernardo, T.M.; Rajic, A.; Young, I.; Robiadek, K.; Pham, M.T.; Funk, J.A. Scoping review on search queries and social media for disease surveillance: A chronology of innovation. *J. Med Internet Res.* **2013**, *15*, e147. [CrossRef] [PubMed]
33. Guy, S.; Ratzki-Leewing, A.; Bahati, R.; Gwadry-Sridhar, F. Social media: A systematic review to understand the evidence and application in infodemiology. *Lect. Notes Inst. Comput. Sci. Soc. Inf. Telecommun. Eng.* **2021**, *91*, 1–8. [CrossRef]
34. Broniatowski, D.A.; Paul, M.J.; Dredze, M. National and Local Influenza Surveillance through Twitter: An Analysis of the 2012–2013 Influenza Epidemic. *PLoS ONE* **2013**, *8*, e83672. [CrossRef]
35. Mohanty, B.; Chughtai, A.; Rabhi, F. Use of Mobile Apps for epidemic surveillance and response–availability and gaps. *Glob. Biosecurity* **2019**, *1*, 37. [CrossRef]
36. Owoyemi, A.; Owoyemi, J.; Osiyemi, A.; Boyd, A. Artificial Intelligence for Healthcare in Africa. *Front. Digit. Health* **2020**, *2*, 6. [CrossRef] [PubMed]
37. Tang, P.; Croxen, M.A.; Hasan, M.R.; Hsiao, W.W.; Hoang, L.M. Infection control in the new age of genomic epidemiology. *Am. J. Infect. Control* **2017**, *45*, 170–179. [CrossRef]
38. Rodríguez-González, A.; Zanin, M.; Menasalvas-Ruiz, E. Public Health and Epidemiology Informatics: Can Artificial Intelligence Help Future Global Challenges? An Overview of Antimicrobial Resistance and Impact of Climate Change in Disease Epidemiology. *Yearb. Med Inform.* **2019**, *28*, 224–231. [CrossRef]
39. Davie, S. Artificial Intelligence in Global Health. *Ethic Int. Aff.* **2019**, *33*, 181–192. [CrossRef]
40. Alessa, A.; Faezipour, M. A review of influenza detection and prediction through social networking sites. *Theor. Biol. Med Model.* **2018**, *15*, 2. [CrossRef] [PubMed]
41. Gluskin, R.T.; Mavinkurve, M.; Varma, J.K. Government leadership in addressing public health priorities: Strides and delays in electronic laboratory reporting in the United States. *Am. J. Public Health* **2014**, *104*, e16–e21. [CrossRef] [PubMed]
42. Szomszor, M.; Kostkova, P.; St Louis, C. Twitter informatics: Tracking and understanding public reaction during the 2009 Swine Flu pandemic. *IEEE Comput. Soc.* **2011**, *1*, 320–323. [CrossRef]
43. Ocampo, A.J.; Chunara, R.; Brownstein, J.S. Using search queries for malaria surveillance, Thailand. *Malar. J.* **2013**, *12*, 390. [CrossRef]
44. Hoyos, W.; Aguilar, J.; Toro, M. Dengue models based on machine learning techniques: A systematic literature review. *Artif. Intell. Med.* **2021**, *119*, 102157. [CrossRef] [PubMed]
45. Abbasi, J. Better Cholera Counts through Machine Learning Models. *JAMA* **2019**, *321*, 1343. [CrossRef]
46. Ginsberg, J.; Mohebbi, M.H.; Patel, R.S.; Brammer, L.; Smolinski, M.S.; Brilliant, L. Detecting influenza epidemics using search engine query data. *Nature* **2009**, *457*, 1012–1014. [CrossRef]
47. Daughton, A.R.; Chunara, R.; Paul, M.J. Comparison of Social Media, Syndromic Surveillance, and Microbiologic Acute Respiratory Infection Data: Observational Study. *JMIR Public Health Surveill.* **2020**, *6*, e14986. [CrossRef]
48. Smolinski, M.S.; Crawley, A.W.; Baltrusaitis, K.; Chunara, R.; Olsen, J.M.; Wójcik, O.; Santillana, M.; Nguyen, A.; Brownstein, J.S. Flu Near You: Crowdsourced symptom reporting spanning 2 influenza seasons. *Am. J. Public Health* **2015**, *105*, 2124–2130. [CrossRef]
49. European Centre for Disease Prevention and Control. *COVID-19 Surveillance Guidance: Transition from COVID-19 Emergency Surveillance to Routine Surveillance of Respiratory Pathogens*; ECDC Technical Report; ECDC: Stockholm, Sweden, 2021; Available online: https://www.ecdc.europa.eu/en/publications-data/covid-19-surveillance-guidance (accessed on 1 December 2021).
50. Lu, F.S.; Hattab, M.W.; Clemente, C.L.; Biggerstaff, M.; Santillana, M. Improved state-level influenza nowcasting in the United States leveraging Internet-based data and network approaches. *Nat. Commun.* **2019**, *10*, 147. [CrossRef]
51. Allam, Z.; Dey, G.; Jones, D. Artificial Intelligence (AI) Provided Early Detection of the Coronavirus (COVID-19) in China and Will Influence Future Urban Health Policy Internationally. *AI* **2020**, *1*, 156–165. [CrossRef]
52. Tong, S. Big Data Predicted the Coronavirus Outbreak and Where It Would Spread. *Marketplace* **2020**. Available online: https://www.marketplace.org/2020/02/04/big-data-predicted-coronavirus-outbreak-where-it-maygo-next/ (accessed on 20 March 2020).
53. Greene, S.K.; McGough, S.F.; Culp, G.M.; Graf, L.E.; Lipsitch, M.; Menzies, N.A.; Kahn, R. Nowcasting for Real-Time COVID-19 Tracking in New York City: An Evaluation Using Reportable Disease Data from Early in the Pandemic. *JMIR Public Health Surveill.* **2021**, *7*, e25538. [CrossRef] [PubMed]

54. Paul, M.J.; Dredze, M.; Broniatowski, D.A.; Generous, N. Worldwide Influenza Surveillance through Twitter. In *Papers of the AAAI, Proceedings of the Workshop on the World Wide Web and Public Health Intelligence, Austlin, TX, USA, 25–26 January 2015*; Shaban-Nejad, A., Buckeridge, D.L., Brownstein, J.S., Eds.; AAAI: Palo Alto, CA, USA, 2015; pp. 6–11. Available online: https://www.aaai.org/Workshops/ws15workshops.php#ws16 (accessed on 1 November 2021).
55. Kostkova, P.; Mano, V.; Larson, H.J.; Schulz, W.S. Vac medi+board: Analysing vaccine rumours in news and social media. In Proceedings of the DH'16: 6th International Conference on Digital Health Conference, Montréal, QC, Canada, 11–13 April 2016; pp. 163–164. [CrossRef]
56. Artus, D.; Larson, H.; Kostkova, P. Role of Social Media in vaccination debate about HPV: The VAC Medi+ plus Board study. *Eur. J. Public Health* **2019**, *29*, ckz185.682. [CrossRef]
57. Szomszor, M.; Kostkova, P.; Quincey, E.D. #Swineflu: Twitter predicts swine flu outbreak in 2009. In *International Conference on Electronic Healthcare*; Springer: Berlin/Heidelberg, Germany, 2010; pp. 18–26.
58. da Silva, C.C.; de Lima, C.L.; da Silva, A.C.G.; Silva, E.L.; Marques, G.S.; de Araújo, L.J.B.; Júnior, L.A.A.; de Souza, S.B.J.; de Santana, M.A.; Gomes, J.C.; et al. COVID-19 Dynamic Monitoring and Real-Time Spatio-Temporal Forecasting. *Front. Public Health* **2021**, *9*, 641253. [CrossRef]
59. Future Society. Responsible AI and AI in Pandemic Response, with the Global Partnership on AI (GPAI). 2020. Available online: https://thefuturesociety.org/2020/12/17/report-release-with-the-global-partnership-on-ai/ (accessed on 20 March 2020).
60. Blasimme, A.; Vayena, E. *The Ethics of AI in Biomedical Research, Patient Care and Public Health) In The Oxford Handbook of Ethics of AI*; Dubber, M.D., Pasquale, F., Das, S., Eds.; Oxford University Press: Oxford, UK, 2020; pp. 703–718. [CrossRef]
61. Murphy, K.; Di Ruggiero, E.; Upshur, R.; Willison, D.J.; Malhotra, N.; Cai, J.C.; Malhotra, N.; Lui, V.; Gibson, J. Artificial intelligence for good health: A scoping review of the ethics literature. *BMC Med. Ethic* **2021**, *22*, 14. [CrossRef]
62. Schwalbe, N.; Wahl, B. Artificial intelligence and the future of global health. *Lancet* **2020**, *395*, 1579–1586. [CrossRef]
63. OECD. OECD. OECD Framework for the Classification of AI systems. In *OECD Digital Economy Papers*; No. 323; OECD Publishing: Paris, France, 2022. [CrossRef]
64. El-Haddadeh, R.; Fadlalla, A.; Hindi, N.M. Is There a Place for Responsible Artificial Intelligence in Pandemics? A Tale of Two Countries. *Inf. Syst. Front. A J. Res. Innov.* **2021**, 1–17. [CrossRef]
65. Fosso Wamba, S.; Queiroz, M.M. Responsible Artificial Intelligence as a Secret Ingredient for Digital Health: Bibliometric Analysis, Insights, and Research Directions. *Inf. Syst. Front.* **2021**, *15*, 1–16. [CrossRef]
66. Floridi, L.; Cowls, J.; Beltrametti, M.; Chatila, R.; Chazerand, P.; Dignum, V.; Luetge, C.; Madelin, R.; Pagallo, U.; Rossi, F.; et al. AI4People-An Ethical Framework for a Good AI Society: Opportunities, Risks, Principles, and Recommendations. *Minds Mach.* **2018**, *28*, 689–707. [CrossRef]
67. Jacob, J.J.; Vasudevan, K.; Pragasam, A.K.; Gunasekaran, K.; Veeraraghavan, B.; Mutreja, A. Evolutionary Tracking of SARS-CoV-2 Genetic Variants Highlights an Intricate Balance of Stabilizing and Destabilizing Mutations. *Mbio* **2021**, *12*, e01188-21. [CrossRef] [PubMed]
68. Zastrow, M. South Korea is reporting intimate details of COVID-19 cases: Has it helped? *Nature* **2020**. [CrossRef] [PubMed]
69. World Health Organization. Ethical Considerations to Guide the Use of Digital Proximity Tracking Technologies for COVID-19 Contact Tracing. *INTERIM Guidance*. 2020. Available online: https://www.who.int/publications/i/item/WHO-2019-nCoV-Ethics_Contact_tracing_apps-2020.1 (accessed on 14 November 2021).
70. Blasimme, A.; Ferretti, A.; Vayena, E. Digital Contact Tracing Against COVID-19 in Europe: Current Features and Ongoing Developments. *Front. Digit. Health* **2021**, *3*, 660823. [CrossRef] [PubMed]
71. Pisa, M. *COVID-19, Information Problems, and Digital Surveillance*; Center for Global Development: Washington, DC, USA, 2020. Available online: https://www.cgdev.org/blog/covid-19-information-problems-and-digital-surveillance (accessed on 15 February 2021).
72. Holzinger, A.; Langs, G.; Denk, H.; Zatloukal, K.; Müller, H. Causability and explainability of artificial intelligence in medicine. *Wiley Interdiscip. Rev. Data Min. Knowl. Discov.* **2019**, *9*, e1312. [CrossRef]
73. Chen, H.; Zeng, D.; Buckeridge, D.L.; Izadi, M.I.; Verma, A.; Okhmatovskaia, A.; Hu, X.; Shen, X.; Cao, Z.; Wang, F.-Y.; et al. AI for Global Disease Surveillance. *IEEE Intell. Syst.* **2009**, *24*, 66–82. [CrossRef]
74. Aiello, A.E.; Renson, A.; Zivich, P.N. Social Media– and Internet-Based Disease Surveillance for Public Health. *Annu. Rev. Public Health* **2020**, *41*, 101–118. [CrossRef]
75. Bardosh, K.; de Vries, D.; Stellmach, D.; Abramowitz, S.; Thorlie, A.; Cremers, L.; Kinsman, J. *Towards People-Centered Epidemic Preparedness and Response: From Knowledge to Action*; Global Research Collaboration for Infectious Disease Preparedness: London, UK, 2019. Available online: https://www.glopid-r.org/wp-content/uploads/2019/07/towards-people-centered-epidemic-preparedness-and-response-report.pdf (accessed on 1 November 2021).
76. Loomba, S.; de Figueiredo, A.; Piatek, S.J.; de Graaf, K.; Larson, H.J. Measuring the impact of COVID-19 vaccine misinformation on vaccination intent in the UK and USA. *Nat. Hum. Behav.* **2021**, *5*, 337–348. [CrossRef]
77. Godinho, M.A.; Borda, A.; Kariotis, T.; Molnar, A.; Kostkova, P.; Liaw, S.-T. Knowledge co-creation in participatory policy and practice: Building community through data-driven direct democracy. *Big Data Soc.* **2021**, *8*. [CrossRef]
78. World Health Organization. Health Data Governance Summit. *Meeting Report 30 June 2021*. Available online: https://www.who.int/publications/m/item/health-data-governance-summit (accessed on 15 February 2021).

79. Van Hout, M.C.; Bigland, C.; Murray, N. Scoping the impact of COVID-19 on the nexus of statelessness and health in Council of Europe member states. *J. Migr. Health* **2021**, *4*, 100053, Published Correction Appears in *J. Migr. Health* **2021**, *4*, 100063. [CrossRef]
80. Ienca, M.; Vayena, E. On the responsible use of digital data to tackle the COVID-19 pandemic. *Nat. Med.* **2020**, *26*, 463–464. [CrossRef]
81. Dietvorst, B.J.; Simmons, J.P.; Massey, C. Algorithm aversion: People erroneously avoid algorithms after seeing them err. *J. Exp. Psychol. Gen.* **2015**, *144*, 114. [CrossRef] [PubMed]
82. Biran, O.; Cotton, C. Explanation and justification in machine learning: A survey. In Proceedings of the IJCAI-17 workshop on explainable AI (XAI), Melbourne, VIC, Australia, 20 August 2017; pp. 8–13.
83. Arrieta, A.B.; Díaz-Rodríguez, N.; Del Ser, J.; Bennetot, A.; Tabik, S.; Barbado, A.; Garcia, S.; Gil-Lopez, S.; Molina, D.; Benjamins, R.; et al. Explainable Artificial Intelligence (XAI): Concepts, taxonomies, opportunities and challenges toward responsible AI. *Inf. Fusion* **2019**, *58*, 82–115. [CrossRef]
84. Roberts, S.; Kostkova, P. Disease Surveillance, Digital Futures, and Data-Sharing in a World 'After' COVID-19. *Global Policy* **2021**. Available online: https://www.globalpolicyjournal.com/articles/science-and-technology/disease-surveillance-digital-futures-and-data-sharing-world-after (accessed on 1 February 2022).

applied sciences

MDPI

Article

The Digital Revolution in the Urban Water Cycle and Its Ethical–Political Implications: A Critical Perspective

Lucia Alexandra Popartan [1,*], Àtia Cortés [2], Manel Garrido-Baserba [3], Marta Verdaguer [1], Manel Poch [1] and Karina Gibert [4]

1 LEQUIA, Institute of the Environment, Universitat de Girona, c/ Maria Aurèlia Capmany 69 Catalonia, 17003 Girona, Spain; marta.verdaguer@udg.edu (M.V.); manuel.poch@udg.edu (M.P.)
2 Barcelona Supercomputing Center, Edifici Omega 201, Jordi Girona 1 and 3, 08034 Barcelona, Spain; atia.cortes@bsc.es
3 inCTRL Solutions Corp., Salt Lake City, UT 84123-2583, USA; mgarrido@inctrl.com
4 Intelligent Data Science and Artificial Intelligence Research Center and Institut de Ciència i Tecnologia de la Sostenibilitat, Universitat Politècnica de Catalunya—BarcelonaTech, 08001 Barcelona, Spain; karina.gibert@upc.edu
* Correspondence: luciaalexandra.popartan@udg.edu

Citation: Popartan, L.A.; Cortés, À.; Garrido-Baserba, M.; Verdaguer, M.; Poch, M.; Gibert, K. The Digital Revolution in the Urban Water Cycle and Its Ethical–Political Implications: A Critical Perspective. *Appl. Sci.* **2022**, *12*, 2511. https://doi.org/10.3390/app12052511

Academic Editor: Jega Veeriah Jegatheesan

Received: 31 January 2022
Accepted: 23 February 2022
Published: 28 February 2022

Abstract: The development and application of new forms of automation and monitoring, data mining, and the use of AI data sources and knowledge management tools in the water sector has been compared to a 'digital revolution'. The state-of-the-art literature has analysed this transformation from predominantly technical and positive perspectives, emphasising the benefits of digitalisation in the water sector. Meanwhile, there is a conspicuous lack of critical literature on this topic. To bridge this gap, the paper advances a critical overview of the state-of-the art scholarship on water digitalisation, looking at the sociopolitical and ethical concerns these technologies generate. We did this by analysing relevant AI applications at each of the three levels of the UWC: technical, operational, and sociopolitical. By drawing on the precepts of urban political ecology, we propose a hydrosocial approach to the so-called 'digital water ', which aims to overcome the one-sidedness of the technocratic and/or positive approaches to this issue. Thus, the contribution of this article is a new theoretical framework which can be operationalised in order to analyse the ethical–political implications of the deployment of AI in urban water management. From the overview of opportunities and concerns presented in this paper, it emerges that a hydrosocial approach to digital water management is timely and necessary. The proposed framework envisions AI as a force in the service of the human right to water, the implementation of which needs to be (1) critical, in that it takes into consideration gender, race, class, and other sources of discrimination and orients algorithms according to key principles and values; (2) democratic and participatory, i.e., it combines a concern for efficiency with sensitivity to issues of fairness or justice; and (3) interdisciplinary, meaning that it integrates social sciences and natural sciences from the outset in all applications.

Keywords: Artificial Intelligence; urban water cycle; hydrosocial urban cycle; urban political ecology

1. Introduction

Digital technologies, particularly those related to Artificial Intelligence, are dramatically transforming our lives and the environment. Urban Water Cycle (UWC) management is a key part of this transformation. The UWC covers all the services related to the extraction, supply and sanitation of waters in a city (see Figure 1). The management of the UWC includes all the water that is present in urban environments: natural surface water, groundwater, drinking water, sewage, stormwater, flood overflow water, and recycled water (stormwater harvesting, managed aquifer recharge, etc.) [1]. In recent years, UWC management has been dramatically transformed by the development and application of new forms of automation and monitoring, as well as the use of AI data sources and knowledge management tools. This so-called "digital revolution" [2,3] has received considerable attention in both research and practitioner circles. The studies focused on this transformation can be grouped into three main categories: first, the tech companies developing digital applications in the water sector present an overwhelmingly positive angle, emphasising the benefits of digitalisation (for an overview, see [3]). Second, scholars coming from water, environmental engineering, and data science fields, who mainly focus on developing and refining existing tools [4–6], also frame digitalisation as a fundamentally positive trend. While some legal and ethical issues are marginally mentioned, this type of analysis presents the intertwining of AI with Big Data science and water management as a 'revolution' in the sector, which opens new ways to analyse, organise, and extract information from large volumes of varying types of data [2,7]. Finally, an emerging literature unpacks the more problematic aspects of water digitalisation [8], but the discussion of the broader sociopolitical implications of this process remains overly sketchy.

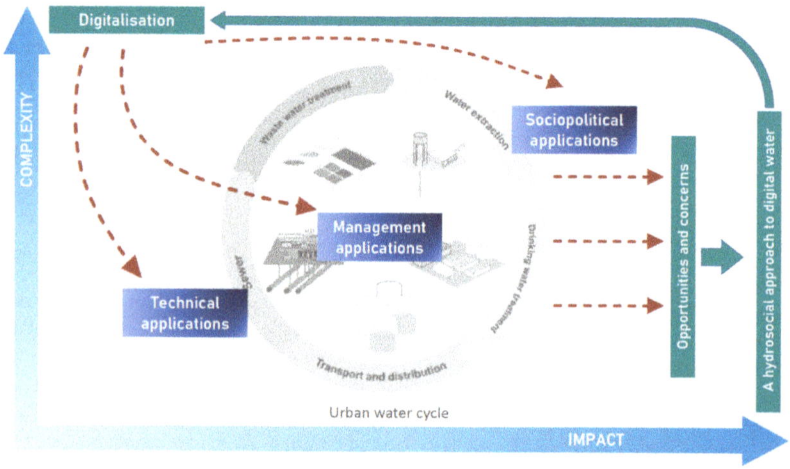

Figure 1. Impact of digitalisation on the UWC.

The lack of critical work on this topic is surprising, given the interest in digitalisation and the proliferation of scientific works devoted to concerns related to AI in general. Scholars of ethics, politics, and technology have unveiled the outright dangers of the unchecked deployment of AI applications, with notorious examples of thinkers such as Shoshana Zuboff [9,10], Noah Yuval Harari [11], and Byung-Chul Han [12] devoting entire volumes to—often dystopian—AI futures. The increase in the references focused on ethics in AI is another indicator; in this sense, see [13–17]. The lack of specific critical attention to the applications of AI in water management is even more puzzling if we consider the importance and ubiquity of water not only as a source of life, but also as a carrier of data and, as an increasingly scarce resource, an object of political conflict. In the context of the overwhelmingly positive accounts of AI application in water management, this paper aims to offer a more nuanced view of the impact of digitalisation in the sector and to look at 'both sides of the coin', which any transformation of this magnitude necessarily possesses. Drawing on the precepts of urban political ecology (Section 2), we propose a hydrosocial approach to digital water management which aims to integrate and overcome the one-sidedness of the state-of-the art approaches to this issue. Thus, the contribution of the article is a new theoretical framework which can be operationalised to analyse the ethical–political implications of the deployment of AI in urban water management.

The paper performs a critical overview of the state-of-the art scholarship on AI applications in water management, looking at both the opportunities and the concerns they generate. However, selecting relevant examples can be problematic, given that the influence of AI and related technologies in the water sector is extremely wide: data science, augmented intelligence techniques, and automation have enabled innumerable applications, from virtual representation of the water system and near-real time flow and quality monitoring to asset management. Therefore, following our methodology (Section 2), we decided to focus on examples that had more potential to result in ethical or political problems. Moreover, we covered each of the three levels of complexity vs. impact in the water sector proposed by Poch et al. [18] and inspired by Funtowicz and Ravetz [19]: the technical, the operational, and the sociopolitical (Figure 1).

The structure of the paper is organised according to these three levels. Thus, in Section 3, we take two examples of technological developments enabled by digitalisation: smart metering and sewer epidemiology. The former refers to the instant measurement of households' water consumption. The latter is the extraction of biological data from wastewater to identify geographical patterns of disease spread, substance consumption, etc., a technique that has become especially popular during the COVID-19 pandemic. The second level corresponds to the design and operation of urban water cycle infrastructures. In Section 4, we refer to Decision Support Systems (DSS), a technology widely used to help in the choice of treatment plants to be included in the urban water cycle. We also look at how AI enables increased optimisation in the water cycle and operation of the infrastructures, but also transforms the workforce in the water sector. The third level is related to wider sociopolitical aspects of the urban water cycle. At this level, we consider two examples (Section 5): the gender issue and water governance more generally. In the discussion section, we explore pathways for the responsible use of AI and other digital technologies in the UWC, making the case for a 'hydrosocial' approach to A. We argue that this approach overcomes the artificial separations between nature, society, and technology, while integrating ethical and sociopolitical considerations from the outset.

2. Theoretical and Methodological Underpinnings

This investigation relied on a three-step method which involved the following: (1) desk research to review the theoretical literature (political ecology and hydrosocial science); (2) the development of criteria for the initial selection of applications and areas of impact; (3) interdisciplinary expert consultations—apart from the authors, consultations included renowned scholars from political science, environmental justice, philosophy, AI ethics, and environmental engineering—to validate the list and discuss the possible implications of digital water management.

The approach proposed in this paper takes its cue from urban political ecology (UPE) [20–23], a theoretical perspective that can be used to help unpack some of the main assumptions of the 'positive' or 'technocratic' approach to water digitalisation. UPE has traditionally explored the entanglements of power with the infrastructural provision in cities: instead of looking at infrastructure as a mere technological and apparently unproblematic issue, UPE scholars explore the political and conflictive dimension of technological and infrastructural projects [24]. Thus, the urban environment can be conceptualised as a socionatural hybrid. It embodies the relations, technological artefacts, networks, and flows that make urban life possible [25]. Building on the same metaphor of the hybrid, the concept of 'hydrosocial' aims to transcend nature–society binaries and envisions the circulation of water as a combined physical and social process [20,23]. Thus, water infrastructures and technologies of any kind—i.e., including the digital—need to be understood as an inseparable intertwining of cultural, political, and economic relations [25]. It follows that an evaluation of the impact of AI on the water sector needs to consider this hybrid nature and, importantly, place these applications in a wider sociopolitical context.

In this sense, urban political ecologists have begun to study the relationships between the deployment of water infrastructures, technologies, and governance within the shifting forms of capitalism and neoliberalism [21,26]. The growing importance of digital technologies in our society has brought about new forms of capital accumulation. Under so-called "cognitive capitalism", the industrial mode of production becomes obsolete and, at the same time, "the object of accumulation consists mainly of knowledge", which is now the primary source of value [27] (p. 57). A dysfunctional form of cognitive capitalism could be considered "surveillance capitalism", a term coined by Shoshana Zuboff, who refers to it as a "fully institutionalised new logic of accumulation" based on the prediction and modification of human behaviour [9]. A fundamental shift in this new regime is that day-to-day practices—therefore water-related ones are included—become a primary target of commercialisation strategies [10]. Importantly, and given the rapid pace at which this trend is advancing, our society is yet to find a way to govern Big Data: how will it be used, to what purpose, who decides, and who decides who decides are fundamental political questions that remain largely unanswered [9].

These questions are also profoundly relevant for how the 'digital revolution' in water will be understood and managed. Digital technologies such as AI, data-mining, and advanced monitoring have the potential to fundamentally transform the way we understand and collect data from water, as the sector becomes riddled with new ethical and political concerns. Bringing together UPE and hydrosocial approaches, the paper focuses on the choreographies of power, nature, and capitalism encapsulated in 'digital water' applications. The article asks: notwithstanding the benefits of these applications, to what extent are they obscuring instances of power and discrimination and what can be done to overcome these concerns?

3. The Impact of AI on UWC: The Technical Level

This section focuses on the technical level of 'digital water' issues and the associated concerns, as exemplified in two applications: sewer epidemiology and smart metering.

3.1. Sewer Epidemiology

Water has the potential to become one of society's key sources of information; according to Garrido-Baserba et al. [2], cities will soon be able to 'mine' their sewer systems for information in order to understand lifestyle habits and the overall health status of the population by measuring human biomarkers. Importantly, by collecting the data available in wastewater, public health decision making will be improved through rapid access to potential sources of epidemiological threat. According to 'optimist' accounts, in exchange for this information, users will be "continuously informed and guided about health (e.g., disease predisposition, recovering follow-ups, tailored health programs, etc.) and lifestyle (e.g., dietary recommendations, personal-suggested activities, sport-recommendation with its corresponding associated products, etc." [2].

However, if we look at this technology and its potentiality from a UPE perspective, several problems arise. As our current experiences with Big Data show us, the relationship between users and the firms which process and commercialise data is more like an extractive process than a fair exchange [9]. It is precisely this one-way process that makes Big Data possible, and it typically occurs in the absence of dialogue or consent. Similarly, in the case of data obtained via sewer mining, optimistic accounts of the benefits of advanced capacity for early detection of diseases do not consider the potential risks of this commodification of data.

Moreover, as the management of COVID-19 has shown, access to intimate data can be used as a tool of advanced surveillance. For instance, the ability to profile different neighbourhoods according to their lifestyle choices raises important threats to privacy, but furthermore, the access to this data can legitimate, even without the threat of health hazards, policies which reproduce inequalities and discrimination against the more vulnerable. For instance, it is known already that the pandemic disproportionately affected poor, black communities, not because of their 'irresponsible' habits, but because of their structural life conditions: dwelling in crowded houses increased the incidence of contagion. The selective lockdown of poor neighbourhoods is just an example of how biopolitics can be spun against these communities [28].

Allowing access to personal lifestyle data can have significant impacts on the lives of the citizens, since in some health systems—notoriously in the USA—lifestyle habits determine the degree of access to health services. A smoker pays up to 50% more for health insurance, depending on the state [29]. Still, so far, the system relies on the 'honour' of the respondents and does not actually perform any checking; in other words, people can hide their habits, even if lying about them constitutes light fraud. A 'smart' system such as sewer epidemiology, which is able to detect eating habits, substance ingest, or other conducts which pose a threat to our health, could not only 'recommend' changes in diets, but also stigmatise and hinder medical autonomy.

Another important concern raised by sewer epidemiology is the potential publication of the databases and the eventual uses of this information. Even if the individual data are not made available—and therefore the European regulations on privacy are safeguarded—the publication of data even at a large geographical aggregation level can have an impact on certain communities. Wastewater can provide information about the level of intake of certain medical drugs in certain areas, which is in fact a valuable commercial information and can help target marketing campaigns for certain 'health' products. These instances require additional cautions that might be beyond the General Data Protection Regulation (GDPR) regulations, since they may be contrary to the principles declared in the recent AI Act of the European Commission from April 2021 (Article 5 of Title II) about the prohibition of any practice oriented to manipulate human behaviour [30].

3.2. Smart Metering

Smart meters installed in every household allow for continuous measuring of water consumption. Through this technique, more accessible data are available to both consumers and companies, and this can impact the way people relate to the service while opening countless possibilities for influencing consumption behaviour. From the utility point of view, it can allow for reductions in peak demands, improved demand forecasting, promotion of efficient appliances, and performance indicators. On the consumer side, smart meters are considered useful as they provide information on how and where water is used to allow the reduction of consumption or quick leak detection.

However, smart meters also present several concerns. The technology can be used to collect and commodify data on consumer behaviour, while the promised reduction in consumption is in part a result of disciplining consumers. Thus, Zetland [31] (p. 126) argues that "[w]ater meters transform water users from passive consumers [...] into active customers entitled to value for money". A more responsible consumer is 'nudged' to consume less water by emphasising the opportunity to save money. The financial value of water is used to urge households to 'take control' over their bill while altering the ways in which they use water. Penetrating the intimacy of the home, smart meters assist in the transformation of the household into a revenue stream and, at the same time, serve to control behaviours and shape practices "associated with the political rule of finance" [32].

Critical literature has also shown that this continuous information about water consumption (combined with pricing mechanisms) may make citizens obsessed with reducing already low and essential water uses instead of focusing on other household expenses [33] (see also Section 5.1 on the gender dimension of this practice). Therefore, it could be argued that smart metering risks consolidating processes of commodification of the urban water supply, even for the most basic uses. At the same time, studies have shown that consumers from vulnerable groups are likely to need more help if they are to obtain the full benefits of smart metering [33].

Moreover, like the sewer epidemiology case, the publication of data can become a problem; even if individual data are strictly preserved and only aggregated data of water consumption at certain territorial levels are made available (at building level, at street level, at district level, etc.), these data can be crossed with other databases such as, for example, TV channel scheduling. This data crossing can yield information about which channel a certain territorial area is predominantly watching, with associated consequences for the advertising policies among others (for instance because there is a direct association between increasing use of water when advertisements happen on TV in the evenings). In Europe, the GRPD is trying to prevent these situations via the informed consent of users, but additional care is required when we are talking about aggregated data which can be analysed in forms that can also have an impact at the individual level [30].

4. The Impact of AI on UWC: The Operational Level

4.1. Decision-Support Systems for Design and Operation of Water Treatment Plants

Selecting the technology mix that can achieve the objectives assigned to urban water infrastructures is a complex task due to the proliferation of new technologies coupled with ever stricter water quality standards [34]. Environmental Decision Support Systems (EDSSs) are AI tools that have been helping operators and policymakers to tackle this problem. Created in the 1980s, the aim of these technologies was to provide decision-making support by combining mathematical models with qualitative knowledge. The use of ontologies and AI techniques specialising in the emulation of human behaviour have allowed EDSSs to be used as systems capable of integrating a wide diversity of knowledge and providing elements of discussion to reach consensual solutions.

While EDSSs have helped to improve decision making across the water cycle [34,35], several pitfalls of this technology are worth reviewing. From an ethical perspective, the fact that they rely on the autonomous nature of AI systems raises issues of the potential loss of human control over decision making, which can yield ethically questionable outcomes. In the same line, the liability for harms resulting from the use of EDSSs remains ambiguous under many legal frameworks.

Application of a political ecology lens to this technique would highlight issues of power inequalities, especially due to how restrictive EDSSs are in terms of use. Thus, while the EDSS as a concept was very promising and there are numerous such systems already in place in the water management sector, the complexity of today's operations poses challenges to their implementation. The integration of expertise, models, statistical analysis, case-based systems, real-time, etc. is key to gaining useful decision-making knowledge. However, because of this diversity of inputs, EDSSs end up difficult to understand and employ for users others than large companies who can afford: (1) to employ data experts, process engineers, and software engineers who can connect different types of data sources, databases, etc.; or (2) software that can create easy-to-use, dynamic interfaces.

4.2. Workforce

The gradual uptake of AI technologies in the water sector will likely affect its workforce, like other economic sectors. From an optimistic perspective, AI has the potential to "free workers from repetitive tasks enabling them to focus on more highly skilled aspects" [2]. At the same time, digital water management will require a requalification of the workforce, including water researchers. According to the same authors, the implementation of effective Big Data exploitation will give birth to a "new generation of researchers/practitioners trained in engineering, statistics, and computer science through the creation of multidisciplinary training programs" [2]. Indeed, one of the objectives of the European strategy on AI is to prepare the society for the socioeconomic changes brought by AI, and in particular to provide support to businesses to strengthen the advanced digital skills of their workforce. Moreover, the Ethics Guidelines for Trustworthy AI of the High-Level Expert Group on AI introduce the education and training of AI stakeholders (including raising awareness of the potential impact of AI and their participatory role in shaping societal development) as a key method through which to ensure the responsible implementation of AI systems [36]. In this sense, the authors [2] are right to stress that the potential of AI comes from training more interdisciplinary researchers and professionals.

While the positive aspects of qualifying staff in managing these new technologies cannot be denied, there are more aspects to consider in this requalification than the purely technical ones. Firstly, the path of deskilling and reskilling involves a set of complexities which are rarely explored in the 'optimist' literature. Not all workers will be able to cope with the reskilling process—or will be willing to—and therefore they will inevitably be affected by redundancy. For most people, jobs are much more than just means to gain a living; they are intimately linked to personal identity, self-esteem, and social status, which will make this process particularly painful for a portion of existing water professionals. Depending on the pace at which AI replaces workers or makes them partially unfit for their positions, compensatory policies such as universal income should be explored [37], representing some form of "intergenerational solidarity between those disadvantaged today and those advantaged tomorrow, to ensure that the disruptive transition between the present and the future will be as fair as possible, for everyone" [13].

Secondly, there is also the danger of perpetuating the technocratic paradigm which already privileges technical skills. The water and AI specialists of tomorrow, with their versatile profiles, may represent an important step forward in understanding the inherent interdisciplinary character of the water sector itself. However, this is not achievable unless interdisciplinarity is understood in its entirety; issues such as values, ethics, and the social and political aspects of AI and of water management need to be part of the basic training of the new water professionals who will govern and manage the "digital water revolution".

5. The Impact of AI on UWC: Sociopolitical Level

In this section, we address the sociopolitical dimension of water digitalisation. We deal with the water governance aspects of water digitalisation and the gender implications.

5.1. Water Governance

The term 'governance' has gained increasing prominence in the scientific literature on water. As an analytical category, the governance perspective asks whether and how institutions can deal with the tasks assigned to them [38]. The governance approach has offered a more flexible (but also more elusive) way to conceptualise the political space, which includes the classic governmental processes—mainly laws, regulation, and policies—but moves beyond it to include more scales, types of public and private actors, and the relationships between them [39].

In the water sector, AI can improve water governance especially by reducing uncertainty; AI capabilities can have a positive impact on policymaking and businesses by offering reliable predictions of climate conditions, water flows, and even envisaged behaviours of communities and actors, thus reducing risks [40]. However, the promise to offer certainty and reduce the complexities of decision making also has several downsides. A political ecology perspective would advise against the danger of excessively relying on AI algorithms for decision making in water governance, since it may consolidate the already reductionist and depoliticised water governance perspective that is dominant nowadays, according to which water is devoid of any cultural, symbolic, and political character. Consequently, its management is conscribed to the realm of technoscience and managerial means. Nevertheless, water is a paradigmatic "wicked problem" [40] imbued with social controversy and needs constant political renegotiations of potential solution paths [39]. Moreover, as Eric Swyngedouw reminds us, water is inherently political, and therefore subject to all manner of tensions, conflicts, and social struggles [22].

The depoliticisation of water is potentially negative because it privileges a dominant way to manage water over alternative ones. As shown in the political ecology literature, depoliticisation goes hand in hand with the commodification of water [41,42] which subordinates the use value of water (i.e., its qualitatively defined characteristics) to the exchange value (monetary price decided by the market). This commodification process implies that the only way to truly 'appreciate' a service is to pay for it (ideally at full market cost). Thus, in the AI-powered expert paradigm, if a value cannot be expressed in monetary terms, it risks not being included in decision making at all [43]. Consequently, service users are increasingly seen (and come to see themselves) as 'customers' instead of 'citizens', with public amenities perceived more like private commodities than public goods, concealing the complex social and labour arrangements behind their exchange price [43,44].

5.2. Water, AI, and Gender

Power imbalances and exploitation are key preoccupations of political ecologists; it is therefore no surprise that gender has been intensively studied, particularly in relation to water supply in developing countries [45–47]. Women are predominantly in charge of providing water for their families, often over very long distances, a fundamental task which does not enjoy social recognition. The responsibility of providing water intensifies the daily work of women and compromises their education and access to leisure, negatively impacting their health and quality of life [48]. These conditions may worsen in a climate change context, which access to water even more difficult [49].

Gender inequality in the water sector is not restricted to developing countries; globally, the water sector continues to employ far less women than men, especially in technical and managerial jobs. A World Bank study shows that women occupy less than a quarter of the total water professionals. Moreover, even if on average, 23 percent of engineers and managers in a utility are female, 32 percent of the sampled utilities had no female engineers and 12 percent had no female managers [50]. As AI penetrates more and more aspects of the water management sector, this will have an impact on how women will cope with digitalisation and will make them more likely to be affected by redundancy. Thus, as AI applications become an essential interface in the work environment, the pressure to reskill will be harder on women; this learning process usually takes place during 'leisure' time, when women carry heavier care burdens than. This is correlated with the gender dimension of the 'digital gap', meaning that girls and young women do not have equal access to technology and digital training, which negatively affects literacy and professional prospects [51].

Another area where the interplay between AI, water, and gender is worth considering is in the decision making in urban water planning interventions. As noted in Section 4.2, AI-based systems using aggregated user data (or instance, about water use in the household through smart meters) are increasingly employed to envisage scenarios and legitimise policies in urban water projects [40,52]. If datasets are not disaggregated by sex and gender (as well as other identities), this means that policies regarding tariffs, water-saving schemes, or infrastructure programs will not factor in women's needs and opinions. This is especially important if we consider that care-related activities, which involve water use, are predominantly the responsibility of women. Using a gendered analysis, planners could gain a more accurate picture of communities, natural resource uses, households, and water users. Understanding the differences among and between women and men (who does what work, who makes which decisions, who uses water for what purpose, who controls which resources, who is responsible for different family obligations, etc.) can yield more effective results [53,54]. Moreover, the fact that only 22 percent of AI professionals are women certainly does not improve the chances that the interpretation of these data includes a gendered perspective [55].

It follows that woman should play a fundamental role in resource management and in the implementation of water technology and infrastructure projects. However, change is not at hand, with studies concluding that the interest of women in advancing within water governance is not well-supported by society, particularly in rural and peri-urban areas [46]. Governments and companies should strengthen women's participation and leadership, promote communication and education with a gender perspective, and ensure the collection of gender-disaggregated data.

6. Discussion: A Hydrosocial Approach to Digital Water

This paper aimed to move beyond the overly optimistic approaches to 'digital water' management and paint a more nuanced picture of this crucial transformation of the sector. Using a hydrosocial and Urban Political Ecology theoretical lens (see also [56]), the analysis highlighted that the gradual penetration of Big Data and AI-related technologies in the water sector can yield opportunities but also have negative effects such as increased surveillance, lack of accountability, workforce exclusion, discrimination, injustice, and loss of democratic quality. These are also areas where more data and research is needed.

How to tackle these concerns? The available ethical frameworks provide some valuable guidance. Thus, the High-Level Expert Group on AI have defined a set of requirements that AI practitioners need to put in place to guarantee a Trustworthy AI: human agency and oversight; technical robustness and safety; privacy and data governance; transparency; diversity, non-discrimination, and fairness; societal and environmental wellbeing; and accountability [36]. However, these principles require further operationalisation for specific AI-based systems in the water domain.

From the overview of opportunities and concerns presented in this paper emerges that a *hydrosocial approach to digital water* is timely and necessary. It envisions AI as a force in the service of the human right to water, the implementation of which needs to be (1) *critical,* in that it takes into consideration gender, race, class, and other sources of discrimination and orients algorithms according to key principles and values; (2) *democratic and participatory,* i.e., it combines a concern for efficiency with sensitiveness to issues of fairness or justice and a shift towards mechanisms that ensure democratic control over the applications, while overcoming the split between experts and society; and (3) *interdisciplinary,* meaning that it integrates social sciences and 'technical'/ natural sciences from the outset, in all applications.

Applying these principles to the risks identified here, the following recommendations emerge:

(1) Risk of surveillance: applications such as sewer epidemiology and smart metering have the capacity to obtain intimate private information which, if not properly protected, can become a tool of increased surveillance and oppression [57]. To this purpose, the direct and indirect uses of data coming from water must be limited, particularly when talking about aggregated data combined with nonwater information. In the EU, the TrustworthyAI ethical guidelines' [36] recommendations and AI Act provide a proper context to guarantee privacy, autonomy, and nondiscrimination; however, they only apply in the EU and they need further operationalisation for the water domain.

(2) Risk of biased decision making: the advance of environmental decision-support systems, agent-based modelling, and Big-Data-powered scenarios for policymaking can be nonrepresentative, biased, and lack accountability. Therefore, there needs to be an increased preoccupation with standards of transparency and access to knowledge. Standards should be complemented by training programmes and integration of a wider variety of disciplines, knowledges, and perspectives in the design of the models, lowering the expertise and gender barrier.

(3) Risk of workforce exclusion: digitalisation will inevitably impact the water workforce, provoking redundancy and the need for reskilling. To avoid the worst consequences for economic means, wellbeing, or dignity of this workforce, there needs to be increased awareness on the part of the policymakers and corporations; the advance of AI should be approached systemically, in an integrated way, and complemented with investments in training, while exploring substantial solidarity programmes such as a universal income. Specific attention needs to be paid to training for women in future professional roles, starting from acknowledgment of the gender dimension of the digital divide.

(4) Risk of democratic quality loss: water digitalisation is largely driven by the private sector, which welcomes the opportunity of reducing risks and complexity of decision-making. However, this transformation should not lead to further commodification of water and the exclusion of communities from decisions in favour of gaining efficiency. Therefore, the digitalisation of the water sector should go hand in hand not only with public control, but also with enhancing mechanisms of direct, democratic participation

in the water governance. Moreover, when we talk about the deployment of new technologies in developing countries, it is especially important to consider the cultural dimension and indigenous knowledges, and to understand how it will affect social relations and what socioeconomic changes need to be foreseen to adopt, adapt to, or reject the water digitalisation processes. It is also necessary to consider possible trade-offs raised from the compliance of all these ethical requirements, where technical concepts such as efficiency or performance might be confronted with sociopolitical requirements. Consequently, responsible AI is as much a question of trustworthy AI techniques as one of accountable governance.

(5) Risk of injustice and gender discrimination: water, gender, and AI are inextricably linked, as the digitalisation of water will likely enhance existing power imbalances in the sector and the lack of representation of women and other gendered identities in water-related interventions. To counteract the bias incorporated in AI applications see also [58], gender should be mainstreamed in planning, implementation, and evaluation of programs. Methodologies such as gender analysis, social mapping, and sex- and gender-disaggregated data should be further encouraged. Promoting an active participation of different gendered people throughout the design, implementation, and evaluation stages can help to ensure that their needs are represented. Additionally, more research correlating data on gender, AI, and water technologies is needed to improve knowledge and awareness on this topic.

7. Conclusions

This paper provides an informed reflection about the ethical and political concerns associated with the deployment of digital technologies in the water sector, with a focus on the Urban Water Cycle. Using the theoretical lens of urban political ecology, and the literature on hydrosocial cycles, we analysed relevant examples of how digitalisation impacts the technical, operational, and sociopolitical levels of the urban water cycle, respectively. The applications were selected using expert consultations. The article did not aim to offer an exhaustive review of the 'digital water' issue, but rather to illustrate how we can—and should—be thinking about this issue. Our results show that while digitalisation can have a positive impact on the urban water cycle, it can also generate concerns in terms of personal and collective rights. In the discussion, we provide paths for further research that approach water digitalisation in a critical, theory-guided and interdisciplinary way. Indeed, interdisciplinarity is key to a safe and fair digital transition, and this article takes this conviction onboard by relying on complementary views of scholars from political science, water engineering, and AI. This first attempt to bring together a diversity of voices can serve as a basis for a timely debate in the water sector that would help to prevent the establishment of dysfunctional ways of functioning, which may be extremely challenging to reverse.

Author Contributions: Conceptualisation, L.A.P. and M.P.; methodology, À.C., M.G.-B., M.V. and K.G.; formal analysis, À.C., L.A.P. and K.G.; writing—original draft preparation, L.A.P., M.P. and M.V.; writing—review and editing À.C., M.G.-B. and K.G.; supervision, M.P. and K.G. All authors have read and agreed to the published version of the manuscript.

Funding: This research received no external funding.

Institutional Review Board Statement: Not applicable.

Informed Consent Statement: Not applicable.

Data Availability Statement: Not applicable.

Conflicts of Interest: The authors declare no conflict of interest.

References

1. Oral, H.V.; Carvalho, P.; Gajewska, M.; Ursino, N.; Masi, F.; van Hullebusch, E.D.; Kazak, J.K.; Exposito, A.; Cipolletta, G.; Andersen, T.R.; et al. A review of nature-based solutions for urban water management in European circular cities: A critical assessment based on case studies and literature. *Blue-Green Syst.* **2020**, *2*, 112–136. [CrossRef]
2. Garrido-Baserba, M.; Corominas, L.; Cortés, U.; Rosso, D.; Poch, M. The Fourth Revolution in the Water Sector Encounters the Digital Revolution. *Environ. Sci. Technol.* **2020**, *54*, 4698–4705. [CrossRef] [PubMed]
3. International Water Association. Digital Water. Industrial Leaders Chart the Transformation Journey. Available online: https://iwa-network.org/publications/digital-water/ (accessed on 12 February 2022).
4. Sun, A.Y.; Scanlon, B.R. How can Big Data and machine learning benefit environment and water management: A survey of methods, applications, and future directions. *Environ. Res. Lett.* **2019**, *14*, 073001. [CrossRef]
5. Corominas, L.; Garrido-Baserba, M.; Villez, K.; Olsson, G.; Corteś, U.; Poch, M. Transforming Data into Knowledge for Improved Wastewater Treatment Operation: A Critical Review of Techniques. *Environ. Model. Softw.* **2018**, *106*, 89–103. [CrossRef]
6. Knüsel, B.; Zumwald, M.; Baumberger, C.; Hadorn, G.H.; Fischer, E.M.; Bresch, D.N.; Knutti, R. Applying big data beyond small problems in climate research. *Nat. Clim. Chang.* **2019**, *9*, 196–202. [CrossRef]
7. Poch, M.; Garrido-Baserba, M.; Corominas, L.; Perelló-Moragues, A.; Monclús, H.; Cermerón-Romero, M.; Melitas, N.; Jiang, S.; Rosso, D. When the fourth water and digital revolution encountered COVID-19. *Sci. Total Environ.* **2020**, *744*, 140980. [CrossRef]
8. Doorn, N. Artificial intelligence in the water domain: Opportunities for responsible use. *Sci. Total Environ.* **2020**, *755*, 142561. [CrossRef]
9. Zuboff, S. Big other: Surveillance Capitalism and the Prospects of an Information Civilization. *J. Inf. Technol.* **2015**, *30*, 75–89. [CrossRef]
10. Zuboff, S. *The Age of Surveillance Capitalism*; Profile Books: London, UK, 2019.
11. Harari, Y.N. Reboot for the AI revolution. *Nature* **2017**, *550*, 324–327. [CrossRef]
12. Han, B.-C. *Non things. Upheaval in the Lifeworld*; Polity Press: Cambridge, UK, 2020.
13. Floridi, L. *The Onlife Manifesto. Being Human in a Hyperconnected Era*; Springer Open: Berlin/Heidelberg, Germany, 2015; p. 264.
14. Jobin, A.; Ienca, M.; Vayena, E. The global landscape of AI ethics guidelines. *Nat. Mach. Intell.* **2019**, *1*, 389–399. [CrossRef]
15. Ryan, M.; Stahl, B. Artificial intelligence ethics guidelines for developers and users: Clarifying their content and normative implications. *J. Inf.* **2015**, *29*, 62–84. [CrossRef]
16. Gupta, A. AI in Smart Cities: Privacy, Trust, and Ethics. New Cities. Available online: https://newcities.org/the-big-picture-ai-smart-cities-privacy-trust-ethics/ (accessed on 12 February 2022).
17. Hagendorf, T. The Ethics of AI Ethics: An Evaluation of Guidelines. *Minds Mach.* **2015**, *30*, 99–120. [CrossRef]
18. Poch, M.; Comas, J.; Rodrıguez-Roda, I.; Sanchez-Marrèb, M.C. Designing and building real environmental decision support systems. *Environ. Model. Softw.* **2003**, *19*, 857–873. [CrossRef]
19. Funtowicz, S.O.; Ravetz, J.R. Science for the post-normal age. *Futures* **1993**, *25*, 739–755. [CrossRef]
20. Boelens, R.; Hoogesteger, J.; Swyngedouw, E.; Vos, J.; Wester, P. Hydrosocial territories: A political ecology perspective. *Water Int.* **2016**, *41*, 1–14. [CrossRef]
21. Loftus, A. *Everyday Environmentalism: Creating an Urban Political Ecology*; University of Minnesota Press: Minneapolis, MN, USA; London, UK, 2015.
22. Swyngedouw, E. Dispossessing H2O: The contested terrain of water privatization. *Capital. Nat. Social.* **2005**, *16*, 81–98. [CrossRef]
23. Swyngedouw, E. *Liquid Power: Contested Hydro-Modernities in 20th Century Spain*; MIT Press: Cambridge, MA, USA, 2015.
24. Heynen, N.; Kaika, M.; Swyngedouw, E. *In the Nature of Cities: Urban Political Ecology and the Politics of Urban Metabolism*; Routledge: London, UK; New York, NY, USA, 2006.
25. Gandy, M. *Concrete and Clay: Reworking Nature in New York City*; MIT Press: Cambridge, MA, USA, 2002.
26. Allen, J.; Pryke, M. Financialising household water: Thames Water, MEIF, and 'ring-fenced' politics. *Camb. J. Reg. Econ. Soc.* **2013**, *6*, 419–439. [CrossRef]
27. Moulier-Boutang, Y. *Cognitive Capitalism*; Polity Press: Cambridge, UK, 2012.
28. Durizzo, K.; Asiedu, E.; Van der Merwe, A.; Van Niekerk, A.; Günther, I. Managing the COVID-19 pandemic in poor urban neighbourhoods: The case of Accra and Johannesburg. *World Dev.* **2021**, *137*, 105175. [CrossRef]
29. Health Markets. What You Need to Know About Smoking and Health Insurance. Available online: https://www.healthmarkets.com/content/smoking-and-health-insurance (accessed on 1 October 2021).
30. European Commission. Laying down Harmonised Rules on Artificial Intelligence (Artificial Intelligence Act) and Amending Certain Union Legislative Acts. 2021. Available online: https://eur-lex.europa.eu/legal-content/EN/TXT/HTML/?uri=CELEX:52021PC0206&from=EN (accessed on 30 January 2022).
31. Zetland, D. The struggle for residential water metering in England and Wales. *Water Altern.* **2016**, *9*, 120–138.
32. Loftus, A.; March, H.; Nash, F. Water Infrastructure and the Making of Financial Subjects in the South East of England. *Water Altern.* **2016**, *9*, 319–335.
33. March, H.; Morote, A.-F.; Rico, A.-M.; Saurí, D. Household smart metering in Spain: Experiences from remote meter reading in Alicante. *Sustainability* **2017**, *9*, 582. [CrossRef]
34. Hadjimichael, A.; Comas, J.; Corominas, L. Do machine learning methods used in data mining enhance the potential of decision support systems? A review for the urban water sector. *AI Commun.* **2016**, *29*, 747–756. [CrossRef]

35. Garrido-Baserba, M.; Reif, R.; Hernández, F.; Poch, M. Implementation of a knowledge-based methodology in a decision support system for the design of suitable wastewater treatment process flow diagrams. *J. Environ. Manag.* **2012**, *112*, 384–391. [CrossRef] [PubMed]
36. High-Level Expert Group on Artificial Intelligence, Ethics Guidelines for Trustworhy AI. Available online: https://www.aepd.es/sites/default/files/2019-12/ai-ethics-guidelines.pdf (accessed on 3 November 2021).
37. Cowls, J.; Floridi, L.; Mariarosaria, T. The Challenges and Opportunities of Ethical AI. Available online: https://digitransglasgow.github.io/ArtificiallyIntelligent/contributions/04_Alan_Turing_Institute.html (accessed on 22 October 2021).
38. Takeda, T.; Kato, J.; Matsumura, T.; Murakami, T.; Abeynayaka, A. Governance of Artificial Intelligence in Water and Wastewater Management: The Case Study of Japan. *Hydrology* **2021**, *8*, 120. [CrossRef]
39. Meisch, S. I Want to Tell You a Story: How Narrative Water Ethics Contributes to Re-theorizing Water Politics. *Water* **2019**, *11*, 631. [CrossRef]
40. Perello-Moragues, A.; Poch, M.; Sauri, D.; Popartan, L.A.; Noriega, P. Modelling domestic water use in metropolitan area using socio-cognitive agents. *Water* **2020**, *13*, 1024. [CrossRef]
41. Bakker, K. From State to Market?: Water Mercantilización in Spain. *Environ. Plan. A* **2002**, *34*, 767–790. [CrossRef]
42. Bakker, K. Water: Political, biopolitical, material. *Soc. Stud. Sci.* **2012**, *42*, 616–623. [CrossRef]
43. McDonald, D.A. To corporatize or not to corporatize (and if so, how?). *Util. Policy* **2016**, *40*, 107–114. [CrossRef]
44. Clarke, J.; Newman, J.; Smith, N.; Vidler, E.; Westmarland, L. *Creating Citizenconsumers: Changing Publics and Changing Public Services*; Sage Press: Thousand Oaks, CA, USA, 2007.
45. Andajani, S.; Chirawatkul, S.; Saito, E. Gender and Water in Northeast Thailand: Inequalities and Women's Realities. *J. Int. Women's Stud.* **2015**, *16*, 200–212.
46. de San, J.A.S.R. Gender and water governance in Mexico. *Manag. Environ. Qual. Int. J.* **2019**, *30*, 695–713.
47. Silva, B.B.; Sales, B.; Lanza, A.C.; Heller, L.; Rezende, S. Water and sanitation are not gender-neutral: Human rights in rural Brazilian communities. *Water Policy* **2020**, *22*, 102–120. [CrossRef]
48. Varua, M.E.; Ward, J.; Maheshwari, B.; Dave, S.; Kookana, R. Groundwater management. *Water* **2018**, *11*, 1959.
49. Eastin, J. Climate change and gender equality in developing states. *World Dev.* **2018**, *107*, 289–305. [CrossRef]
50. World Bank. Women in Water Utilities. Breaking Barriers. 2019. Available online: https://openknowledge.worldbank.org/bitstream/handle/10986/32319/140993.pdf (accessed on 1 January 2022).
51. UNICEF. What We Know about the Gender Digital Divide for Girls: A Literature Review. 2021. Available online: https://www.unicef.org/eap/media/8311/file (accessed on 22 January 2022).
52. Smith, G.; Rustagi, I. When Good Algorithms Go Sexist: Why and How to Advance AI Gender Equity, Standford Social Innovation Review. 2021. Available online: https://ssir.org/articles/entry/when_good_algorithms_go_sexist_why_and_how_to_advance_ai_gender_equity# (accessed on 12 February 2022).
53. UNDP. Resource Guide for Mainstreaming Gender in Water Management. Available online: https://www.undp.org/publications/resource-guide-mainstreaming-gender-water-management (accessed on 12 August 2021).
54. Beebeejaun, Y. Gender, urban space, and the right to everyday life. *J. Urban Aff.* **2016**, *39*, 323–334. [CrossRef]
55. Young, E.; Wajcman, J.; Sprejer, L. Where are the Women? Mapping the Gender Job Gap in AI. Policy Briefing: Full Report. The Alan Turing Institute. 2021. Available online: https://www.turing.ac.uk/news/where-are-women-mapping-gender-job-gap-ai (accessed on 30 January 2022).
56. Kostakis, V.; Roos, A.; Bauwens, M. Towards a political ecology of the digital economy: Socio-environmental implications of two competing value models. *Environ. Innov. Soc. Transitions* **2016**, *18*, 82–100. [CrossRef]
57. Noble, S.U. *Algorithms of Oppression*; New York University Press: New York, NY, USA, 2018. [CrossRef]
58. Ferrer, X.; van Nuenen, T.; Such, J.M.; Cote, M.; Criado, N. Bias and Discrimination in AI: A Cross-Disciplinary Perspective. *IEEE Technol. Soc. Mag.* **2021**, *40*, 72–80. [CrossRef]

applied sciences

MDPI

Article

Use of Deep Learning to Improve the Computational Complexity of Reconstruction Algorithms in High Energy Physics

Núria Valls Canudas *, Míriam Calvo Gómez *, Elisabet Golobardes Ribé * and Xavier Vilasis-Cardona *

Data Science for the Digital Society (DS4DS) Research Group, Engineering Department, La Salle-Universitat Ramon Llull, Sant Joan de La Salle 42, 08022 Barcelona, Spain

* Correspondence: nuria.valls@salle.url.edu (N.V.C.); miriam.calvo@salle.url.edu (M.C.G.); elisabet.golobardes@salle.url.edu (E.G.R.); xavier.vilasis@salle.url.edu (X.V.-C.)

Citation: Valls Canudas, N.; Calvo Gómez, M.; Golobardes Ribé, E.; Vilasis-Cardona, X. Use of Deep Learning to Improve the Computational Complexity of Reconstruction Algorithms in High Energy Physics. *Appl. Sci.* **2021**, *11*, 11467. https://doi.org/10.3390/app112311467

Academic Editors: Aida Valls and Karina Gibert

Received: 15 November 2021
Accepted: 1 December 2021
Published: 3 December 2021

Abstract: The optimization of reconstruction algorithms has become a key aspect in the field of experimental particle physics. Since technology has allowed gradually increasing the complexity of the measurements, the amount of data taken that needs to be interpreted has grown as well. This is the case with the LHCb experiment at CERN, where a major upgrade currently undergoing will considerably increase the data processing rate. This has presented the need to search for specific reconstruction techniques that aim to accelerate one of the most time consuming reconstruction algorithms in LHCb, the electromagnetic calorimeter clustering. Together with the use of deep learning techniques and the understanding of the current reconstruction algorithm, we propose a method that decomposes the reconstruction process into small parts that can be formulated as a cellular automaton. This approach is shown to benefit the generalized learning of small convolutional neural network architectures and also simplify the training dataset. Final results applied to a complete LHCb simulation reconstruction are compatible in terms of efficiency, and execute in nearly constant time with independence on the complexity of the data.

Keywords: deep learning; convolutional neural network; cellular automaton; reconstruction; complexity; optimization; high energy physics

1. Introduction

In the field of high energy physics (HEP) there are many computational challenges regarding data. In most collider experiments the first step that needs to be done, when a collision happens inside a detector, is reconstructing the data. This process consists of an algorithm that transforms these data into information of what physical phenomena happened inside the detector, so that later on physicists can use it for a detailed analysis. The LHCb experiment at CERN [1] is not an exception. Due to the collision rate generated at the large hadron collider (LHC) accelerator [2], the throughput rate is so high that data cannot be stored before processing. In this case, together with all of the experiments at LHC, the reconstruction process of data has a very strict time constraint in order to obtain determinant information, to decide whether or not to store the data. More details on the overview of data processing are provided in Section 2.

At this point, optimization of reconstruction techniques is needed. Moreover, the vision of future upgrades [3] enhances the importance of developing scalable and flexible software methodologies. In the case of LHCb, the current time performance analysis of the reconstruction process points the calorimeter reconstruction as the second most computational expensive process, representing almost 25% of the LHCb reconstruction time [4].

The LHCb electromagnetic calorimeter (ECAL) [5] is a subdetector designed to measure the energy of particles as they interact with the detector material. It has a rectangular shape of 7.8×6.3 m and is placed perpendicular to the accelerator beam pipe. Moreover, the purpose of the LHCb detector is to study heavy quark hadron decays. Those decays are known to produce particles coming out almost parallel to the beam pipe. Hence, the incident angle of particles striking the calorimeter is between 1° and 21°, with respect to the beam axis. The calorimeter detector structure is segmented into individual square-shaped modules that perform the energy measurement. Each module is made from lead absorber plates interspaced with scintillator tiles as active material and has a variable number of readout cells, depending on the position. To avoid confusion between the calorimeter cells and the cells from a cellular automaton, the calorimeter cells will be referred to as readout cells. The output data obtained from the calorimeter are the values from each readout cell concerning the accumulated energy deposited by incident particles in a single collision. Since particles may deposit energy in more than one readout cell, the energy deposits need to be reconstructed and clustered together with the ones belonging to the same particle.

Such a challenge, under very tight execution time constraints, invites one to think of neural network techniques that can provide the capability of learning complex problems and a fast inference. Considering its increasing popularity in recent years, there have been many improvements in the optimization of reconstruction algorithms. Specially, deep learning models are able to solve many complex issues at very high speeds, only at the cost of increasing the time and complexity of the training. However, most of these proposals approach the whole scenario at once, forcing the networks to process hundreds of thousands of data samples and understand their insights, to be able to provide a complete solution at the output. As problems become more challenging, deeper networks need to be trained with more data. Although nowadays, computational time is not an inconvenience in most cases, data acquisition is, especially in the HEP case, where data are simulated with time-consuming algorithms.

Beginning with the comprehension of the current implementation of the reconstruction algorithm, it is based on a cellular automaton [6]. Further details are provided in Section 3.1. Due to the typical square-shaped modular structure of the calorimeter, its geometry can be easily mapped into a two-dimensional grid. Therefore, the cellular automaton strategy has long been used in calorimeters for high energy physics [7,8]. Such a method provides an efficient reconstruction of the clusters, although the classical formulation of the cellular automaton requires several iterative processes along the readout cell values that are programmed as loops in the algorithm's code. As this causes a strong dependency of the algorithm's complexity on the number of clusters in the data, other approaches have attempted to avoid this dependency by exploiting the architecture similarities between cellular automata and neural networks [9,10]. However, these approaches focus on a proof of concept rather than providing a specific reconstruction solution for the LHCb calorimeter. Within recent years, the evolution of deep learning models has encouraged the use of image processing techniques for this challenge. An early stage project has shown promising results when approaching the LHCb calorimeter reconstruction with a deep neural network structure [11] based on the YOLOv3 network [12]. This particular case uses the order of 100,000 simulation samples to train a network of the order of 65 million parameters.

In this article, we propose a different formulation of the LHCb calorimeter reconstruction problem to improve the generalized learning of small deep learning architectures [13]. Those networks have been trained using artificially generated sets of data and preprocessed pieces of simulated LHCb data. With this methodology, we aim to have an efficient cluster reconstruction algorithm that performs at a nearly constant speed with independence of the event complexity.

2. Overview of Data Processing in Modern Particle Detectors

The physics analysis conducted in the modern HEP field starts with a particle accelerator. In the case of LHC, two streams of proton particles are accelerated up to $0.99999999c$ and collide at four specific detection points. The LHCb experiment is placed in one of these four locations and has a subset of eight dedicated detectors to acquire data from the particles generated in the collisions. The electromagnetic calorimeter is one of them. Complementing the previous definition, the calorimeter detector's general structure is segmented in three different rectangular-shaped regions, as can be seen in Figure 1.

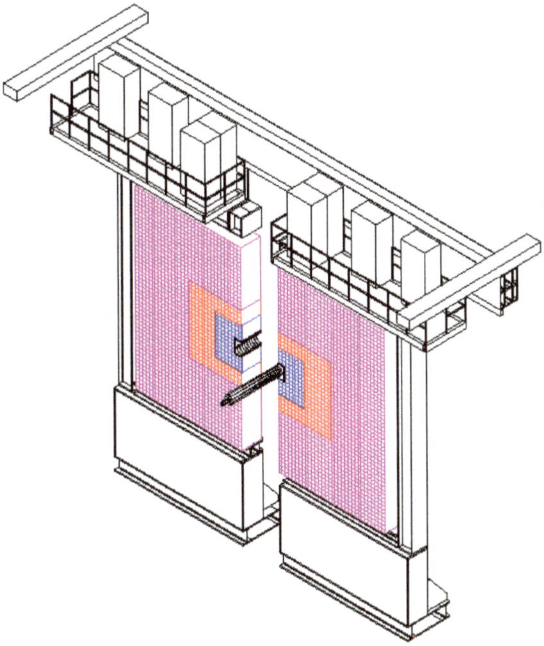

Figure 1. The electromagnetic calorimeter 3d-view from behind the detector towards the interaction point [5].

Although all modules have the same size of 12 × 12 cm, the number of readout cells on a module depends on the region. The inner region is the closest to the accelerator pipe and has nine readout cells of 4 × 4 cm per module. It has the highest granularity among the three regions since it is the region with the highest occupancy of incident particles. The middle region surrounds the inner and has four readout cells of 6 × 6 cm per module, and the outer region has a single readout cell of 12 × 12 cm per module. Since the occupancy decreases with the distance from the accelerator pipe, the shape of the readout cells for middle and outer regions is increased following a trade-off between precision and cost. Regarding the size of the inner readout cells, it is calculated following the *Molière* radius. Therefore, a particle is expected to deposit all of its energy in one inner readout cell if it falls on its center. However, to ensure that all the energy from a particle is captured, and to simplify the reconstruction algorithm, the definition of a calorimeter cluster stands for a 3 × 3 block of readout cells around an energy peak. Some studies have been done regarding the cluster shapes [14] where 2 × 2 clusters show promising performance for high luminosity, although the 3 × 3 cluster is used as a base for masking other shapes on clusters. Therefore, for simplicity in the reconstruction process, the definition of 3 × 3 readout cell clusters is maintained through all the regions of the detector. Since in this approach we focus on improving the reconstruction performance rather than particle identification, we can only evaluate comparisons between the reconstruction results and simulation data.

In all particle physics experiments, data are not analyzed directly from the detector readout, what is called raw data. Instead, domain experts need information of what happened inside the detectors in order to make the physics' analysis. This information is extracted through the reconstruction process. In the case of the calorimeter detector, a sample of raw data contains the list of readout cells that have an energy deposit from one collision. This needs to be reconstructed as all the groups of 3×3 readout cells with the energy, belonging to the same particle colliding into the calorimeter. Up until now, the amount of data generated by all the detectors at LHCb reached the order of 35 GB/s. However, the processing rate of each datum sample was decreased by a factor 40 to a throughput of 1 MHz using a hardware selector known as level 0 trigger [15]. In the upcoming upgrade of the experiment, the data processing rate is going to increase to 30 MHz, meaning the full 35 GB/s will be processed. Since this throughput rate cannot be stored with current technology, the trigger system will be upgraded into a full software trigger called the high level trigger (HLT), where complete data reconstruction needs to take place. Even so, since the number of collisions will also increase in the upgrade, the occupancy of the LHCb detector is going to increase in a factor five. This enhances the need to optimize and accelerate the current reconstruction algorithms in order to fit in the processing time constraints and throughput rate of the HLT.

At the time of building and testing reconstruction algorithms, there is no point in working with raw data from the detector, since there is no "truth information" of the particles that generated the raw sample. Instead, simulation data are produced using the Monte Carlo method, where particles are simulated together with its interaction with the detector. With these simulations, the reconstruction algorithms have a supervised set of data to compare the reconstruction output to the real particles impacting the detector.

3. Methods

In this section, the proposal is explained in detail, starting with an explanation of the fundamental principles applied.

3.1. Fundamentals

The current implementation for data reconstruction in the LHCb calorimeter consists of a cellular automaton based clustering [6]. A cellular automaton (CA) [16] is a computational system used to describe the evolution of a discrete system under a set of rules through discrete steps in time. The system is defined as a grid of cells of any dimension where each cell can have a finite number of states. At each step of time, every cell updates simultaneously its own state depending on the state of its neighbor cells following a defined set of rules.

Moving forward to the cluster reconstruction, the algorithm can be segmented into three different steps. The first one consists of a local maxima finder algorithm to identify the potential centers of particle clusters. The second step is the proper cellular automaton, which iteratively tags each of the cells to the closest maximum and enhances those cells that may have contributions from more than one cluster. The final step consists of an iterative algorithm that resolves those particular cases, from now on known as overlapping cells.

Each of the three steps mentioned are indeed iterative algorithms that analyze the grid of calorimeter readout cells using a set of specific rules to give a certain condition in the end. Going further, all steps can be defined as a CA. Given the universality theorem of the CAs, there exists a set of rules and a set of states that can model the behavior of the mentioned algorithms as a dynamic system.

In this article, we propose modeling each of the three steps of the reconstruction process as a CA with an ad-hoc formulation. With this we can benefit from the fact that convolutional neural networks have been proven to learn the generalized rules of cellular automata [17]. Hence, we will train a convolutional neural network with the rules of each of the reconstruction steps to build a pipeline of networks that perform the full reconstruction of the calorimeter.

3.2. Local Maxima Formulation

Starting with the local maxima finder, we want to identify cells that are a local maxima among the others. Hence, its cellular automata characteristics can be defined as follows:

- States: two. One (1) if the cell is a maximum and zero (0) otherwise.

- Neighborhood: eight cells. In order to check if it is a local maxima, the surrounding cells at distance one need to be checked. In a two-dimensional grid of cells, the number of distance one neighbors is eight.
- Ruleset function: in order to define the condition of a cell to be a maximum: (1) its value needs to be higher or equal to its neighbors. Although the equal condition is not obvious, it is needed to represent a specific pair of particles from a π^0 that strike the calorimeter at a very short distance. Therefore Equation (1) defines the ruleset, where $c^t_{i,j}$ stands for the value of each cell at time t and the case $M = 0, N = 0$ is excluded. The initial states of the grid cells concerning $t = 0$ are the values of the calorimeter readout cells.

$$c^{t+1}_{i,j} = \begin{cases} 1, & \text{if } c^t_{i,j} >= c^t_{i+M,j+N} \text{ for } M \in \{-1,0,1\},\ N \in \{-1,0,1\} \\ 0, & \text{otherwise.} \end{cases} \quad (1)$$

By looking at the function, it can be seen that the only operation is a comparison between the value of the central cell and one of its neighbors. Yet this comparison will perform the same way with independence of the values we have to compare. Hence, there is no dependence on the numerical scale value of the cells to the application of this ruleset function.

Therefore, the dataset used for the training of the first network needs to reflect significantly the two possible cases of the ruleset function in any numerical value scale. Hence, each input test sample is generated as a two-dimensional grid of the same size as the calorimeter, with uniformly distributed random values from 0 to 99. This range of values was chosen, taking into account the statistical number of ones that appear on a sample. Regarding the maximum number of local maxima that can fit in a sample, which is one fourth of the total cells, we consider one sixth of the total cells as a good estimation on the number of local maxima that can be on a sample to make sure that some maxima occurrences happen to be adjacent in some cases. The random generation with values from 0 to 99 happen to match these conditions. For reference, the range of values of the calorimeter goes from 0 to 10,240 MeVs. Each expected output test sample is then generated with the application of the ruleset function of the CA on all the cells of the input sample. As a visual example, Figure 2 shows a sample of the input and expected output the network is trained with. Given that we want the network to learn to reproduce the CA ruleset on the calorimeter data, the testing samples are raw data samples of the calorimeter with the corresponding readout cell values. The expected output of the testing samples is obtained again with the application of the CA on the training input samples. The last image (d) on Figure 2 shows the image generated by the trained neural network when it is given image c at the input.

Given that the three regions of the calorimeter have different cell sizes, samples from each region have different shapes. Hence, we will train one network for each region. All of them have the same structure that consists of a two-dimensional convolutional layer followed by two/three dense layers, depending on the region, and finally, an output dense layer of two neurons, since the network is trained as a classifier understanding the output class as the state of a cell in the CA formulation. Table 1 shows a summary of the network parameters and characteristics obtained in the training. To evaluate the network performance, we will use the accuracy metric, since we want to maximize the number of correct classifications. The accuracy is measured as the number of correct classifications over the total number of cells.

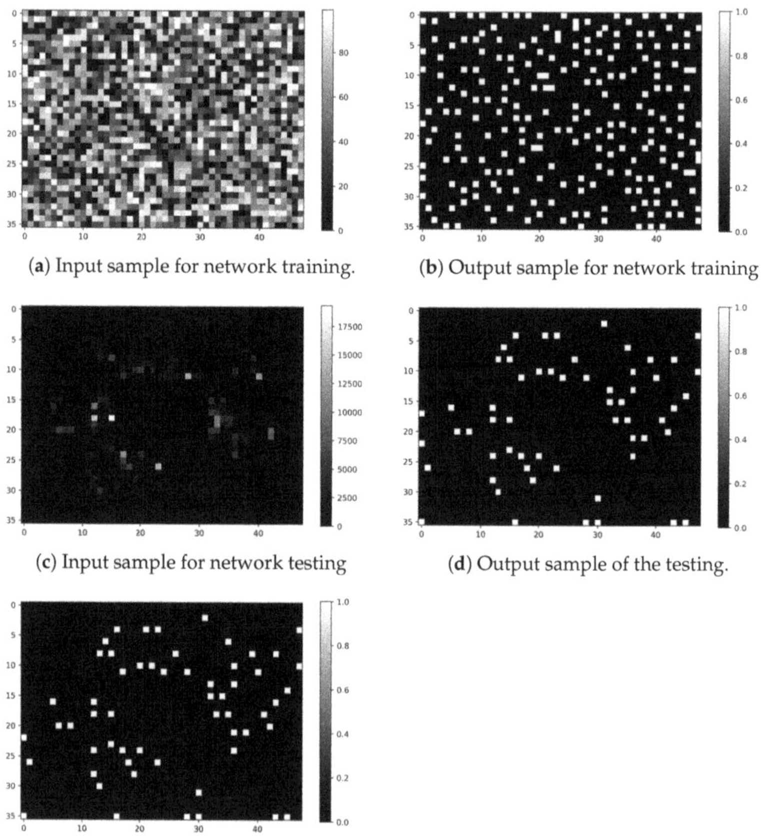

(**a**) Input sample for network training.

(**b**) Output sample for network training

(**c**) Input sample for network testing

(**d**) Output sample of the testing.

(**e**) Output sample of the network.

Figure 2. Samples of the data used for the training and testing of the local maxima finder network for the inner region of the LHCb calorimeter. The training input sample (**a**) is artificially generated and the output sample for training (**b**) is obtained applying the CA rules on the (**a**) sample. The testing input sample (**c**) is an LHCb simulation and the testing output sample (**d**) again generated applying the CA rules on sample (**c**) and represent the expected output from the network. Then image (**e**) is the real output obtained from the network trained to reproduce the CA when sample (**c**) is on the input, and is compared to sample (**d**) to obtain the accuracy value.

Table 1. Parameter summary of the local maxima finder neural networks.

Region	Image Shape	Training Samples	Neurons Per Layer	Parameters	Training Time	Accuracy
Outer	64×52	10,000	[20, 20, 20, 10, 2]	1272	1354.7 s	99.96%
Middle	64×40	10,000	[20, 20, 20, 10, 2]	1272	1052.3 s	99.92%
Inner	48×36	10,000	[10, 10, 10, 2]	342	461.8 s	99.93%

3.3. Clustering Formulation

The following step of the reconstruction process is proper clustering, which indeed was already formulated as a cellular automaton in the classical implementation. In this case, the algorithm needs to identify the cells that belong to a cluster and the overlap cases where a cell can belong to more than one cluster over the rest. Where the rest of the cells can only be a local maximum already identified by the first reconstruction step or an energy deposit located too far from a local maximum to be taken into account. We can formulate the mentioned CA characteristics as follows:

- States: three. Since we need to identify three types of cells: cells that belong to one cluster (1), overlapping cells (−1), and the rest of them (0).
- Neighborhood: eight cells. In order to differentiate cells as overlapping or belonging to a cluster from the others, the neighborhood at one cell distance needs to be checked.
- Ruleset function: in order to identify the cell states, the algorithm needs to check how many local maxima are in the neighborhood of a cell. Cells that belong to a certain cluster are those that have a single local maximum in its neighborhood of distance one. In the same way, overlap cases are identified when there is more than one local maximum in its neighborhood. If there are no local maxima in a cell neighborhood then it is either a local maximum itself or not relevant for the clustering. This is formulated in Equation (2), where K is defined as the number of local maxima in the neighborhood of a cell.

$$c_{i,j}^{t+1} = \begin{cases} 1, & \text{if } K == 1 \\ -1, & \text{if } K > 1 \\ 0, & \text{otherwise.} \end{cases} \quad (2)$$

At the time of designing the formulation of the third step of the reconstruction, we realized that the information provided by this second step was redundant. Since, in order to solve the overlap cases, the number of local maxima also need to be calculated and cannot be transmitted to the third step, we propose formulating the clustering and the overlap solver as a single step in this approach.

3.4. Clustering and Overlap Formulation

For the case of the last step of the reconstruction, the overlap algorithm needs to resolve the cases where a cell belongs to more than one cluster. To do so, the energy of that cell needs to be distributed among the involved clusters, depending on the total energy of each cluster. To be specific, since the energy measured in an overlapping cell may come from the addition of two different particles, one fraction of the overlapping cell energy belongs to one particle and the rest of the fraction to the other particle. Therefore, the desired output for this step is, given an input cell with overlap, the part of the energy designated to each of its contributing clusters. In case the cell does not have overlap, the output needs to be the same energy value of the input.

To correctly perform the distribution of the energy in case of overlap, the algorithm needed to know which cluster is the fraction of energy going to belong. Since there is a maximum of four possible clusters contributing to a cell by construction (see Figure 3), each fraction operation needs to be related to one of the clusters involved in the overlap. Two conclusions can be extracted at this point for the overlap cases:

- The neighborhood window for this step needs to be big enough to identify all the cells belonging to the possible clusters causing overlap. This number corresponds with 24 neighbors inside a 5×5 square window around the given cell, as can be seen in Figure 3.
- Extra information needs to be added to the 5×5 window, indicating the cluster to which the energy fraction is going to be part of. Since each fraction of energy needs to be assigned to a certain cluster, in the overlap cases, the same 5×5 window must be evaluated as many times as the number of clusters involved, but each time selecting a cluster that the fraction is going to contribute. This dominant cluster, within a 5×5 window, will be called central cluster.

In order to provide the central cluster information, before transforming the input image into blocks of 5×5 windows, there is a first transformation into windows of 7×7. Within this 7×7 window, we can identify if the cell located at the centre is a local maximum and generates another stream of data to indicate that, for this particular window, the central cluster is the one in the middle of the window. An example representation can be seen in Figure 3 where the blue cross represents a local maxima and in this case would be marked as a central cluster as well. Regarding the identification of the central cluster, these data are generated with a masking of the local maxima stream of each 7×7 window. Where the mask is a 7×7 matrix of zeros and a single one on the central position where we expect to find the central cluster.

Once we include this third stream of information in a window of 7×7, we can subsample all nine possible 5×5 windows that can fit inside the 7×7. At this time, we have, on a single 5×5 sample, the data regarding the readout cells of the calorimeter, the local maxima information and the central cluster information. It is prepared to generate a prediction of the designated energy partition of the cell located at the centre of the window concerning the central cluster.

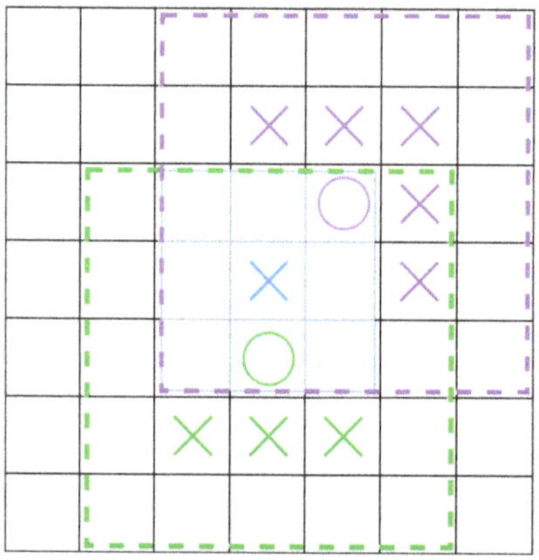

Figure 3. Diagram representing the possible cluster centre positions overlapping with the central cluster in a 7×7 window. The maxima positions are marked with crosses and the two overlap cells that need to be predicted are marked with a circle.

Gathering the previous concepts, the cellular automaton formulation of this step has the following characteristics:

- States: 10,240. Since the algorithm needs to give, as output, a value concerning the energy of a cell, the CA must have enough states to model the full calorimeter sensitivity, which is of 12 bits on the ADC lecture with a gain of 2.5 MeVs per ADC value ($2^{12} \times 2.5$).
- Neighborhood: 24 cells. As explained above, the window around a cell to predict its value needs to be of 5×5 cells.
- Ruleset function: at this point of the reconstruction, we have three streams of information:
  - Original data sample. Obtained from the calorimeter simulation with values from 0 to 10,240.
  - Local maxima information. Obtained from the output of the local maxima finder network. With values from 0 to 1.
  - Central cluster information. Obtained from the masking of the central cell on each 7×7 window from the image. With values from 0 to 1.

  At this point, the ruleset function for this step defines the fractioning of a cell's energy in case of overlap in Equation (3). In the same equation, *num_clusters* refers to the number of local maxima that could be causing overlap. As an example, if we look at the purple circle in Figure 3 to account for *num_clusters*, the positions marked with a purple X should be checked.

$$c_{i,j}^{t+1} = \begin{cases} \frac{c_{i,j}^{t} C_0}{\sum_{k=0}^{K} C_k}, & \text{for } K = num\_clusters \text{ if } K > 1 \\ c_{i,j}^{t}, & \text{otherwise,} \end{cases} \quad (3)$$

where

$$C_k = \sum_{m=-1}^{1} \sum_{n=-1}^{1} c_{o+m,p+n}^{t} \quad (4)$$

and variables o and p stand for the local maxima coordinates of cluster k in the 5×5 image. For $k = 0$ the cluster is specifically the marked as central cluster, the one in the center of the 7×7 window.

Before starting with the network training, there is a key aspect on the ruleset function that affects the learning capacity of a convolutional network, like the ones used on the first step. It can be seen that, for the second condition, there is a division that transforms the function into a non-linear behavior. Following the universal approximation theorem [18], there needs to be at least one hidden layer to approximate non-linear functions. However, the previous used convolutional architectures had in fact two and three hidden layers, yet the convolutional layer itself does not have a hidden layer structure on the convolution operation. The convolution is performed through the multiplication between the data and a linear kernel of parametric values; hence, there is no chance for this convolution to be able to learn the desired non-linear behavior before losing the neighborhood information on the dense layers. Therefore, the strategy for the network architecture is to use a multilayer perceptron (MLP) structure and train it to be the kernel of the convolution.

Following this strategy, one MLP network is enough for the three regions, since the sampling into 5×5 windows normalizes the input beyond the different region granularity. In this case, the network architecture consists of four dense layers and the output layer is a single neuron representing the predicted value of the central input cell. Hence, the network will be trained as a regressor. The output values need to be aggregated in groups of nine concerning the predicted values from a cluster at all cells of the calorimeter. As an example of the data used for the training of this network, see Figure 4.

Regarding the training dataset generation, the same numerical value independence from the first formulation is observed in this ruleset function. However, in this case, the conditions of the function are not so simple to achieve in a homogeneous scenario using randomly generated samples. At this point, we decided to take a set of only 2000 samples of LHCb simulation data and take a selected subset of approximately 30,000 windows of 7×7 centered on a cluster. The selection has taken into account the balancing of the dataset between six different cases specified in Table 2. Case 6 was included to enhance the training of the fraction operation when clusters have a big energy difference between them because on these cases big clusters tend to mask completely smaller clusters on its surroundings. Since case 5 has the lowest number of samples, we chose to increase its samples rotating each window by 90°, 180° and 270°, reaching more than 5000 different samples for that case. Finally, the balanced dataset was constructed, collecting 5000 samples from each of the six cases and subsampling nine windows of 5×5 for each of them, reaching a combined number of 270,000 samples for the training.

As it can be seen in Table 2, there is an RMSE value for each datum case. This gives an overview of the precision of the network as a function of the data complexity. Since case 1 stands for samples in which there are no clusters around, it is expected to have the lowest RMSE value. Once the samples start increasing in number of clusters involved, the RMSE value goes up accordingly. It is also expected to observe an increase in the error with the increment on data complexity. A good "symptom" is to see the case 6 samples, which have increased complexity regarding the energy difference between clusters, showing to have good performance. Even though the RMSE values are considerably low inside the full calorimeter range, it must be stated that the average energy value seen in the dataset used is of 1202.64 MeVs. With respect to that value, the maximum RMSE obtained from group 5 represents 20.3%. However, it must be taken into account that the RMSE metric is sensitive to out layers.

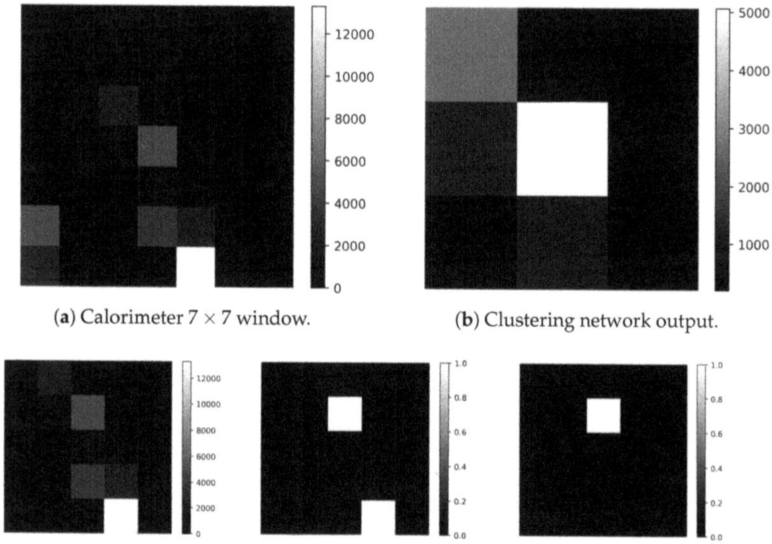

(**a**) Calorimeter 7 × 7 window. (**b**) Clustering network output.

(**c**) Calorimeter input. (**d**) Local maxima input. (**e**) Central cluster input.

Figure 4. Samples of the data used for the training of the clustering and overlap network. The three streams of data (**c**–**e**) are extracted from the analysis of a 7 × 7 window (**a**). Image (**b**) shows the output predicted by the network representing the reconstructed cluster located at the centre of the (**a**) sample.

Table 2. Case characteristics in the balanced dataset for the MLP training. Case 6 is a selection of samples with overlap with at least one cluster, but where the energy difference between clusters is bigger than an order of magnitude. The RMSE values were extracted, comparing the network predicted values and the samples generated with the application of the CA rule.

Case	Number of Clusters (K)	Overlap with Central Cell	Samples on 2k Events	RMSE
1	0	No	153,519	96.281
2	1	No	121,316	135.616
3	2	Yes (1 cluster)	45,066	147.501
4	3	Yes (2 clusters)	9937	199.644
5	4+	Yes (3+ clusters)	1367	244.312
6	>1	Yes (energy difference of 1 order of magnitude)	6816	181.693

A summary of the network parameters for this reconstruction step is provided in Table 3. Given the regressive nature of this network, results, in terms of training performance, are measured with the RMSE metric, comparing all the values predicted by the network to the results of applying the CA rules to the training input samples. Considering the reference of the mean energy value seen in the training samples, the RMSE of the network represents 14% of the average energy.

Table 3. Parameter summary of the cluster and overlap neural network.

Region	Image Shape	Training Samples	Neurons Per Layer	Parameters	Training Time	RMSE
All	5×5	270,000	[64, 64, 64, 32]	108,993	2130.4 s	168.884

Aiming to provide a detailed overview of the structured proposal, gathering the relations between the neural networks and the data, Figure 5 shows the entire dataflow of the reconstruction process for the inner calorimeter region. The diagram starts with the list of energy cells transformed into an image and ends with the list of reconstructed clusters. For the other two regions, the structure is maintained, but the shapes of the constructed images adapt to each region size.

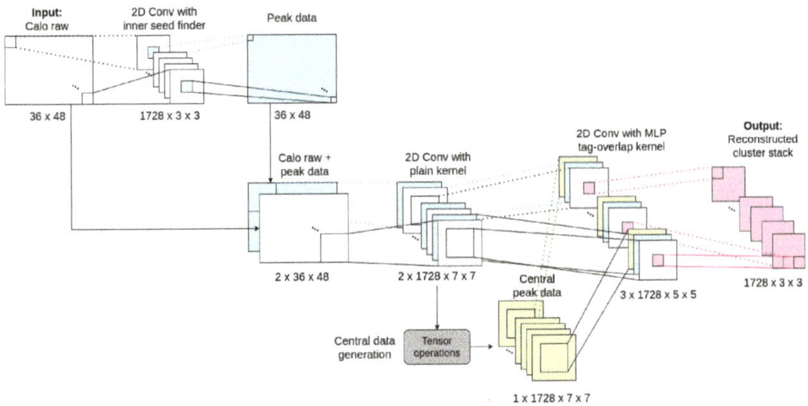

Figure 5. Detailed scheme of the proposed reconstruction dataflow for the inner calorimeter region.

4. Results

The results provided in this paper were obtained operating on a computer with the following properties: memory of 15.5 GB, processor Intel Core i7-6500U CPU @ 2.50 GHz ×4, disk capacity of 512.1 GB with Ubuntu 20.04.1 LTS 64-bit OS. The algorithms are coded in Python 3.8 and the explained neural networks have been built and trained using the TensorFlow 2.3.0 library.

Aiming to make a fair comparison, in terms of computational performance between the proposed reconstruction algorithm and the original implementation of LHCb in a local environment, a version of the original calorimeter reconstruction implemented in LHCb was replicated in Python with the same computational complexity, referred to as the Python version of LHCb algorithm.

To make sure the comparison between both algorithms was fair, we defined a metric of efficiency on the reconstruction as relative error. This relative error calculated the difference of energy between reconstructed clusters and the true clusters reconstructed by the original LHCb reconstruction algorithm implemented in the LHCb framework. Using the relative error metric in the reconstructed clusters from the proposed deep learning algorithm gives as a result the first entry in Table 4. In the same table, we find the relative error for the reconstructed clusters obtained with the Python version of the LHCb algorithm. The values shown, concerning the two compared algorithms in relative error, were obtained as the mean values from over 200 simulation samples from the full LHCb calorimeter. It can be observed that the proposed deep learning approach shows, in general, a lower relative error value than the original version. Although this needs to be studied in detail with further experimentation, these results give us a hint that the two algorithms behave as expected and further comparisons will be fair.

Table 4. Results concerning the mean value and standard deviation of the relative error measured as the difference of energy reconstructed per cluster from a total of 200 simulated events.

Algorithms	Mean of Relative Error	STD of Relative Error
Deep Learning	0.056	0.105
Python version of LHCb algorithm	0.079	0.159

Results, in terms of computational performance, are measured as the time in seconds elapsed between the reading of the readout cell values, and the generation of the list of clusters for the three regions of the calorimeter. The execution of both methods was done using a single thread in each case. Figure 6 shows a plot of the computational time for the number of energy cells (hits) on a single sample of a calorimeter simulation (event). The time plotted is measured as a mean of one hundred iterations on the same event for 200 different events. Looking at the curve from the Python version of the LHCb algorithm (iterative), it performs really fast in events with a low number of energy cells. However, it shows a clear quadratic growth with the number of cells activated. On the other hand, the deep learning approach (DL total) shows nearly constant behavior towards the number of cells. Although it has a small positive slope of 4.97×10^{-6}, the tendency shows to be linear. Even so, around 72% of the events processed in the testing have less than 2575 energy cells and, therefore, stay under the time performance curve of the deep learning approach. Even so, we achieved a constant computational time with independence of the event complexity.

In addition, as stated in Section 3.2, for the first step of the proposed reconstruction process, the information from the three regions of the calorimeter needs to be treated separately. Given that the reconstruction process is the same for each region, excepting the ad-hoc local maxima neural network, another way of accelerating the execution time is executing the reconstruction of the three regions in parallel. To approximate the behavior of such execution, Figure 6 shows the execution time measured by each of the three region reconstructions independently (DL inner, DL middle, and DL outer). It is observed that, in this parallel condition, the maximum time is achieved by the outer reconstruction, since it has the highest number of readout cells. Although more studies should be made in this direction, the proposed deep learning algorithm shows that it benefits from a parallel execution.

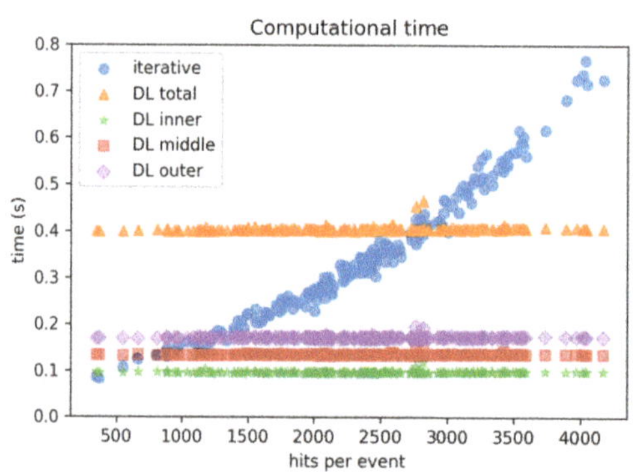

Figure 6. Scatter plot of the mean computational time over the number of readout cells per event from LHCb simulations. Comparing the Python version of the LHCb algorithm (iterative) and the proposed deep learning implementation (DL total) with executions segmented by regions (DL inner, DL middle, DL outer). Hits refer to the readout cells with energy on a sample.

5. Discussion

Within the development of this proposal, there are several things that have been learned. We have seen that, for the specific problem of calorimeter reconstruction in LHCb, segmenting the reconstruction steps can help in simplifying the development of a deep learning solution. Moreover, as seen in Section 3.2, data can be artificially generated as long as they correctly equally represent all cases to be learned. For more complex functions, such as the one seen in Section 3.4, the understanding of the rules also leads to a simplification of the dataset, where we were able to extract thousands of samples from only 2000 full LHCb simulated events. Understanding that there is no need to work with full simulation data to train specific networks can simplify the training dataset generation on further deep learning developments for the calorimeter reconstruction. In other words, we trained neural networks on the rules that solve a general formulation of the problem. It was proven that the network learns the application of the formulated rules in a generalized context. The complexity reduction on the training data was also reflected into a fast training process and the simplification of the networks, in terms of architecture and the number of parameters, compared with previous deep learning approaches.

Comparing the results with the state-of-the-art, we improved the relation between the network's complexity and the amount of training data. Furthermore, the proposed model is validated by construction, since the same reconstruction steps as the current method are being reproduced.

As a proof of concept, the performance comparison in this paper is done with the current implementation of the reconstruction self-implemented in Python. In terms of computational time, there is a clear gain in the reconstruction complexity with the proposed approach. However, the execution time could possibly be reduced with a vectorized implementation of the proposal. Apart from that, the proposed implementation clearly benefits from parallel execution, reducing the computational time by nearly a factor three. Moreover, its convolutional formulation could benefit even more from a GPU architecture without conditioning the efficiency, as the insight neural networks and convolutional operators are highly parallelizable.

Despite the results, there is still room for improvement in terms of performance. Due to the region-independent strategy used in this approach; clusters that fall in the boundary regions of the calorimeter are now reconstructed partially as two separate clusters in each region. There is the idea of using a graph neural network (GNN) with similar training as the MLP, in order to perform the reconstruction in the boundary regions, as GNNs can model irregular neighborhoods. Another aspect that needs to be worked is the identification of π^0 particles. By nature, the two photons of the decay of energetic π^0 particles arrive at the calorimeter as two very close similar energy particles, but need to be reconstructed as a single cluster. Hence, the window that surrounds a pair of photons is bigger than the defined 3×3. With the current training of the MLP in our proposal, the network wrongly reconstructs these specific photons as two very close overlapping clusters. There is the idea of improving this shortcoming, re-training the network of the current implementation to identify the defined photons and make an ad-hoc reconstruction for those cases.

In conclusion, we implemented and tested an alternative approach to the LHCb calorimeter reconstruction. It adapts the current reconstruction steps to a formulation that can be learned by simple deep learning structures. With this, we make sure that the reconstruction process is correct as it mimics the implementation of the current algorithm, gaining in computational complexity. Results from the testing show interesting behavior, in terms of computational time, which could be promising for a full calorimeter reconstruction implementation on GPUs.

Author Contributions: Conceptualization, N.V.C., M.C.G., E.G.R. and X.V.-C.; methodology, N.V.C., M.C.G., E.G.R. and X.V.-C.; software, N.V.C.; validation, N.V.C., M.C.G., E.G.R., and X.V.-C.; formal analysis, N.V.C., M.C.G., E.G.R. and X.V.-C.; investigation, N.V.C., M.C.G., E.G.R. and X.V.-C.; resources, N.V.C., M.C.G., E.G.R. and X.V.-C.; data curation, N.V.C., M.C.G., E.G.R. and X.V.-C.; writing—original draft preparation, N.V.C., M.C.G., E.G.R. and X.V.-C.; writing—review and editing,

N.V.C., M.C.G., E.G.R. and X.V.-C.; visualization, N.V.C., M.C.G., E.G.R. and X.V.-C.; supervision, M.C.G., E.G.R. and X.V.-C.; project administration, M.C.G., E.G.R. and X.V.-C.; funding acquisition, M.C.G., E.G.R. and X.V.-C. All authors have read and agreed to the published version of the manuscript.

Funding: This research was funded by Ministerio de Ciencia e Innovación grant number PID2019-106448GB-C32.

Institutional Review Board Statement: Not applicable.

Informed Consent Statement: Not applicable.

Data Availability Statement: No new data were created or analyzed in this study. Data sharing is not applicable to this article.

Conflicts of Interest: The authors declare no conflict of interest.

References

1. Alves, A.A., Jr.; Andrade Filho, L.; Barbosa, A.; Bediaga, I.; Cernicchiaro, G.; Guerrer, G.; Lima, H., Jr.; Machado, A.; Magnin, J.; Marujo, F.; et al. The LHCb detector at the LHC. *J. Instrum.* **2008**, *3*, S08005.
2. Evans, L.; Bryant, P. LHC Machine. *JINST* **2008**, *3*, S08001. [CrossRef]
3. Bediaga, I.; Torres, M.C.; De Miranda, J.; Gomes, A.; Massafferri, A.; Rodriguez, J.M.; dos Reis, A.; Aoude, R.; Amato, S.; Akiba, K.C.; et al. Physics case for an LHCb Upgrade II-Opportunities in flavour physics, and beyond, in the HL-LHC era. *arXiv* **2018**, arXiv:1808.08865.
4. LHCb Collaboration. *Throughput and Resource Usage of the LHCb Upgrade HLT*; Technical Report; LHCB-FIGURE-2020-007; CERN: Geneva, Switzerland, 2020.
5. Omelaenko, O.; Dalpiaz, P.; Guzik, Z.; Spiridenkov, E.; Jarron, P.; Semenov, V.; Ocariz, J.; Khan, A.; Perret, P.; Schneider, O.; et al. *LHCb Calorimeters: Technical Design Report*; Technical Report; LHCb-TDR-002; CERN: Geneva, Switzerland, 2000.
6. Breton, V.; Brun, N.; Perret, P. *A Clustering Algorithm for the LHCb Electromagnetic Calorimeter Using a Cellular Automaton*; Technical Report; CERN-LHCb-2001-123; CERN: Geneva, Switzerland, 2001.
7. Breton, V.; Fonvieille, H.; Grenier, P.; Guicheney, C.; Jousset, J.; Roblin, Y.; Tamin, F. Application of neural networks and cellular automata to interpretation of calorimeter data. *Nucl. Instrum. Methods Phys. Res. Sect. A Accel. Spectrometers Detect. Assoc. Equip.* **1995**, *362*, 478–486. [CrossRef]
8. Casolino, M.; Picozza, P. A cellular automaton to filter events in a high energy physics discrete calorimeter. *Nucl. Instrum. Methods Phys. Res. Sect. A Accel. Spectrometers Detect. Assoc. Equip.* **1995**, *364*, 516–523. [CrossRef]
9. Denby, B. Neural networks and cellular automata in experimental high energy physics. *Comput. Phys. Commun.* **1988**, *49*, 429–448. [CrossRef]
10. Baldanza, C.; Bisi, F.; Bruschi, M.; D'Antone, I.; Meneghini, S.; Rizzi, M.; Zuffa, M. A cellular neural network for peak finding in high-energy physics. In Proceedings of the 2000 6th IEEE International Workshop on Cellular Neural Networks and their Applications (CNNA 2000) (Cat. No. 00TH8509), Catania, Italy, 25 May 2000; pp. 443–448.
11. Mazurek, M. Deep Learning Solutions for 2D Calorimetric Cluster Reconstruction at LHCb. In Proceedings of the 4th Inter-Experiment Machine Learning Workshop, Zürich, Switzerland, 19–23 October 2020.
12. Redmon, J.; Farhadi, A. Yolov3: An incremental improvement. *arXiv* **2018**, arXiv:1804.02767.
13. Canudas, N.V.; Cardona, X.V.; Gómez, M.C.; Ribé, E.G. Deep Learning approach to LHCb Calorimeter reconstruction using a Cellular Automaton. *EPJ Web Conf. EDP Sci.* **2021**, *251*, 04008. [CrossRef]
14. Abba, A.; Caponio, F.; Cusimano, A.; Geraci, A.; LHCb Collaboration. *LHCb Particle Identification Upgrade: Technical Design Report*; CERN: Geneva, Switzerland, 2013.
15. Bediaga, I.; Chanal, H.; Hopchev, P.; Cadeddu, S.; Stoica, S.; Calvo Gomez, M.; T'Jampens, S.; Machikhiliyan, I.V.; Guzik, Z.; Alves, A.A., Jr.; et al. *Framework TDR for the LHCb Upgrade: Technical Design Report*; Technical Report; LHCb-TDR-012; CERN: Geneva, Switzerland, 2012.
16. Neumann, J.; Burks, A.W. *Theory of Self-Reproducing Automata*; University of Illinois Press: Urbana, IL, USA, 1966; Volume 1102024.
17. Gilpin, W. Cellular automata as convolutional neural networks. *Phys. Rev. E* **2019**, *100*, 032402. [CrossRef] [PubMed]
18. Cybenko, G. Approximation by superpositions of a sigmoidal function. *Math. Control. Signals Syst.* **1989**, *2*, 303–314. [CrossRef]

applied sciences

MDPI

Article

Recognition of the Mental Workloads of Pilots in the Cockpit Using EEG Signals †

Aura Hernández-Sabaté [1,2,*], José Yauri [1,2], Pau Folch [3,4], Miquel Àngel Piera [4] and Debora Gil [1,2]

1 Computer Vision Center, Universitat Autònoma de Barcelona, Bellaterra, 08193 Barcelona, Spain; jyauri@cvc.uab.cat (J.Y.); debora@cvc.uab.cat (D.G.)
2 Departament de Ciències de la Computació, Universitat Autònoma de Barcelona, Bellaterra, 08193 Barcelona, Spain
3 Aslogic, Parc de Recerca UAB, Bellaterra, 08193 Barcelona, Spain; pau.folch@uab.cat
4 Telecommunications and Systems Engineering Department, Universitat Autònoma de Barcelona, Bellaterra, 08193 Barcelona, Spain; miquelangel.piera@uab.cat
* Correspondence: aura@cvc.uab.cat

† This paper is an extended version of our paper published in 23rd International Conference of the Catalan Association for Artificial Intelligence, Lleida, Spain, 20–22 October 2021.

Abstract: The commercial flightdeck is a naturally multi-tasking work environment, one in which interruptions are frequent come in various forms, contributing in many cases to aviation incident reports. Automatic characterization of pilots' workloads is essential to preventing these kind of incidents. In addition, minimizing the physiological sensor network as much as possible remains both a challenge and a requirement. Electroencephalogram (EEG) signals have shown high correlations with specific cognitive and mental states, such as workload. However, there is not enough evidence in the literature to validate how well models generalize in cases of new subjects performing tasks with workloads similar to the ones included during the model's training. In this paper, we propose a convolutional neural network to classify EEG features across different mental workloads in a continuous performance task test that partly measures working memory and working memory capacity. Our model is valid at the general population level and it is able to transfer task learning to pilot mental workload recognition in a simulated operational environment.

Keywords: cognitive states; mental workload; EEG analysis; neural networks; multimodal data fusion

Citation: Hernández-Sabaté, A.; Yauri, J.; Folch, P.; Piera, M.À.; Gil, D. Recognition of the Mental Workloads of Pilots in the Cockpit Using EEG Signals. *Appl. Sci.* **2022**, *12*, 2298. https://doi.org/10.3390/app12052298

Academic Editors: Aida Valls and Karina Gibert

Received: 22 December 2021
Accepted: 6 February 2022
Published: 22 February 2022

1. Introduction

A fundamental aspect of multiple task management is to attend to new stimuli and integrate associated task requirements into an ongoing task set—that is, to engage in interruption management [1]. Interruptions often negatively affect human performance. Specifically, most laboratory and applied experiments demonstrate that interruptions increase post-interruption performance times [2] and error rates [3], increase perceived workload [4], and motivate compensatory behavior [5].

The commercial flightdeck is a naturally multi-tasking work environment, one in which interruptions are frequent and of various forms. Further, interruptions have been cited as a contributing factor in many aviation incident reports. External and aircraft events, and interactions with other operators, compete for pilots' attention and require pilots to integrate performance requirements associated with these unexpected prompts with ongoing flightdeck tasks.

For that, the study of workload is essential to prevent accidents, since it could compromise human task performance [6]. Since workload involves cognitive, neuro-physiologic, and perceptual processes to resolve a task, it is affected by individual capabilities, motivation, and physical and emotional state [7]. Although this multifaceted nature of workload prevents one from studying workload directly, it is feasible to infer it from various quantifiable variables [8]. There exist many proposals for recognizing workload based on physiological features, such as hearth rate, eye movement and dilation, electroencephalogram (EEG), and electrocardiogram (ECG) [9,10]. The recent emergence of low cost EEG headsets has driven new researches (such as interaction with home devices, teaching-learning educative methods, and mentally control robotic arms) further than the medical screening of neurological disorders. In the particular case of cognitive state assessment, EEG alone is becoming the preferred sensor for addressing its characterization [11–13]. However, there is not enough evidence in the literature to validate how well models generalize to new subjects performing tasks of a workload similar to the ones included during the model's training.

The goal of this study was to characterize the mental workloads of airplane pilots in the cockpit from the analysis of EEG signals.

The remainder of this paper is organized as follows: Section 2 summarizes the state-of-the-art related work. Section 3 details the ground truth generation. Section 4 explains the models used to recognize the different levels of workload. Section 5 presents the experimental design. Section 6 is devoted to the experimental results. Finally, Section 7 outlines the conclusions and future work.

2. Related Work

The most generalized mechanisms to measure workload can be split into two main categories [9,14,15]: subjective measures based on the subject perception and objective scores based on physiological responses.

On the one hand, subjective measures are still the most used to assess mental workload, the NASA Task Load Index (TLX) [16] being the most prominent test used to gain insights about the perceived workload levels while a subject works with various human–machine interface systems [17,18]. This questionnaire measures the mental workload based on a weighted average of six sub-variables: mental demand, physical demand, temporal demand, performance, effort, and frustration. It is widely used in aviation to assess mental workload of pilots while interacting with plane controls [19,20].

On the other hand, physiological measures provide more reliable data of workload by measuring physiological dynamic changes which cannot be controlled consciously, so they have been becoming more popular among researchers in recent years [21–23]. The most common sensors/measurements used to record physiological data are: electrocardiogram (ECG) to register heart electrical activity, electromyograph to read skeletal muscles' electrical activity, electroencephalogram (EEG) to detect electrical activity in the brain, photoplethysmography to register volumetric changes in the blood flow, respiration rate sensors, electro-dermal activity (EDA) to read skin surface temperature, oxygen density in the blood in the brain, and eye movement trackers, among others [24]. TLX surveys allow one to assess the perceived workload [16], but it is highly subjective. However, physiological data occur spontaneously, and together with TLXs, provide more reliable information [9,17,21].

The combination of several physiological sensors to classify workload states gives better results than using a single one. The approach proposed in [25] combines EEG, ECG, and electrooculography (EOG); and results show the best predictive power for their combination (80%) rather than the analysis of each one independently (70%). In addition, the study in [10] reports an accuracy average of 85.2 (±4.3%) combining EEG, ECG, respiration rate, and EDA to classify four mental states. The work in [26] still shows better results combining EEG, ECG, and EDA than using only EEG signals from classifying four mental states, although results from the single sensor are promising (86.66%).

Deep learning (DL) approaches are gaining ground over more classical machine learning techniques due to their ability to automatically extract features [24,27,28].

The application fields fall into five general groups: emotion recognition [29], motor imagery [30], seizure detection [31], sleep scoring [32], and mental workload. Saadati et al. [33] combined functional Near Infrared Spectroscopy (fNIRS) and adapted a CNN architecture to allow fNIRS-EEG input to the CNN with promising results (89/5 of correct classification). The study in [6] proposed a concatenated structure of deep recurrent and 3D convolutional neural networks to combine both raw and spectral EEG data and assess two degrees of mental workload, reporting an average accuracy of 88.9% in a cross-task assessment. In the same fashion, Kwak et al. [34] proposed a LSTM based temporal attention technique to simultaneously extract EEG features containing both local and global structure information, obtaining an accuracy of 90.8% on their own dataset. Chakladar et al. presented a new framework using the grey wolf optimizer algorithm and deep BLSTM-LSTM neural model for estimating different levels of mental workload, achieving 86.33% and 82.57% classification accuracy for "No task" and "multitasking activity" experiments, respectively. None of them transfer learning to another type of task.

Contributions

AI methods characterizing WL from EEG signals must face several challenges. First, in order to properly be trained and tested, it is mandatory to have data with unambiguous annotations (known as ground truth, GT). The collection of this annotated data is complex because the concept of WL itself is multifaceted and difficult to determine in an objective, systematic manner. Second, for optimal performance of the system, it should properly combine the signals recorded from the different EEG electrodes. Finally, a main issue that a machine learning (ML) system involving humans should consider is its generalization power, including reproducibility of results and the capability of transfer learning—that is, to what extent a general model trained over a set of individuals can successfully predict a new unseen individual performing a different task than the ones used for training the system [35].

This work contributes to the three challenges as follows:

1. Unambiguous Annotated Dataset. In order to generate data with unambiguous annotation, we have designed serious games and flight scenarios in an A320 simulator. The serious game was a modified n-back-test [36] with increasing memory demand. The level of difficulty of the test is our GT for training models. Such level of difficulty was cross-checked with the difficulty perceived by the player assessed using NASA TLX questionnaire. Models were trained using n-back-test data recorded from a population that did not include pilots. The task and population transfer of systems were validated in cockpit simulation exercises designed to have different levels of complexity, and unexpected unsaved situations known to substantially drop pilots' performance.
2. Models able to recognize two levels of workload with high generalization capability. Two different architectures are proposed for the fusion of EEG sensor signals (channels) at two different levels [37]: input data (labeled input projector model) and convolutional feature (labeled feature projector model) models. Both architectures consist of an input unit managing fusion at the input level, a convolutional unit, and an output unit for fusion of convolutional features. For each architecture, several classification problems (including an increasing number of WL classes) were trained on n-back-test data using a one-subject-out scheme and tested in binary problem for detection of WL on flight simulations.

The results show that between the two models, projecting convolutional feature channels achieved higher performance, with 76.25% sensitivity and 87.81% specificity in WL detection in n-back-test leave-one-out subject evaluation, and good task transfer with the detected WL increasing with the number of interruptions.

3. Data Annotation and Ground Truth Generation

In this paper, we provide two different automatically annotated datasets that served to train, validate, and verify the learning and population transfer of models. The first dataset was recorded from a group of non-pilot subjects playing a memory demanding, serious game with an increasing WL. The second dataset was recorded from pilots flying scenarios of different complexity on an A320 flight simulator.

3.1. Dual N-Back Test

N-Back-tests are memory demanding games requiring the resolution of tasks according to a stimulus presented N trials before. We used three variants of the n-back-tests to induce low, medium, and high mental workloads:

1. Position 1-back for low workload. A square appears every few seconds in one of eight different positions on a regular grid over the screen. Players must press a keyboard key when the position of the square on the current screen is the same as the square of the previous grid.
2. Arithmetic 1-back for medium workload. An integer between 0 and 9 appears every few seconds on the screen while an audio message says an arithmetic operation (plus, minus, multiply, or divide). Players have to solve this operation using the current number and number that appeared prior.
3. Dual arithmetic 2-back for high workload. This test combines the two previous ones. An integer between 0 and 9 appears every few seconds in one of eight different positions on a regular grid. At the same time, for each number that appears on screen, an operator is presented with an audio message. As before, players have to solve this operation using the current number and the number that appeared two instances before. In addition, players have to press a key if the position of the current number is the same as the position of the number shown two screens before.

The neurophysiological response of a subject against mentally demanding tasks depends on his baseline state, which is prone to vary across time. In order to account for differences in the baseline states of subjects, prior to the n-back-tests, participants watched a relaxing video for 10 min. For each experiment (1—low, 2—medium, and 3—high workload), we had a video watching stage, a baseline phase, and the n-back-test, the workload phase. Thus, we call BL1, BL2, and BL3 the baseline phases of the experiments; and WL1, WL2, and WL3 are devoted to the workload phases of the experiments.

After the game, participants answered a TLX questionnaire to collect their subjective perceptions of game difficulty and workload. Results presented in [38] showed that the level of difficulty of the games was correlated to the performance of players and also to the subjective perception of WL computed using NASA-TLX questionnaire.

A total of 20 subjects participated in the experiment. Subjects were adults between 20 and 60; all of them were healthy without any condition that might have caused an imbalance in the data recorded. The sequence of tasks was randomly assigned to subjects, and recording of each session was on different days and hours.

3.2. Flight Simulations

The experiments were designed considering the importance of collecting experimental data that could be useful to quantifying the impact of a task load increment to pilots through operational interruptions by an air traffic controller (ATC), cabin crew (TCP), and electronic centralized aircraft monitor (ECAM) warnings, in order to assess to what extent the system presented to discriminate between low and high workload can be transferred to a more complex environment.

Four scenarios with different levels of complexity were designed, all of them assuming pilot monitoring (PM) incapacitation in order to check how interruptions can overload pilot flying (PF).

- Flight 1. It is based in a nominal standard flight. This experiment is used to take reference parameters. Thus, nominal flight without considering any interrupting event from abnormal procedures due to system failure nor ATC vectoring instructions. In this scenario, ATC provides a minimum number of instructions which the pilots are used to. This scenario the lowest complexity and is considered as the BL class.
- Flight 2. It also relies on the approach phase and it is modified from the nominal scenario, by three different interruptions which increase the PF workload. This scenario has an overall high WL demand.
- Flight 3. This scenario is based on the previous experiment with similar interruptions, but they are slightly advanced or delayed to times at which the PF workload is low and the pilot can attend the interruption without a negative performance impact. Given that interruptions were issued at the most appropriate times, this scenario has a lower level of WL demand than Flight 2.
- Flight 4. This last scenario is based on the previous experiment with the same interruptions, but they are fired at times in which PF is attending to concurrent actions, considerably increasing the workload and impacting the PF performance. This scenario has a similar or greater WL than Flight 2.

The functional resonance analysis method (FRAM) [39] is an agent based modeling framework to identify those factors that affect the performance of the pilot in cockpit functionalities considering different socio-technical operational conditions. According to this agent, the impact of an interruption on the PF workflow depends largely on the time at which the interruption occurs. Consequently, FRAM provides a reliable measure of the workload that will be faced by the pilot, and thus, it was used to design simulation scenarios with interruptions triggered at times when the pilot had a low and high WL peaks, and thus to provide realistic flying situations of controlled difficulty. In addition, FRAM output (both, number of tasks and its complexity) was used to assess the ability of ML models to detect WL peaks associated with highly demanding tasks. In this case a single pilot flew the 4 scenarios.

Figure 1 illustrates a volunteer during a session for the dual n-back test task (a) and a pilot during a simulated flight session (b).

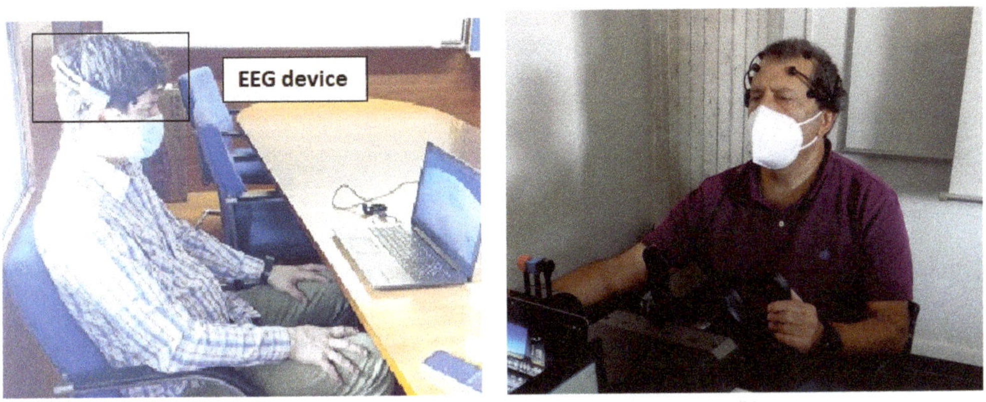

(a) (b)

Figure 1. Data collection with Emotiv Epoc+ headset. (**a**) A volunteer during a n-back-test and (**b**) a pilot during simulated flight session.

4. Workload Recognition

In this section, we present our models, able to recognize between two levels of workload. Each method consists of two stages: First, raw input data are extracted from EEG signals and preprocessed to obtain the proper input data. Later, these signals are fed into the network model to automatically extract the features that will be further combined in a classifier step to discriminate among the number of classes previously determined (baseline vs. workload in our case).

4.1. Extracting Input Data from EEG Signals

For EEG recording, an EMOTIV EPOC+ headset [40] was used, which has 14 electrodes placed according to the 10/20 system. This sensor provides both raw data and power spectra for the main brain frequencies (θ, α, β_{low}, β_{high}, and γ). Given that proposed n-back tasks are memory demanding stressing games and baseline phases consist in watching a relaxing video, the theta wave [41] is the best candidate for discriminating the different mental loads of our experimental phases. In this work, we used the power spectrum of theta wave (4–8 Hz) sampled at 8 Hz.

Eye blinking and sudden head movements introduce abrupt sharp peaks of large amplitude in the power spectra wave that should be filtered before using them as predictors of a mental state [21]. In particular, we used an interquartile range (IQR) [42] filtering strategy to detect outlier values associated with muscular movement wave peaks. Our IQR filtering was based on setting the value of the 99% percentile of the distribution to all points above it.

To ensure a high quality of signals, we further filter data according to the quality of the EEG during recordings provided by the headset itself. For each sensor and recorded sample, Emotiv reports the quality of the recording in a discrete scale with values in the range 0–4 indicating how good the contact between sensor and head is: 4 for optimal—0 for none. For the sake of data with the highest possible quality while keeping a reasonable sample size, signals with a 25% of bad recordings were discarded (<3). Further, since there is no evidence about what are the most discriminating sensors that best correlate with the detection of mental workload, the whole phase was discarded if the signals of two or more of the sensors were low quality. Finally, a subject was discarded if either all its base line or its workload phases were discarded, since, in this case, there were not enough data to define the binary classification. After this quality filtering, only 16 of the 20 subjects were selected for models training and testing.

In order to feed data to models, θ signals were cut in temporal windows. Notice that the size and overlap of the temporal windows might be a critical issue in order to properly include workload peaks [43]. For that we have used several window widths with different overlaps, obtaining the best results with 40 s windows overlapped 30 s. Thus, the input data of the networks were the concatenations of 40 s windows for the 14 EEG sensors ($14 \times 40 = 560$-dimensional feature space). In order to account for the difference in units and magnitudes, input data were standardized using the mean and standard deviation of the training set.

4.2. Network Architectures

The spatio-temporal representation of EEG signals is an issue that any classification ML system has to face. The simplest question is when to combine the signals: before or after extracting features? As Figures 2 and 3 show, we propose two architectures that differ in the moment when EEG sensor signals (channels) are projected: one projects input EEG sensors (input projector model) and the other one projects the convolutional features extracted from each EEG sensor (feature projector model). Each model has one input unit projecting EEG channels (if applicable), a convolutional unit equal for both models, and an output unit projecting the convolutional features extracted from each EEG sensor (if applicable). This output unit has a fully connected layer with sigmoid activation and output the number of classes. To account for different window lengths, we apply an average pooling before the classification layer. All convolutional layers use kernels of size 3 and stride 1 and have Relu activation.

The convolutional unit has 3 blocks consisting of one convolutional layer with max pooling and having 16, 32, and 64 neurons for each convolutional layer, respectively. The classification layer has 256 neurons. For the input projector model, the projection unit has one convolutional layer with 16 neurons. For the feature projector model, the output unit has 2 blocks consisting of one convolutional layer before the classification layer. The first one has 64 neurons, the second one projects convolutional features also using 64 neurons.

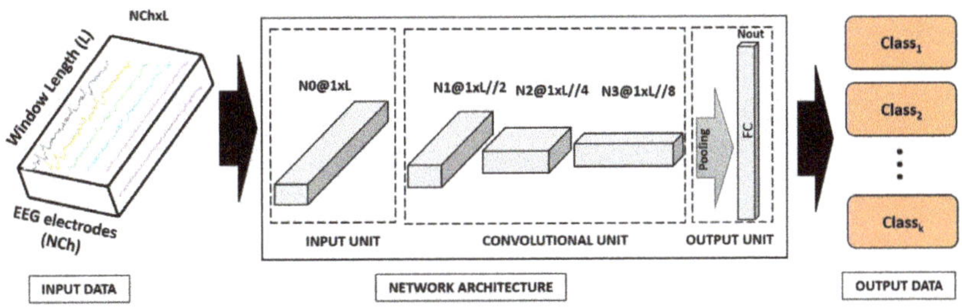

Figure 2. Architecture of the input projector model.

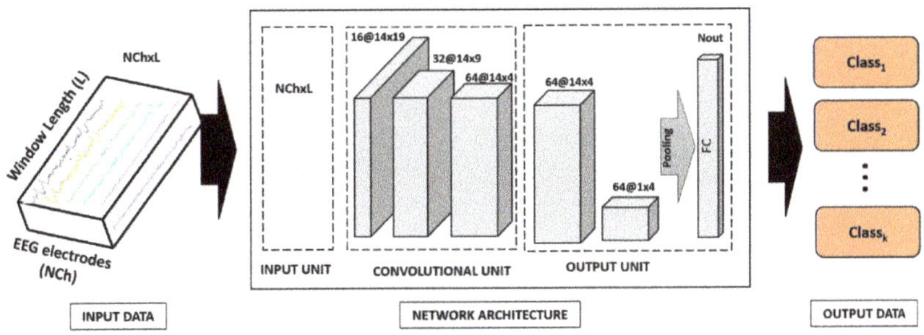

Figure 3. Architecture of the feature projector model.

Although our main problem is a binary one, to ensure generalization capabilities of the classifier (including task transfer), we increased the diversity of the classifier by increasing the number of classes used to train the network. That is, our architecture was trained as a classifier to discriminate between a BL and WL classes using 4 different grouping of the data recorded from the 3 n-back tests:

1. Binary problem (noted BLs-WL2) given by BL = (BL1, BL2, BL3) and WL2. That is, the BL class is defined by aggregating the baselines for the 3 games and WL class defined by the workload phase of the second experiment.
2. Three class problem 1 (noted BLs-WL2-WL3) given by BL = (BL1, BL2, BL3), WL2 and WL3. That is, a BL class defined as before and two WL classes given by the workload phase of the second and third experiments.
3. Three class problem 2 (noted WL1-WL2-WL3) given by WL1, WL2 and WL3. That is, a BL class defined by the workload phase of the first experiment and two WL classes given by the phase 2 of the second and third experiments.
4. Four class problem (noted BLs-WL1-WL2-WL3) given by BL = (BL1, BL2, BL3), WL1, WL2 and WL3. That is, a BL class defined as in the first configuration and also defined by the workload phase of the first experiment and two WL classes given by the workload phase of the second and third experiments.

Unlike binary problems, in multiclass settings, the classifier does not predict the probability of belonging to each class. It rather gives a score of belongingness. It follows that the class predicted is not the one having a score above 0.5 (as is the case in binary problems), but the one having the largest value of the score predicted by the classifier.

In our case, since the final class prediction is binary, we compute the binary class labels in the multiclass settings by binarizing first the output probabilities and then taking the maximum between the two as the final class label. The transformation between classifier output and BL-WL classes scores is as follows:

1. BLs-WL2-WL3: The probability of BL is directly the probability of the train BL class, whereas the probability of the class WL it is the maximum of the probabilities of the WL2 and WL3 classes.
2. WL1-WL2-WL3: The probability of the class BL is given the probability of the class WL1, whereas for the class WL it is the maximum of the probabilities of the WL2 and WL3 classes.
3. BLs-WL1-WL2-WL3: The probability of the class BL is the maximum probability of the BL and WL1 classes, whereas for the class WL it is the maximum of the probabilities of the WL2 and WL3 classes.

5. Experimental Design

In order two validate the proposed models, two experiments were conducted:

5.1. Training and Validation Using N-Back-Test Data

To assess to what extent a model trained over a set of individuals can successfully predict a new unseen individual, we have used a generalist population model, where a single model using all subjects was trained to assess whether inter subject variability can be properly modeled. The validation of the capability for modeling a population was tested using a leave-one-out scheme to allow statistical analysis. Models were trained using a batch size of 750, a weighted cross-entropy loss to compensate unbalances between baseline and workload phases, Adam [44] as optimization method, 100 epochs, and a learning rate of 0.0001.

The performances of the different approaches for detection of mental workload were assessed using the accuracy (or sensitivity) for each class:

$$\text{Sensitivity} = \frac{\text{TP}}{\text{TP} + \text{FN}}$$

where TP = number of true positives and FN = number of false negatives. Sensitivity measures the ability of the system to detect BL and WL classes. Since we have a binary classification problem with WL the positive class, the sensitivity for BL corresponds to the specificity of the model.

5.2. Task Transfer Verification Using Flight Simulator Data

To assess the capability of our model for transfer learning, experiments were devoted to showing that the model trained to detect WL in a memory demanding task (n-back test) can detect an increase of WL associated with multitask procedures with interruptions decreasing performance.

The EEG signals of the flight dataset explained in Section 3 are intended to assess:

1. Correlation of WL recognition with the number of tasks carried out by the pilot. Since we expected that the proportion of samples classified by our model as medium–high WL would be higher in the intervals where the PF performed more tasks, we show the percentages of predictions for BLs and each WL in correspondence with the number of tasks demanded.
2. Correlation of WL recognition to flight complexity. Flights 2 and 4 were designed to have higher workloads than Flight 3 (Flight 1 is considered the baseline) so that the hypothesis is that the proportion of samples classified by the model as medium–high WL will be higher than in flight 3.

6. Results

In this section, we show and discuss the results obtained.

6.1. Training and Validation Using N-Back-Test Data

Tables 1 and 2 summarize the recalls of baseline (BL) and workload (WL2) for the binarized models trained on different class problems for, respectively, the input and feature projector models. Tables show ranges for WL and BL detection computed for the 16 subjects after removing three outlying cases (80% of population) that all approaches failed to correctly predict.

For all cases, performance was more robust for the three-class problem, although specificity was better in the 2-class and 4-class problems. Regarding projection approaches, models projecting features achieved higher performance. In particular, the binary class feature projector model achieved an average detection of BL of 87.81% and a WL of 76.25%.

Table 1. Input projector model binarized.

		All Population	80% of Population
BL-WL2	BL	85.72 ± 7.52	84.15 ± 7.50
	WL	76.22 ± 17.64	82.81 ± 11.73
BLs-WL2-WL3	BL	78.16 ± 10.83	75.5 ± 10.29
	WL	78.62 ± 16.59	84.35 ± 10.87
WL1-WL2-WL3	BL	72.94 ± 18.08	70.58 ± 19.29
	WL	77.34 ± 16.72	82.85 ± 11.48
BLs-WL1-WL2-WL3	BL	80.75 ± 9.87	79.42 ± 10.07
	WL	76.44 ± 16.81	80.96 ± 13.16

Table 2. Feature projector model binarized.

		All Population	80% of Population
BL-WL2	BL	87.81 ± 7.07	86.65 ± 7.33
	WL	76.25 ± 19.27	82.73 ± 14.85
BLs-WL2-WL3	BL	79.00 ± 9.22	77.11 ± 9.13
	WL	80.94 ± 16.21	85.96 ± 11.68
WL1-WL2-WL3	BL	81.34 ± 15.76	81.27 ± 15.21
	WL	82.47 ± 15.78	86.54 ± 11.81
BLs-WL1-WL2-WL3	BL	84.75 ± 8.88	83.96 ± 9.24
	WL	76.34 ± 15.78	80.65 ± 12.49

6.2. Task Transfer Verification Using Flight Simulator Data

Barplots in Figures 4 and 5 show the percentages of WL detection as a function of the number of interruptions (0, 1, or 2). The expected pattern was the percentage of WL detection increasing with the number of interruptions. For both projection models, the 3-class problem WL1_WL2_WL3 is the only model that does not follow the expected increasing pattern. For the remaining problems, both architectures seem to behave equally.

Figures 6 and 7 show the barplots for the number of BL and WL predictions for the four flights. The expected pattern was to have the most detections in Flight 1, Flight 2, and Flight 4 (similar amounts of WL detected) and for Flight 3 to present a decrease in detected WL with respect these flights. Only the feature projector model follows the pattern expected. The most significant differences between flights are evident in the 3-class problem

BLs_WL2_WL3, followed by the 4-class problem. The 3-class problem WL1_WL2_WL3 does not apparently detect any difference among Flight 3 and Flight 4.

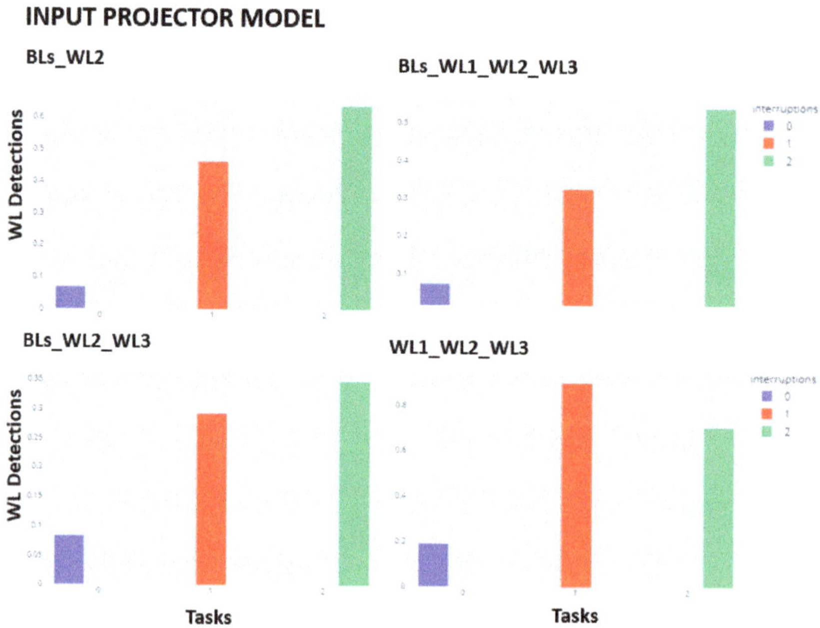

Figure 4. FRAM tasks barplots of WL predictions for the input projector model.

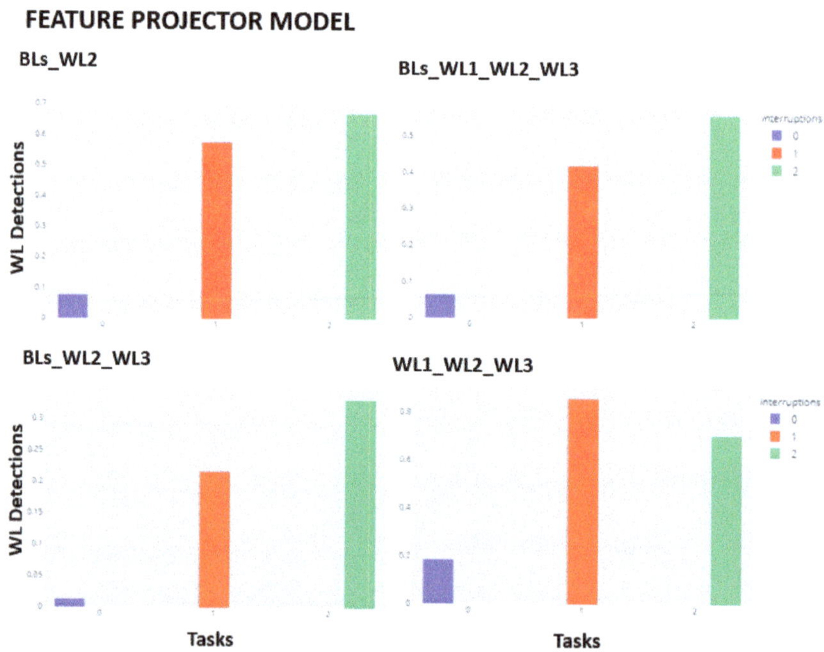

Figure 5. FRAM tasks barplots of WL predictions for the feature projector model.

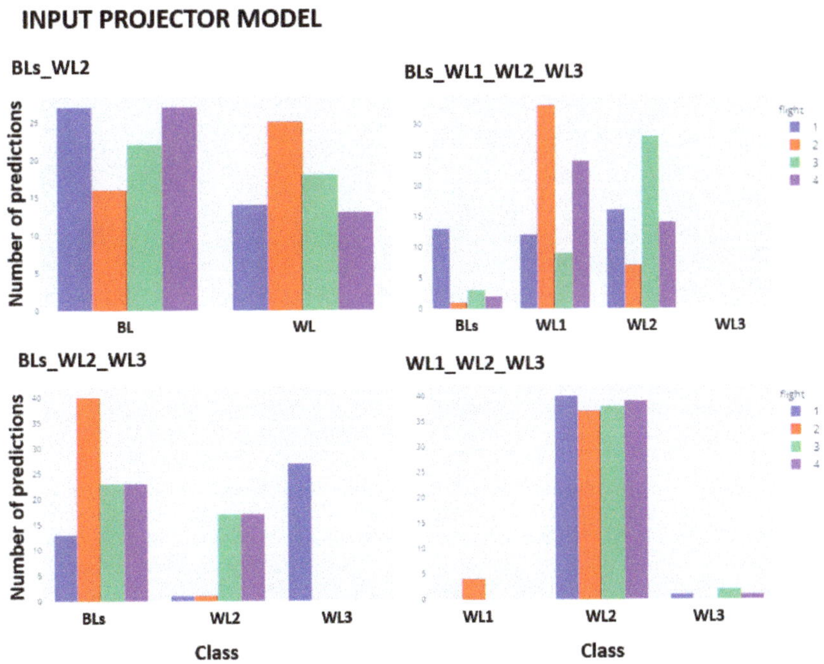

Figure 6. Flight test barplots of WL predictions for the input projector model.

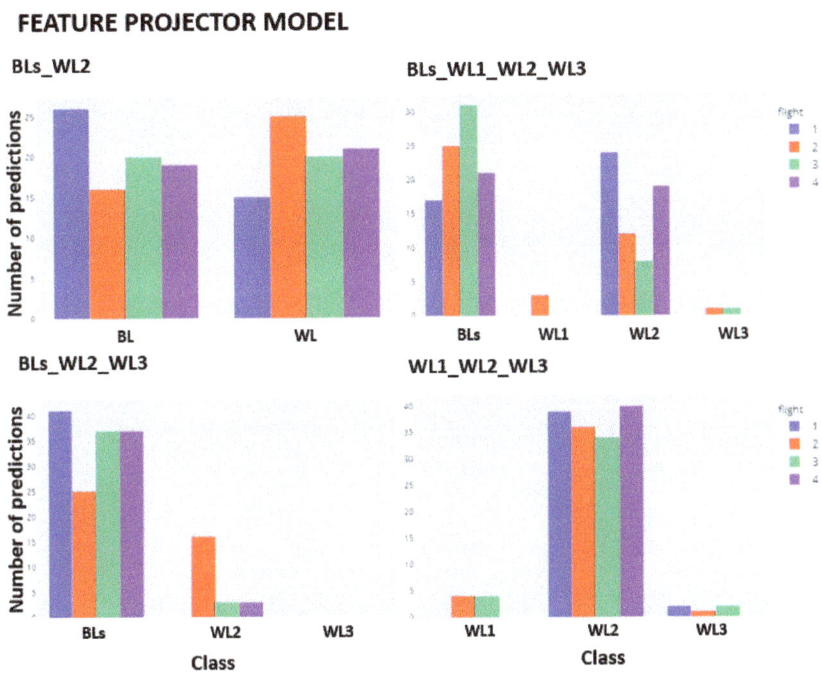

Figure 7. Flight test barplots of WL predictions for the feature projector model.

7. Conclusions

In this paper, we have presented two different approaches to the fusion of EEG sensor signals. Models were trained and validated on self-designed games (one serious game and one flight simulator with specific scenarios) to ensure unambiguous annotations. Models were trained and validated on the serious game using a one-subject-out scheme; and simulator data gathered from a subject not included in the training data were used to evaluate transfer capability.

Results show that between the two models, projecting convolutional feature channels achieved higher performance, with 76.25% sensitivity and 87.81% specificity in WL detection in n-back-test leave-one-out subject evaluation and good task transfer with the detected WL increasing with the number of interruptions. Although these results provide evidence of the ability of the EEG sensor to discern between more and less demanding tasks—increasing the evidence the robustness of the EEG and its ability to transfer tasks—the fact that the 3-class problem BLs_WL2_WL3 does not correlate with flight complexity suggests the following improvements.

A delicate issue that has an impact on the performances of methods is the filtering of signals required to remove muscular motion peaks and other artifacts. EEG pre-processing approaches have not been standardized, and even small changes in the artifact removal strategy may result in differences with large effects on particular portions of the signal. In this study, we have adopted a filtering approach based on signal probabilistic distribution for outlier removal in the temporal space. We consider that muscular motion could be filtered calibrating muscular signals before test recording to set either the values or the frequency ranges associated with muscular motion.

Some studies claim the importance of considering multiple aspects of a user's state when developing cognitive state detection algorithms [45]. Consequently, affective state should be considered.

Given that the way EEG sensors are fused has a direct impact in performance of models, alternative architectures should be further investigated. In this context, a direct improvement would be to consider ensemble models processing each sensor separately with own-learned weights. Furthermore, more recent architectures such as convolutional/LSTM and Lambda Nets that include attention modeling should be also studied.

Author Contributions: Conceptualization, D.G. and M.À.P.; methodology, P.F., A.H.-S. and D.G.; software, P.F. and J.Y.; validation, J.Y., P.F., A.H.-S. and D.G.; formal analysis, A.H.-S. and D.G.; investigation, P.F., A.H.-S. and D.G.; resources, M.À.P.; data curation, J.Y. and P.F.; writing—original draft preparation, J.Y. and A.H.-S.; writing—review and editing, A.H.-S. and D.G.; visualization, J.Y. and A.H.-S.; supervision, D.G., A.H.-S. and M.À.P.; project administration, M.À.P. and D.G.; funding acquisition, M.À.P. All authors have read and agreed to the published version of the manuscript.

Funding: This research was partially funded by Cleansky, grant number 831993, Ministerio de Ciencia e Innovación (MCI), Agencia Estatal de Investigación (AEI) and Fondo Europeo de Desarrollo Regional (FEDER), RTI2018-095209-B-C21 (MCI/AEI/FEDER, UE); Agència de Gestió d'Ajuts Universitaris i de Recerca grant numbers 2017-SGR-1597 and 2017-SGR-1624; and CERCA Programme/Generalitat de Catalunya.

Institutional Review Board Statement: Ethical review and approval were waived for this study, because data collected was not identifiable human material (the only data collected are anonymised EEG signals without any other information). Even so, participants were provided with information on the purpose, and on the content of the research. They were also given the choice to participate by agreeing to this information, or to not participate, and could quit the experiments at any moment. All this is in line with regulations on the use of personal information in scientific research in Spain.

Informed Consent Statement: Informed consent was obtained from all subjects involved in the study.

Data Availability Statement: Publicly available datasets were analyzed in this study. These data can be found here: http://iam.cvc.uab.es/portfolio/e-pilots-dataset/ accessed on 7 February 2022.

Acknowledgments: Authors would like to thank Carles Sánchez for his help in the revision of the paper. DGil is a Serra Hunter Fellow. DGil would like to dedicate this work to her mother Esther Resina Enfedaque, the best woman ever.

Conflicts of Interest: The authors declare no conflict of interest.

References

1. Latorella, K.A. *Investigating Interruptions: Implications for Flightdeck Performance*; NASA: Washington, DC, USA, 1999; Volume 99.
2. Foroughi, C.K.; Werner, N.E.; McKendrick, R.; Cades, D.M.; Boehm-Davis, D.A. Individual differences in working-memory capacity and task resumption following interruptions. *J. Exp. Psychol. Learn. Mem. Cogn.* **2016**, *42*, 1480. [CrossRef]
3. Oulasvirta, A.; Saariluoma, P. Long-term working memory and interrupting messages in human–computer interaction. *Behav. Inf. Technol.* **2004**, *23*, 53–64. [CrossRef]
4. Kirmeyer, S.L. Coping with competing demands: Interruption and the type A pattern. *J. Appl. Psychol.* **1988**, *73*, 621. [CrossRef]
5. Cellier, J.M.; Eyrolle, H. Interference between switched tasks. *Ergonomics* **1992**, *35*, 25–36. [CrossRef]
6. Zhang, P.; Wang, X.; Zhang, W.; Chen, J. Learning Spatial-Spectral-Temporal EEG Features With Recurrent 3D Convolutional Neural Networks for Cross-Task Mental Workload Assessment. *IEEE Trans. Neural Syst. Rehabil. Eng.* **2019**, *27*, 31–42. [CrossRef]
7. Li, D.; Wang, X.; Menassa, C.C.; Kamat, V.R. Understanding the impact of building thermal environments on occupants' comfort and mental workload demand through human physiological sensing. In *Start-Up Creation*; Elsevier: Amsterdam, The Netherlands, 2020; pp. 291–341.
8. Hendy, K.C.; Liao, J.; Milgram, P. Combining time and intensity effects in assessing operator information-processing load. *Hum. Factors* **1997**, *39*, 30–47. [CrossRef]
9. Heine, T.; Lenis, G.; Reichensperger, P.; Beran, T.; Doessel, O.; Deml, B. Electrocardiographic features for the measurement of drivers' mental workload. *Appl. Ergon.* **2017**, *61*, 31–43. [CrossRef]
10. Han, S.Y.; Kwak, N.S.; Oh, T.; Lee, S.W. Classification of pilots' mental states using a multimodal deep learning network. *Biocybern. Biomed. Eng.* **2020**, *40*, 324–336. [CrossRef]
11. Zhang, P.; Wang, X.; Chen, J.; You, W.; Zhang, W. Spectral and Temporal Feature Learning with Two-Stream Neural Networks for Mental Workload Assessment. *IEEE Trans. Neural Syst. Rehabil. Eng.* **2019**, *27*, 1149–1159. [CrossRef]
12. Lee, D.H.; Jeong, J.H.; Kim, K.; Yu, B.W.; Lee, S.W. Continuous EEG Decoding of Pilots' Mental States Using Multiple Feature Block-Based Convolutional Neural Network. *IEEE Access* **2020**, *8*, 121929–121941. [CrossRef]
13. Wu, E.Q.; Peng, X.; Zhang, C.Z.; Lin, J.; Sheng, R.S. Pilots' fatigue status recognition using deep contractive autoencoder network. *IEEE Trans. Instrum. Meas.* **2019**, *68*, 3907–3919.
14. Averty, P.; Collet, C.; Dittmar, A.; Athènes, S.; Vernet-Maury, E. Mental workload in air traffic control: An index constructed from field tests. *Aviat. Space Environ. Med.* **2004**, *75*, 333–341.
15. da Silva, F.P. Mental Workload, Task Demand and Driving Performance: What Relation? *Procedia-Soc. Behav. Sci.* **2014**, *162*, 310–319. [CrossRef]
16. Hart, S.G. NASA-task load index (NASA-TLX); 20 years later. In *Proceedings of the Human Factors and Ergonomics Society Annual Meeting*; Sage Publications: Los Angeles, CA, USA, 2006; Volume 50, pp. 904–908.
17. Borghini, G.; Astolfi, L.; Vecchiato, G.; Mattia, D.; Babiloni, F. Measuring neurophysiological signals in aircraft pilots and car drivers for the assessment of mental workload, fatigue and drowsiness. *Neurosci. Biobehav. Rev.* **2014**, *44*, 58–75. [CrossRef]
18. Hart, S.G.; Staveland, L.E. Development of NASA-TLX (Task Load Index): Results of empirical and theoretical research. In *Advances in Psychology*; Elsevier: Amsterdam, The Netherlands, 1988; pp. 139–183.
19. Wickens, C.D. Situation awareness and workload in aviation. *Curr. Dir. Psychol. Sci.* **2002**, *11*, 128–133. [CrossRef]
20. Parasuraman, R.; Sheridan, T.B.; Wickens, C.D. Situation awareness, mental workload, and trust in automation: Viable, empirically supported cognitive engineering constructs. *J. Cogn. Eng. Decis. Mak.* **2008**, *2*, 140–160. [CrossRef]
21. Wang, Z.; Yang, L.; Ding, J. Application of heart rate variability in evaluation of mental workload. *Chin. J. Ind. Hyg. Occup. Dis.* **2005**, *23*, 182–184.
22. Stanton, N.; Salmon, P.M.; Rafferty, L.A. *Human Factors Methods: A Practical Guide for Engineering and Design*; Ashgate Publishing, Ltd.: Farnham, UK, 2013.
23. Jang, E.H.; Park, B.J.; Kim, S.H.; Chung, M.A.; Park, M.S.; Sohn, J.H. Classification of human emotions from physiological signals using machine learning algorithms. In Proceedings of the Sixth International Conference on Advances in Computer-Human Interactions 2013 (ACHI 2013), Nice, France, 24 February–1 March 2013; Citeseer: Princeton, NJ, USA, 2013; pp. 395–400.
24. Rim, B.; Sung, N.J.; Min, S.; Hong, M. Deep learning in physiological signal data: A survey. *Sensors* **2020**, *20*, 969. [CrossRef]
25. Ziegler, M.D.; Russell, B.A.; Kraft, A.E.; Krein, M.; Russo, J.; Casebeer, W.D. Computational Models for Near-real-time Performance Predictions Based on Physiological Measures of Workload. In *Neuroergonomics*; Elsevier: Amsterdam, The Netherlands, 2019; pp. 117–120.
26. Secerbegovic, A.; Ibric, S.; Nisic, J.; Suljanovic, N.; Mujcic, A. Mental workload vs. stress differentiation using single-channel EEG. In *CMBEBIH 2017*; Springer: Berlin/Heidelberg, Germany, 2017; pp. 511–515.
27. Ismail Fawaz, H.; Forestier, G.; Weber, J.; Idoumghar, L.; Muller, P.A. Deep Learning for Time Series Classification: A Review. *Data Min. Knowl. Discov.* **2019**, *33*, 917–963. [CrossRef]

28. Faust, O.; Hagiwara, Y.; Hong, T.J.; Lih, O.S.; Acharya, U.R. Deep learning for healthcare applications based on physiological signals: A review. *Comput. Methods Programs Biomed.* **2018**, *161*, 1–13. [CrossRef]
29. Zhang, Y.; Chen, J.; Tan, J.H.; Chen, Y.; Chen, Y.; Li, D.; Yang, L.; Su, J.; Huang, X.; Che, W. An investigation of deep learning models for EEG-based emotion recognition. *Front. Neurosci.* **2020**, *14*, 1344. [CrossRef]
30. Venkatachalam, K.; Devipriya, A.; Maniraj, J.; Sivaram, M.; Ambikapathy, A.; Iraj, S.A. A Novel Method of motor imagery classification using eeg signal. *Artif. Intell. Med.* **2020**, *103*, 101787.
31. Zhao, W.; Wang, W. SeizureNet: A model for robust detection of epileptic seizures based on convolutional neural network. *Cogn. Comput. Syst.* **2020**, *2*, 119–124. [CrossRef]
32. Zhang, X.; Xu, M.; Li, Y.; Su, M.; Xu, Z.; Wang, C.; Kang, D.; Li, H.; Mu, X.; Ding, X.; et al. Automated multi-model deep neural network for sleep stage scoring with unfiltered clinical data. *Sleep Breath.* **2020**, *4*, 581–590. [CrossRef]
33. Saadati, M.; Nelson, J.; Ayaz, H. Convolutional Neural Network for Hybrid fNIRS-EEG Mental Workload Classification. In *Advances in Neuroergonomics and Cognitive Engineering*; Ayaz, H., Ed.; Springer International Publishing: Cham, Switzerland, 2020; pp. 221–232.
34. Kwak, Y.; Kong, K.; Song, W.J.; Min, B.K.; Kim, S.E. Multilevel feature fusion with 3d convolutional neural network for eeg-based workload estimation. *IEEE Access* **2020**, *8*, 16009–16021. [CrossRef]
35. Ziegler, M.D.; Kraft, A.; Krein, M.; Lo, L.C.; Hatfield, B.; Casebeer, W.; Russell, B. Sensing and assessing cognitive workload across multiple tasks. In *International Conference on Augmented Cognition*; Springer: Cham, Switzerland, 2016; pp. 440–450.
36. Jaeggi, S.M.; Buschkuehl, M.; Jonides, J.; Perrig, W.J. Improving fluid intelligence with training on working memory. *Proc. Natl. Acad. Sci. USA* **2008**, *105*, 6829–6833. [CrossRef]
37. Bokade, R.; Navato, A.; Ouyang, R.; Jin, X.; Chou, C.A.; Ostadabbas, S.; Mueller, A.V. A cross-disciplinary comparison of multimodal data fusion approaches and applications: Accelerating learning through trans-disciplinary information sharing. *Expert Syst. Appl.* **2021**, *165*, 113885. [CrossRef]
38. Yauri, J.; Hernández-Sabaté, A.; Folch, P.; Gil, D. Mental Workload Detection Based on EEG Analysis. In *Artificial Intelligence Research and Development*; IOS Press: Amsterdam, The Netherlands, 2021; pp. 268–277.
39. Piera, M.A.; Ramos, J.J.; Muñoz, J.L. A socio-technical holistic agent based model to assess cockpit supporting tools performance variability. *IFAC-PapersOnLine* **2019**, *52*, 122–127. [CrossRef]
40. Emotiv. *EMOTIV EPOC+ 14-Channel Wireless EEG Headset*; Emotiv: San Francisco, CA, USA, 2021.
41. Addante, R.J.; Watrous, A.J.; Yonelinas, A.P.; Ekstrom, A.D.; Ranganath, C. Prestimulus Theta Activity Predicts Correct Source Memory Retrieval. *Proc. Natl. Acad. Sci. USA* **2011**, *108*, 10702–10707. [CrossRef]
42. Wasserman, L. *All of Statistics: A Concise Course in Statistical Inference*; Springer: Berlin/Heidelberg, Germany, 2010.
43. Gupta, S.S.; Taori, T.J.; Ladekar, M.Y.; Manthalkar, R.R.; Gajre, S.S.; Joshi, Y.V. Classification of cross task cognitive workload using deep recurrent network with modelling of temporal dynamics. *Biomed. Signal Process. Control* **2021**, *70*, 103070. [CrossRef]
44. Kingma, D.P.; Ba, J. Adam: A Method for Stochastic Optimization. In Proceedings of the 3rd International Conference on Learning Representations, ICLR 2015-Conference Track Proceedings, Banff, AB, Canada, 14–16 April 2014.
45. Bagheri, M.; Power, S.D. EEG-based detection of mental workload level and stress: The effect of variation in each state on classification of the other. *J. Neural Eng.* **2020**, *17*, 056015. [CrossRef]

applied sciences

MDPI

Article

Bootstrap–CURE: A Novel Clustering Approach for Sensor Data—An Application to 3D Printing Industry

Shikha Suman *,†, Ashutosh Karna † and Karina Gibert †

Knowledge Engineering and Machine Learning Group at Intelligent Data Science and Artificial Intelligence Research Centre, Universitat Politecnica de Catalunya, 08034 Barcelona, Spain; ashutosh.karna@upc.edu (A.K.); karina.gibert@upc.edu (K.G.)

* Correspondence: shikha.suman@estudiantat.upc.edu

† These authors contributed equally to this work.

Abstract: The agenda of Industry 4.0 highlights smart manufacturing by making machines smart enough to make data-driven decisions. Large-scale 3D printers, being one of the important pillars in Industry 4.0, are equipped with smart sensors to continuously monitor print processes and make automated decisions. One of the biggest challenges in decision autonomy is to consume data quickly along the process and extract knowledge from the printer, suitable for improving the printing process. This paper presents the innovative unsupervised learning approach, **bootstrap–CURE**, to decode the sensor patterns and operation modes of 3D printers by analyzing multivariate sensor data. An automatic technique to detect the suitable number of clusters using the dendrogram is developed. The proposed methodology is scalable and significantly reduces computational cost as compared to classical CURE. A distinct combination of the 3D printer's sensors is found, and its impact on the printing process is also discussed. A real application is presented to illustrate the performance and usefulness of the proposal. In addition, a new state of the art for sensor data analysis is presented.

Keywords: CURE; hierarchical clustering; cluster validity indices; Calinski–Harabasz index; bootstrapping; Industry 4.0; 3D printing

Citation: Suman, S.; Karna, A.; Gibert, K. Bootstrap–CURE: A Novel Clustering Approach for Sensor Data—An Application to 3D Printing Industry. *Appl. Sci.* **2022**, *12*, 2191. https://doi.org/10.3390/app12042191

Academic Editor: Anming Hu

Received: 14 December 2021

Accepted: 17 February 2022

Published: 19 February 2022

1. Introduction

Industry 4.0 [1] has been revolutionizing the manufacturing practices with a strong influence on mechanization and automation. Inadvertently, this also brings up the importance of sensors and their pattern analysis. A sensor is a physical device that detects or measures an external signal and records it or responds to it. A comprehensive definition and types of sensors are explained in the work [2]. New generation machines are now equipped with dozens of sensors to extract data at high temporal resolution. An exhaustive analysis of the data provided by sensors can help obtain crucial information about the health of the overall system, as well as developing tools for faster knowledge extraction and automation. Sensor data are typically unlabeled and thus demand an unsupervised methodology to characterize their impact on a machine and also explainable AI techniques to interpret the results from a semantic point of view. When a 3D printer works in the real production environment under smart manufacturing [3], most of its activities are completely automated and governed by an electronic control system. There can still be situations when the printer stops a print job or results in some fault, and this leads to a lot of open questions. A proper understanding of machine operations under various conditions can help answer what factors cause such problems.

The main goal of this paper is to use data to study the behavior of the 3D printer machine to better understand its operation and to obtain insights with respect to the control systems that govern the printing process in a real production environment. Understanding what factors result in a *successful* or *unsuccessful* job and how they are expressed through sensor data is crucial for the future development of 3D printer technologies.

It is important to note that this behavioral study is conducted from the perspective of aiding the printer manufacturer in predictive maintenance as well as gaining a superior understanding of the printing subsystems and their operation modes by solely using the multivariate sensor data.

This research addresses the enterprise-grade multi-jet fusion 3D printers working in a real production environment. As these machines operate in the customer's environment with strict confidentiality agreements, neither confidential data (such as print-layer images or video) nor any external modification to the printer are possible, and this is why this approach is based on the mere use of sensor data. This situation also applies to other domains outside 3D printing, such as wastewater treatment plants [4], gas turbines [5], aero-generators [6], etc. Therefore, the methodology proposed in this paper may be useful in other Industry 4.0 application domains as well.

The long-term goal of this research is to build an intelligent diagnosis system that can understand what leads to a failure while the machine is operating and can react in real time to restore the normal behavior of the machine. In the long term, it is also expected to predict such failures in advance and react preventively. The focus of this paper is a preliminary step to reach these goals. This step consists of identifying and understanding the main operation modes exhibited by the machine and eventually association with different kinds of failures.

Thus, the main goal of the proposal is to provide a tool that identifies states of operation in the 3D printing machine without needing any prior hypotheses on the number of clusters and is also suitable for analyzing a large amount of data in a rather short computation time. This is directly related to the use of hierarchical clustering methods, where the real number of existing clusters is not required as an input parameter. However, hierarchical clustering is of quadratic complexity and does not scale up to large datasets automatically.

Although the research is focused on analyzing sensor data from 3D printers, the clustering methodology itself is agnostic to the domain and can be used in any generic application wherein data from sensors are used. Thus, the final and fourth contribution of this paper is to provide a general conceptualization of the field of sensor data.

The structure of the paper is as follows. First, a summary of the motivation behind the research, including the high-level goals, is presented in Section 1. A brief overview of 3D printing is shared in Section 2, followed by the literature review of state-of-the-art models in Section 3. The contributions of the paper are explicitly mentioned in Section 4. Section 5 describes the methodology and continues to Section 6, providing an application from a real-life dataset from 3D printing. Section 7 discusses the proposed methodology in an industrial setting of 3D printers. The paper finally rests with the conclusion in Section 8 along with the future lines of work in the research.

2. 3D Printing Process

The term, *additive manufacturing* (commonly known as *3D printing*), has been drawing attention from different sections of the industries by allowing the digitization of physical models. In [7], a review of the main 3D printing technologies can be found.

One of the major advantages of 3D printing is the ability to produce a monolithic structure of complex geometry. This is possible by building parts through a stack of thin cross sections in an additive manner. Hence, this type of printing also saves print material, which is often wasted in traditional subtractive manufacturing. Some of the key 3D printing technologies are as follows [8]:

1. **Fused deposition modeling**: Machine lets the plastic filament melt and extrude through nozzles onto the bed platform, where it is cooled and solidified.
2. **Binder jetting**: Machine distributes a layer of powder onto a build platform, and a bonding agent helps to fuse the parts. The process keeps repeating until the parts are built up in the powder bed.

3. **Laser sintering**: Uses a laser as a power source to aim at points in 3D space to sinter the powder material and create a solid part.
4. **Laser melting**: Machine uses laser(s) to melt metal powder.
5. **Stereo-lithography**: Machine builds parts out of liquid photopolymer through polymerization activated by a UV laser.

This paper focuses on finding interpretable models to describe the work done on anonymized sample sensor measurements from *HP Multi-Jet 3D Fusion* printers. According to the overview provided in *additive manufacturing* [7], the *jet-fusion* technology is a further improvement developed by HP, on the fused deposition modeling and polyjet technologies. A brief technical introduction of multi-jet fusion technology is provided in the work [9]. Such 3D printers are equipped with dozens of sensors placed across different components to measure and control factors, such as temperature, humidity, velocity, pressure, etc. A 3D printing process is a complex job, where many parameters must align among themselves for the success of the job. A brief technical introduction of multi-jet fusion technology is provided in the work by [9].

The data collected through the sensors network of the printer are produced in a streaming mode and sent to a cloud-based server for further processing. Analyzing the sensor-based network is challenging in the sense that it is required to run the analysis on the fly and be able to ingest large amount of sensor data rather quickly.

3. State-of-the-Art

The literature review is divided into two major parts, focusing, respectively, on data science applications in sensor analysis (Section 3.1) and the methods to scale hierarchical clustering and identify the best number of clusters (Section 3.2.1).

3.1. Data Science Applications in Sensor Analysis

The field of sensor analysis has been quite active in recent years, with both academics and industry contributing to the research alike.

Although one can find innumerable data science applications of sensor data (including the one in 3D printing) in the literature, and it is difficult to write an exhaustive survey, the authors identify the following main research areas.

- Anomaly detection;
- Automatic reporting and visualization;
- Pattern analysis;
- Process control;
- Predictive maintenance.

The authors were thus able to synthesize the map displayed in Figure 1. The authors conclude that most of the literature in this field can be grouped into five broad categories as shown in Figure 1. This picture helps to organize the contents of this section.

3.1.1. Anomaly Detection

A majority of works in recent times focus on using sensor data to detect anomalies in real time or offline data. This itself is a fairly general framework with applications in a wide variety of domains, ranging from healthcare, autonomous vehicles, printing, internet-of-things, etc. The literature can further be subdivided based on the kind of algorithms used for the modeling. Deep learning-based methods are most popular in this category [10], sometimes in their classical supervised version [11,12], and sometimes in an unsupervised deep learning approach, such as Ref. [13] applying autoencoders to detect anomalies in an aircraft system, and Parllaku et al. [14] and Karna et al. [15] finding anomalies in 3D printing processes. Although deep learning techniques generally have high accuracy, there is a high variety of architectures depending on each concrete application. This implies that the sensor data behave differently as per the application field. In addition, the inability of neural network models, in general, to explain anomalies is a big industrial challenge for research purposes.

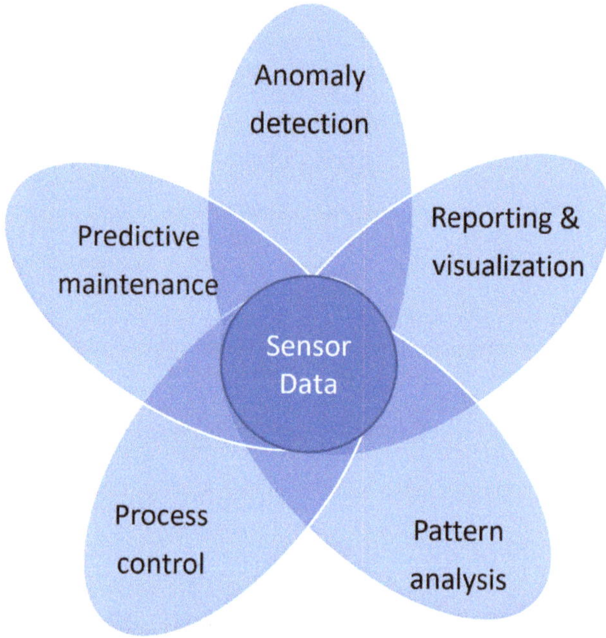

Figure 1. Sensor analyses goals.

Although relatively fewer, there are still several studies using other machine learning methods, most of them hybrid approaches combining different techniques: Ref. [16], where hybrid random forest with wavelet transform is developed to recognize vehicle steering mode and further detect abnormal behavior; some references based on the combination of support vector machines with other techniques, such as deep belief networks [17] and DBScan [18]; or statistical applications, such as kernel PCA [19]. These works propose a hybrid scheme of using a previously unsupervised method to identify anomalies and build a predictive model on top of using SVM. Although SVM provides more explainable predictions than deep learning, their learning times are not scalable to big-data-based real industrial scenarios yet.

3.1.2. Automatic Reporting and Visualization

Here, the authors find works contributing mainly to visualizing traffic coming from sensors and interpreting their patterns. From analyzing EEG (*ElectroEncephaloGram*) [20,21], to visualizing traffic data [22], a good visualization tool can allow interacting with anomalies in real time [23] or combine common system abnormalities with simple descriptive statistics [24]. In this field, interactivity and visual interfaces provide business dashboards and tools to track key business process indicators. This field focuses more on the human–computer interaction than intensive algorithmic models.

3.1.3. Pattern Analysis

A typical data challenge in industrial applications is to discover or recognize patterns from sensor data. A large number of works use clustering techniques and related methods to discover patterns or characterize features. In [25], a theoretical approach to density estimation clustering on sensor networks is found. Although the literature is limited with respect to pattern recognition research in Industry 4.0, some important works have been found in the following. In [26], the unsupervised deep convolutional network was used to recognize patterns of cyber attacks in industrial wastewater treatment plants, whereas [27] used Markov models on sensor data to identify patterns of energy consumption and comfort management in intelligent buildings. Some noteworthy work in the characterization of sensor signals using *k-means* clustering was done by Hromic et al. [28] on sensor data for air quality, and Loane et al. [29] used clustering to find patterns of households based on domestic sensors. Ref. [30] used k-means and support vector machines (SVM) to first identify patterns of print conditions in *selective laser melting* machines and then identified them with an SVM model.

Some of these works are approaching the dynamics of the system. In [26], a univariate approach of time series modeling was followed, while in [27], a multivariate approach was used for characterizing the relationships among the sensors along time. All other works are based on clustering, finding groups of similar objects under a multivariate approach. K-means is used frequently, although it requires many runs and finding the optimization criteria to identify the right number of clusters. Only one work used hierarchical clustering [29], showing evidence that the optimal number of clusters is obtained by running several k-means and optimizing the silhouette index results in the same number of clusters given by the dendrogram of the hierarchical clustering. Specific to 3D printing, only one work is found that combined k-means with a classifier to identify types of operation with a predictive model. The authors also want to use clustering to identify types of operation of 3D printers and thus understand the relationships among sensors. A detailed overview of the related works in clustering is provided in Section 3.2.

3.1.4. Process Control

Process control and monitoring sensors is highly relevant for Industry 4.0. The literature is relatively limited and mostly comes from the field of time series forecasting or supervised machine learning based on images.

A number of references describe supervised learning applications especially on small-scale 3D printers (commonly based on fused-deposition-modeling principle). Some notable works that help in detecting quality of 3D printed part are presented in [31–36]. Although these works provide valuable contributions to 3D printing especially in detecting and monitoring part quality, they assume the possibility of using tagged data to train the models, and they are mainly based on images of video processing. These approaches are out of the scope of this research for several reasons:

- The need to obtain supervised data every time a model has to be trained is unrealistic in real Industry 4.0 manufacturing.
- In a customer environment with a fleet of 3D printers installed, obtaining global knowledge of the machines' health from the manufacturer's perspective would require analyzing data anonymously, as any confidential data (involving 3D print design or the print images) other than the sensor records cannot be shared or stored outside the customer site.
- Additionally, some of these works assume the possibility of installing additional cameras into the 3D printers; however, in an industrial setting, this is not always possible, as only customers have the sole decision-making authority to make any changes.

The time series approach based on sensor data seems a more suitable reference for our research. In this regard, some works merit a reference. In [37], an application from the iron and steel industries processes sensor data through a discrete-time extended *Kalman* filter for both process state estimation and sensor data fusion. Understanding the sensors' patterns and invoking real-time process control can tremendously improve print job success on a whole. In the context of 3D printing, process control helps monitor the evolution of the part by printing layers and guarantees the overall success of the job. In process control, some works can be found in the field of 3D printing, although they are not suitable for our research goals. In [38], exponential weighted moving average (EWMA) charts are used to monitor the process parameters and detect shifts in the fused deposition printing process and thereby assess geometrical deformations of the printed part. In [39], Zang et al. present a novel scheme of simulating surface properties of printed part by defining in-control and out-of-control specifications and use bootstrap sampling to estimate the control limit of several control charts. Both of these works consider small-scale 3D printers. In the context of a large-scale 3D printer, such as multi-jet fusion, such approaches do not apply, as the machine is far more complex, so process monitoring cannot be limited to just one surface of the part, and external cameras are also not applicable. However, it is interesting to see the idea of introducing bootstrap techniques in the processing to scale up the modeling.

3.1.5. Predictive Maintenance

Predictive maintenance [40,41], is the latest trend in Industry 4.0 to provide superior customer experience by using an automated and highly interconnected network of sensors to predict faults at the early stage and minimize industrial downtime. Predictive maintenance is a wider approach and mostly includes anomaly detection by detecting faults at an early stage in addition to estimating the remaining useful life and time-to-failure of equipment in industrial processes. A general framework of designing predictive maintenance solution is described in [42] with a case study from steel industries to estimate the degradation of coiler drums, using the discrete Bayesian filter. Bonci et al. in [43] used the wavelet decomposition model to analyze and predict faults in Cartesian robots based on the motor current signal, while Lin et al. in [44] presented an ARIMA-based time series prediction approach to predict the remaining useful life of the target device in the factory settings based on aging features. Gibert et al. [45] used clustering techniques to assess the health of the wind turbines based on sensor data and introduced some post-processing techniques to interpret the meaning of the clusters and the associated healthy state they represent.

Deep learning techniques are also frequently used in predictive maintenance. In [46], autoencoders and deep belief networks were proposed for early fault detection in digital manufacturing. In [47], dynamic predictive maintenance framework was proposed for the turbofan engine by using deep learning. In [48], deep learning was used on IoT for predictive maintenance of a *Porsche* car based on the sound recorded on different parts of the engine.

Shi et al. [49] concluded that multi-core CPU machines do not scale sufficiently well for training deep neural networks and still require a dedicated GPU. This is a a serious limitation for those Industry 4.0 applications with machines embedded with computation-intensive hardware, such as 3D printers, as adding a GPU implies increasing the printing time of jobs. This also adds extra cost to the total value of the machine. Additionally, deep learning models are generally assumed to be black-box in nature and cannot explain the proposed predictions, which is critical in a decision-making context, as predictive maintenance often is.

Thus, there is no existing method known to be effective in analyzing and controlling the manufacturing process in the context of multi-jet large-scale 3D printers [50]. A machine like a multi-jet 3D printer, running non-stop and generating tons of data, would thus require appropriate statistical techniques to ingest data and process them rather rapidly. Hence, for a suitable insertion in real production systems, near-real-time and non-supervised approaches are required; classic deep learning approaches for predictive maintenance might not be suitable.

In this paper, the authors are attempting to establish telemetry monitoring and analysis, using an unsupervised learning approach (hierarchical clustering in particular) with a focus on explainable artificial intelligence methods.

3.2. Clustering Strategies

In general, clustering refers to the task of grouping similar objects. Xu et al. [51] provided a comprehensive survey of various clustering algorithms. While hard-clustering algorithms (e.g., hierarchical and k-means) uniquely assign an object into a single cluster, soft-clustering methods, such as fuzzy-clustering, rather assign each object with a certain membership degree to belong to different clusters. Fuzzy clustering provides a fuzzy partition as a result and requires defuzzification in the post-processing step.

Gupta et al. [52] presented a new approach based on evolutionary multi-objective optimization in fuzzy clustering to identify clusters at different levels of fuzziness. Lahmar et al. [53] provided a self-adaptive fuzzy c-means method to find the number of clusters; however, scalability to large datasets is not guaranteed in the paper. Shirkhorshidi et al. [54] provided an evolving fuzzy-clustering approach, where a small subset of data is clustered in every epoch, centroids are generated, and a global clustering takes place using k-means on these centroids. Although this work also elaborates on the idea of finding previous clusters and making a further clustering over centroids, the use of k-means demands a clear hypothesis on the number of clusters in each epoch, which is not possible in the real 3D printer management application field that the authors are targeting.

Among non-fuzzy clustering algorithms, *k-means* is one of the most popular for its ability to work on large datasets. Sebastian et al. [55] provided an interesting application of k-means clustering in the characterization of snore signals in the context of upper-airway collapse. Chakraborty et al. [56] proposed the *Lass-weighted k-means* algorithm specifically useful for a high-dimensional dataset, such as gene-expression data. Another interesting algorithm was developed by Gondeau et al. [57], using *object-weighting* in k-means clustering. The algorithm therein can help as a data preprocessing step to deal with outliers, especially in the case of noisy data. This method attempts to increase the weights of the outliers instead of removing them from the study. The results of this algorithm on both simulated and real-life datasets are quite promising too. However, the context of the current research is far from the situation where the real number of clusters to be built is known a priori, and all partitioning methods, including k-means, require the number of clusters to be found as an input parameter. For this reason, the authors will not work with partitioning methods.

In some recent works, the use of evolutionary and meta-heuristic optimization is also seen in the context of clustering. Li et al. [58] used gravitational search to optimize the number of clusters (obtained using *DBSCAN* approach) in multiple iterations. The final number of clusters is, however, decided manually. The approach seems scalable to large datasets; however, the requirement of running the algorithm in multiple iterations limits the application on cases where near-real-time data are to be analyzed rather quickly, such as the data from 3D printers' sensors. In the current research, the 3D printing sensor data require quicker analysis on a large amount of data, and running multiple iterations may not be acceptable from the practical point of view. Liu et al. [19] proposed another meta-heuristic optimization to improve the *maximum-entropy* clustering method. The applications of such an algorithm on large-scale data would be interesting to check.

In the current age of deep learning, the *graph-convolutional network* algorithm was provided by Zhao et al. [59] for incremental clustering. The algorithm was tested on face images and showed promising results. However, in the context of this paper, the authors deal only with the unsupervised sensor data, and no print images are provided for the analysis. The authors' goal is to generate a tool to support the learning and management of 3D printers at the manufacturer's site; the images are available only at the customer site and are bound by privacy clauses. Indeed, all customer images are confidential and not available to the company making the printers.

Some other recent works in the literature also include the work done in subspace clustering to deal with high dimensional datasets. Menon et al. [60] proposed a parameter-free approach in the subspace clustering, where the data are clustered based on statistical distribution within a subspace. The performance of the algorithm is also shown on various public datasets. In our context, distributional assumptions might be a challenge.

3.2.1. Hierarchical Clustering and Automatic Identification of the Number of Clusters

Hierarchical clustering creates a tree-like structure, called the *dendrogram*, to disclose the internal multivariate structure of data by moving from a pairwise distance matrix toward nested partitions on data. Unlike other clustering algorithms, such as k-means, it does not require prior knowledge about the number of clusters, as the clusters result from the appropriate horizontal cut of the dendrogram (see Section 5.2).

In [9], the authors presented an analysis to discover print profiles based on a random sample using hierarchical clustering with Ward's method. However, hierarchical clustering has higher space and time complexity than partitioning methods (as k-means does) and seems not very reliable for real applications comprising sensor data. Indeed, the standard algorithm is $O(n^2)$, in both space and time [61] and although some implementations in class $O(nlog(n))$ are available, it is still prohibitive for large datasets. Rafsanjani et al. [62] discussed different hierarchical clustering algorithms along with their comparison.

3.2.2. Clustering Using Representation (CURE)

CURE is a clustering algorithm that combines hierarchical clustering with sampling to deal with bigger datasets at smaller computation costs. Guha et al. [63] presented the *CURE* (clustering using representation) algorithm that can scale hierarchical clustering up to large datasets. The CURE algorithm can broadly be divided into two phases: initialization and completion. Initialization begins with taking a random sample without replacement from the original population and applying hierarchical clustering. Both the number of representative points and the shrinkage factor can be optimized as part of hyperparameter tuning. In the completion step, for each resulting cluster, CURE chooses a small set of representative points which are well scattered inside the cluster (so representing the shape of the cluster itself) and after shrinking toward the cluster centroid, a *KNN* scheme is used to assign the cluster to all remaining objects by computing distances toward all those cluster representatives. Due to its robust approach, the CURE algorithm can even recognize arbitrary shaped geometries while being robust against outliers. Another important advantage of the algorithm is that it is linear in space complexity, of order $O(n)$; however, the time complexity is still of order $O(n^2log(n)$ [63]. The algorithm also does not need any strict assumption about the distribution of the data inside the clusters. However, in the context of big data [64], this method is still not sufficiently scalable. In this paper, a modification of CURE is proposed to increase the model capacity based on bootstrap strategies.

3.2.3. Detection of the Number of Clusters

The most common strategies used in deciding the number of clusters in a hierarchical setting are cross-validation, resampling, and finding the *knee* or *elbow* of an error curve. The approach of cross-validation aims at computing the regression coefficients on the $v - 1$ portion of the dataset and validating the same on the vth portion, which is not used in building the regression model. The approach, however, becomes time intensive for large data and is not recommended. In [65], Kawamoto et al. proposed a criterion to detect the number of clusters in a modular network, using a leave-one-out cross-validation approach. Fu et al. [66] proposed a cross-validation-based approach to determine the number of clusters in a k-means type of clustering. Hence, the idea of using cross validation in clustering is to iteratively test which cluster solution yields the lowest error rate and consider that the best clustering solution. However, as pointed out by McIntyre et al. in [67], cluster analysis has no such linear coefficients which can be applied to multiple random samples and true cluster membership is rather unknown. Consequently, the cross-validation strategy cannot be used globally for all cases. This was also highlighted by Krieger et al. in [68], where the authors demonstrated how the cross-validation technique fails in hierarchical clustering by doing an *Monte-Carlo* simulation-based study of cross-validation techniques under different conditions.

Several works related to using resampling methods to decide the number of clusters in data can be found in the literature. In [69], Overall and Magee presented a replication-based stopping rule in which a replication defined by higher-order clustering helps identify the distinct underlying populations (clusters) in a multidimensional space. An improved version of this criterion was presented in [70], where Tonidandel et al. used the bootstrapping procedure along with the increase in the size of the resampled dataset with respect to the primary dataset and thus showed an increase in accuracy of the clustering solution. Related work is found in [71], where Fang et al. discussed a bootstrapping-based approach to estimate the clustering instability and then select the best number of clusters.

Another popular method of finding the optimal number of clusters is to seek the local maxima (or minima) corresponding to the *knee* or *elbow* of a curve that plots values of some clustering evaluation criteria on a range of numbers. Sevilla et al. in [72] reviewed several CVIs and their association with the type of data. Tibshirani et al. [73] also provided a measure called the *gap statistic*. It tests the hypothesis that the model has a single cluster ($K = 1$) and tries to reject it with an alternative hypothesis ($K > 1$). This method compares the total within intra-cluster variation for different values of k with their expected values under null reference distribution of the data, i.e., when there is no underlying clustering. However, the rejection of the null hypothesis only indicates insufficient evidence in support of the null hypothesis and does not really make it true. The underlying methodology in *elbow* or *knee* is, in fact, agnostic to the cluster validity index being used.

In [74], Jung et al. proposed *clustering gain* to find the right number of clusters in hierarchical clustering. Clustering gain is designed to have a maximum value when the intra-cluster similarity is maximized and inter-cluster similarity is minimized. The optimal number of clusters is then chosen based on the maximum point in the clustering gain curve.

Similarly, in [75], Zhou et al. proposed another criterion, called *CSP* (compact separation proportion) wherein the optimal number of clusters is estimated corresponding to the maximum average value of the *CSP* index. The *CSP* is used as a substitute for the *Calinski–Harabasz* index, and it functions in a similar relationship between intracluster homogeneity and intercluster separability. However, part of the CSP is focused on building a minimum spanning tree, and the proposal is of $O(n^3)$, which is prohibitive in large dataset contexts.

The *Calinski–Harabasz* index [76] is often regarded as the most suitable criterion to determine the number of clusters in hierarchical clustering. Milligan in [77] conducted an extensive experiment on 30 different CVIs and concluded the *Calinski–Harabsz* index to be the most consistent one. In the recent work by Karna et al. [78], the authors performed empirical analysis on several real-life datasets and presented an improved CVI, called $\Delta_{K_{cond}}$, that maximizes the difference of successive *Calinski–Harabasz* indices over a range of K clusters ($k = 1, 2, \ldots K$) and suggests the conditional criterion to select between 2 and the next best clustering solution. However, several instances were found where the number of clusters by the proposed $\Delta_{K_{cond}}$ criterion did not match correctly against the expert's judgment based on the dendrogram.

Additionally, several unique solutions to correctly determine the number of clusters are mentioned in the literature. In [79], Cowgill et al. proposed the genetic-algorithm-based method, *COWCLUS*, which optimizes a fitness function defined in terms of *within-cluster cohesion* and *between-cluster isolation* which itself is a *Calinski–Harabasz* criterion. However, such a method does not appear to be scalable for large datasets and, hence, not very suitable for real-life applications.

Similarly, in [80,81], Bruzzese et al. proposed a permutation test-based approach to determine the optimal number of clusters in hierarchical clustering. This too appears to be complex to execute for large-scale datasets in real time.

In [82], the reachability plot (derived from the density-based clustering methods) is used as preprocessing to identify the number of clusters of the hierarchical tree. The process of the heights of tree nodes is complex and iterative in this technique.

In [83], a single-height similarity threshold is applied, using a dynamic slider to identify the main clusters. Continuing this, in [84], Vogogias et al. used the height of the nodes to identify interesting branches of the tree. However, the concept of using non-horizontal cuts of the trees violates the ultra-metric properties of the dendrogram itself and is thus not very aligned with the objective of this current research work by the authors.

To the extent of interpreting cluster patterns, several works are studied. In [85], the authors presented an approach to interpret cluster patterns in real datasets. Gibert et al. [86] presented the *KLASS* clustering system to measure similarity in ill-structured domains. In [5], using mixed metrics was proposed while clustering complex messy datasets. Gibert et al. in [87] introduced the concept of semantic variables and generalized *Gibert's* mixed metrics for clustering heterogeneous data matrices. Another related work can be seen in [4], where the authors presented a real-life application from a wastewater treatment plant, using a clustering-based approach.

In the recent study by Suman et al. [88], a new cluster-validity-criteria was proposed that operates directly on the linkage matrix of the hierarchical clustering to detect the suitable number of clusters similar to how human experts perceive clusters visually in a dendrogram.

4. Contributions

This paper presents three novel contributions:

- A modification of the CURE approach consists of substituting the first phase of the original CURE approach by a bootstrap process that generates several small samples (S samples) from the original dataset and runs some clustering and super-classification processes to create the centroids that constitute the input of the second step of the CURE strategy. This contribution permits to scale up a hierarchical clustering process to large datasets and also reduces the CPU time drastically. Section 5.1.1 provides the details on it.
- As a consequence of using the *bootstrap–CURE* strategy in real large dataset applications, a new challenge appears of developing an automatic criterion to cut the resulting dendrograms ($S + 1$) in order to identify the number of clusters in such a way that the number of clusters manually proposed by an expert is properly approached (see Section 5.2).
- A third contribution is the proposal of an entire data science process that inserts *bootstrap–CURE* with the automatic criterion to cut the dendrograms in a process, including the steps from the preprocessing to the interpretation-oriented tools. This automatically interprets the clusters emerging from this process, in line with the works in [89,90] and with the emerging field of *explainable AI* [91]. The proposal is described in Section 5.

Therefore, this work contributes to automatically profile the operation modes of 3D printers, providing an understandable description of profiles. This bridges the gap between obtaining the profiles with advanced clustering methods and connecting them with actionable knowledge, supporting decision making. The proposal is extremely useful in a real production process to better manage 3D printers, as it can automatically detect the operational modes of 3D printers by analyzing their sensor data and is also potentially linkable to an intelligent alert system, for example. Although the methodology presented is developed in the context of 3D printing, it is easily adaptable to other situations, where a machine is monitored through sensors and thus has a wide range of applications, contributing to the maintenance and follow up of machines in industrial settings as well as contributing to solving the open problem of real-time management of the machines.

5. Methodology

As mentioned previously, this paper proposes a data science approach [92] to discover states of operation in a machine monitored through several sensors in a 3D printing scenario. As stated before, the main contribution of the paper is to introduce a modification of the CURE algorithm [63] that scales the hierarchical clustering up to big data and automatically cuts the hierarchical dendrogram so that the operation states of a certain population of printers can be identified. The proposed approach also introduces interpretation-oriented tools to provide a conceptual description of the discovered clusters to assist the printer operators in real-time decision making.

5.1. Preprocessing

The importance of *preprocessing* was highlighted in [93] along with the proposal of a general-purpose methodology for preprocessing in data science. In the current research work, sensor data consist of all numerical measurements, and the preprocessing step includes only a few steps from the general methodology. The very first step is to re-label the sensor names to support direct conceptualization and knowledge production. The actual names of the sensors are anonymized due to confidentiality issues. All variables represent sensor measurements captured as numerics but with different scales and units. Normalization is applied across sensors to bring them to a uniform scale.

The sensor data addressed in this work come from multi-jet fusion 3D printers in a smart factory setting that feeds data continuously to the cloud environment, which reduces the probability of missing-data instances to be negligibly small.

In all the experiments performed throughout the research, no missing values are observed yet. This eliminates the need to perform any missing value imputation in the current state of the research. Thus, the missing data treatment, which is based on multivariate interpolation techniques, is delayed to further steps of the research. As a matter of fact, the missing data issue will become important in the context of processed log files, which is out of the scope of this paper.

The proposed approach relies on using raw log files instead of processed ones since the authors pretend that the proposal works well in real time when introduced in an industrial manufacturing process and the method must be able to work with raw and unprocessed data. Thus, no feature extraction steps are considered to deal with transformed data. The hierarchical clustering-based methodologies are also useful since they are robust to the presence of outliers and there is no need for prior outlier detection.

Following the approach described in [94] and the main goals of this research, clustering methods are most suitable to find patterns in sensor data. Clustering is one of the widely used unsupervised distance-based learning methods that aims at deciphering the hidden patterns in the data. Some key measures of distance were discussed in [95]. Jain et al. [96] presented a good overview of clustering algorithms and their applicability in the real world. In this work, since no previous hypotheses are available about the number of real existing patterns, hierarchical clustering is used as a starting point.

5.1.1. The Proposed Bootstrap–CURE Approach

As said before, the CURE algorithm accelerates the hierarchical clustering processes so as to increase the capacity to deal with bigger datasets, but this is still not sufficient to deal with sensor datasets that provide millions of readings per second.

In this work, a modification of CURE is proposed to overcome this limitation. (Figure 2 shows an overview of the proposal.)

The proposed modification is based on the well-known mathematical fact that, given a data set of size n, decomposed in S disjunct subsets of size n_s such that

$$n = \sum_{s=1}^{S} n_s \tag{1}$$

then the square of the total size is bigger than or equal to the sum of squares of individual components

$$n^2 = (\sum_{s=1}^{S} n_s)^2 \gg \sum_{s=1}^{S} {n_s}^2 \tag{2}$$

Consequently, any quadratic algorithm running on the entire dataset of size n will be more expensive from the computational point of view than its replication on the S samples. There is no need to run any experiment to demonstrate this property since it is based on an analytical geometrical property of the sum of squares.

This means that applying the bootstrap technique to hierarchical clustering by repeating clustering of small samples from the original dataset several times will surely decrease the complexity of the total process as compared to the global clustering of the entire dataset.

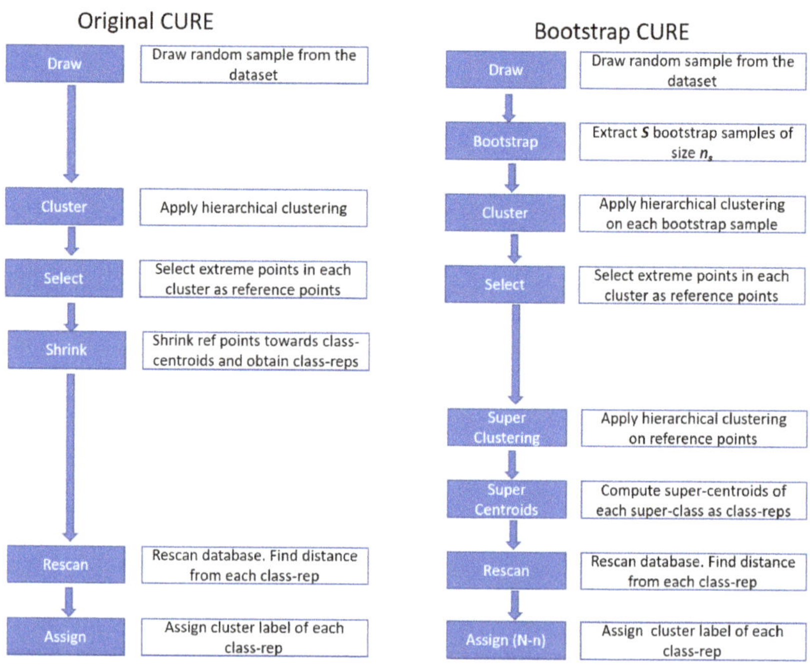

Figure 2. Comparison between original and *bootstrap–CURE* processing.

The authors introduce the *bootstrap* technique in the process in the following. As mentioned earlier, the CURE initialization step in the original definition of the method involves taking a small sample of data, applying hierarchical clustering on it, and further identifying a set of representative points to be used to extend the clusters to the whole dataset. The whole CURE algorithm is based on the idea that the initialization sample is representative of the total population and that other objects assigned in the completion step follow the same cluster structure as the one discovered in the sample. This is quite a strong hypothesis that might not necessarily hold when we are sampling a subset of data, sufficiently small to process hierarchical clustering (quadratic) in a big data dynamic environment. Hence, the authors propose to use the *bootstrap* principle to extract S small samples $(n_s \ll n)$, with n being the size of the original dataset and n_s being the size of the sth sample from the initial dataset for the initial clustering step.

This proposal is based on the idea that each sample is clustered under a hierarchical method independently, thus resulting in S dendrograms. A subsequent intermediate step of super classification is proposed as well to encompass the results of independent sample clusters. The super classification performs over the set of class representatives (centroids) coming from all the samples processed and thus the final set of clusters is more robust than the original CURE algorithm and eventually, has better representation and lower computational cost. To assign class labels to the remaining objects, the following steps are followed:

1. Draw S random samples of a single reference dataset I, each of same size n_s, without replacement.
2. Subject each sample to hierarchical clustering (in this work with *Euclidean* distance and *Ward's* method), and obtain the S dendrograms.

3. Cut the S sample dendrograms and retrieve a set of clusters for each sample. In Section 5.2, a method to automatically determine the number of clusters in each dendrogram is proposed.
4. Compute the centroid of all clusters found in the previous step and build a final dataset with all centroids.
5. Super-classification step: Apply hierarchical clustering on the centroids dataset.
6. Cut the resulting dendrogram by using the automatic criterion proposed in Section 5.2 and find the set of centroids belonging to each super-class.
7. Compute the super-centroids of each super-class.
8. Retrieve the list of original points belonging to each super-class by finding the centroids belonging to the super-class and the original elements used for each centroid.
9. Assign the label of the corresponding super-class to all elements included in the S samples used.
10. For all the elements that were not part of the S samples, compute the distances to each of the super-centroids.
11. Assign each element the class of the nearest super-centroid.

Figure 2 compares the steps in the original CURE and the proposed *bootstrap–CURE* methods.

5.2. Determining the Number of Clusters: Calinski–Harabasz Index

In any hierarchical clustering algorithm, the number of clusters emerges after plotting the dendrogram and optimizing both the homogeneity and the distinguishability of the clusters. There is an abundance of works in the literature available, dealing with the evaluation and performance of clustering [97–99]. A robust evaluation measure was provided by the Calinski–Harabasz index [76], also popularly known as the *variance-ratio* method. The Calinski–Harabasz index, for a clustering solution P consisting of k clusters from a dataset having n rows, is calculated as follows:

$$CH(k) = \frac{B_k/(k-1)}{W_k/(n-k)} \tag{3}$$

where k is the number of clusters. B_k is the between classes variability, defined as

$$B_k = \sum_{C \in P} n_C \, \mathrm{d}(\bar{i}_C, \bar{i})^2 \tag{4}$$

and W_k is the within classes variability, defined as

$$W_k = \sum_{C \in P} \sum_{\mathrm{i} \in \mathrm{C}} \mathrm{d}(\mathrm{i}, \bar{i}_C)^2 \tag{5}$$

$\bar{i}_C$ is the centroid of the cluster C, n_C is the size of cluster C, and $\bar{i}$ is the centroid of the whole dataset.

In theory, the number of clusters obtained by using the *Calinski–Harabasz* method should match with the number of clusters deduced by the experts looking at the dendrogram. In the earlier work by Karna et al. [78], the authors evaluated five different criteria based on the *Calinski–Harabasz* index over 100 datasets of varying size and assessed the validity of those five criteria in regards to visual inspection by human experts. In practice, a human expert usually finds the best cut of the dendrogram, finding the branches with a wider vertical gap between consecutive nodes. The final clusters correspond to the branches of the tree isolated by the horizontal cut. The result of the experiments performed in [88], however, indicated all criteria based on the Calinski–Harabasz index to be underperforming. It was also evidenced that the *Calinski–Harabasz* index does not align with real expert practices well enough.

Thus, Suman et al. proposed two new criteria to overcome this limitation based on the heights of the internal nodes of the dendrogram that better follow the real practices performed by experts and prove that the criterion $\Delta_{H_{cond}}$ is the one that better approaches that of experts. This is the criteria proposed to be included in the general methodology proposed in Section 5.1.1

Let $h_v, v \in 1 : n-1$ be the height of node v in a given dendrogram built over a dataset I. The values of h_v depend on the linkage method used in the hierarchical process that generates the dendrogram. The $\Delta_{H_{cond}}$ [88] is defined as follows:

$$K^*_{\Delta_{H_{cond}}} = \begin{cases} K^*_{2\Delta_H} & \text{if } K^*_{\Delta_H} = 2 \text{ and } (h_2/h_{root}) > 1/3 \\ K^*_{\Delta_H} & \text{otherwise} \end{cases} \quad (6)$$

where h_{root} and h_2 represent the heights of the two highest nodes of the dendrogram and the $K^*_{\Delta_H}$ and $K^*_{2\Delta_H}$ are the maximum and second maximum of the Δ_H criterion as defined in [78]:

$$K^*_{\Delta_H} = \underset{2 \leq k \leq K}{\operatorname{argmax}}(\Delta_{H_k}); k \in (2, 3, \ldots K-1) \quad (7)$$

where

$$\Delta_{H_k} = h_k - h_{k+1}; k \in (2, 3, \ldots K-1) \quad (8)$$

It is shown empirically that the Δ_H criterion underperforms where the best cut of the tree is two clusters, as experts are biased toward this scenario and use a heuristic to skip the two clusters' cut sometimes. The modified criterion, defined as, The $\Delta_{H_{cond}}$ incorporates this heuristic by introducing the concept of the *height factor* as a ratio between the heights of the two highest nodes of the dendrogram. This can be visualized in Figure 3. Here, the tree structure on the right (Figure 3b) represents the annotated dendrogram that reveals that the biggest gap between nodes occurs at $K = 2$, where the difference in height between involved nodes is (85.5 − 27.1 = 58.4). The second-best clustering solution occurs at $K = 4$ with the height difference as (25.9 − 19.5 = 6.4). In this case, the expert does not go for a solution of $K = 4$, but a solution of $K = 2$.

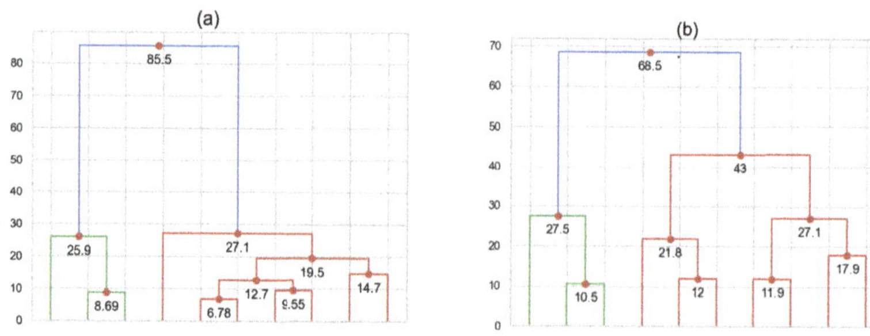

Figure 3. (**a**) Dendrogram of sample 10; (**b**) dendrogram of sample 31.

At the same time, in Figure 3a, the annotated dendrogram reveals the biggest gap of the tree as occurring at solution $K = 2$, the maximum $\Delta_{h1} = 68.5 - 43 = 25.5$, and the second best solution is seen at $K = 3$, $\Delta_{h2} = 43 - 27.5 = 15.5$. Thus, following the *height-factor* notion, the second best solution coincides with the expert's understanding of three clusters.

This corresponds to a practical situation where the cut in two clusters often provides general results, which may not be very interesting from the application point of view, and the second-best cut tends to be preferred by stakeholders, as it is more informative.

The proposed method by Suman et al. [88] shows an impressive result with the $\Delta_{H_{cond}}$ criterion matching the expert criterion in 93% of tested datasets and performs significantly better than all other criteria, including the proposals in [78].

5.3. Post-Processing: Toward Explaining the Clusters

Once the dataset is clustered, specific assessment tests are used to identify the significant sensors in the clusters according to the methodologies presented in [90]. The *class panel graphs* (CPG) [89] are used to visualize the behavior of the sensors through the different clusters; properly interpret the results of the tests given in [90] as well as to show how sensors are related with one another; and eventually to induce conceptualization to the experts. The CPG is used to analyze the conditional distribution of sensor data versus clusters. It also helps understand some key patterns among the clusters with respect to sensor behavior. Illustrative variables, which were not part of the original data used in the clustering, are also used to enrich the interpretations.

The proposal of using conditional distributions of sensors toward clusters as a basis to the automatic interpretation of the clusters is based on the post-processing methodologies discussed in [89,100,101], which is directly aligned with the introduction of a knowledge production step on the data mining process as proposed by Fayyad et al. [102]. This approach of interpreting clusters is also aligned with the emerging field of *explainable AI* [103] by providing conceptual explanations of the discovered patterns. This confers explanatory capabilities to the AI methods, which is one of the fundamental requirements for integrating a data-driven method in an intelligent decision support system.

5.4. Validation of the Proposal

The validation is performed with regards to two different aspects. Firstly, the discovered clusters are validated by adding external information from the job summary that provides a status message for each job as well as the reasons of failure (if available), which is captured separately, analyzing how this external information distributes with the discovered clusters.

Secondly, regarding the benefits of *bootstrap–CURE* with respect to scalability, the methodology is also validated in terms of computational cost. The total CPU time it takes to cluster a large dataset by the classical hierarchical clustering method and original CURE clustering method is compared against the proposed *bootstrap–CURE* approach. Hence, the scalability of the algorithm is directly tested by running several experiments of varying dataset sizes, both by the traditional and proposed approaches. The experimental results are discussed in Sections 6.5 and 6.6.

6. Applications to 3D Printer Data

6.1. Data Collection

The data from eight anonymized HP multi-jet 3D printers were collected for over 300 printing jobs and appended together to create a large dataset comprising sensors' behaviors toward various internal processes and sub-systems, including pressure, temperature, humidity, etc. A dataset of approximately 562,000 records was thus generated, containing the behavior of the machines in different phases of a print job. For the current study, the focus was given to the print phase only. For experiments, *Python 3.6* was used along with standard scientific libraries (*Pandas, NumPy, Scipy, Scikit-learn,* and *Matplotlib*). All experiments were conducted on a GPU-enabled, four-core processor, Windows computer with 32 GB memory.

6.2. Data Preprocessing

The data analysis starts with the computation of descriptive statistics and preprocessing of the variables. Preprocessing follows the scheme provided in Section 5.1. After eliminating the redundant and non-informative sensor features, the final dataset contains 46,821 sensor records across 41 sensors. Table 1 contains the final list of sensors considered in the study [9] with anonymous names to maintain confidentiality. Each sensor rules as a variable of the data matrix (one column of the dataset), whereas rows of the dataset represent measurements at a certain time interval (in seconds).

Table 1. List of sensors.

Variables	Description
Timestamp	Timestamp of the sensor recording
Sensor_1	To measure pressure in Air release system
Sensor_2	To measure Ambient temperature
Sensor_3	To measure temperature in cooling system-1
Sensor_4	To measure temperature in cooling system-2
Sensor_5	To measure temperature in cooling system-3
Sensor_6	To detect glass breakage on left fusing system
Sensor_7	To detect glass breakage on right fusing lamp
Sensor_8	To measure temperature in carriage back
Sensor_9	To measure temperature in carriage front
Sensor_10	To measure temperature in carriage middle
Sensor_11	Internal camera reading
Sensor_12	To measure the reference temperature in subsystem–back
Sensor_13	To measure the reference temperature in subsystem–front
Sensor_14	To measure the reference temperature in subsystem–middle
Sensor_15	To measure the temperature in the subsystem–back
Sensor_16	To measure the temperature in the subsystem–front
Sensor_17	To measure the temperature in the subsystem–middle
Sensor_18	To measure the reference temperature in subsystem–back
Sensor_19	To measure the reference temperature in subsystem–front
Sensor_20	To measure the reference temperature in subsystem-middle
Sensor_21	To measure the temperature in the subsystem–back
Sensor_22	To measure the temperature in the subsystem–front
Sensor_23	To measure the temperature in the subsystem–middle
Sensor_24	To check obstruction in pressure system-left
Sensor_25	Temperature coefficient sensor
Sensor_26	Temp. coefficient for fusing system1–left
Sensor_27	Temp. coefficient for fusing system1–right
Sensor_28	Temp. coefficient for fusing system2–left
Sensor_29	Temp. coefficient for fusing system2–right
Sensor_30	Temp. coefficient for camera system
Sensor_31	Temp. coefficient for Cooling left air exit
Sensor_32	Temp. coefficient for Top heating
Sensor_33	Temp. coefficient for right air exit
Sensor_34	Humidity sensor for subsystem XX
Sensor_35	Temperature sensor for subsystem XX
Sensor_36	Connectivity check sensor for fusing system-left
Sensor_37	Connectivity check sensor for fusing system–right
Sensor_38	To check obstruction in pressure system-right
Sensor_39	Sensor in subsystem
Sensor_40	Temperature Sensor in subsystem_z1
Sensor_41	Temperature Sensor in subsystem_z2

6.3. Hierarchical Clustering

As proposed in Section 5.1.1, the study began with a random sample of 10,000 records drawn without replacement to subject it to hierarchical clustering with Ward's method. The result was shown in the earlier research work of [9]. Figure 4 shows the corresponding dendrogram.

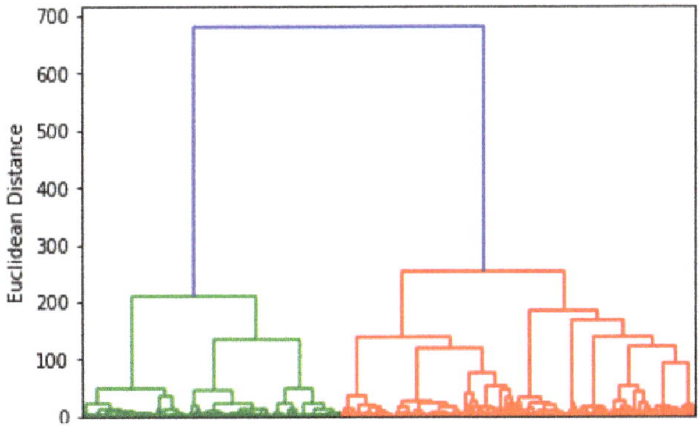

Figure 4. Classical dendrogram suggesting four potential clusters.

The four clusters so revealed in the dendrogram were further studied in detail. The use of the class panel graph (Figure 5) helped identify the redundancy among the sensor measurements and drop the redundant variables further to enrich the interpretation.

- **Cluster 0**: Specific sensors not meeting the required conditions to print without error.
- **Cluster 1**: Shows many sensors reaching the acceptable threshold (to initiate printing).
- **Cluster 2**: Conditions associated with print job to fail.
- **Cluster 3**: Imbalance in the internal cooling which would result in system error.

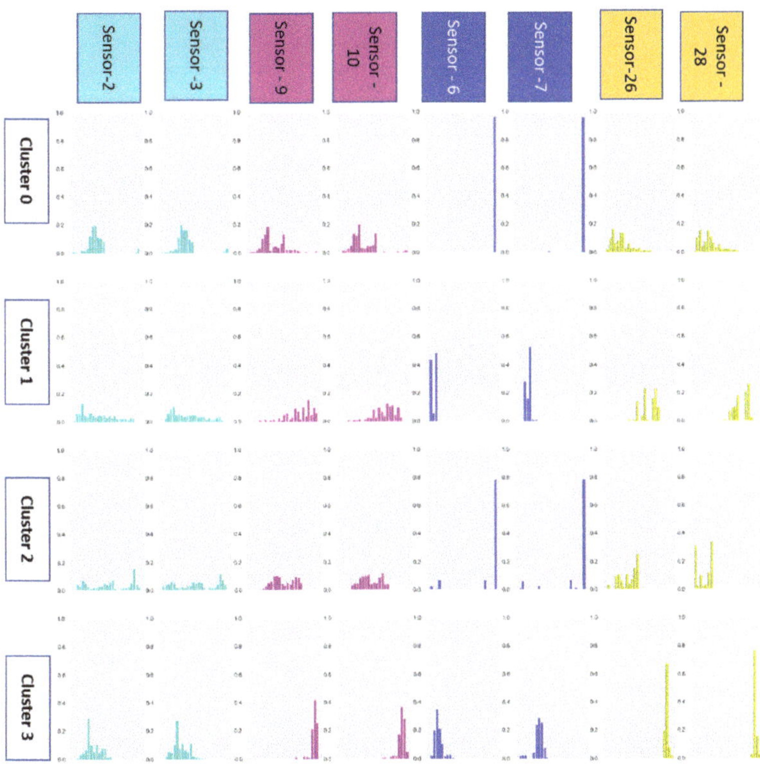

Figure 5. CPG of redundant variables.

6.4. CURE and Bootstrap–CURE

The first step toward the CURE strategy is to shuffle the original data of size n records into a sample of size $n' = pn$, with p being a certain proportion of n. For the *bootstrap–CURE*, this n' size sample is split into $r = 10$ datasets of size $s = n'/r$ records each, drawn at random without replacement. The index of each record from the original dataset was preserved to use them at the later stage of the clustering and to keep the reproducibility of experiments. All 10 sample datasets share the same set of variables as used in work [9] to enable comparisons.

This work continues the application of both CURE and *bootstrap–CURE* to data according to Section 5.1.1.

For $n = 46821$ and $s = 2000$ (and $r = 10$, so that almost half of the data size is involved in the bootstrap phase and the bootstrap samples have sufficient variability). The dendrograms of some of the samples in the bootstrap process are shown in Figure 6. The proposed approach resulted in a total of 39 centroids (3 centroids from sample_0 and 4 centroids from each of the rest). These 39 centroids merge in a new dataset of final representative points of the clusters. This new dataset of centroids would be subject to the super-clustering step (see Figure 7) in such a way that all centroids across all samples representing the same cluster group are merged. Without going into detail, it is interesting to see that all samples provide more or less similar structures, which are aligned with a good sample representation, even with smaller sample sizes as compared to the size of the complete dataset. As is evident from the dendrograms (Figure 6), all samples (1–9) exhibit 4 cluster structures, except sample_0, which shows 3 clusters.

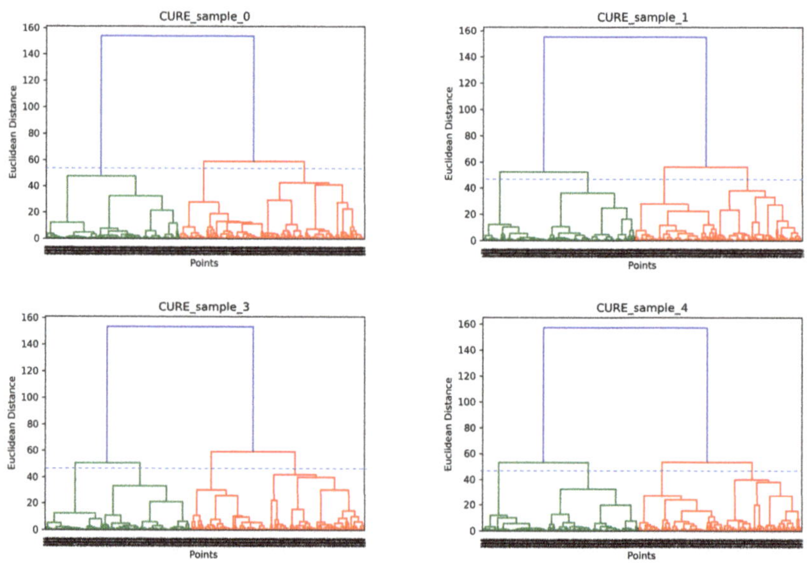

Figure 6. Sample dendrograms in bootstrap method.

The final *bootstrap–CURE* dendrogram (Figure 7) indicates a two-cluster solution; however, as a general practice, a human expert in such a situation often considers the second-best solution which shows four clusters. This strengthens the fact that a four-cluster solution (also established in [9]) is sufficient to represent the sensor behavior in the 3D printer.

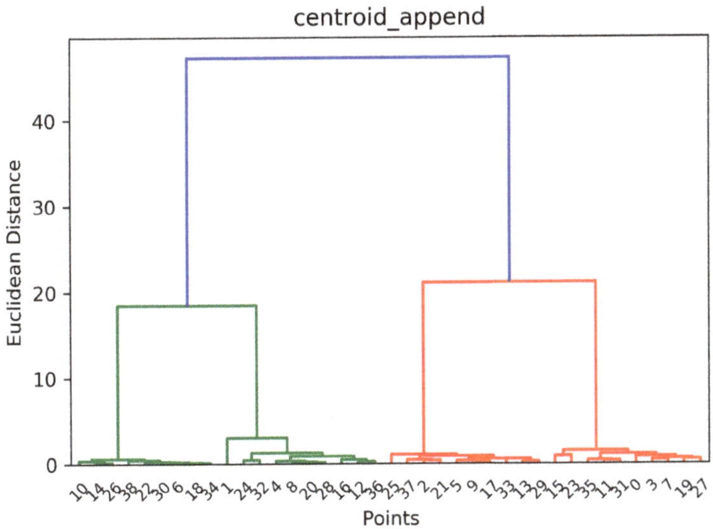

Figure 7. Dendrogram of centroids.

6.5. Comparison of Original CURE against Bootstrap–CURE

In order to evaluate the *bootstrap–CURE* implementation, the algorithm was tested on eight datasets from 3D printers of different sample sizes n, fixed $p = 0.5$ and $r = 10$ and the total time to execute CURE and *bootstrap–CURE*, as well as the hierarchical clustering algorithms, were computed, as shown in Table 2, and a graphical summary was provided (Figure 8).

Table 2. CPU time (sec) comparison among hierarchical, CURE and bootstrap–CURE methods.

Data Size	Hierarchical Clustering	CURE	Bootstrap–CURE
5000	36.296	2.922	0.578
10,000	68.875	4.125	1.406
15,000	101.094	8.297	2.094
20,000	139.625	14.25	3.156
25,000	257.375	22.125	4.187
30,000	262.718	32.422	5.328
35,000	274.109	47.469	6.75
40,000	330.391	62.703	8.062
45,000	443.578	76.078	10.141

It is quite evident from Table 2 that as the data grows in size, the original CURE implementation [63] takes quadratically longer CPU time (*TCPU*). In fact, a quadratic regression between *TCPU* and sample size provides a single significant coefficient (to the quadratic term) and significant goodness-of-fit coefficient $R^2 = 0.9982$.

The proposed bootstrap implementation processes the data quite rapidly, even for a large sample size, and appears to exceed 7× speed for bigger datasets when compared with the classical implementation of the CURE algorithm. The same appears to be up to 40× faster than the hierarchical clustering algorithm in the experiments. This property seems to be well aligned with the goals of Industry 4.0, which require quick automation and analysis of a huge amount of data.

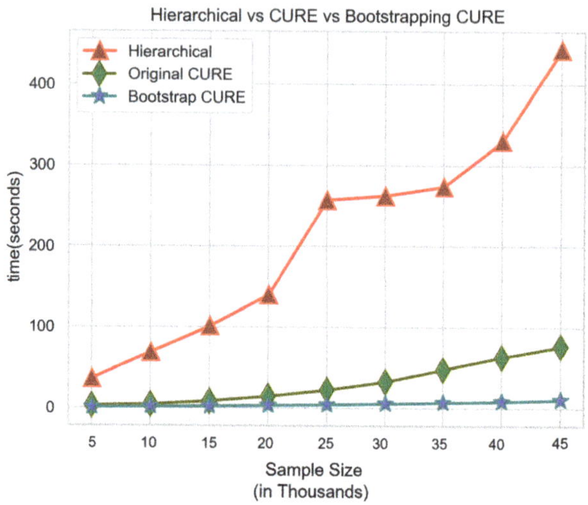

Figure 8. CPU time comparison.

It is also interesting to remark that the automatic criterion to cut the dendrograms developed in [88] was introduced into the process to help automate the entire *bootstrap–CURE* methodology, which is a great advantage.

From another perspective, the quality of the clusters does not change between CURE and *bootstrap–CURE* by construction, since the data used for the sampling step are the same in both algorithms; one is clustered together while the other is clustered, divided into 10 samples, and thus is only impacting on *TCPU* savings but not on the quality of the resulting clusters.

6.6. Post-Processing in Bootstrap–CURE Method

Post-processing of bootstrap–CURE dataset is accompanied with the class panel graph. Figure 9 shows part of the class panel graph from final data (with vertical scales of all cells normalized to a range of [0–1]). The conditional distribution of sensor_1 to sensor_5 exhibits a similar behavior as observed in the work [9].

To compare hierarchical clustering with the *bootstrap–CURE* strategy, the authors focus on the 10,000 elements common between both experiments. Table 3 shows that class 0 of both clustering approaches is distributed among the other classes, thereby showing some differences in the results for both strategies. The objects included in each cluster are not exactly the same. However, 81.36% of the printing jobs data match in the same class for both hierarchical clustering and *bootstrap–CURE*, strengthening our proposal.

Table 3. Cluster evaluation between hierarchical and bootstrap–CURE.

		Bootstrap–CURE			
		Cluster 0	**Cluster 1**	**Cluster 2**	**Cluster 3**
Hierarchical Clustering	Cluster_0	1476	549	770	133
	Cluster 1	157	2324	0	0
	Cluster 2	201	0	2639	0
	Cluster 3	54	0	0	1697

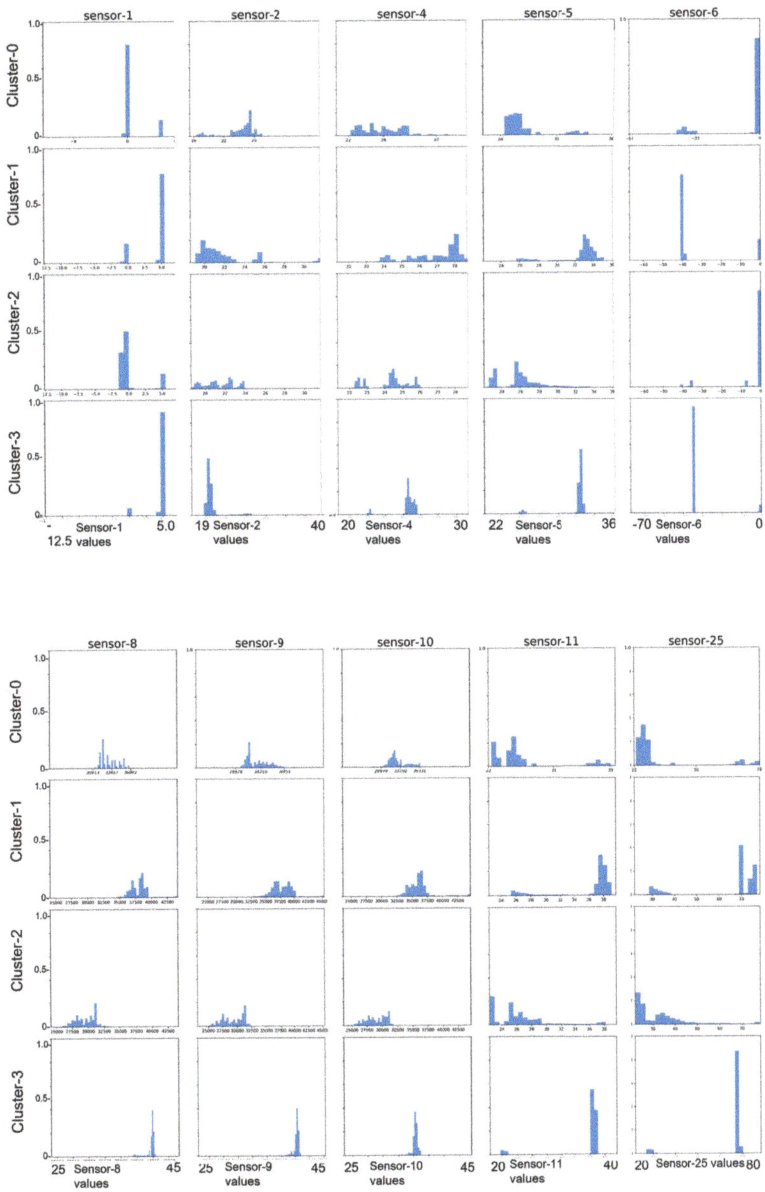

Figure 9. Class panel graph for selected sensors.

Table 4 compares the *Calinski–Harabasz (CH)* indices between the original sample using the classical hierarchical algorithm, original CURE and the *bootstrap–CURE* proposal. The *CH* index obtained on the *bootstrap–CURE* clustering is much lower than the same obtained in [9], using hierarchical clustering. It is to be noted that the *bootstrap–CURE* method is computationally much less expensive, as the assignment of labels does not need the whole set of rows to be used in clustering. This is particularly useful to approximate large data by a substantially lesser number of records to process in clustering.

Hence, the profiles discovered by both of the approaches maintain stable clusters; however, the ones obtained by the *bootstrap–CURE* method are more relevant for the big data framework, as they are based on the larger dataset size and most of the clusters show smaller standard deviations as well. Illustrative variables are also used to help understand the structure of the clusters (see Figure 10 for job status by clusters and Figure 11 for fail reasons by clusters).

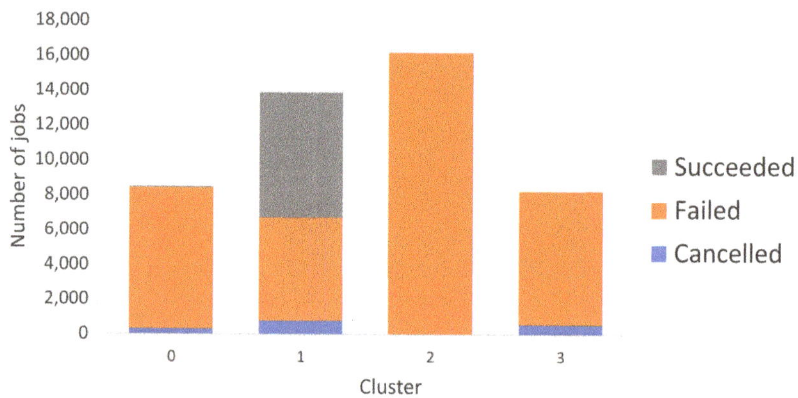

Figure 10. Job status across clusters.

Out of 41 sensors, 12 were found to be non-significant subsystems (those referring to sensor_12 through sensor_23), and both sensor_24 and sensor_38 pressure values were found to be constant and, thus, not providing any insight about the printing process. For the remaining 25 sensors, some pairs were found to be redundant according to the clusters, and thus only one of them is considered (sensor_2–sensor_3, sensor_6–sensor_7, and sensor_26–sensor_28) for further study. The remaining 24 variables were analyzed to be following three basic profiles that can be interpreted in detail by analyzing the distribution of sensors.

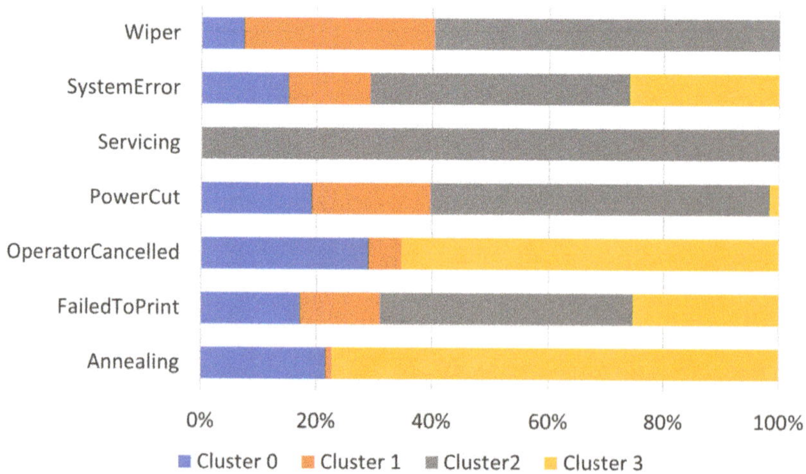

Figure 11. Job failure reasons.

Table 4. Calinski–Harbasz index comparison.

Attributes	Hierarchical Clustering	Original CURE	Bootstrap–CURE
Sample size	10,000	10,000	10,000
No. of clusters	4	4	4
Calinski–Harabasz	9208.0117	2885.2101	5198.4981

Cluster 0 (non-conformable jobs):

The main characteristics of this cluster are as follows. Firstly, sensor_1 does not reach the recommended pressure for most of the instances and, hence, the job does not start. Sensor_3 shows a higher than expected temperature, which might result in system error. However, sensor_39 shows the measurement within specifications. Thus, the cluster should lead to unsuccessful printing. Further, the cluster is also found to contain several failed jobs with few successful ones, with the most common reasons being *failed-to-print* and *system-error*. This aligns with our interpretation based on unsupervised analysis of the cluster data.

Cluster 1 (successful jobs):

This cluster is characterized as follows. Sensor_1 has the recommended pressure reading to initiate the job. The temperature measurement in sensor_3 is also adequate as per the domain knowledge, as is sensor_39. In general, the behavior of sensor_39 and sensor_1 should match what is seen in this cluster. Sensor_35 also has the correct measurements; however, sensor_34 does not match the behavior all the time. Hence, this cluster is expected to lead to successful jobs, but there may be a few failures due to a mismatch in the behavior of sensor_35 and sensor_34. Further, analysis of the data reveals that the cluster indeed has mostly successful jobs with few that are 'failed' or 'operator-canceled'. The main reason found for the failure of the job is 'power-cut', while some failed to print because an operator canceled the jobs.

Cluster 2 (failed jobs due to malfunction of components):

In this cluster, sensor_1 does not reach the recommended pressure value to start the job and this should result in system error. Although the measurements of sensor_3 are within spec, the behavior of sensor_1 and sensor_39 does not match (the pressure values should be reversed to each other), which is evident almost throughout the data. Additionally, sensor_34 shows measurements that are higher than normal. Overall, the cluster shows quite unfavorable readings that would hamper a job to start or would lead to failure. Further analysis revealed that most of the jobs found in this cluster failed with the main reason being 'system error'.

Cluster 3 (failed jobs due to imbalance of sensors):

This cluster is peculiar, as sensor_1 and sensor_3 both are found to be within the recommended specifications, as are sensor_4 and sensor_5. This indicates sufficient conditions for a print job to start. Further, the measurements of sensor_1 and sensor_39 also correspond to each other, which is necessary for a print job to continue. Sensor_34 shows very high readings and thus points to a higher level of humidity; sensor_32 and sensor_33, which control the overall airflow in the machine, run at a lower temperature. This creates an imbalance in the internal cooling environment, which might result in system error. Further verification of the jobs in this cluster disclosed that most of them failed due to 'system error'.

7. Discussion

This paper presents a sensor data science contribution in the field of Industry 4.0 and comprises all steps from the very first of identifying operation modes in a machine monitoring scenario, where large multivariate sensor data are continuously reported. Understanding these operation modes and identifying the patterns of failures can be of interest for further development of statistical models for predictive maintenance, and also to analyze the relationships among other sensors that suggest future improvements in the design of the 3D printer.

It is important to note that the methodology proposed here is general for any large-scale sensor data environment, and this applies to other Industry 4.0 domains, such as gas turbines or air quality in smart cities.

The specific application presented here is with regards to the industrial 3D printers, where customers' print-layer images or videos are considered confidential, and the method has to rely solely on the sensor data to obtain a global understanding of machine behavior. This is only possible through unsupervised analysis.

In fact, in a real industry context, rapid product innovation makes it difficult to prepare supervised data continuously to learn models. When unsupervised learning is faced, there is no possibility to compute either the misclassification rates or the confusion matrices. To introduce the proposed methodology in a real production environment, the technique must be capable of executing large datasets rapidly in a smart factory setting. The current research identified a bootstrapping-based CURE technique that indeed executes much faster than the traditional methods.

With the help of the proposed methodology, sensor data from the real industrial large 3D printers were analyzed in lesser computational time, and four potential clusters of sensor profiles were discovered. Three of them were associated with print job failure due to different problems, and specific sensors reporting abnormal values were observed. The job summary captured separately is a good resource to validate whether the patterns interpreted from the proposed methodology point to the actual scenarios in the print jobs. Thus, along with the domain knowledge, one can find useful patterns of sensor behavior leading to the overall quality of the print job.

This proposed methodology is unsupervised and scalable to big data frameworks in real-time applications. Compared with the CURE algorithm, without including the proposed bootstrap-based modification, this approach was also found to be very efficient with minimal CPU cost and an increase in speed of up to two orders of magnitude for large datasets. This seems quite promising for Industry 4.0, where such a technique could help retrieve insights from machines or printers with limited computational power rather quickly.

8. Conclusions

In this paper, a new data-science-based intelligent methodology is proposed to assist the management of 3D printers. The work presented in this paper includes three novel contributions.

The main contribution of the proposal is a new scalable hierarchical clustering method, called *bootstrap–CURE*, which modifies the first step of the original CURE algorithm by introducing a combination of a bootstrap strategy and a super-classification method so that large datasets can be hierarchically clustered. Apart from scaling up the process to large datasets, the impact on the CPU time reduction is also drastic. This provides a new hierarchical-clustering-based algorithm that suitably scales up to the large sensor data and becomes applicable to additive manufacturing. Hierarchical clustering has many advantages, as it is a non-supervised learning method that does not require human intervention. In particular, the hierarchical clustering family does not require any prior hypotheses for the number of clusters, which is extremely useful in additive manufacturing, as, in real cases, users have no a priori idea of the real number of clusters, and it is much better to leave the algorithm to discover it.

Since hierarchical clustering is quadratic, it is not suitable in its original form for large datasets. The proposed *bootstrap–CURE* method overcomes this limitation and reduces the computational cost drastically by keeping all advantages of the hierarchical clustering.

In addition, a second contribution is a novel criterion to automatically determine the number of clusters from the dendrogram in such a way that the results properly approach what clustering experts find by manual inspection. The proposed criterion is based on the analysis of the level indices of the internal nodes of the dendrograms. Being computationally cheap, this method can be introduced in the *bootstrap–CURE* strategy without incrementing the computational cost significantly.

This provides a clustering method that is fully automated and can be used with sensor data coming directly from the 3D printers to identify the operational modes of the machines. Finally, the *bootstrap–CURE* was introduced in a complete data-science methodology to identify those operational modes by obtaining an automatic interpretation of the clusters' meaning. This provides a very useful decision support tool in the management of 3D printers or complex sensor-equipped machines in general (aero-generators, gas turbines, etc.).

The proposal was tested on real data coming from real 3D printers with very promising results.

For the future, the integration of the proposed method in an intelligent decision support system that can interpret the operation of an entire fleet of machines in real time is in progress. This will help manufacturers to build better predictive maintenance policies or identify possibilities to improve the 3D printers' design as well. Regarding the proposed algorithm itself, in the current formulation of the *bootstrap–CURE* approach, the shrinking step (as used in original CURE) is skipped before the integral clustering of all objects since a super-classification process is used with the bootstrap centroids. However, the improvements of implementing the shrinking on top of them will be analyzed in the mid-term. Once the behavior of 3D printers is profiled and interpreted, in the next steps of the research, a reduced subset of significant sensors associated with the different types of failures will be identified, and predictive models will be developed to identify anomalies in advance.

Author Contributions: Conceptualization: A.K. and K.G.; methodology: S.S. and K.G.; validation: S.S. and K.G. and A.K.; resources: A.K.; visualization: S.S.; supervision: K.G. All authors have read and agreed to the published version of the manuscript.

Funding: This work was supported in part by KEMLG-at-IDEAI (UPC) under Grant SGR-2017-574 from the Catalan government.

Conflicts of Interest: The authors declare no conflict of interest. The funders had no role in the design of the study; in the collection, analyses, or interpretation of data; in the writing of the manuscript; or in the decision to publish the results.

Abbreviations

The following abbreviations are used in this manuscript:

CVI	Cluster validity indices
K_f	Optimal number of clusters using criterion f
h_i	Height of ith node in dendrogram
h_{root}	Height of the root node in dendrogram
CPG	Class panel graph
CH	Calinksi–Harabasz index
$TCPU$	Total CPU time in seconds
$\Delta_{H_{cond}}$	Proposed CVI based on dendrogram height

References

1. Rüßmann, M.; Lorenz, M.; Gerbert, P.; Waldner, M.; Justus, J.; Engel, P.; Harnisch, M. Industry 4.0: The future of productivity and growth in manufacturing industries. *Boston Consult. Group* **2015**, *9*, 54–89.
2. Sureshkumar, P.; Rajesh, R. The Analysis of Different Types of IoT Sensors and security trend as Quantum chip for Smart City Management. *IOSR J. Bus. Manag. (IOSR-JBM)* **2018**, *20*, 55–60.
3. Kang, H.S.; Lee, J.Y.; Choi, S.; Kim, H.; Park, J.H.; Son, J.Y.; Kim, B.H.; Do Noh, S. Smart manufacturing: Past research, present findings, and future directions. *Int. J. Precis. Eng. Manuf.-Green Technol.* **2016**, *3*, 111–128. [CrossRef]
4. Gibert, K.; Rodríguez-Silva, G.; Rodríguez-Roda, I. Knowledge discovery with clustering based on rules by states: A water treatment application. *Environ. Model. Softw.* **2010**, *25*, 712–723. [CrossRef]
5. Gibert, K.; Nonell, R. Impact of mixed metrics on clustering. In *Iberoamerican Congress on Pattern Recognition*; Springer: Berlin/Heidelberg, Germany, 2003; pp. 464–471.
6. Marti-Puig, P.; Blanco-M, A.; Cárdenas, J.J.; Cusidó, J.; Solé-Casals, J. Effects of the pre-processing algorithms in fault diagnosis of wind turbines. *Environ. Model. Softw.* **2018**, *110*, 119–128. [CrossRef]
7. Wong, V.K.; Hernandez, A. A Review of Additive Manufacturing. *ISRN Mech. Eng.* **2012**, *2012*, 1–10. [CrossRef]
8. Nale, S.B.; Kalbande, A.G. A Review on 3D Printing Technology. *Int. J. Innov. Emerg. Res. Eng.* **2015**, *2*, 2394–5494.
9. Karna, A.; Gibert, K. Using Hierarchical Clustering to Understand Behavior of 3D Printer Sensors. *Adv. Intell. Syst. Comput.* **2020**, *976*, 150–159. [CrossRef]
10. Chalapathy, R.; Chawla, S. Deep learning for anomaly detection: A survey. *arXiv* **2019**, arXiv:1901.03407.
11. Malhotra, P.; Ramakrishnan, A.; Anand, G.; Vig, L.; Agarwal, P.; Shroff, G. LSTM-based encoder-decoder for multi-sensor anomaly detection. *arXiv* **2016**, arXiv:1607.00148.
12. van Wyk, F.; Wang, Y.; Khojandi, A.; Masoud, N. Real-Time Sensor Anomaly Detection and Identification in Automated Vehicles. *IEEE Trans. Intell. Transp. Syst.* **2019**, *21*, 1264–1276. [CrossRef]
13. Sakurada, M.; Yairi, T. Anomaly detection using autoencoders with nonlinear dimensionality reduction. In Proceedings of the MLSDA 2014 2nd Workshop on Machine Learning for Sensory Data Analysis, Gold Coast, QLD, Australia, 2 December 2014; pp. 4–11.
14. Parllaku, F.; Zaman, A.; Shah, F.; Karna, A.; de Pena, S. Using computational intelligence for smart device operation monitoring. In Proceedings of the 2019 International Conference on Computational Intelligence and Knowledge Economy (ICCIKE), Dubai, United Arab Emirates, 11–12 December 2019; pp. 124–129.
15. Karna, A.; Shah, F. Machine Learning Based Approach to Process Characterization for Smart Devices in 3D Industrial Manufacturing. In Proceedings of the 2020 International Conference on Electrical, Communication, and Computer Engineering (ICECCE), Istanbul, Turkey, 12–13 June 2020; pp. 1–6.
16. Ouyang, Z.; Niu, J.; Guizani, M. Improved vehicle steering pattern recognition by using selected sensor data. *IEEE Trans. Mob. Comput.* **2017**, *17*, 1383–1396. [CrossRef]
17. Erfani, S.M.; Rajasegarar, S.; Karunasekera, S.; Leckie, C. High-dimensional and large-scale anomaly detection using a linear one-class SVM with deep learning. *Pattern Recognit.* **2016**, *58*, 121–134. [CrossRef]
18. Emadi, H.S.; Mazinani, S.M. A novel anomaly detection algorithm using DBSCAN and SVM in wireless sensor networks. *Wirel. Pers. Commun.* **2018**, *98*, 2025–2035. [CrossRef]
19. Liu, L.; Guo, Q.; Liu, D.; Peng, Y. Data-driven remaining useful life prediction considering sensor anomaly detection and data recovery. *IEEE Access* **2019**, *7*, 58336–58345. [CrossRef]
20. Wulsin, D.; Blanco, J.; Mani, R.; Litt, B. Semi-Supervised Anomaly Detection for EEG Waveforms Using Deep Belief Nets. In Proceedings of the 2010 Ninth International Conference on Machine Learning and Applications, Washington, DC, USA, 12–14 December 2010; pp. 436–441. [CrossRef]
21. Salem, O.; Naseem, A.; Mehaoua, A. Epileptic seizure detection from EEG signal using Discrete Wavelet Transform and Ant Colony classifier. In Proceedings of the 2014 IEEE International Conference on Communications (ICC), Sydney, Australia, 10–14 June 2014; pp. 3529–3534. [CrossRef]
22. Wibisono, A.; Jatmiko, W.; Wisesa, H.A.; Hardjono, B.; Mursanto, P. Traffic big data prediction and visualization using fast incremental model trees-drift detection (FIMT-DD). *Knowl.-Based Syst.* **2016**, *93*, 33–46. [CrossRef]
23. Riveiro, M.; Falkman, G. Interactive Visualization of Normal Behavioral Models and Expert Rules for Maritime Anomaly Detection. In Proceedings of the 2009 Sixth International Conference on Computer Graphics, Imaging and Visualization, Tianjin, China, 11–14 August 2009; pp. 459–466. [CrossRef]
24. Salehi, A.; Jimenez-Berni, J.; Deery, D.M.; Palmer, D.; Holland, E.; Rozas-Larraondo, P.; Chapman, S.C.; Georgakopoulos, D.; Furbank, R.T. SensorDB: A virtual laboratory for the integration, visualization and analysis of varied biological sensor data. *Plant Methods* **2015**, *11*, 53. [CrossRef] [PubMed]
25. Nowak, R.D. Distributed EM algorithms for density estimation and clustering in sensor networks. *IEEE Trans. Signal Process.* **2003**, *51*, 2245–2253. [CrossRef]
26. Kravchik, M.; Shabtai, A. Detecting Cyber Attacks in Industrial Control Systems Using Convolutional Neural Networks. In *Proceedings of the 2018 Workshop on Cyber-Physical Systems Security and PrivaCy, Toronto, ON, Canada, 19 October 2018*; Association for Computing Machinery: New York, NY, USA, 2018; pp. 72–83. [CrossRef]

27. Dong, B.; Andrews, B. Sensor-based occupancy behavioral pattern recognition for energy and comfort management in intelligent buildings. In Proceedings of the Eleventh International IBPSA Conference, Glasgow, Scotland, 27–30 July 2009; International Building Performance Simulation Association: Vancouver, BC, Canada, 2009; pp. 1444–1451.
28. Hromic, H.; Le Phuoc, D.; Serrano, M.; Antonić, A.; Žarko, I.P.; Hayes, C.; Decker, S. Real time analysis of sensor data for the internet of things by means of clustering and event processing. In Proceedings of the 2015 IEEE International Conference on Communications (ICC), London, UK, 8–12 June 2015; pp. 685–691.
29. Loane, J.; O'Mullane, B.; Bortz, B.; Knapp, R.B. Interpreting presence sensor data and looking for similarities between homes using cluster analysis. In Proceedings of the 2011 5th International Conference on Pervasive Computing Technologies for Healthcare (PervasiveHealth) and Workshops, Dublin, Ireland, 23–26 May 2011; pp. 438–445.
30. Uhlmann, E.; Pontes, R.P.; Laghmouchi, A.; Bergmann, A. Intelligent pattern recognition of a SLM machine process and sensor data. *Procedia Cirp* **2017**, *62*, 464–469. [CrossRef]
31. Grasso, M.; Colosimo, B.M. Process defects and in situ monitoring methods in metal powder bed fusion: a review. *Meas. Sci. Technol.* **2017**, *28*, 044005. [CrossRef]
32. Grasso, M.; Colosimo, B. A statistical learning method for image-based monitoring of the plume signature in laser powder bed fusion. *Robot. Comput.-Integr. Manuf.* **2019**, *57*, 103–115. [CrossRef]
33. Mani, M.; Feng, S.; Lane, B.; Donmez, A.; Moylan, S.; Fesperman, R. Measurement science needs for real-time control of additive manufacturing powder bed fusion processes. *Int. J. Prod. Res.* **2017**, *55*, 1400–1418. [CrossRef]
34. Repossini, G.; Laguzza, V.; Grasso, M.; Colosimo, B.M. On the use of spatter signature for in-situ monitoring of Laser Powder Bed Fusion. *Addit. Manuf.* **2017**, *16*, 35–48. [CrossRef]
35. Colosimo, B.M.; Grasso, M. Spatially weighted PCA for monitoring video image data with application to additive manufacturing. *J. Qual. Technol.* **2018**, *50*, 391–417. [CrossRef]
36. Yuan, B.; Guss, G.M.; Wilson, A.C.; Hau-Riege, S.P.; DePond, P.J.; McMains, S.; Matthews, M.J.; Giera, B. Machine-Learning-Based Monitoring of Laser Powder Bed Fusion. *Adv. Mater. Technol.* **2018**, *3*, 1800136. [CrossRef]
37. Salahshoor, K.; Mosallaei, M.; Bayat, M. Centralized and decentralized process and sensor fault monitoring using data fusion based on adaptive extended Kalman filter algorithm. *Measurement* **2008**, *41*, 1059–1076. [CrossRef]
38. He, K.; Zhang, Q.; Hong, Y. Profile monitoring based quality control method for fused deposition modeling process. *J. Intell. Manuf.* **2019**, *30*, 947–958. [CrossRef]
39. Zang, Y.; Qiu, P. Phase I monitoring of spatial surface data from 3D printing. *Technometrics* **2018**, *60*, 169–180. [CrossRef]
40. March, S.T.; Scudder, G.D. Predictive maintenance: strategic use of IT in manufacturing organizations. *Inf. Syst. Front.* **2019**, *21*, 327–341. [CrossRef]
41. Poór, P.; Basl, J.; Zenisek, D. Predictive Maintenance 4.0 as next evolution step in industrial maintenance development. In Proceedings of the 2019 International Research Conference on Smart Computing and Systems Engineering (SCSE), Colombo, Sri Lanka, 28 March 2019; pp. 245–253.
42. Ruiz-Sarmiento, J.R.; Monroy, J.; Moreno, F.A.; Galindo, C.; Bonelo, J.M.; Gonzalez-Jimenez, J. A predictive model for the maintenance of industrial machinery in the context of industry 4.0. *Eng. Appl. Artif. Intell.* **2020**, *87*, 103289. [CrossRef]
43. Bonci, A.; Longhi, S.; Nabissi, G.; Verdini, F. Predictive Maintenance System using motor current signal analysis for Industrial Robot. In Proceedings of the 2019 24th IEEE International Conference on Emerging Technologies and Factory Automation (ETFA), Zaragoza, Spain, 10–13 September 2019; pp. 1453–1456.
44. Lin, C.; Hsieh, Y.; Cheng, F.; Huang, H.; Adnan, M. Time Series Prediction Algorithm for Intelligent Predictive Maintenance. *IEEE Robot. Autom. Lett.* **2019**, *4*, 2807–2814. [CrossRef]
45. Gibert, K.; Marti-Puig, P.; Cusidó, J.; Solé-Casals, J. Identifying health status of wind turbines by using self organizing maps and interpretation-oriented post-processing tools. *Energies* **2018**, *11*, 723.
46. Luo, B.; Wang, H.; Liu, H.; Li, B.; Peng, F. Early Fault Detection of Machine Tools Based on Deep Learning and Dynamic Identification. *IEEE Trans. Ind. Electron.* **2019**, *66*, 509–518. [CrossRef]
47. Nguyen, K.T.; Medjaher, K. A new dynamic predictive maintenance framework using deep learning for failure prognostics. *Reliab. Eng. Syst. Saf.* **2019**, *188*, 251–262. [CrossRef]
48. der Mauer, M.A.; Behrens, T.; Derakhshanmanesh, M.; Hansen, C.; Muderack, S. Applying sound-based analysis at porsche production: Towards predictive maintenance of production machines using deep learning and internet-of-things technology. In *Digitalization Cases*; Springer: Berlin/Heidelberg, Germany, 2019; pp. 79–97.
49. Shi, S.; Wang, Q.; Xu, P.; Chu, X. Benchmarking state-of-the-art deep learning software tools. In Proceedings of the 2016 7th International Conference on Cloud Computing and Big Data (CCBD), Macau, China, 16–18 November 2016; pp. 99–104. [CrossRef]
50. HP Jet Fusion 3D 4200 Printer Review 2018 | Industrial 3D Printer Reviews, 0. Available online: https://www.3dbeginners.com/hp-jet-fusion-3d-4200-review/ (accessed on 19 October 2019).
51. Xu, R.; Wunsch, D. Survey of clustering algorithms. *IEEE Trans. Neural Netw.* **2005**, *16*, 645–678. [CrossRef]
52. Gupta, A.; Datta, S.; Das, S. Fuzzy clustering to identify clusters at different levels of fuzziness: An evolutionary multiobjective optimization approach. *IEEE Trans. Cybern.* **2019**, *51*, 2601–2611. [CrossRef]

53. Lahmar, I.; Zaier, A.; Yahia, M.; Bouallegue, R. A New Self Adaptive Fuzzy Unsupervised Clustering Ensemble Based On Spectral Clustering. In Proceedings of the 2020 17th International Multi-Conference on Systems, Signals & Devices (SSD), Sfax, Tunisia, 20–23 July 2020; pp. 1–5.
54. Shirkhorshidi, A.S.; Wah, T.Y.; Shirkhorshidi, S.M.R.; Aghabozorgi, S. Evolving Fuzzy Clustering Approach: An Epoch Clustering That Enables Heuristic Postpruning. *IEEE Trans. Fuzzy Syst.* **2021**, *29*, 560–568. [CrossRef]
55. Sebastian, A.; Cistulli, P.A.; Cohen, G.; de Chazal, P. Characterisation of Upper Airway Collapse in OSA Patients Using Snore Signals: A Cluster Analysis Approach. In Proceedings of the 2020 42nd Annual International Conference of the IEEE Engineering in Medicine & Biology Society (EMBC), Montreal, QC, Canada, 20–24 July 2020; pp. 5124–5127.
56. Chakraborty, S.; Das, S. Detecting meaningful clusters from high-dimensional data: A strongly consistent sparse center-based clustering approach. *IEEE Trans. Pattern Anal. Mach. Intell.* **2020**. [CrossRef]
57. Gondeau, A.; Aouabed, Z.; Hijri, M.; Peres-Neto, P.; Makarenkov, V. Object weighting: a new clustering approach to deal with outliers and cluster overlap in computational biology. *IEEE/ACM Trans. Comput. Biol. Bioinform.* **2019**, *18*, 633–643. [CrossRef]
58. Li, K.; Cao, X.; Ge, X.; Wang, F.; Lu, X.; Shi, M.; Yin, R.; Mi, Z.; Chang, S. Meta-heuristic optimization-based two-stage residential load pattern clustering approach considering intra-cluster compactness and inter-cluster separation. *IEEE Trans. Ind. Appl.* **2020**, *56*, 3375–3384.
59. Zhao, X.; Wang, Z.; Gao, L.; Li, Y.; Wang, S. Incremental face clustering with optimal summary learning via graph convolutional network. *Tsinghua Sci. Technol.* **2021**, *26*, 536–547. [CrossRef]
60. Menon, V.; Muthukrishnan, G.; Kalyani, S. Subspace clustering without knowing the number of clusters: A parameter free approach. *IEEE Trans. Signal Process.* **2020**, *68*, 5047–5062. [CrossRef]
61. Firdaus, S.; Uddin, M. A Survey on Clustering Algorithms and Complexity Analysis. *Int. J. Comput. Sci. Issues (IJCSI)* **2015**, *12*, 62.
62. Kuchaki Rafsanjani, M.; Asghari Varzaneh, Z.; Emami Chukanlo, N. A Survey Of Hierarchical Clustering Algorithms. *J. Math. Comput. Sci.* **2012**, *05*, 229–240. [CrossRef]
63. Guha, S.; Rastogi, R.; Shim, K. CURE: An efficient clustering algorithm for large databases. *Inf. Syst.* **2001**, *26*, 35–58. [CrossRef]
64. Jagadish, H.; Gehrke, J.; Labrinidis, A.; Papakonstantinou, Y.; Patel, J.M.; Ramakrishnan, R.; Shahabi, C. Big data and its technical challenges. *Commun. ACM* **2014**, *57*, 86–94. [CrossRef]
65. Kawamoto, T.; Kabashima, Y. Cross-validation estimate of the number of clusters in a network. *Sci. Rep.* **2017**, *7*, 3327. [CrossRef]
66. Fu, W.; Perry, P.O. Estimating the number of clusters using cross-validation. *J. Comput. Graph. Stat.* **2020**, *29*, 162–173. [CrossRef]
67. McIntyre, R.M.; Blashfield, R.K. A nearest-centroid technique for evaluating the minimum-variance clustering procedure. *Multivar. Behav. Res.* **1980**, *15*, 225–238. [CrossRef]
68. Krieger, A.M.; Green, P.E. A cautionary note on using internal cross validation to select the number of clusters. *Psychometrika* **1999**, *64*, 341–353. [CrossRef]
69. Overall, J.E.; Magee, K.N. Replication as a rule for determining the number of clusters in hierarchial cluster analysis. *Appl. Psychol. Meas.* **1992**, *16*, 119–128. [CrossRef]
70. Tonidandel, S.; Overall, J.E. Determining the number of clusters by sampling with replacement. *Psychol. Methods* **2004**, *9*, 238. [CrossRef] [PubMed]
71. Fang, Y.; Wang, J. Selection of the number of clusters via the bootstrap method. *Comput. Stat. Data Anal.* **2012**, *56*, 468–477. [CrossRef]
72. Sevilla-Villanueva, B.; Gibert, K.; Sànchez-Marrè, M. Using CVI for understanding class topology in unsupervised scenarios. In *Proceedings of the Spanish Association for Artificial Intelligence, Salamanca, Spain, 14–16 September 2016*; Springer: Berlin/Heidelberg, Germany, 2016; pp. 135–149.
73. Tibshirani, R.; Walther, G.; Hastie, T. Estimating the number of clusters in a data set via the gap statistic. *J. R. Stat. Soc. Ser. B (Stat. Methodol.)* **2001**, *63*, 411–423. [CrossRef]
74. Jung, Y.; Park, H.; Du, D.Z.; Drake, B.L. A decision criterion for the optimal number of clusters in hierarchical clustering. *J. Glob. Optim.* **2003**, *25*, 91–111. [CrossRef]
75. Zhou, S.; Xu, Z.; Liu, F. Method for determining the optimal number of clusters based on agglomerative hierarchical clustering. *IEEE Trans. Neural Netw. Learn. Syst.* **2016**, *28*, 3007–3017. [CrossRef] [PubMed]
76. Caliński, T.; Harabasz, J. A dendrite method for cluster analysis. *Commun. Stat.-Theory Methods* **1974**, *3*, 1–27. [CrossRef]
77. Milligan, G.W. A Monte Carlo study of thirty internal criterion measures for cluster analysis. *Psychometrika* **1981**, *46*, 187–199. [CrossRef]
78. Karna, A.; Gibert, K. Automatic identification of the number of clusters in hierarchical clustering. *Neural Comput. Appl.* **2021**, *34*, 119–134. [CrossRef]
79. Cowgill, M.C.; Harvey, R.J.; Watson, L.T. A genetic algorithm approach to cluster analysis. *Comput. Math. Appl.* **1999**, *37*, 99–108. [CrossRef]
80. Bruzzese, D.; Vistocco, D. Cutting the dendrogram through permutation tests. In Proceedings of the COMPSTAT'2010, Paris, France, 22–27 August 2010; pp. 847–854.
81. Bruzzese, D.; Vistocco, D. DESPOTA: DEndrogram slicing through a pemutation test approach. *J. Classif.* **2015**, *32*, 285–304. [CrossRef]

82. Sander, J.; Qin, X.; Lu, Z.; Niu, N.; Kovarsky, A. Automatic extraction of clusters from hierarchical clustering representations. In *Proceedings of the Pacific-Asia Conference on Knowledge Discovery and Data Mining, Seoul, Korea, 30 April–2 May 2003*; Springer: Berlin/Heidelberg, Germany, 2003; pp. 75–87.
83. Vogogias, A.; Kennedy, J.; Archaumbault, D.; Smith, V.A.; Currant, H. Mlcut: Exploring multi-level cuts in dendrograms for biological data. In *Proceedings of the Computer Graphics and Visual Computing Conference (CGVC), London, UK, 10–11 September 2016*; Eurographics Association: Bournemouth, UK, 2016.
84. Vogogias, A.; Kennedy, J.; Archambault, D.W. Hierarchical Clustering with Multiple-Height Branch-Cut Applied to Short Time-Series Gene Expression Data. In *EuroVis (Posters)*; 2016; pp. 1–3. Available online: https://diglib.eg.org/handle/10.2312/eurp20161127 (accessed on 19 October 2019).
85. Sevilla-Villanueva, B.; Gibert, K.; Sànchez-Marrè, M. A methodology to discover and understand complex patterns: Interpreted Integrative Multiview Clustering (I2MC). *Pattern Recognit. Lett.* **2017**, *93*, 85–94. [CrossRef]
86. Gibert, K.; Cortés García, C.U. Weighting quantitative and qualitative variables in clustering methods. *Mathw. Soft Comput.* **1997**, *4*, 3.
87. Gibert, K.; Valls, A.; Batet, M. Introducing semantic variables in mixed distance measures: Impact on hierarchical clustering. *Knowl. Inf. Syst.* **2014**, *40*, 559–593. [CrossRef]
88. Suman, S.; Karna, A.; Gibert, K. Towards Expert-nspired Automatic Criterion to Cut a Dendrogram for Real-Industrial Applications. *Artif. Intell. Res. Dev.* **2021**, 339, 235.
89. Gibert, K.; García-Rudolph, A.; Rodríguez-Silva, G. The Role of KDD Support- Interpretation Tools in the Conceptualization of Medical Profiles: An Application to Neurorehabilitation. *ACTA Inform. Medica* **2008**, *16*, 178–182.
90. Gibert, K.; Sevilla-Villanueva, B.; Sànchez-Marrè, M. The role of significance tests in consistent interpretation of nested partitions. *J. Comput. Appl. Math.* **2016**, *292*, 623–633. [CrossRef]
91. Adadi, A.; Berrada, M. Peeking Inside the Black-Box: A Survey on Explainable Artificial Intelligence (XAI). *IEEE Access* **2018**, *6*, 52138–52160. [CrossRef]
92. Gibert, K.; Horsburgh, J.S.; Athanasiadis, I.N.; Holmes, G. Environmental Data Science. *Environ. Model. Softw.* **2018**, *106*, 4–12. [CrossRef]
93. Gibert, K.; Sànchez-Marrè, M.; Izquierdo, J. A survey on pre-processing techniques: Relevant issues in the context of environmental data mining. *AI Commun.* **2016**, *29*, 627–663. [CrossRef]
94. Gibert, K.; Izquierdo, J.; Sànchez-Marrè, M.; Hamilton, S.H.; Rodríguez-Roda, I.; Holmes, G. Which method to use? An assessment of data mining methods in Environmental Data Science. *Environ. Model. Softw.* **2019**, *110*, 3–27. [CrossRef]
95. Choi, S.S.; Cha, S.H.; Tappert, C.C. A Survey of Binary Similarity and Distance Measures. *J. Syst. Cybern. Inform.* **2010**, *8*, 43–48.
96. Jain, A.K. Data Clustering: 50 Years Beyond K-means. In *Machine Learning and Knowledge Discovery in Databases*; Springer: Berlin/Heidelberg, Germany, 2010; pp. 3–4. [CrossRef]
97. Maulik, U.; Bandyopadhyay, S. Performance evaluation of some clustering algorithms and validity indices. *IEEE Trans. Pattern Anal. Mach. Intell.* **2002**, *24*, 1650 –1654. [CrossRef]
98. Gurrutxaga, I.; Muguerza, J.; Arbelaitz, O.; Perez, J.M.; Martin, J.I. Towards a standard methodology to evaluate internal cluster validity indices. *Pattern Recognit. Lett.* **2011**, *32*, 505–515. [CrossRef]
99. Salvador, S.; Chan, P. Determining the Number of Clusters/Segments in Hierarchical Clustering/Segmentation Algorithms. In Proceedings of the 16th IEEE International Conference on Tools with Artificial Intelligence, Boca Raton, FL, USA, 15–17 November 2004; pp. 576–584.
100. Gibert, K.; Conti, D.; Vrecko, D. Assisting the end-user in the interpretation of profiles for decision support. an application to wastewater treatment plants. *Environ. Eng. Manag. J.* **2012**, *11*, 931–944. [CrossRef]
101. Pérez-Bonilla, A.; Gibert, K. Towards automatic generation of conceptual interpretation of clustering. In *Iberoamerican Congress on Pattern Recognition*; Springer: Berlin/Heidelberg, Germany, 2007; pp. 653–663.
102. Fayyad, U.; Piatetsky-Shapiro, G.; Smyth, P. From data mining to knowledge discovery in databases. *AI Mag.* **1996**, *17*, 37.
103. Gunning, D. Explainable artificial intelligence (xai). *Def. Adv. Res. Proj. Agency (DARPA)* **2017**, *2*, 2. [CrossRef] [PubMed]

applied sciences

MDPI

Article

CAIT: A Predictive Tool for Supporting the Book Market Operation Using Social Networks

Jessie C. Martín Sujo *,†, Elisabet Golobardes i Ribé † and Xavier Vilasís Cardona †

Research Group of Data Science for the Digital Society, La Salle-Ramon Llull University, 08024 Barcelona, Spain; elisabet.golobardes@salle.url.edu (E.G.i.R.); xavier.vilasis@salle.url.edu (X.V.C.)
* Correspondence: jessiecaridad.martin@salle.url.edu
† These authors contributed equally to this work.

Abstract: A new predictive support tool for the publishing industry is presented in this note. It consists of a combined model of Artificial Intelligence techniques (CAIT) that seeks the most optimal prediction of the number of book copies, finding out which is the best segmentation of the book market, using data from the networks social and the web. Predicted sales appear to be more accurate, applying machine learning techniques such as clustering (in this specific case, KMeans) rather than using current publishing industry expert's segmentation. This identification has important implications for the publishing sector since the forecast will adjust more to the behavior of the stakeholders than to the skills or knowledge acquired by the experts, which is a certain way that may not be sufficient and/or variable throughout the period.

Keywords: artificial intelligence; machine learning; segmentation; clustering; forecasting; book copies; social networks; publishing industry

Citation: Martín Sujo, J.C.; Golobardes i Ribé, E.; Vilasís Cardona, X. CAIT: A Predictive Tool for Supporting the Book Market Operation Using Social Networks. *Appl. Sci.* **2022**, *12*, 366. https://doi.org/10.3390/app12010366

Academic Editor: Emanuele Carpanzano

Received: 15 November 2021
Accepted: 25 December 2021
Published: 31 December 2021

1. Introduction

When a book is launched, the publisher faces a big problem: how many books should be printed? This is known in the industry as print runs. Books that are not sold in stores are returned to the warehouse. Naturally, publishers do not want returns, nor do they want books that languish in the warehouse. For the publisher, this implies increased cost of promotion, correction time, legal procedures, etc.; for the environment, this leads to higher impact due to paper consumption to procedure surplus copies of the book. For example, during 2018, up to 40% of the 225 million titles published in Spain were returned [1].

Currently, the publishing industry uses the knowledge of marketing experts as a predictive measure, classifying sales (or the number of copies to print) into four fundamental segments. This classification is not using, however, the information coming from social networks and internet searches and mentions. As Fishbein and Azjen [2] wrote, "the best individual predictor of an individual's behavior will be a measure of his intention to perform that behavior". Since social network analysis can provide and insight on the market pulse, we are driven to pose the following research question: Will the identification of the market segments of the book using the stakeholders' behavior in social networks and the web improve the prediction of copies to print?

In this way, the main objectives that we set ourselves with this research are:

- Find a better segmentation method.
- Adjust the prediction of copies to print to each segment found.
- Create a combined model of Artificial Intelligence techniques (CAIT), which will serve as a support tool to predict the number of books copies contributes to increasing revenue (publishers only).

Starting from the base of our research question and the objectives set, we hypothesize that an automatic segmentation can predict and/or improve current results with the experts' segmentation.

The article is organized as follows. First, in Section 2, we review the work related to the topic that we will propose to work on. Then, Section 3, we will describe our research design depicting the data and the method used. Next, the experimentation results are presented in Section 4. Finally, in Section 5, we discuss our results and conclude by describing the general contributions of the research, the ethical aspects of being mindful of the use of Artificial Intelligence, as well as future directions.

2. Background

The estimation of the number of copies to be printed of a new book is at present determined by the editor's experience and ability to gauge the potential reader's interest. In order to assess the viability of a decision support tool, relying on information of social networks and internet mentions, let us review first three key aspects such as (a) The influence of social networks on sales, (b) The success of a book based on sales, and (c) Sales segmentation forecast techniques.

2.1. *The Influence of Social Networks on Sales*

From the perspective of social influence, no study in the publishing sector considers the impact of social networks on the product. Still, there are many similar cases in different retail industries. Fashion, for example, ref. [3], shows us that a strong presence in social networks is important for the sale of a product, rather than being under the advertising standards by the industry; mobile telephony [4], they indicate us how these communication channels can be used to predict the income of a product or entertainment [5], reveal that beyond social networks, valuable information can be extracted from famous blogs. Specifically, this paper [6] proposes to use the number of blog references as an indicator of the success of the sale of a book. Another work [7] understands the effects of social networks in the interactions of the seller and the consumer, offering us new insights on how social influence improves the impacts of consumption and contributes positively to the performance of the retailers and consumer loyalty. Following this line of thought, we found another study [8] based on the customer's commitment to a product or brand, which reinforces the importance that this feature may have for future predictions. Recent publications such as [9] serve as the basis to begin to understand the mechanics of reader preference. Still, they are only based on the sale of bestsellers, thus leaving a large gap in the sale of the remaining books. Another article like [10] makes us reflect on the effect that sharing the famous "likes" of consumers on social networks has on sales, exerting great social pressure in the community.

2.2. *The Success of a Book Based on Sales*

From the point of view of the success of the book based on sales, we reference this work [11] which analyzes the characteristics that make a book a bestseller, using statistical techniques and data analysis. However, the work is aimed exclusively at authors already recognized by readers. Another study [12] is only based on historical book sales data and some attributes collected from Amazon. Up to this point of the investigation, none of the sources includes social networks in the publishing sector to predict sales. Still, they help us consider which characteristics to input for the prediction model.

2.3. *Sales Segmentation Forecast Techniques*

The goal of segmentation is to partition heterogeneous groups into homogeneous subgroups based on similarities. One of the most widely used statistical techniques for this purpose is quartiles [13], which divide populations into four groups, or quarters, of equal size. The following work [14] allows us to analyze the use of quartiles for market segmentation, but it is only focused on the purchaser. The consumer market in this paper is segmented by price, where it is shown that wealthy purchasers pay moderately higher prices for pills and injectables. Another study [15] reveals a negative wage gap in the lowest quartile from the wage distribution of an employment balance in Russia. These articles are useful, as they provide us with a segmentation technique that we will use to compare with the expert's segmentation (currently used) in the publishing sector. Finally, although following this line of research, revising state of the art, these works [16–18] where pattern matching techniques are applied to time series data are interesting because they use the similarity of historical data as a basis for grouping time series. All this allows us to visualize an idea of grouping the historical information of the books, but we still need to incorporate the data from social networks and the web.

Finally, no study considers the impact of the presence of social networks in the prediction of the number of copies to print of a book (or what is the same sales) and, especially, that it is done automatically.

3. Research Design and Method

This section describes the proposal of a new decision support tool to help publishers determine the number of copies to print. Using as a basis three methods of segmentation of the book market: (1) a priori segmentation based on quartiles, the most basic proposal, (2) segmentation based on the criteria of experts, current segmentation, and (3) grouping the data according to their behavior patterns, automatic segmentation.

3.1. Description of the Dataset

The data used is provided by (from now on, we will call it) "The Editorial", respecting the anonymity of the medium; and correspond to the data of GFK [19] on the sale of books. It contains book titles, authors, release dates, price, and other features. The data is limited to the territory of Spain during the 2018–2020 period and to the literary genres Non-Fiction and Children/Youth. As a sample, 1169 books have been used, which, according to the experts of the publishing industry, are considered more sensitive to the popularity of the author. The analysis is based on the characteristics of a book, the author's public social networks, and web mentions, which are shown in Figure 1.

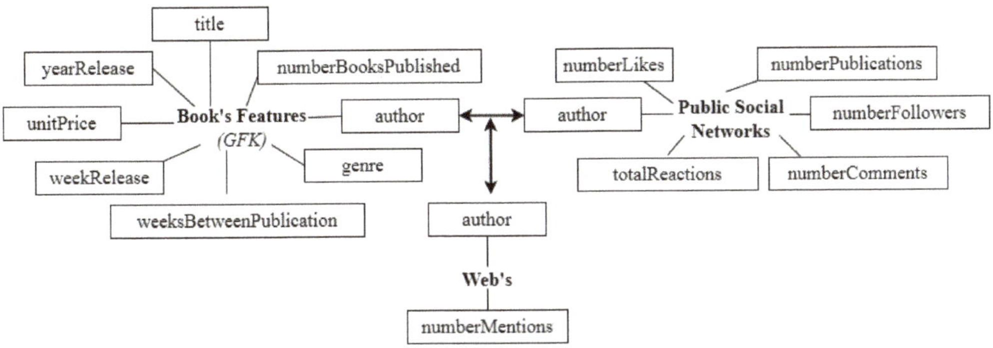

Figure 1. The entity–relationship diagram of the various sources works to form the input dataset to the proposed system. The two-way arrows indicate that author is the key to the union between the different sources.

For a better understanding of the experimentation that we will carry out in Section 4, in Figure 2 we will observe the three different segmentations that we can perform on the data. After obtaining the classification label to test whether our hypothesis is true or false, we pass the new data through a combined classification and regression model to determine the precise number of copies for each segmentation.

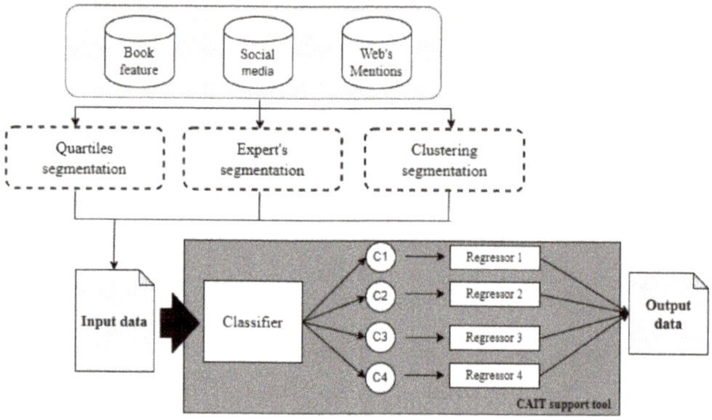

Figure 2. Research structure diagram. The input data in the combined model of Artificial Intelligence techniques (CAIT) support tool are tested with the three different segmentations. Upon entering into the system, they will be classified into the defined segmentations, and individual regressors will be applied for each segmentation. Finally, an output report is generated with the predicted data. The 4 classifications represented within CAIT are given: (1) the quartiles are influenced from the perspective of the experts; (2) by the expert´s segmentation there are 4; and (3) in the case of clustering segmentation, there could be more, but coincidentally the k groups gives a value of 4, as seen in the Section 4.

3.2. *CAIT: The Proposal*

Once we have the input data, it contains the characteristics of the book, the data from social networks, the number of mentions on the Web, and the segmentation according to the book market. Then, they are entered into the support tool to verify the accuracy of the predictions. In this subsection, we will present the composition of the CAIT and the considerations that were considered for the selection of the optimal algorithm in the two parts of this combined model, according to the processed data. The use of classifiers *(Part A)* is due to the previously performed segmentation, based on the objective variable (number of copies or, what is the same, book sales). This analysis will be detailed in Section 4 and the use of regressors *(Part B)* because, thanks to its linear adjustment, it can bring us closer to the dependency relationship between the independent and dependent variables. Next, we will describe each of its components and, finally, the proposed predictive support system.

3.2.1. Part A: The Classifier

Three classification algorithm shall be considered for the implementation of the classifier. These are,

- Decision Tree: It is a representation in the form of a tree whose branches branch according to the values taken by the variables and which end in a specific action. It is generally used when the number of conditions is not very large in this study. See [20,21] for a detailed description of this algorithm.
- Random Forest: It is a combination of predictor trees such that each tree depends on the values of a random vector tested independently and with the same distribution for each of these. It is implemented in data mining to classify or forecast a target variable. See [22,23] for a detailed description of this algorithm.
- K-Nearest Neighbors: It is a classification method used to estimate the density function of the predictors for each class. See [24,25] for a detailed description of this algorithm.
- XGBoost: Part of the decision tree that is implemented in data mining to classify or forecast on a target variable (book copies), through machine learning that is performed on a set of data, using several weak classifiers. In this case, they are the decision trees, but enhancing the results of these, due to the sequential processing of the data with a loss or cost function, minimizes the error iteration after iteration, thus making

it a strong predictor. However, this will depend on the level of adjustment of the parameters used in the function. See [26,27] for a detailed description of this algorithm.

For our specific dataset, as will be detailed in Section 4, the best algorithm is XGBoost. We can observe the implementation of this algorithm in Algorithm 1, in which we will enter the input data that is made up of the characteristics of the book, the author's social media data, plus the mentions it has on the web (X_i). Finally, the output is given by the different segmentations into which books can be divided based on the number of copies of a book (Y_i).

Algorithm 1 Classifier phase

Split: D_{total} in D_{train} and D_{test} (from K-Fold stratified cross-validation, in this case $k = 10$)
Input: $D_{train} = (X_j, Y_j)$ Where the target variable will be the segmentation
1: An initial tree F_0 is obtained to predict the objective variable Y_j, the residual is associated with the difference $(Y_j - F_0)$.
2: A new tree "h_1" is obtained that adjusts the error to the previous weight.
3: The results of F_0 and h_1 are combined to obtain the tree F_1, where the mean square error of F_1 will be less than that of F_0
4: $F_1 x < -F_0 x + h_1(x)$
5: This process is continued iteratively until the error is minimized as much as possible in the following way:
6: $F_m x < -F_m - 1x + h_m(x)$
7: The classifier is tested with D_{test} using Accuracy, Precision, Recall, F1Score, and MAE as the evaluation metrics.
Output: Y_{class}, predicted segmentation.

3.2.2. Part B: The Regressors

In the second part, we use the regressor algorithms; with them, the aim is to study the effect of one or more independent variables on a single dependent variable. The dependent variable (Y) will be the one we seek to survey through statistical regression to understand how it adapts when modifying the independent variables (X_i). After mathematically describing what has just been explained, we can obtain the following formula:

$$Y = 0 + B_1 * X_1 + B_2 * X_2 + \ldots + B_n * X_n + \varepsilon \quad (1)$$

where Y represents the dependent variable that is being studied or trying to predict, $X_1, X_2 \ldots X_n$ are all the independent variables that influence or can affect the dependent variable Y. The function of ϵ is to explain the possible variability of the data that cannot be presented through the linear relationship of the formula; in other words, it represents the possible existing error.

Knowing the objective and operation of the regressor algorithms, we show below those selected for the competition, looking for the one that best suits the input data:

- Gradient Boosting: It is a machine learning technique [28,29] which produces a predictive model in the form of a set of weak prediction models (typically decision trees). When building the model, it is done in a stepwise manner (as boosting methods do), and it generalizes them, allowing the arbitrary optimization of a differentiable loss function.
- XGBoost: Described in previous section.
- LightGBM: It is a distributed gradient impulse framework for machine learning. It is based on decision tree algorithms but does not grow at the tree level but in leaves. Therefore, by choosing the one will produce the greatest decrease in loss. See [30] for a detailed description of this algorithm.

For our specific dataset, as will be detailed in Section 4, the best algorithm is XGBoost. The implementation of this algorithm can be observed in Algorithm 2, in which we will enter as input data the characteristics of the book, the author's social network data, plus the mentions that this has on the web (X_j). The output is our objective variable that will be given by the number of copies (Y_j).

Algorithm 2 Regressor phase

Split: D_{total} in D_{train} and D_{test} (from K-Fold stratified cross-validation, in this case $k = 10$)
Input: $D_{train} = (X_j, Y_j)$ Where the target variable will be the number of copies

1. An initial tree F_0 is obtained to predict the objective variable Y_j, the residual is associated with the difference $(Y_j - F_0)$.
2. A new tree "h_1" is obtained that adjusts the error to the previous weight.
3. The results of F_0 and h_1 are combined to obtain the tree F_1, where the mean square error of F_1 will be less than that of F_0
4. $F_1 x < -F_0 x + h_1(x)$
5. This process is continued iteratively until the error is minimized as much as possible in the following way:
6. $F_m x < -F_m - 1x + h_m(x)$
7. The regressor is tested with D_{test} using R^2 as the evaluation metric.

Output: $Y_{predicted}$, predicted number of copies.

3.2.3. CAIT: A Predictive Support Tool

Once we have described each component of our proposed predictive support tool, we can observe in the Algorithm 3 its implementation, where we introduce the characteristics of the book, the author's social network, web mentions, and the segmentation of said data (X_k). These data will go through the classification function, obtaining as a result which segmentation group each book can belong to. Given this classification, the data corresponding to each group will be divided, and the regressors will be applied individually through hyperparameters optimization. The details of the hyperparameters used can be seen in Section 4.

Algorithm 3 CAIT algorithm

Input: X_k

class = Classifier phase (X_k)
if class == 1:
    Regressor phase (X_k)
elseif class == 2:
    Regressor phase (X_k)
elseif class == 3:
    Regressor phase (X_k)
elseif class == 4:
    Regressor phase (X_k)

Output: $Y_{nbcopies}$, number of book copies to print according to its segmentation in the market.

Finally, this subsection has detailed the composition of the predictive support tool created. Demonstrating the effect of obtaining its best advantages from each part and joining them in one helps us improve the precision of the number of copies.

4. Results

This section shows the different distributions that a priori data can have with one of the segmentations to be evaluated. First, the most basic segmentation is carried out, quartiles; then expert segmentation is used, and finally, automatic segmentation is given by pattern matching. Subsequently, each of them is tested in the support tool, showing the comparison results of each of the parts referred to in Section 3, seeking the improvement of the predictions. All calculations were performed in Python [31] on Intel (R) Core ™ i5-9400F @ 2.90GHz PC CPU.

4.1. Quartiles Segmentation (the Most Basic Segmentation)

The first segmentation to test is the quartiles. Quartiles have been selected influenced by the segmentation of the experts after the analysis of the project requirement. For the dataset used, the numbers of book copies are grouped by author. If we observe Figure 3a, we can detect the existence of outliers, and they are eliminated above 1.5 of the interquartile range. Finally, we are left with Figure 3b, in which we can easily identify the 4 quartiles into which the number of copies can be segmented, being: less than 1808 (Q1—low sales); between 1808 and 4229 (Q2—low intermediate sales); between 4229 and 12 781 (Q3—high intermediate sales); and finally greater than 12781 (Q4—high sales) copies. This last quartile is the so-called Bestseller books. We will refer to the quartiles as classes so that it is easier later to compare them with the rest of the segmentations to be analyzed. A better understanding of the distribution of volume of data by the different segmentations of the quartiles can be seen in Figure 4.

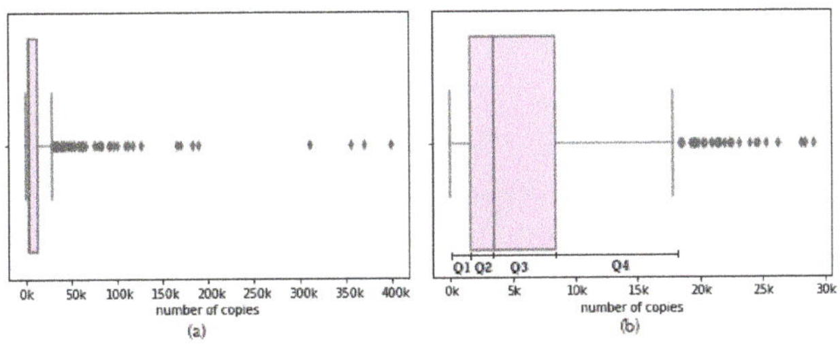

Figure 3. Results of quartiles segmentation. The boxplot represents the mean (center lines), standard deviation (box), range (dotted lines), and outliers (crosses) of the number of copies of books. (**a**) The quartiles can hardly be appreciated given the number of existing outliers. They are eliminated above 1.5 of the interquartile range, and the quartiles in (**b**) are appreciated where it is observed that in Q1 there will be less than 1808, in Q2 between 1808 and 4229, in Q3 between 4229, and 12,781 and Q4 greater than 12,781 number of copies.

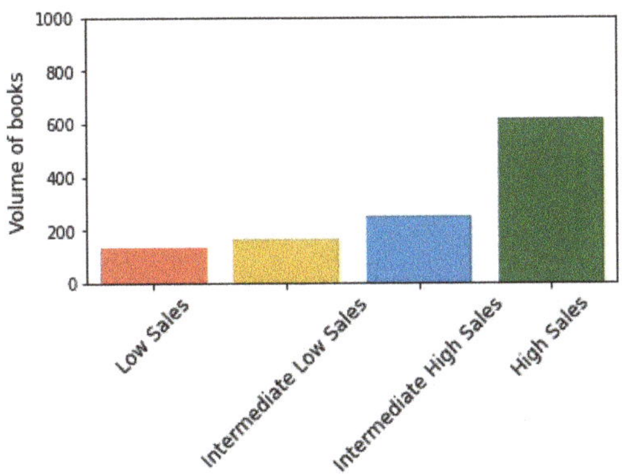

Figure 4. Distribution of the volume of books for each of the segmentations by quartiles. The results show a significant data imbalance between the different segmentations.

4.2. Expert's Segmentation (the Current Segmentation)

The experts provide the segmentations after the analysis of the project requirement. They will be identified as class 1—low sales (C1), class 2—low intermediate sales (C2), class 3—high intermediate sales (C3), and class 4—high sales (C4). The reason for the segmentation carried out by the experts will not be detailed, but in Figure 5 the volume of data due to the different segmentations will be observed.

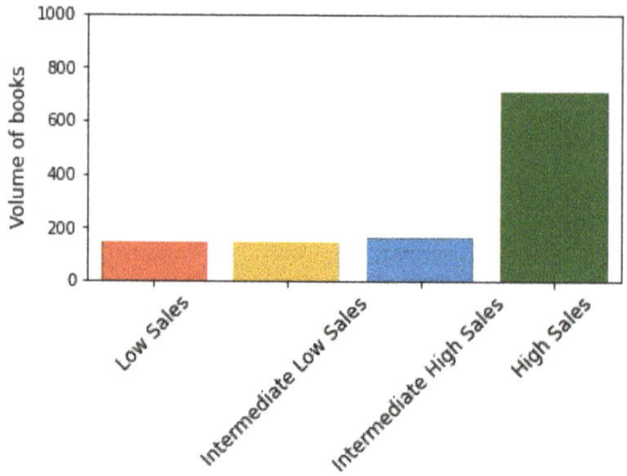

Figure 5. Distribution of the volume of books for each of the expert's segmentations. The results show a significant data imbalance between the different segmentations.

4.3. Clustering (the Automatic Segmentation)

Before applying unsupervised learnings techniques for pattern matching, it is necessary to establish the optimal number of k groups. For this, we use the Elbow curve, which consists of plotting the sum of squared distance between each point and the centroid in a cluster (Wcss). As the number of clusters increases, the Wcss value will decrease. The rule applied to choose $k = 4$, as we can see in Figure 6. These results confirm that the number of segments suggested by the experts from the start is correct but that they can be obtained regardless of their knowledge.

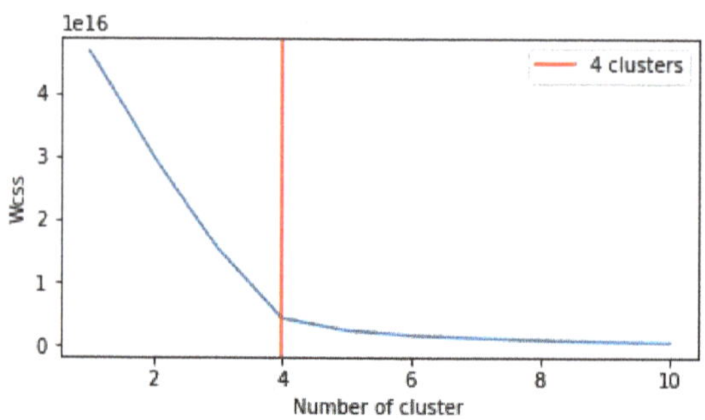

Figure 6. Elbow curve graph. The blue line indicate the within-cluster sums of squares values. The more this value decreases, the greater the number of clusters. The red line indicates the exact point where the "elbow" occurs, which indicates the optimal number of clusters to choose from, in our case 4.

Once we have selected the optimal number of clusters to group our data based on their behavior, we use the K-means, the simplest and fastest training method. In Figure 7 we can see the volume of data presented by the 4 clusters detected by KMeans. They will be identified as class 1—low sales (C1), class 2—low intermediate sales (C2), class 3—high intermediate sales (C3), and class 4—high sales (C4).

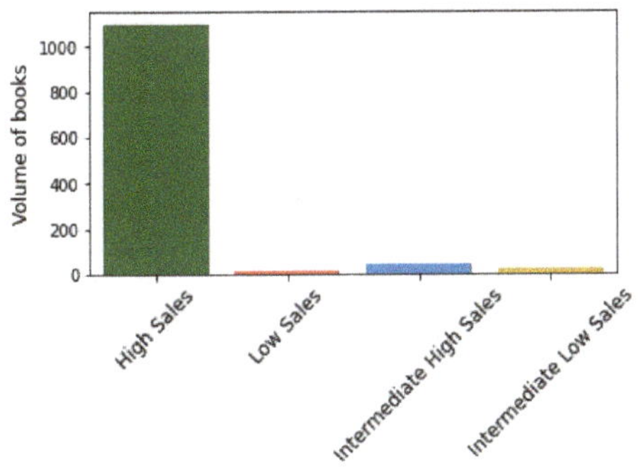

Figure 7. Distribution of the volume of books for each of the clustering segmentations. The results show a significant data imbalance between the different segmentations.

If we analyze Figures 4, 5 and 7, some of the classes have a fairly low volume. Therefore, before using any prediction method such as the one we will use with the predictive support tool CAIT, the data is balanced.

4.4. Performance Evaluation Methods: For Classification Part

- K-Fold stratified cross-validation, this validation seeks to ensure that each k group is representative in all data strata. It is intended to ensure that each class is (roughly represented equally in each test fold) and thus avoid overtraining. In this specific case, our variable $k = 10$
- *Accuracy* (Equation (2)), which refers to how close a sample statistic is to a population parameter, being *TP* (true positive value), *TN* (true negative value), *FP* (false positive value), *FN* (false negative value).
- *Precision* (Equation (3)) with which this algorithm hits each of the classes is also analyzed.
- *Recall* (Equation (4)), represents the model's ability to correctly predict the positives out of actual positives.
- *F1-Score* (Equation (5)), this gives a weighted average of the precision and recall metrics. It is the best metric for averaging out and balancing all the evaluation metrics as a whole.
- *Mean Absolute Error (MAE)* is used, which is a measure of the difference between two continuous variables (Equation (6)). Where y_i is the prediction, $\hat{y}$ and the true value.

$$Accuracy = \frac{TP + TN}{TP + TN + FP + FN} \quad (2)$$

$$Precision = \frac{TP}{TP + FP} \quad (3)$$

$$Recall = \frac{TP}{TP + FN} \quad (4)$$

$$F1\text{-}Score = \frac{Precision * Recall}{Precision + Recall} \quad (5)$$

$$MAE(y, \hat{y}) = \frac{1}{n_{samples}} \sum_{i=0}^{n_{samples}} |y_i - \hat{y}_i| \quad (6)$$

Algorithm Comparison Analysis

All the values of the metrics discussed above are combined into a single overall scorecard for each classifier for each of the segmentations.

As can be observed in Tables 1 and 2, all show that the XGBoost classifier algorithm outperforms the decision tree, KNN, and random forest. Observing Table 3, we could also select the KNN, but the XGBoost is chosen as the optimal one, given its flexibility as a parallelizable algorithm, and that in general in the three segmentations, the scores of this algorithm are higher than other classifiers. The selected classifier can provide very good consistency across all classes. Let us analyze the obtained results from the expert's segmentation (Table 2) concerning the quartiles segmentation (Table 1). We do not see a great difference with what can be interpreted, as the use of this type of segmentation does not improve or worsen the expert´s segmentation. However, if we compare the expert´s segmentation with the clustering segmentation (Table 3), we observe that the classification improves significantly since it fully adjusts to the patterns found in the data.

Table 1. Comparison of classifiers algorithms with quartiles segmentation. This table shows that the XGBoost algorithm is the optimal one of the four to be compared since it presents the best accuracy, with the lowest mean absolute error.

	Decision Tree	K-Neares Neighbors	Random Forest	XGBoost
Accuracy	0.86	0.86	0.89	**0.91**
MAE	0.21	0.23	0.16	**0.12**
Precision	0.86	0.86	0.89	**0.91**
Recall	0.86	0.86	0.89	**0.91**
F1-Score	0.86	0.86	0.89	**0.91**

Table 2. Comparison of classifiers algorithms with expert's segmentation. This table shows that the XGBoost algorithm is the optimal one of the four to be compared, since it presents the best accuracy, with the lowest mean absolute error.

	Decision Tree	K-Nearest Neighbors	Random Forest	XGBoost
Accuracy	0.89	0.87	0.90	**0.93**
MAE	0.19	0.21	0.16	**0.11**
Precision	0.89	0.87	0.90	**0.93**
Recall	0.89	0.87	0.90	**0.93**
F1-Score	0.89	0.87	0.90	**0.93**

Table 3. Comparison of classifiers algorithms with clustering segmentation. This table shows that both the XGBoost and KNN algorithms can be the most optimal of the four to be compared, given that they present the best accuracy, with the lowest mean absolute error.

	Decision Tree	K-Nearest Neighbors	Random Forest	XGBoost
Accuracy	0.99	**1.00**	0.99	**1.00**
MAE	0.01	**0.00**	0.02	**0.00**
Precision	0.99	**1.00**	0.99	**1.00**
Recall	0.99	**1.00**	0.99	**1.00**
F1-Score	0.99	**1.00**	0.99	**1.00**

4.5. Performance Evaluation Methods: For Regression Part

In the case of the regressors part, the determination coefficient (R^2) is used, a statistical metric in the regression models that allows determining the proportion of variance in the dependent variable, which is explained by the independent variable. In other words, R^2 [32] (Equation (7)) shows us how well the data fit the regression model (the goodness of fit). Where $SS_{regression}$ is the sum of squares due to regression *(explained sum of squares)* and SS_{total} is the total sum of squares.

$$R^2 = \frac{SS_{regression}}{SS_{total}} \quad (7)$$

Algorithm Comparison Analysis

To better understand the comparison of the regressors mentioned Section 3, for each of the segmentation carried out, remember that class 1 (C1) corresponds to low sales, class 2 (C2) to low intermediate sales, class 3 (C3) to high medium sales, and finally class 4 (C4) to increased sales.

As can be seen, once again, the most optimal algorithm is the XGBoost Regressor since it is the one with the best results of the three. If we make a comparison between the different segmentations, we observe that the segmentation presents the lowest prediction values by quartiles (Table 4). Clearly, the prediction with clustering segmentation (Table 5) is the same or better than with expert segmentation (Table 6).

Table 4. Comparison of regressors algorithms using (R^2) as the evaluation metric. This table shows that the XGBoost algorithm is the one that best predicts the number of copies of books for each of the segmentations obtained with segmentation by quartiles.

	Class 1	**Class 2**	**Class 3**	**Class 4**
GBoosting	0.75	0.72	0.51	0.16
XGBoost	**0.93**	**0.96**	**0.98**	**0.94**
LGBM	0.87	0.72	0.51	0.43

Table 5. Comparison of regressors algorithms using (R^2) as the evaluation metric. This table shows that the XGBoost algorithm is the one that best predicts the number of copies of books for each of the segmentations obtained with clustering segmentation.

	Class 1	**Class 2**	**Class 3**	**Class 4**
GBoosting	0.32	0.34	0.77	0.06
XGBoost	**0.94**	**0.96**	**1.00**	**0.96**
LGBM	0.20	0.00	0.00	0.38

Table 6. Comparison of regressors algorithms using (R^2) as the evaluation metric. This table shows that the XGBoost algorithm is the one that best predicts the number of copies of books for each of the segmentations obtained with the expert's segmentation.

	Class 1	**Class 2**	**Class 3**	**Class 4**
GBoosting	0.73	0.73	0.13	0.16
XGBoost	**0.95**	**0.97**	**1.00**	**0.96**
LGBM	0.90	0.87	0.86	0.41

The hyperparameters optimized by each class used with the algorithm selected as optimal are described below:

- Regressor 1: lambda = 3; booster = "gblinear", alpha = 5, feature selector = "shuffle"
- Regressor 2: lambda = 5; booster = "gblinear", alpha = 18, feature selector = "cyclic"
- Regressor 3: lambda = 4; booster = "gblinear", alpha = 12, feature selector = "cyclic"
- Regressor 4: lambda = 8; booster = "gblinear", alpha = 2, feature selector = "shuffle"

5. Discussion and Conclusions

In this section, we will highlight our contribution to the publishing sector. Furthermore, the ethical and social considerations that must be according in Artificial Intelligence solutions are also valued. Finally, in the conclusions, we will summarise the main points addressed during this work and propose further work along the same lines.

5.1. General Discussion

Despite the limitations of the data provided by "The Editorial", we can observe that the results are promising. Nevertheless, we are aware that to reach the final validation of the improvement of the number of copies to print predictions through the proposed segmentation, the following is required: (a) A greater amount of data, (b) Other literature genre, (c) Increase the analysis period, (d) Other types of networks social, since these depend specifically on the period in which they are found, and (e) Expand the scope of the work, without being limited to a single publisher or country.

5.2. Ethical and Social Considerations

As the last part of our discussion (and not least), it is necessary to highlight that the analysis carried out not only identifies which segmentation is the most optimal to improve predictions, but also validates that social networks are becoming a double-edged tool if we do not know how to handle it with ethical principles. They can create a human profile based on their tastes, which leaves us without the main tool of the living being, reasoning. This leads us to ask ourselves different questions: Where are the limits of Artificial Intelligence? Is the sale of books, given the author's influence, directly proportional to the book's literary quality? How can we include ethics in the behavior of a model?

Finally, with this section, we highlight the benefits that this tool can provide, but we also consider continuous improvement by applying our ethical sense in this type of work.

5.3. Conclusions and Futher Work

In conclusion, the results have shown that the prediction of the number of copies to be printed improves significantly if automatic segmentation methods are used. The hypothesis that an automatic segmentation can predict and/or improve current results with expert´s segmentation is tested and validated. Another main finding of our work is the promising results shown after using our proposed support tool. Once this system can be validated with more data, more sustainable consumption and production patterns can be guaranteed under the action plan to implement the 2030 Agenda [33].

Many different adaptations, tests, and experiments have been left for the future due to time constraints (that is, experiments with real data are often time-consuming and take even days to complete a single run). Future work concerns a more in-depth analysis of new proposals. For example, the following ideas could be tested: (a) Add segmentation automation in CAIT, as the results are promising; (b) Put this study into production once it has been validated with sufficient data; (c) Validate CAIT's capabilities through MLOps [34] with ways to expand their use in other retail fields such as textiles, music and film, etc.

Author Contributions: Conceptualization, J.C.M.S., X.V.C. and E.G.i.R.; methodology, J.C.M.S. and E.G.i.R.; software, J.C.M.S.; validation, J.C.M.S., X.V.C. and E.G.i.R.; formal analysis, J.C.M.S. and X.V.C.; investigation, J.C.M.S. and E.G.i.R.; resources, J.C.M.S. and X.V.C.; writing—original draft preparation, J.C.M.S.; writing—review and editing, J.C.M.S., E.G.i.R. and X.V.C.; supervision, E.G.i.R. All authors have read and agreed to the published version of the manuscript.

Funding: This research received no external funding.

Institutional Review Board Statement: Not applicable. This studies not involving humans or animals.

Informed Consent Statement: Not applicable. This studies not involving humans.

Data Availability Statement: Not applicable. The data for this study provided by the GFK application. Data sharing is not applicable to this article.

Acknowledgments: It has also been possible thanks to "The Editorial" and DS4DS research group at La Salle—Ramon Llull University.

Conflicts of Interest: The authors declare no conflict of interest.

References

1. ElPais. Available online: https://elpais.com/cultura/2018/07/09/actualidad/1531163370_371133.html#:~:text=Babelia%C3%9Altimas%20noticias-,Hasta%20un%2040%25%20de%20los%20225%20millones%20de,editados%20en%20Espa%C3%B1a%20se%20devuelve (accessed on 8 November 2021).
2. Fischbein, M.; Ajzen, I. *Belief, Attitude, Intention and Behavior*; Addison-Wesley: Boston, MA, USA, 1975.
3. Park, J.; Ciampaglia, G.L.; Ferrara, E. Style in the age of instagram: Predicting success within the fashion industry using social media. In Proceedings of the 19th ACM Conference on Computer-Supported Cooperative Work & Social Computing, San Francisco, CA, USA, 27 February–2 March 2016; pp. 64–73.
4. Lassen, N.B.; Madsen, R.; Vatrapu, R. Predicting iphone sales from iphone tweet. In Proceedings of the 2014 IEEE 18th International Enterprise Distributed Object Computing Conference, Ulm, Germany, 1–2 September 2014; pp. 81–90.
5. Abel, F.; Diaz-Aviles, E.; Henze, N.; Krause, D.; Siehndel, P. Analyzing the blogosphere for predicting the success of music and movie products. In Proceedings of the 2010 International Conference on Advances in Social Networks Analysis and Mining, Odense, Denmark, 9–11 August 2010; pp. 276–280.
6. Moon, G.C.; Kikuta, G.; Yamada, T.; Yoshikawa, A.; Terano, T. Blog information considered useful for book sales prediction. In Proceedings of the 7th International Conference on Service Systems and Service Management, Tokyo, Japan, 28–30 June 2010; pp. 1–5.
7. Rapp, A.; Beitelspacher, L.S.; Grewal, D.; Hughes, D.E. Understanding social media effects across seller, retailer, and consumer interactions. *J. Acad. Mark. Sci.* **2013**, *41*, 547–566. [CrossRef]
8. Guesalaga, R. The use of social media in sales: Individual and organizational antecedents, and the role of customer engagement in social media. *Ind. Mark. Manag.* **2016**, *54*, 71–79. [CrossRef]
9. Wang, X.; Yucesoy, B.; Varol, O.; Eliassi-Rad, T.; Barabási, A.L. Success in books: Predicting book sales before publication. *EPJ Data Sci.* **2019**, *8*, 31. [CrossRef]
10. Namil, K.I.M.; Wonjoon, K.I.M. Do your social media lead you to make social deal purchases? Consumer-generated social referrals for sales via social commerce. *Int. J. Inf. Manag.* **2018**, *39*, 38–48.
11. Yucesoy, B.; Wang, X.; Huang, J.; Barabási, A.L. Success in books: a big data approach to bestsellers. *EPJ Data Sci.* **2018**, *7*, 7. [CrossRef]
12. Feng, T.Q.; Choy, M.; Laik, M.N. Predicting book sales trend using deep learning framework. *Int. J. Adv. Comput. Sci. Appl.* **2020**, *11*, 28–39. [CrossRef]
13. Rew, H. Francis Galton. *J. R. Stat. Soc.* **1922** *85*, 293–298.
14. Winfrey, W.; Heaton, L. *Market Segmentation Nalysis of the Indonesian Family Planning Market: Consumer, Provider and Product Market Segments and Public Sector Procurement Costs of Family Planning under*; USAID: Washington, DC, USA, 1996.
15. Lehmann, H.; Zaiceva, A. Informal Employment in Russia: Incidence, Determinants and Labor Market Segmentation. 2013. Available online: https://ssrn.com/abstract=2330214 (accessed on 15 January 2021).
16. Duncan, G.T.; Gorr, W.L.; Szczypula, J. Forecasting analogous time series. In *Principles of Forecasting*; Springer: Boston, MA, USA, 2001; pp. 195–213.
17. Maharaj, E.A.; Inder, B.A. Forecasting time series from clusters. In *Monash Econometrics and Business Statistics Working Papers*; Department of Econometrics and Business Statistics, Monash University: Melbourne, Australia, 1999.
18. Mitchell, R. Forecasting Electricity Demand using Clustering. In *Proceedings of 21st IASTED International Conference on Applied Informatics*; UNSPECIFIED: Innsbruck, Austria, 2003; pp. 225–230.
19. GFK. Available online: https://www.gfk.com/home (accessed on 6 February 2019).
20. Quinlan, J.R. Induction of decision trees. *Mach. Learn.* **1986**, *1*, 81–106. [CrossRef]
21. Quinlan, J.R. Decision trees and decision-making. *IEEE Trans. Syst. Man Cybern.* **1990**, *20*, 339–346. [CrossRef]
22. Ho, T.K. Random decision forests. In Proceedings of the 3rd International Conference on Document Analysis and Recognition, Montreal, QC, Canada, 14–16 August 1995; Volume 1, pp. 278–282.
23. Breiman, L. Random forests. *Mach. Learn.* **2001**, *1*, 5–32. [CrossRef]
24. Guo, G.; Wang, H.; Bell, D.; Bi, Y.; Greer, K. KNN model-based approach in classification. In Proceedings of the OTM Confederated International Conferences on the Move to Meaningful Internet Systems, Catania, Italy, 3–7 November 2003; Springer: Berlin/Heidelberg, Germany, 2003; pp. 986–996.
25. Cover, T.; Hart, P. Nearest neighbor pattern classification. *IEEE Trans. Inf. Theory* **1967**, *13*, 21–27. [CrossRef]
26. Chen, T.; Guestrin, C. Xgboost: A scalable tree boosting systems. In Proceedings of the 22nd ACM Sigkdd International Conference on Knowledge Discovery and Data Mining, San Francisco, CA, USA, 13–17 August 2016; pp. 785–794.
27. Nielsen, D. Tree Boosting with Xgboost-Why Does Xgboost Win Every Machine Learning Competition? Master's Thesis, NTNU, Taipei, Taiwan, 2016.
28. Freund, Y.; Schapire, R.E. Experiments with a new boosting algorithm. *ICML* **1996**, *96*, 148–156.

29. Friedman, J.H. Stochastic gradient boosting. *Comput. Stat. Data Anal.* **2002**, *38*, 367–378. [CrossRef]
30. Ke, G.; Meng, Q.; Finley, T.; Wang, T.; Chen, W.; Ma, W.; Liu, T.Y. Lightgbm: A highly efficient gradient boosting decision tree. *Adv. Neural Inf. Process. Syst.* **2017**, *30*, 3146–3154.
31. Python. Available online: https://www.python.org/ (accessed on 6 February 2019).
32. Nagelkerke, N.J. A note on a general definition of the coefficient of determination. *Biometrika* **1991**, *78*, 691–692. [CrossRef]
33. Agenda2030. Available online: https://www.agenda2030.gob.es/recursos/docs/METAS_DE_LOS_ODS.pdf (accessed on 8 November 2021).
34. Alla, S.; Adari, S.K. What Is MLOps? In *Beginning MLOps with MLFlow*; Apress: Berkeley, CA, USA, 2021; pp. 79–124.

applied sciences

MDPI

Article

A Comparative Study of Two Rule-Based Explanation Methods for Diabetic Retinopathy Risk Assessment

Najlaa Maaroof [1,*], Antonio Moreno [1], Aida Valls [1], Mohammed Jabreel [1] and Marcin Szeląg [2]

[1] ITAKA Research Group, Department of Computer Science and Mathematics, Universitat Rovira i Virgili, 43007 Tarragona, Spain; antonio.moreno@urv.cat (A.M.); aida.valls@urv.cat (A.V.); mhjabreel@gmail.com (M.J.)

[2] Institute of Computing Science, Poznań University of Technology, 60-965 Poznań, Poland; marcin.szelag@cs.put.poznan.pl

* Correspondence: najlamaroof2007@gmail.com or najlaamaaroofwahib.al-ziyadi@urv.cat

Abstract: Understanding the reasons behind the decisions of complex intelligent systems is crucial in many domains, especially in healthcare. Local explanation models analyse a decision on a single instance, by using the responses of the system to the points in its neighbourhood to build a surrogate model. This work makes a comparative analysis of the local explanations provided by two rule-based explanation methods on RETIPROGRAM, a system based on a fuzzy random forest that analyses the health record of a diabetic person to assess his/her degree of risk of developing diabetic retinopathy. The analysed explanation methods are C-LORE-F (a variant of LORE that builds a decision tree) and DRSA (a method based on rough sets that builds a set of rules). The explored methods gave good results in several metrics, although there is room for improvement in the generation of counterfactual examples.

Keywords: explainable AI; machine learning; fuzzy rules; dominance-based rough set approach; diabetic retinopathy

Citation: Maaroof, N.; Moreno, A.; Valls, A.; Jabreel, M.; Szeląg, M. A Comparative Study of Two Rule-Based Explanation Methods for Diabetic Retinopathy Risk Assessment. *Appl. Sci.* **2022**, *12*, 3358. https://doi.org/10.3390/app12073358

Academic Editor: Vincent A. Cicirello

Received: 21 February 2022
Accepted: 22 March 2022
Published: 25 March 2022

1. Introduction

Healthcare costs are continuously raising, due to the increase of life expectancy, the improvements in the management of chronic diseases and the development of new treatments. Diabetes Mellitus (DM), suffered by 382 million adults worldwide, is one of the most important chronic diseases. DM patients are estimated to increase up to 592 million adults by 2035 [1]. Moreover, specialists estimate that around 46% of diabetic patients have not been diagnosed [1]. DM has been growing steadily in the last few years. In Spain, the National Health Survey (NHS) detected that diabetes increased from 4.1% of the population in 1993 to 6.4% in 2009. Specialists predict an incidence of more than 3 million DM patients in Spain by 2030 [2].

Diabetic Retinopathy (DR) is an ocular disease related to DM. It is the main cause of blindness and visual impairment worldwide and the most common among working-aged adults [3]. Overall, DR affects 30% of diabetic patients, 11% show some degree of vision loss (sight-threatening diabetic retinopathy [4]), and 4% lose their sight completely. However, early detection through periodic screening can reduce this risk by as much as 95%.

Artificial Intelligence (AI) techniques may improve the screening quality by identifying the patient's risk of developing DR using information from the Electronic Health Record. In the healthcare domain, it is common to build clinical decision support systems (CDSS) using Machine Learning (ML) tools and algorithms. These intelligent CDSS assist clinicians in diagnosing diseases and choosing treatment decisions.

Recently, we observed significant and continuous success in the development of ML-based systems. Such success is attributed to many factors, including the presence of vast amounts of data, the advances in ML and the affordability of more advanced computer equipment. As a result, ML-based systems have become ubiquitous in our society and regular life, and a vital component of multiple applications in many domains, including healthcare.

In line with this progress, our research group and the Ophthalmology Unit of the University Hospital Sant Joan (Reus, Tarragona) have developed a CDSS called RETIPROGRAM [5,6], that helps clinicians to estimate the personalised risk of developing DR as early as possible. The AI core of RETIPROGRAM is a Fuzzy Random Forest (FRF) composed of 100 Fuzzy Decision Trees (FDTs). A FDT is a hierarchical structure that classifies patients based on the values of a set of attributes related to DR risk factors. Each node of the tree represents an attribute. A branch of a node is associated to a possible value of that attribute. Finally, the tree leaves assign patients to two categories: patients with/without risk of developing DR. Each branch is a rule, that provides a result if the attributes have certain values. Experimental results have shown that the system could be incorporated in DR screening programs and improve the quality of screening models [7].

However, it is known that, in domains such as healthcare, high precision is not enough to convince society to trust the decisions of ML-based systems. When a decision is made or suggested by an automated system, it is crucial for practical, social, and increasingly legal reasons to explain the rationale behind that decision to users, developers or regulators. In the case of RETIPROGRAM, a patient may want to know why she has a high risk of developing DR, so that she can try to change her lifestyle to reduce it.

As a result, we started to develop methods to derive explanations for the predictions of the RETIPROGRAM system. In the literature, we can find a variety of techniques to achieve this goal. The simplest solution is to directly inspect the components of the models—e.g., the activated path in a fuzzy decision tree. However, such a naive solution is not feasible in the case of random forests with hundreds of trees. The internal structure of such models is complex, and it is not easy to be inspected. Alternatively, we can develop post-hoc explanation methods [8]. Such methods aim to study the relationship between the input and output produced by the system to explain and extract post-hoc explanations. Most of the explanation methods that follow this approach generate a set of inputs, analyse the answers provided by the system to be explained and then create a simpler model from which we can infer an explanation [9–11].

The most well-known examples of post-hoc ML explanations are Local Interpretable Model-agnostic Explanations (LIME) and Local Rule-Based Explanations (LORE). LIME provides local explanations for a classifier's prediction by fitting a linear regression model locally around the data point of which the prediction is to be explained. LORE generates a set of neighbours of the input point, applies the classification system to them, and builds a decision tree on these results. The explanation obtained by the LORE method is composed of two parts: the activated path, i.e., the *rule* used to produce the decision, and a set of *counterfactual rules* which represent the minimal number of changes in the feature values of the instance that would change the conclusion of the system. So, the main difference between LIME and LORE is the neighbourhood generation procedure and the type of explanations each derives.

In a previous work [12] we proposed Guided-LORE, an adaptation of the LORE method in which the neighbourhood generation, which is the key in obtaining a solid explanation, was formalised as a search problem and solved using Uniform Cost Search. Such adaptation allowed us, to some extent, to make the generation process more informed. However, we found that both LORE and Guided-LORE did not take into consideration explicitly the case in which the attributes that define the objects are fuzzy. For example, when we use Guided-LORE, a neighbour of a point is generated by adding (or subtracting) a fixed amount (which is called step) to the value of an attribute. We addressed these shortcomings by proposing C-LORE-F (Contextualized LORE for Fuzzy attributes) [13]. If we know that an attribute is fuzzy and we have the information on its fuzzy labels and their associated fuzzy sets, we can make a more focused neighbourhood generation. Based on the definition of the fuzzy sets of each attribute, we can generalise its step from being a fixed value to being a function that depends on that knowledge. In that way, the proposed method is more general. It works in cases where the fuzzy sets associated with the linguistic labels are uniformly or non-uniformly distributed. To the best of our knowledge, that work was the first one that considered such knowledge to develop explanation methods for ML systems based on fuzzy logic, e.g., Fuzzy Decision Trees and Fuzzy Random Forests [14].

In this work, we study two different ways of generating rules in the C-LORE-F method. On one hand, we use the classic crisp decision trees. On the other hand, we propose the construction of preferential decision rules based on rough sets (using the Dominance-Based Rough Set Approach, DRSA, [15]). Both methods are used to generate explanations for the RETIPROGRAM classifier.

The rest of this article is structured as follows. Section 2 provides an overview of the related works. In Section 3, we present the two rule explanation methods used in this comparative work. Section 4 presents a general framework for generating counterfactual-based explanations for the RETIPROGRAM classifier. The focus is on the methods for obtaining the decision rules that are used to generate the explanation. In Section 5 we describe the experimentation, including several tests to evaluate and compare the performance of both methods. Finally, in Section 6, we conclude the paper and list some points for future work.

2. Literature Review

Although it is possible to consider machine learning-based systems as reliable, their effectiveness is restricted by the lack of explanation of their decisions and actions to end-users. So, in the literature, we find an increasing body of work on interpretable and transparent Machine Learning algorithms in general, especially applied to sensitive domains such as healthcare. This section provides a brief literature review about the work on explanation methods.

Recently, the research of methods for explaining the output of black-box decision systems has got critical attention [16], and there has been an extraordinary amount of articles in ML interpretability in the last years. We can categorise the works on ML interpretability into those based on features' importance (Section 2.1), counterfactual examples (Section 2.2) and visualisation mechanisms (Section 2.3).

2.1. Methods Based on Measuring the Importance of Features

There are two main directions for developing explanation methods based on the importance of features, global and local explanation methods. Global methods try to explain the entire model behaviour using surrogate models, whereas local models explain a single prediction. It can be helpful, in some scenarios, to understand the global logic of a model. However, the major issue with such approaches is that, as the explanations are extracted from simpler surrogate models, there is no guarantee that they are faithful to the original model [8,11,17].

Local explanation methods are arguably the fundamental approaches to the construction of post-hoc explanations. LIME [11], already mentioned in the introduction, is a well-known example. It is independent of the type of data and the black box to be explained. Given a black box model b, an instance x, and a decision y produced by b on x, LIME constructs a simple linear model that approximates b's input-output behaviour to justify why b predicts y. It generates some neighbours of x randomly in the feature space centred on x. Such an approach is becoming a conventional method. We can find now LIME implementations in multiple popular packages, including Python, R and SAS.

The authors of LIME observed that it does not measure its fidelity. As a result, the local behaviour of a notably non-linear model may lead to faulty linear approximations. Hence, they were motivated to work on a new model-agnostic method, Anchors, based on if-then rules [18]. This method highlights the part of the input that is adequate for the classifier to make the prediction, delivering more intuitive and easy-to-understand explanations.

SHAP [19] is a method that provides an explanation of the prediction of the output of a black box for an instance x by estimating each feature's contribution to that prediction.

These contributions are collected by measuring the Shapley values from coalitional game theory. The features act like players in a coalition. Each player can be formed by a single feature or a subset of features. The Shapley values show the payout distribution of the prediction among the features.

2.2. Methods Based on Counterfactual Examples

Another important category of explanations is based on the generation of *counterfactuals*. The methods of this approach seek minimal changes to the feature values such that the model's predicted outcome changes. Such kind of explanations can be helpful in different scenarios (e.g., an applicant for a bank loan might want to know which part of her application could be changed to get her application approved).

The work presented in [20] is considered to be the first one that employed the counterfactual examples to provide an explanation for the decisions of a classifier. It used a heuristic best-first search to develop a model-agnostic method for finding evidence counterfactuals that can explain the predictions of any classification model. LORE [21] is another example of this approach that is more related to our work, although it can also be seen as an example of the determination of features' importance. It constructs a decision tree c based on a synthetic neighbourhood of the input point generated by a genetic algorithm. Then, an explanation e, composed of a decision rule and a set of counterfactual rules, based on some extracted counterfactual examples, is obtained from the logic of c. The authors in [22] developed a general optimisation framework to generate sets of diverse counterfactual examples for any differentiable Machine Learning classifier. Russell proposed in 2019 what is called a "mixed polytope", a set of constraints that can be used with integer programming solvers to extract counterfactual explanations without making a brute-force enumeration [23].

2.3. Methods Focused on Visualisation

Visualisation-based interpretation methods play an essential role to show comprehensible explanations. We can find several visualisation methods proposed in the literature to help ML engineers and domain experts to understand, debug, and refine ML models. For example, the work presented in [24] proposed an interactive visualisation method to help users, even those without expertise in Machine Learning, to understand, explore and confirm predictive models. This method by Ming et al. extracts a set of rules that approximates a classifier's prediction and visualises them using an interactive visual interface. Neto and Paulovich proposed Explainable Matrix (ExMatrix) [25], a visualisation method to interpret Random Forests. They used a matrix as a visual metaphor in which rows represent rules, columns are features, and cells are rules predicates. They showed that their method is capable of offering global and local explanations of Random Forest models.

3. Preliminaries

This section provides a brief background about Contextualised LORE for Fuzzy attributes (C-LORE-F), which uses classic decision rules, and the Dominance-based Rough Set Approach (DRSA), that builds a set of rules taking into account the preference direction of the variables.

3.1. Contextualised LORE for Fuzzy Attributes (C-LORE-F)

C-LORE-F is a variant of the LORE and Guided-LORE methods for fuzzy-based ML models. It provides an explanation for the decision assigned to a specific instance based on some contextual information (e.g., the type of attribute and the fuzzy sets associated to the linguistic values of the fuzzy attributes). The inputs of C-LORE-F are a trained fuzzy-based ML model, b, and an example x. Algorithm 1 shows the main steps of C-LORE-F. First, we apply b to x to get a decision y. We obtain a set of neighbours of x, D, and a rule-based model t is built by considering the output of b in these points. From this model

t it is possible to derive an explanation, that contains the rule r used to classify x, a set of counterfactual rules δ and a set of counterfactual instances $\mathbb{C}$.

Algorithm 1: C-LORE-F

input : x: an instance to explain, T: an auxiliary set, b: a black-box model, L: maximum level of exploration, and KB: knowledge base.
output: E: the explanation of the decision of b on x

$y \leftarrow b(x)$;
$\mathcal{D}^+ \leftarrow GetNeighbours(x, y, b, L, KB)$;
$x^-, y^- \leftarrow FindDiffExample(x, y, b, T)$;
$\mathcal{D}^- \leftarrow GetNeighbours(x^-, y^-, b, L, KB)$;
$\mathcal{D} \leftarrow \mathcal{D}^+ \cup \mathcal{D}^-$;
$t \leftarrow BuildModel(\mathcal{D}, b)$;
$r = (p \rightarrow y) \leftarrow ExtractRule(x, t)$;
$\delta \leftarrow ExtractCounterfactualRules(x, r, t)$;
$\mathbb{C} \leftarrow ExtractCounterfactuals(x, \delta, t)$;
$E \leftarrow (r, \delta, \mathbb{C})$;

The set $\mathcal{D}$ is obtained by merging two subsets, $\mathcal{D}^+$ and $\mathcal{D}^-$. The first one is called the positive set, and it contains a set of instances that belong to the same class of x. The second one, the negative set, contains examples with a different class. We obtain $\mathcal{D}^-$ by looking at an auxiliary set T and finding the closest example to x , i.e., x^-, that has a different label than y. T can be the training set used to train the black-box model, if accessible, or any other data set from the same distribution.

The neighbours generation step is the key point in C-LORE-F and similar methods. As a first change with respect to LORE and Guided-LORE, we have defined the following types of attributes.

- Attributes with a fixed value (e.g., sex).
- Attributes whose value increases in time (e.g., age).
- Attributes whose value decreases in time (e.g., years left until retirement).
- Variable attributes, that can change positively and negatively (e.g., weight).

The motivation of such definitions is to generate useful and actionable explanations. For example, nobody can reduce his/her age, so it is not useful to give a counterfactual rule saying "if you were 10 years younger, your risk of developing DR would be much lower".

We define the neighbourhood generation as a search problem in which we explore the neighbourhood space of a point x by applying a Uniform Cost Search based on the Heterogeneous Value Difference Metric (HVDM, [26]), using some contextual information about the attributes (e.g., the attribute type and the fuzzy set definitions). This search problem can be formulated as follows:

- **State Space**: the set of all possible examples S.
- **Initial State**: (x, y), where x is the instance of which we want to generate its neighbours and y is the label of this instance obtained by the black-box b.
- **Actions**: Modifications of the value of a single attribute (feature). These actions leverage some contextual information about the feature to make the desired changes to generate new neighbours. In our case, we define two types of actions, next and prev, described later.
- **Transition Model**: returns a new instance in which the value of a feature is changed by applying all actions.
- **Goal Test**: We check, for each generated individual, if, according to the black box, it has the same label as the root, y. If that is the case, we generate its neighbours in the same way (i.e., applying one positive/negative change in the value of a single attribute). Otherwise, we have found an individual close to x that belongs to another class; thus, we have reached a boundary of y, and we terminate the search from that instance.

- **Path Cost**: The path cost of each example is calculated by measuring the HVDM distance between the generated example and x.

Hence, the generation of the closest neighbours of an instance x is a tree search procedure. The search process starts from this instance, and the available actions to move from one instance to another are applied. For each feature f, the number of possible actions can be zero (f is Fixed), one (either next if the feature is temporally increasing or prev if it is temporally decreasing) or both, if f is variable. Each action only changes the value of one feature.

The candidate node to be expanded, n, is the one closest to x, based on the path cost. If the outcome of the black-box model for the instance in n is different from y, then it is a leaf of the tree. Otherwise, we expand that node. Consequently, for each node in the second level, we would have changes in two attributes or double changes in the same attribute, and so on. The expanding process terminates when we reach a predefined max-level, or when there are no more nodes to be expanded (all the leaves have led to changes in the initial classification). Repeated instances are ignored to avoid cycles.

The expanding process is done by cloning the instance of the node to be expanded and applying the next and/or prev actions. After that, we pass the obtained instance to the black-box model b to obtain its corresponding label.

To apply the actions step and prev for a given attribute we consider some separate zones based on its fuzzy sets, which are defined as shown in Figure 1, taking into account the intersection point between two consecutive fuzzy sets and the intervals of maximum activation.

In Figure 1 the zones would be 0–5, 5–10, 10–15, 15–20, 20–25, 25–40, 40–50, 50–60, 60–75, 75–90 and 90–100. Given the value of the attribute, we locate its zone, and then we take the middle of the previous zone as the lower neighbour (the result of the prev action), and the middle of the next zone as the upper neighbour (the result of the next action). Figure 1 shows an example. The input value is 22, which belongs to the zone 20–25. Thus, the middle of the previous zone is the lower neighbour, (15 + 20)/2 = 17.5, and the middle of the next zone is the upper neighbour, (25 + 40)/2 = 32.5. We might end up applying only either the next action, if the located zone was the first one, or the prev action, if it was the last one.

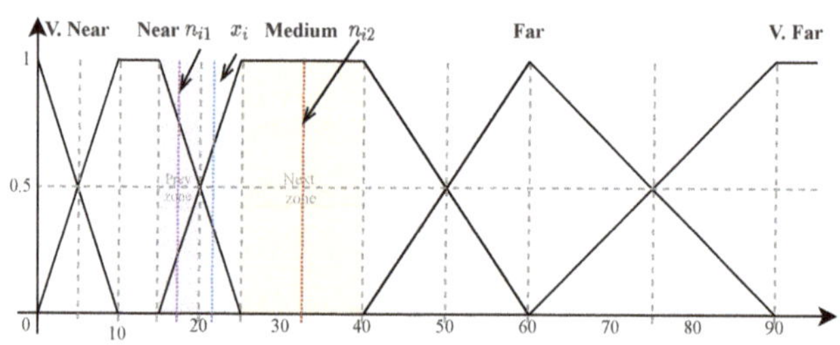

Figure 1. Illustration of the next and prev actions.

As soon as we obtain the neighbours $\mathcal{D}$, we use the *YaDT* system [27], an implementation of the C4.5 classification algorithm with multi-way splits of categorical attributes, to build the decision tree. Finally, from the tree we can extract a set of decision rules, one for each branch in the tree.

The rules obtained are conjunctive and use inequality conditions that check if a certain numerical input value v_j for attribute f_j is above or below a threshold. In case of categorical attributes, the condition is an equality. So, rules have the following format:

$$IF\ f_i\ op\ t_i\ [\text{AND} f_j\ op\ t_j\ \ldots [\text{AND}\ f_z\ op\ t_z]]\ \text{THEN}\ y\ =\ y_k$$

Here, f_i, f_j and f_z are attributes, op is the logical condition and can be either $\leq$ or $>$ if the attribute is continuous, and $=$ if it is categorical, and t_i, t_j and t_z are thresholds.

3.2. Dominance-Based Rough Set Approach (DRSA)

The DRSA method [15] can be used to construct a set of decision rules in line 6 of Algorithm 1. The main difference of these rules with respect to those obtained from a decision tree is that they are classification rules with a set of conditions that take into account the preference directions of the input variables. In this subsection, we explain how the rules are constructed from the set of examples $\mathcal{D}$.

Rough set theory (RST) [28] is a formal theory derived from fundamental research on the logical properties of information systems. The main goal of the rough set analysis is the approximation of concepts. In addition, it offers mathematical tools to discover patterns hidden in data. As a result, it has a wide range of applications, including feature and pattern extraction, data reduction and decision rules generation (our goal in this work). It can also identify partial or total dependencies in data, among other things.

Rough set analysis concerns data stored in a table known as an Information Table. Each row represents an object x_i, evaluated with respect to multiple attributes representing different points of view; the information table is defined as a pair $(X, \mathcal{F})$, where X is a non-empty finite set of objects and $\mathcal{F}$ is a non-empty finite set of attributes. A special kind of information table is a Decision Table $(X, \mathcal{F} \cup Dec)$, where the attributes are divided into *condition* attributes $\mathcal{F}$ and *decision* attributes Dec. The former are related to features of objects, while the latter relate to decisions about objects. Often there is just a single decision attribute $\mathcal{Y}$. Distinct values y_k of this attribute, called class labels, induce a partition of the set of objects into so-called decision classes Cl_k.

DRSA is an extension of RST, suitable for analysis of decision tables where both condition attributes from $\mathcal{F}$ and the output decision variable (decision attribute) $\mathcal{Y}$ are ordinal, and there exist monotonic relationships between attributes from $\mathcal{F}$ and $\mathcal{Y}$. A positive relationship means that the greater the value of the condition attribute, the higher the class label. A negative relationship means that the greater the value of the condition attribute, the lower the class label. Both types of relationships are captured by induced decision rules. In general, in DRSA the number of decision classes can be more than two. Then, one has to consider upward and downward unions of decision classes. However, in the case of RETIPROGRAM, we only have two classes: 0 for the absence of DR risk and 1 for the presence of DR risk. Thus, using DRSA, we calculate rough approximations and induce decision rules for exactly these two classes.

Rules are constructed using elementary building blocks, known as *dominance cones*, with origins in each object in the attribute space. Based on the rough set concept, rules for a lower or/and an upper approximation of each decision class are obtained from a training set ($\mathcal{D}$ in our case) [29]. The choice of DRSA for explainability in the Diabetic Retinopathy disease is motivated by the fact that the values of the attributes are mainly ordinal, and a change from one value to another may be an indicator of the risk of developing DR. Moreover, using the VC-DomLEM algorithm [30], one can induce a set of rules being a minimal cover of consistent objects from both classes. This enables to efficiently distinguish between the two possible decision outputs [30], which is one of the aims of a surrogate model. Two types of rules may be distinguished:

1. $\mathcal{Y} \geq$ decision rules, providing lower profile descriptions for objects belonging at least to class Cl_k (so they belong to Cl_k or a better class, $Cl_{k+1}, Cl_{k+2}, \ldots$):
   IF $f_1 \geq v_1$ AND $f_2 \geq v_2$ AND $\ldots f_n \geq v_n$ THEN $y \geq y_k$
2. $\mathcal{Y} \leq$ decision rules, providing upper profile descriptions for objects belonging at most to class Cl_k (so they belong to Cl_k or a lower class, $Cl_{k-1}, Cl_{k-2}, \ldots$):
   IF $f_1 \leq v_1$ AND $f_2 \leq v_2$ AND $\ldots f_n \leq v_n$ THEN $y \leq y_k$.

In this notation, we must take into account that all condition attributes in $\mathcal{F}$ are considered to be maximisation functions (the higher the value, the higher the class label), which are called *Gain* attributes. In case an attribute has to be minimised, it is called a *Cost* attribute, and the lower its value, the higher the class label. It is also possible to introduce a criterion as both Cost and Gain. In this case, the attribute may appear twice in the rule and define an interval of values.

An important feature of the DRSA method coupled with the VC-DomLEM algorithm is the fact that particular rules are minimal (without redundant conditions) and the whole set of rules is non-redundant (if any rule would be removed, some consistent objects would not be covered by any rule).

Algorithm 2 shows the main steps of the DRSA method, assuming that only certain decision rules are considered (as in our case).

Algorithm 2: DRSA method

input : $\mathcal{D}$ – training set of objects (decision table)
output: γ – quality of classification,
$\mathcal{R}$ – set of decision rules generated on $\mathcal{D}$

1 $\mathcal{X}^{\geq} \leftarrow CalculateUpwardClassUnions(\mathcal{D})$;
2 $\mathcal{X}^{\leq} \leftarrow CalculateDownwardClassUnions(\mathcal{D})$;
3 **foreach** $X \in \mathcal{X}^{\geq} \cup \mathcal{X}^{\leq}$ **do**
4 | $X.LowerApproximation \leftarrow CalculateLowerApproximation(X, \mathcal{D})$;
5 **end**
6 $\gamma = CalculateQualityOfClassification(\mathcal{X}^{\geq}, \mathcal{X}^{\leq}, \mathcal{D})$;
7 $\mathcal{R}^{\geq} \leftarrow VC\text{-}DomLEM(\mathcal{X}^{\geq})$;
8 $\mathcal{R}^{\leq} \leftarrow VC\text{-}DomLEM(\mathcal{X}^{\leq})$;
9 $\mathcal{R} \leftarrow \mathcal{R}^{\geq} \cup \mathcal{R}^{\leq}$;

In lines 1–2, all upward and downward unions of decision classes are identified, depending on the class labels of $\mathcal{Y}$. In the loop defined in the following lines 3–5, for each upward/downward union its lower approximation is calculated. These approximations are stored inside objects representing particular unions of classes. In line 6, the quality of classification is calculated. This is a typical rough set descriptor related to consistency of data, defined as a ratio of the number of consistent objects and all objects in $\mathcal{D}$. During calculation of γ, one takes into account the lower approximations calculated previously. In line 7, the VC-DomLEM algorithm is invoked for the upward unions of classes to induce decision rules. It generates rules describing objects from the lower approximations of subsequent unions, iterating from the most specific to the least specific union to control rule minimality. Suppose decision attribute $\mathcal{Y}$ has labels $1, 2, 3, 4, 5$, and the higher the label, the more preferred the respective decision class. Then, VC-DomLEM will first generate rules for class Cl_5, then for upward union of classes $Cl_4^{\geq} = Cl_4 \cup Cl_5$, then for upward union $Cl_3^{\geq}$, and finally for upward union $Cl_2^{\geq}$. Obviously, considering union $Cl_1^{\geq}$ does not make sense (set of all objects). In line 8, the VC-DomLEM algorithm is invoked to induce decision rules for the downward unions of classes. This is realized analogously, with the only difference that this time first class Cl_1 will be taken into account, then downward union of classes $Cl_2^{\leq} = Cl_1 \cup Cl_2$, next downward union $Cl_3^{\leq}$, and finally downward union $Cl_4^{\leq}$. Remark that VC-DomLEM algorithm was introduced for the Variable Consistency DRSA (VC-DRSA), being an extension of the classical DRSA. In [30], there are four input parameters: set of upward or downward unions of classes, rule consistency measure, set of consistency thresholds for particular unions, and object covering option s (strategy). When invoking the algorithm, we set measure $\widehat{\epsilon}$ [31] for rule consistency measure, supply a set of consistency thresholds all equal to zero (which forces the classical DRSA), and choose 1 for object covering option (indicating that a rule induced for any upward/downward union of classes is allowed to cover only objects from the lower approximation of that union). Moreover, in our problem (binary classification) there is just one upward union $Cl_1^{\geq} = Cl_1$ and one downward union $Cl_0^{\leq} = Cl_0$. Finally, in line 9 the resulting set of decision rules is built by adding sets of rules induced for upward and downward unions of classes.

In the experiments described in this paper, we used the implementations of the DRSA method and the VC-DomLEM algorithm available in the open source ruleLearn library (https://github.com/ruleLearn/rulelearn, last access: 21 March 2022).

4. Explanation Generation System

In this section, we define the explanation generation methodology. Figure 2 shows the proposed architecture. Given an input x, i.e., a patient record, we first pass it to the RETIPROGRAM to obtain a class, y. Then, we pass the input x and its corresponding output y to the explanation unit (shown in the bottom part of Figure 2) to extract an explanation for that decision. The explanation unit (in blue) is composed of three parts: the neighbours' generation module, the training module and the explanation extraction module. Finally, we forward the obtained results to the evaluation part (explained in Section 5) of the system to obtain insights into its performance.

In this article, we focus on describing the explanation and evaluation parts. We explain in detail each of these parts and their sub-modules below. For more information about the RETIPROGRAM and its development and evaluation, we refer the reader to our previous papers [14,32,33].

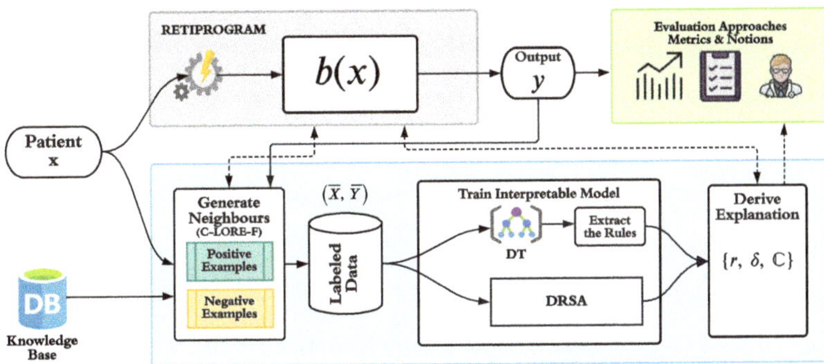

Figure 2. Architecture of the proposed explanation generation methodology.

As we presented in Section 3, the neighbours' generation module applies a Uniform Cost Search based on the HVDM [26] distance metric, using some contextual information about the features to generate positive neighbours (instances with the same label as x) and negative neighbours (instances with different labels than x). All of these examples are labelled using the RETIPROGRAM system, combined into one set $\mathcal{D}$ and fed to the interpretable model training module to build a surrogate model to mimic the behaviour of the RETIPROGRAM locally on $\mathcal{D}$. This model can be any simple model such as Linear Regression. However, as stated earlier, we consider two types of models in this work: a Decision Tree and Ordinal Decision Rules.

To construct the desired explanation, we first extract all decision rules from the interpretable model in the following form: IF $condition_1$ AND $condition_2 \ldots$ AND $condition_n$ THEN $decision_k$. In the case of the Ordinal Rules, we use DRSA and we obtain a minimal set of rules as output of the model training. In the case of the Decision Tree, we can derive such decision rules just following the conditions of each branch of the tree. In the case of decision trees, the selection criteria of the nodes during their construction ensures that the set of rules is small and that they use an small number of conditions. So in both cases, the generated models are much simpler than the original Fuzzy Random Forest (which has 100 trees, each one with several rules).

In the last step, we extract the explanation from these small sets of decision rules in the form of a triplet $(r, \delta, \mathbb{C})$, where:

- r is the decision rule(s) that covers the instance x. This rule tells which are the necessary conditions to be satisfied by the object for being classified as y, so they indicate the minimal reasons for belonging to that class. When using DRSA, we can have more than one applicable rule with different combinations of conditions.

- δ is the set of counterfactual rules that lead to an outcome different than the one of x. They indicate the minimal number of conditions that should be simultaneously changed in the object for not being in class y.
- $\mathbb{C}$ is a set of counterfactual instances that represent examples of objects that belong to a different class and have the minimum changes with respect to the original input object x.

The counterfactual rules extraction procedure is described in Algorithm 3. It looks for all rules leading to a decision different than y and it chooses the ones with the minimum number of conditions not satisfied by x (returned by the function nf).

Algorithm 3: Extraction of counterfactual rules

```
input  : R: The set of decision rules, x: instance to explain, and y: the decision of x
output : δ: set of counterfactual rules
Q ⟵ GetRulesWithDifferentDecision(R, y) ;
δ ⟵ ∅;
min ⟵ +∞ ;
foreach rule q ∈ Q do
    qlen ⟵ nf(q, x) ;
    if qlen < min then
        δ ⟵ q ;
        min ⟵ qlen
    else
        if qlen = min then
            δ ⟵ δ ∪ q ;
        end
    end
end
return δ ;
```

The counterfactual examples are useful for the decision maker to understand what changes in the values of the attributes produce a change in the classification label. In Medicine, this knowledge is particularly interesting, as it can tell a patient how she could move to a better category. These instances are obtained from the counterfactual rules δ and the original input x. Given a counterfactual rule $r : q \to y$, and x, we find the instance that needs the minimum changes in x to satisfy the conditions q. We look at all the attributes in the conditions q that are not satisfied by x, and then we make the smallest change (up or down) to the values of these attributes to satisfy the conditions in q.

Let us take as an example the explanation for the patient x shown in the first row of Table 1. Using RETIPROGRAM this patient is assigned to class 1. Then, the explanation system will construct the rules in Figure 3.

Table 1. Patient example and counterfactual instances.

Age	Sex	EVOL	TTM	HbA1c	CDKEPI	MA	BMI	HTAR
71.0	1	14.0	2	7.4	90.07	0.0	31.05	1
—	—	—	—	6.5	—	—	—	—
—	—	—	0	—	—	—	—	—
—	—	—	1	—	—	—	—	—

$$R1: \ \{HbA1c \leq 6.5\} \ \rightarrow \ \{y = 0\}$$
$$R2: \ \{HbA1c > 6.5 \ \& \ TTM = 0\} \ \rightarrow \ \{y = 0\}$$
$$R3: \ \{HbA1c > 6.5 \ \& \ TTM = 1\} \ \rightarrow \ \{y = 0\}$$
$$R4: \ \{HbA1c > 6.5 \ \& \ TTM = 2 \ \& \ HTAR = 0 \ \& \ EVOL \leq 9.0\} \ \rightarrow \ \{y = 0\}$$
$$R5: \ \{HbA1c > 6.5 \ \& \ TTM = 2 \ \& \ HTAR = 0 \ \& \ EVOL > 9.0\} \ \rightarrow \ \{y = 1\}$$
$$\boxed{R6: \ \{HbA1c > 6.5 \ \& \ TTM = 2 \ \& \ HTAR = 1\} \ \rightarrow \ \{y = 1\}}$$

Figure 3. The constructed rules using the explanation system for x.

The activated rule r is $R6$ (the applicable one on the given patient). There are four rules that lead to the opposite decision, $Q = \{R1, R2, R3, R4\}$. Being $nf(R1, x) = 1, nf(R2, x) = 1$, $nf(R3, x) = 1, nf(R4, x) = 2$, the set of counterfactual rules δ are $\{R1, \ R2, \ R3\}$. The final step in the explanation extraction process is to construct the set of counterfactual examples $\mathbb{C}$. First we take the rule $R1$ and as its condition $(HbA1c \leq 6.5)$ is false for patient x, we change its values to the smallest below the upper bound of HbA1c, which is 6.5. We repeat this process for the rest of rules in δ, obtaining two other examples that only make one change in the TTM variable (as the rest of conditions are already satisfied by x). The obtained counterfactual instances are shown in Table 1 (rows 2–4), in which the empty cells mean that the initial value of that attribute is not changed. We can see that the number of changes in the counterexamples is small. In this example, changing the treatment type would decrease the risk of DR (to class 0); the other option is a decrease of the HbA1c variable, which is the glycated hemoglobin, an indicator of a bad control of the diabetes which affects the blood.

5. Experiments and Results

5.1. Experimental Setup

We evaluated the proposed explanation system on a private data set to assess the risk of developing diabetic retinopathy for diabetic patients. It is composed of 2323 examples of binary classification. The Diabetic-Retinopathy data set was used to develop a fuzzy random forest-based system, called RETIPROGRAM, which is currently being used in the Hospital de Sant Joan in Reus (Tarragona). Each instance in the data set is defined by nine attributes: current age, sex, years since diabetes detection, type of diabetes treatment, good or bad control of arterial hypertension, HbA1c level, glomerular filtrate rate estimated by the CKD-EPI value, microalbuminuria, and body mass index. The data was split into a training set of 1212 examples and a test set of 1111 examples. The classification model used in RETIPROGRAM achieves an accuracy of 80%, with a sensitivity of 81.3% and specificity of 79.7% [32]. We used the test set in all our experiments to evaluate the effectiveness of the proposed explanation system.

5.2. Evaluation of the Explanation Results

As we mentioned above, the explanation contains two main parts: first, the explanation decision rule(s), r, and second, a set of counterfactual rules, δ from which we can derive the counterfactual examples, $\mathbb{C}$. These components are obtained from a set of rules, that we call the explanation model. In this section we want to compare the quality of the rules generated by the two methods. We will denote as C-LORE-F the method using typical decision trees, and we will name as DRSA the version of the same method using rules generated with Dominance-based Rough Sets. The following evaluation metrics are used to measure the quality in both cases, for the RETIPROGRAM black-box method.

- **Hit**: this metric computes the similarity between the output of the explanation model and the black-box, b, for all the testing instances. It returns 1 if they are equal and 0 otherwise.

- **Fidelity**: this metric measures to which extent the explanation model can accurately reproduce the black-box predictor for the particular case of instance x. It answers the question of how good is the explanation model at mimicking the behaviour of the black-box by comparing its predictions and the ones of the black-box on the instances that are neighbours of x, which are in $\mathcal{D}$.
- **l-Fidelity**: it is similar to the fidelity; however, it is computed on the subset of instances from $\mathcal{D}$ covered by the explanation rule(s), r. It is used to measure to what extent this rule is good at mimicking the black-box model on similar data of the same class.
- **c-Hit**: this metric compares the predictions of the explanation model and the black-box model on all the counterfactual instances of x that are extracted from the counterfactual rules in $\mathbb{C}$.
- **cl-Fidelity**: it is also similar to the fidelity; however, it is computed on the set of instances from $\mathcal{D}$ covered by the counterfactual rules in δ.

Table 2 shows the means and standard deviations of the metrics for the C-LORE-F and DRSA explanation methods on the test set. It may be seen that C-LORE-F outperforms DRSA in all metrics. Let us look at the Fidelity and l-Fidelity for the DRSA method. We can find a difference of 10% in favour of l-Fidelity, which means that most of the disagreements between RETIPROGRAM and DRSA occurred with the examples with a different outcome than the original input. So, the rules describing the opposite classes are worse in DRSA than in C-LORE-F. We can also observe that in both C-LORE-F and DRSA, the cHit and cl-Fidelity show lower performance than the other metrics. This can be attributed to the quality of the generated counterfactual examples (which are evaluated in more depth in Section 5.4).

Table 2. Evaluation results of the C-LORE-F and DRSA explanation methods.

	Hit	Fidelity	l-Fidelity	cHit	cl-Fidelity
C-LORE-F	1.00 ± 0.00	0.99 ± 0.002	0.99 ± 0.002	0.89 ± 0.290	0.88 ± 0.282
DRSA	0.97 ± 0.152	0.831 ± 0.32	0.93 ± 0.176	0.830 ± 0.315	0.83 ± 0.298

5.3. Evaluating the Locality of the Methods

The proposed explanation system is local, because it focuses on the behaviour of RETIPROGRAM around the specific instance x. The Fidelity metrics defined above validate the models' performance in terms of locality with respect to the generated neighbours and the instance to be explained. Assuming that we have access to the test set used to evaluate the black-box model, we can validate the locality of the model with respect to the test set by defining a new metric, the **xt-Fidelity**. It is the fidelity measure computed on the set of instances from the test set with a distance to the instance x less than or equal a threshold t. The overall xt-Fidelity on a set X given a threshold t is computed by taking the average of xt-Fidelity for all $x \in X$. We use it to measure the locality vs the globality of the explanation method. It is expected that a local method shows a degradation in its performance with large thresholds, as a significant number of the selected instances will belong to subspaces different than the one used to build the explanation model.

We compared the xt-Fidelity results of both C-LORE-F and DRSA under different thresholds as illustrated in Figure 4. In general, we can find that, as the threshold increases, the overall score decreases, which means we lose the models' locality. In other words, the input space turns out to be more global, and the model fails to cover that space. The degradation in the performance is obvious in the case of DRSA. On the other hand, the C-LORE-F method mostly preserves the performance (the degradation is minor than in DRSA). We can attribute that to the fact that the multiple and small decision trees of C-LORE-F were trained on subspaces of the global space and formed a random forest model. Such a model can show comparable performance on the test set to the performance of RETIPROGRAM (a fuzzy random forest model built from multiple fuzzy decision trees).

Hence, if the locality of the explanation model is more important than the globality, we can choose DRSA. Otherwise, C-LORE-F is ideal.

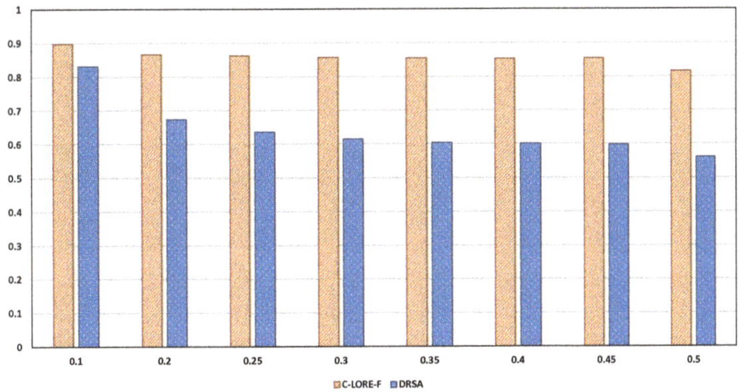

Figure 4. Degree of locality vs globality of the explanation models. The x-axis represents the thresholds and the y-axis represents the xt-Fidelity.

5.4. Evaluation of the Counterfactual Examples

It is well-known that counterfactual examples help to understand what changes are needed to obtain a different outcome. This is particularly interesting in health-care applications. Hence, it is important to have counterfactual examples that balance a wide range of suggested modifications (diversity) and the relative facility of adopting those modifications (proximity to the actual input). Moreover, counterfactual examples must be actionable, e.g., people can not reduce their age or change their race.

In this subsection, we evaluate the generated counterfactual examples using the following evaluation metrics [22]:

- **Validity**: is the number of counterfactual examples with a different outcome than the original input, i.e., x, divided by the total number of counterfactual examples.

$$\text{Validity} = \frac{|\hat{x} \in \mathbb{C}\ s.t. b(x) \neq b(\hat{x})|}{|\mathbb{C}|} \quad (1)$$

  Here $\mathbb{C}$ refers to the set of returned counterfactual examples and b is the black-box model.

- **Proximity**: is the mean of feature-wise distances between a counterfactual example c and the original input x. We considered two different proximity measures, the Euclidean distance for the continuous features and the Hamming distance for the categorical ones.

$$\text{Continuous Proximity} = -\frac{1}{|\mathbb{C}|} \sum_{c \in \mathbb{C}} dist\_cont(c, x) \quad (2)$$

$$\text{Categorical Proximity} = 1 - \frac{1}{|\mathbb{C}|} \sum_{c \in \mathbb{C}} dist\_cat(c, x) \quad (3)$$

- **Sparsity**: it measures the average of changes between a counterfactual example and the original input.

$$\text{Sparsity} = 1 - \frac{1}{|\mathbb{C}| * |\mathbb{F}|} \sum_{c \in \mathbb{C}} \sum_{f \in \mathbb{F}} \mathbb{1}[c_f \neq x_f] \quad (4)$$

  Here, $\mathbb{F}$ is the set of features, and $\mathbb{1}$ is the indicator function.

- **Diversity**: it is similar to proximity. However, instead of computing the feature-wise distance between the counterfactual example and the original input, we compute it between each pair of counterfactual examples. As in the proximity measure, we considered two different diversity versions, one for the continuous and the other for the categorical features.

Figure 5 shows the results of the Validity and Sparsity metrics for C-LORE-F and DRSA. C-LORE-F generates better valid counterfactual examples than DRSA. For both of them, when they generate a single counterfactual example, it is valid (near 100%). Notice that C-LORE-F has never generated more than 5 counterfactual examples. On the contrary, DRSA is able to generate more counterfactuals but the validity decreases and sparsity keeps similar.

Both C-LORE-F and DRSA show outstanding performance on Sparsity (DRSA is slightly better) with an average of 0.9.

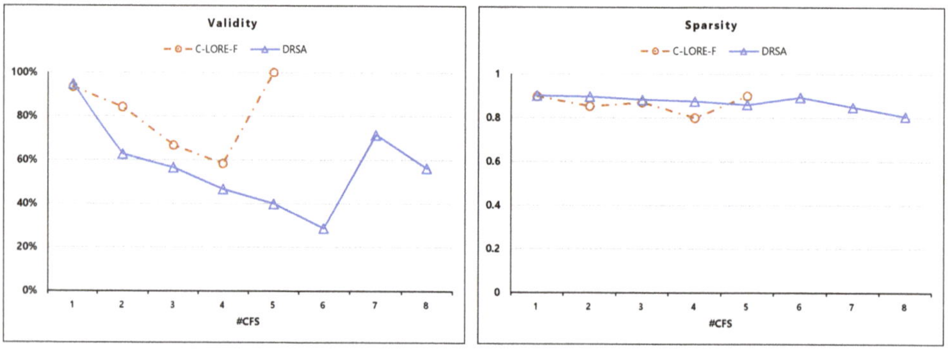

Figure 5. Validity and Sparsity of counterfactual examples. The numbers in the y-axis represent the metrics values (the higher values the better performance), while the x-axis represents the number of generated counterfactual instances.

Looking at the Diversity results (Figure 6), we can find that C-LORE-F generates more diverse examples with respect to the categorical features and diversity increases as the number of counterfactual examples increases. In the case of continuous features, DRSA is slightly better than C-LORE-F.

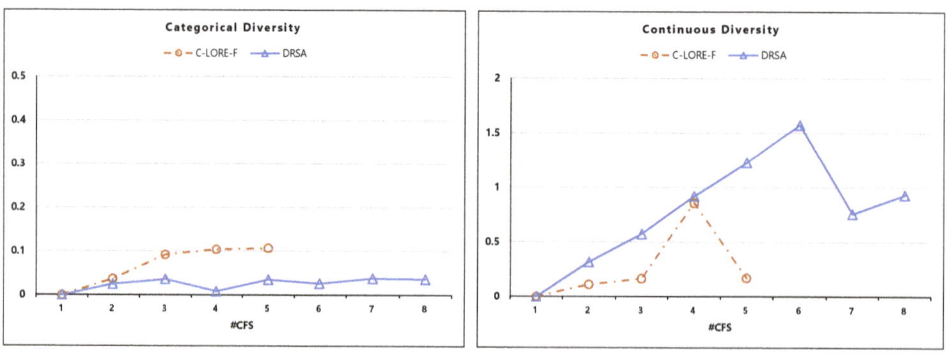

Figure 6. Diversity of counterfactual examples, in Categorical and Continuous attributes. The numbers in the y-axis represent the metrics values (the higher values the better performance), while the x-axis represents the number of generated counterfactual instances.

The Proximity results are shown in Figure 7. C-LORE-F generates counterfactual examples with lower proximity than DRSA for both the categorical and continuous features. Moreover, the inherent trade-off between diversity and proximity metrics can be observed in the case of categorical features.

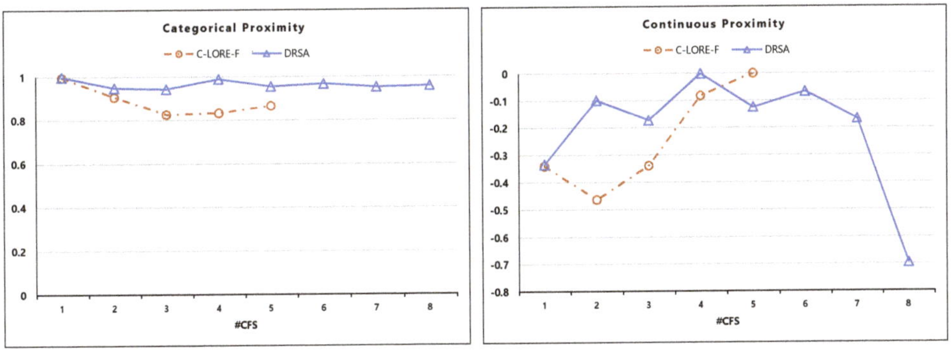

Figure 7. Proximity of counterfactual examples, in Categorical and Continuous attributes. The numbers in the y-axis represent the metrics values (the higher values, the better performance), while the x-axis represents the number of generated counterfactual instances.

5.5. Analysis of Computational Complexity

Computational resources are crucial in any practical application. In this subsection, we analyse the time complexity of the system. The most costly parts of the system are the generation of neighbours, the construction of the decision tree and the construction of the rules in the DRSA method. So, we first present the theoretical analysis of the time complexity of each part, and then we show the experimental setup and the running time in seconds.

We solve the generation of neighbours as a searching problem using the Uniform Cost Search algorithm. Hence, the total running time complexity of this part of the system is $O(b^{1+\lfloor C^*/\epsilon \rfloor})$, where b is the branching factor and C^* is the cost of the optimal solution, assuming that every action costs at least ϵ [34]. The time complexity for the decision tree algorithm is $O(n) + O(m \cdot n \cdot log_2 n) + O(n \cdot log_2 n)$ [35]. The time complexity of the DRSA method is $O(m^2 \cdot n^2)$. Here, m is the number of examples, and n is the number of attributes.

The experiments were carried out on a 64- bit computer, with AMD Ryzen 7 3700U (4 Cores, 2.3 GHz) and 16 GB RAM, running Windows 11 operating system. The number of examples in the test set is 1111. For each example, we generate 800 neighbours and use them to derive the explanation. Table 3 shows the elapsed time of generating the neighbours, building the decision trees and constructing the rules in seconds. As expected, the most expensive part of the system is the generation of the neighbours. The decision tree is slightly faster than the DRSA, which is expected as the DRSA complexity is quadratic, whereas the decision tree complexity is logarithmic. The estimated time to obtain an explanation is 9.051 s using the C-LORE-F method and 9.268 s using the DRSA method.

Table 3. Running time comparisons.

	Min	Max	Average
Neighbours Generation	5.453	19.252	8.882
C-LORE-F	0.093	1.149	0.169
DRSA	0.125	2.556	0.386

6. Conclusions

We have proposed a methodology to derive an explanation for the decision made by the RETIPROGRAM system, which was developed in our research group to estimate the personalised risk of developing diabetic retinopathy as early as possible. RETIPROGRAM is in use at a regional hospital in city of Reus (Spain). The paper is focused on comparing two different explanation methods: one based on decision trees (C-LORE-F) and the other one based on decision rules (DRSA). These methods are post-hoc explanation methods and rely on the generation of neighbours around the instance to be explained, which are used to train an explainable model (either a decision tree or decision rules), and finally they generate an explanation from that local model. The explanation result is formed by the rule(s) applicable to the instance, a set of counterfactual rules and a set of counterfactual instances. C-LORE-F with DT was previously published in [13], where was compared with other state of the art methods. The current paper shows that DRSA is also a valid method for generating explanatory rules. After comparing the obtained explanation from the C-LORE-F and DRSA methods using multiple evaluation metrics, we found that both of them generate an adequate explanation. C-LORE-F is slightly better in hid and fidelity indicators, but its sparsity is a bit smaller than the one of DRSA. We have also shown that the time needed for the generation of the explanation is of 9 s, which is an acceptable time for real use for medical physicians when visiting a patient.

As a weak point, we found that the counterfactual instances are not sufficiently good. In our future work, we will focus on resolving this issue by improving the generation of counterfactual examples [22].

It is worth to mention that the methods proposed for constructing explanations based on rules are general even if they have been studied for the Diabetic Retinopathy problem. They could be applied to other fields and with any other black box classifier.

Author Contributions: Conceptualization, N.M., A.M., A.V.; methodology, N.M., M.J., A.M., A.V., M.S.; software, N.M., M.J., M.S.; validation, N.M., M.J., A.M.; formal analysis, N.M., A.V., M.S.; investigation, N.M., M.J., A.M., A.V.; resources, A.M., A.V.; data curation, N.M., A.V., M.S.; writing—original draft preparation, N.M.; writing—review and editing, M.J., A.M., A.V., M.S.; visualization, N.M.; supervision, A.M., A.V.; project administration, A.V.; funding acquisition, A.V. All authors have read and agreed to the published version of the manuscript.

Funding: This work has been funded by the Spanish FIS projects PI21/00064 and PI18/00169 (ISCIII and Fondos FEDER) and the URV grant 2019-PFR-B2-61. It was also partially supported by TAILOR, a project funded by EU Horizon 2020 research and innovation programme under GA No.952215. The first author is funded by a URV Martí Franquès predoctoral grant.

Institutional Review Board Statement: Not applicable.

Informed Consent Statement: Not applicable.

Data Availability Statement: Not applicable.

Acknowledgments: The authors would like to thank the collaboration of Oriol Villaró for his initial implementation work. We also thank the collaboration of the Ophthalmology department of Hospital Sant Joan de Reus in Catalonia.

Conflicts of Interest: The authors declare no conflict of interest.

Abbreviations

The following abbreviations are used in this manuscript:

DM	Diabetes Mellitus
DR	Diabetic Retinopathy
NHS	National Health Surveys
AI	Artificial Intelligence
CDSS	Clinical Decision Support System
ML	Machine Learning
FDT	Fuzzy Decision Tree
FRF	Fuzzy Random Forest
LIME	Local Interpretable Model-agnostic Explanations
LORE	Local Rule-Based Explanations
C-LORE-F	Contextualized LORE for Fuzzy Attributes
DRSA	Dominance-based Rough Set Approach
HVDM	Heterogeneous Value Difference Metric
CKD-EPI	Chronic Kidney Disease Epidemiology index

References

1. Aguiree, F.; Brown, A.; Cho, N.H.; Dahlquist, G.; Dodd, S.; Dunning, T.; Hirst, M.; Hwang, C.; Magliano, D.; Patterson, C.; et al. *IDF Diabetes Atlas*; International Diabetes Federation: Brussels, Belgium, 2013.
2. Shaw, J.E.; Sicree, R.A.; Zimmet, P.Z. Global estimates of the prevalence of diabetes for 2010 and 2030. *Diabetes Res. Clin. Pract.* **2010**, *87*, 4–14. [CrossRef] [PubMed]
3. Nair, A.T.; Muthuvel, K.; Haritha, K. Effectual Evaluation on Diabetic Retinopathy. In *Information and Communication Technology for Competitive Strategies (ICTCS 2020)*; Springer: Singapore, 2022; pp. 559–567.
4. López, M.; Cos, F.X.; Álvarez-Guisasola, F.; Fuster, E. Prevalence of diabetic retinopathy and its relationship with glomerular filtration rate and other risk factors in patients with type 2 diabetes mellitus in Spain. DM2 HOPE study. *J. Clin. Transl. Endocrinol.* **2017**, *9*, 61–65. [CrossRef] [PubMed]
5. Romero-Aroca, P.; Valls, A.; Moreno, A.; Sagarra-Alamo, R.; Basora-Gallisa, J.; Saleh, E.; Baget-Bernaldiz, M.; Puig, D. A clinical decision support system for diabetic retinopathy screening: Creating a clinical support application. *Telemed. e-Health* **2019**, *25*, 31–40. [CrossRef] [PubMed]
6. Saleh, E.; Valls, A.; Moreno, A.; Romero-Aroca, P.; Torra, V.; Bustince, H. Learning fuzzy measures for aggregation in fuzzy rule-based models. In *International Conference on Modeling Decisions for Artificial Intelligence*; Springer: Cham, Switzerland, 2018; pp. 114–127.
7. Romero-Aroca, P.; Verges-Pujol, R.; Santos-Blanco, E.; Maarof, N.; Valls, A.; Mundet, X.; Moreno, A.; Galindo, L.; Baget-Bernaldiz, M. Validation of a diagnostic support system for diabetic retinopathy based on clinical parameters. *Transl. Vis. Sci. Technol.* **2021**, *10*, 17. [CrossRef] [PubMed]
8. Burkart, N.; Huber, M.F. A Survey on the Explainability of Supervised Machine Learning. *J. Artif. Intell. Res.* **2021**, *70*, 245–317. [CrossRef]
9. Strumbelj, E.; Kononenko, I. An efficient explanation of individual classifications using game theory. *J. Mach. Learn. Res.* **2010**, *11*, 1–18.
10. Krause, J.; Perer, A.; Ng, K. Interacting with predictions: Visual inspection of black-box machine learning models. In Proceedings of the 2016 CHI Conference on Human Factors in Computing Systems, San Jose, CA, USA, 7–12 May 2016; pp. 5686–5697.
11. Ribeiro, M.T.; Singh, S.; Guestrin, C. Why Should I Trust You?: Explaining the Predictions of Any Classifier. In Proceedings of the 22nd ACM SIGKDD International Conference on Knowledge Discovery and Data Mining, San Francisco, CA, USA, 13–17 August 2016; pp. 1135–1144.
12. Maaroof, N.; Moreno, A.; Valls, A.; Jabreel, M. Guided-LORE: Improving LORE with a Focused Search of Neighbours. In *International Workshop on the Foundations of Trustworthy AI Integrating Learning, Optimization and Reasoning*; Springer: Cham, Switzerland, 2020; pp. 49–62.
13. Maaroof, N.; Moreno, A.; Jabreel, M.; Valls, A. Contextualized LORE for Fuzzy Attributes. In *Artificial Intelligence Research and Development*; IOS Press: Amsterdam, The Netherlands, 2021; pp. 435–444.
14. Saleh, E.; Moreno, A.; Valls, A.; Romero-Aroca, P.; de La Riva-Fernandez, S. A Fuzzy Random Forest Approach for the Detection of Diabetic Retinopathy on Electronic Health Record Data. In *Artificial Intelligence Research and Development*; Frontiers in Artificial Intelligence and Applications; IOS Press: Amsterdam, The Netherlands, 2016; Volume 288, pp. 169–174.
15. Greco, S.; Matarazzo, B.; Slowinski, R. Rough sets theory for multicriteria decision analysis. *Eur. J. Oper. Res.* **2001**, *129*, 1–47. [CrossRef]
16. Guidotti, R.; Monreale, A.; Ruggieri, S.; Turini, F.; Giannotti, F.; Pedreschi, D. A survey of methods for explaining black box models. *ACM Comput. Surv.* **2018**, *51*, 1–42. [CrossRef]
17. Molnar, C. *Interpretable Machine Learning*; Lulu.com: Morrisville, NC, USA, 2020.

18. Ribeiro, M.T.; Singh, S.; Guestrin, C. Anchors: High-Precision Model-Agnostic Explanations. *AAAI* **2018**, *18*, 1527–1535.
19. Lundberg, S.M.; Lee, S.I. A unified approach to interpreting model predictions. *Adv. Neural Inf. Process. Syst.* **2017**, *30*, 4765–4774.
20. Martens, D.; Provost, F. Explaining data-driven document classifications. *MIS Q.* **2014**, *38*, 73–100. [CrossRef]
21. Guidotti, R.; Monreale, A.; Ruggieri, S.; Pedreschi, D.; Turini, F.; Giannotti, F. Local rule-based explanations of black box decision systems. *arXiv* **2018**, arXiv:1805.10820.
22. Mothilal, R.K.; Sharma, A.; Tan, C. Explaining machine learning classifiers through diverse counterfactual explanations. In Proceedings of the 2020 Conference on Fairness, Accountability, and Transparency, Barcelona, Spain, 27–30 January 2020; pp. 607–617.
23. Russell, C. Efficient search for diverse coherent explanations. In Proceedings of the Conference on Fairness, Accountability, and Transparency, Atlanta, GA, USA, 19–31 January 2019; pp. 20–28.
24. Ming, Y.; Qu, H.; Bertini, E. Rulematrix: Visualizing and understanding classifiers with rules. *IEEE Trans. Vis. Comput. Graph.* **2018**, *25*, 342–352. [CrossRef]
25. Neto, M.P.; Paulovich, F.V. Explainable Matrix—Visualization for Global and Local Interpretability of Random Forest Classification Ensembles. *IEEE Trans. Vis. Comput. Graph.* **2020**, *27*, 1427–1437. [CrossRef] [PubMed]
26. Wilson, D.R.; Martinez, T.R. Improved heterogeneous distance functions. *J. Artif. Intell. Res.* **1997**, *6*, 1–34. [CrossRef]
27. Ruggieri, S. YaDT: Yet another Decision Tree Builder. In Proceedings of the 16th IEEE International Conference on Tools with Artificial Intelligence (ICTAI 2004), Boca Raton, FL, USA, 15–17 November 2004; pp. 260–265.
28. Pawlak, Z. *Rough Sets: Theoretical Aspects of Reasoning about Data*; Springer Science & Business Media: Dordrecht, The Netherlands, 1991; Volume 9.
29. Słowiński, R.; Greco, S.; Matarazzo, B. Rough set methodology for decision aiding. In *Springer Handbook of Computational Intelligence*; Springer: Berlin/Heidelberg, Germany, 2015; pp. 349–370.
30. Błaszczyński, J.; Słowiński, R.; Szeląg, M. Sequential covering rule induction algorithm for variable consistency rough set approaches. *Inf. Sci.* **2011**, *181*, 987–1002. [CrossRef]
31. Błaszczyński, J.; Słowiński, R.; Szeląg, M. Induction of Ordinal Classification Rules from Incomplete Data. In *Rough Sets and Current Trends in Computing*; Yao, J., Yang, Y., Słowiński, R., Greco, S., Li, X., Mitra, S., Polkowski, L., Eds.; Springer: Berlin/Heidelberg, Germany, 2012; Volume 7413, pp. 56–65.
32. Saleh, E.; Maaroof, N.; Jabreel, M. The deployment of a decision support system for the diagnosis of Diabetic Retinopathy into a Catalan medical center. In *Proceedings of the 6th URV Doctoral Workshop in Computer Science and Mathematics*; Universitat Rovira i Virgili: Tarragona, Spain, 2020; p. 45.
33. Blanco, M.E.S.; Romero-Aroca, P.; Pujol, R.V.; Valls, A.; SaLeh, E.; Moreno, A.; Basora, J.; Sagarra, R. A Clinical Decision Support System (CDSS) for diabetic retinopathy screening. Creating a clinical support application. *Investig. Ophthalmol. Vis. Sci.* **2020**, *61*, 3308.
34. Russell, S.; Norvig, P. *Artificial Intelligence: A Modern Approach*, 4th ed.; Pearson: London, UK, 2020.
35. Sani, H.M.; Lei, C.; Neagu, D. Computational complexity analysis of decision tree algorithms. In *International Conference on Innovative Techniques and Applications of Artificial Intelligence*; Springer: Cham, Switzerland, 2018; pp. 191–197.

applied sciences

MDPI

Article

AWEU-Net: An Attention-Aware Weight Excitation U-Net for Lung Nodule Segmentation

Syeda Furruka Banu [1,*], Md. Mostafa Kamal Sarker [2], Mohamed Abdel-Nasser [1,3], Domenec Puig [1] and Hatem A. Raswan [1]

1 Departament d'Enginyeria Informàtica i Matemàtiques, Universitat Rovira i Virgili, 43007 Tarragona, Spain; mohamed.abdelnasser@urv.cat (M.A.-N.); domenec.puig@urv.cat (D.P.); hatem.abdellatif@urv.cat (H.A.R.)
2 National Subsea Centre, Robert Gordon University, Aberdeen AB10 7GJ, UK; m.kamal.sarker@gmail.com
3 Department of Electrical Engineering, Aswan University, Aswan 81542, Egypt
* Correspondence: syedafurruka.banu@estudiants.urv.cat

Abstract: Lung cancer is a deadly cancer that causes millions of deaths every year around the world. Accurate lung nodule detection and segmentation in computed tomography (CT) images is a vital step for diagnosing lung cancer early. Most existing systems face several challenges, such as the heterogeneity in CT images and variation in nodule size, shape, and location, which limit their accuracy. In an attempt to handle these challenges, this article proposes a fully automated deep learning framework that consists of lung nodule detection and segmentation models. Our proposed system comprises two cascaded stages: (1) nodule detection based on fine-tuned Faster R-CNN to localize the nodules in CT images, and (2) nodule segmentation based on the U-Net architecture with two effective blocks, namely position attention-aware weight excitation (PAWE) and channel attention-aware weight excitation (CAWE), to enhance the ability to discriminate between nodule and non-nodule feature representations. The experimental results demonstrate that the proposed system yields a Dice score of 89.79% and 90.35%, and an intersection over union (IoU) of 82.34% and 83.21% on the publicly available LUNA16 and LIDC-IDRI datasets, respectively.

Keywords: artificial intelligence; computer-aided diagnosis; computed tomography; lung cancer; deep learning; lung nodule detection; lung nodule segmentation

Citation: Banu, S.F.; Sarker, M.M.K.; Abdel-Nasser, M.; Puig, D.; Rashwan, H.A. AWEU-Net: An Attention-Aware Weight Excitation U-Net for Lung Nodule Segmentation. *Appl. Sci.* **2021**, *11*, 10132. https://doi.org/10.3390/app112110132

Academic Editor: Francesco Bianconi

Received: 29 September 2021
Accepted: 22 October 2021
Published: 28 October 2021

1. Introduction

According to the World Health Organization (WHO), lung cancer is the leading cause of cancer deaths in 2020 (1.80 million deaths) [1]. The estimated number of new cases has risen to 2.89 million, and the number of deaths may reach 2.45 million worldwide by 2030 [2]. These deaths could be avoidable by an early diagnosis with a proper treatment plan. The National Lung Screening Trial (NLST) showed that the mortality of lung cancer is reduced by 20% by emphasizing the significance of nodule detection and assessment [3]. Many studies have shown the efficacy of computed tomography (CT) screening for lung cancer diagnosis and the detection of subsolid nodules, as well as suspected/unsuspected lung cancer nodules [4].

CT imaging technology helps make an efficient investigation to discover pulmonary nodules. CT imaging technology generates a hundred images of the lung within a second by a single scan. It is difficult for radiologists to manually detect and segment the lung nodules from such a high number of images. In this context, computer-aided diagnosis (CAD) systems have assisted radiologists in the automated diagnosis of lung cancer and pulmonary diseases over the last several years. The authors of [5] noted that the use of an accurate lung nodule CAD system accelerates the entire diagnosis and radiotherapy process, such that patients can perform the required radiation or photon therapy on the same day. CAD systems mainly depend on the detection and segmentation of various pulmonary parts. Computer-aided detection (CADe) systems identify the region of interest

(ROI) in the lung nodule, while computer-aided segmentation (CASe) systems segment the nodule region and determine its boundaries.

Automated analysis of lung CT images is essential to measure the properties of lung nodules for identifying malignancy in a tumor. Lung nodule segmentation systems can determine malignancy by analyzing nodule size, shape, and change rate [6]. Although many automated nodule detection/segmentation systems have been presented in the last years [7–9], their accuracies are not high due to several challenges, such as the heterogeneity of CT images and the variation present in nodule size, shape, and location.

In the last years, deep convolutional neural networks (CNNs) have been widely used to handle the lung nodule detection and segmentation problem, achieving promising results [7–13]. CNNs can learn complex features to detect and segment the nodule accurately. However, existing nodule segmentation systems use deep learning models to segment nodules from the whole-input CT images. This reduces the segmentation precision because the input images are usually resized before feeding into the deep learning model. Such resizing processes yield artifacts that badly affect the objects' boundaries and details.

In an attempt to handle the challenges that accompany the automated segmentation of lung nodules, in this article, we propose the AWEU-Net method for a lung nodule detection and segmentation system based on deep learning. AWEU-Net is a fully automated deep learning-based framework that includes two cascaded stages. In the first stage, AWEU-Net automatically detects lung nodules based on a fine-tuned Faster R-CNN model. In the second stage, AWEU-Net automatically delineates lung nodules from the ROI which results from the first stage based on a U-Net architecture with two powerful blocks via position attention-aware weight excitation (PAWE) and channel attention-aware weight excitation (CAWE). Both blocks help model the correlation between the spatial and channel features and encourage the CNN to learn the most relevant features that enhance its ability to discriminate between nodule and non-nodule feature representations. The contributions of this article can be listed as follows:

- A fully automated deep learning-based framework called AWEU-Net is proposed for improving the accuracy of lung nodule detection and segmentation;
- PAWE and CWEU mechanisms are proposed to model the correlation between the spatial and channel features and encourage the CNN model to learn the most relevant features that enhance its ability to discriminate between nodule and non-nodule feature representations;
- A comparative study of different nodule detection models and nodule segmentation models is presented using two publicly available datasets, namely LUNA16 and LIDC-IDRI.

Section 2 of this article discusses the existing lung nodule segmentation systems based on classical computer vision and deep learning techniques. Section 3 introduces the proposed system workflow and model architecture. Section 4 presents and discusses experimental results. Finally, Section 5 concludes the article and highlights a future extension of this research.

2. Related Work

In the literature, several lung nodule detection and segmentation systems have been presented based on classical computer vision and deep learning techniques. Table 1 lists some common lung nodule segmentation techniques. Below, we present and discuss classical computer vision-based and deep learning-based lung nodule segmentation methods.

Table 1. Summary of existing lung nodule segmentation methods. The undeclared information is marked with dashes (-) in the referred literature. Aug., morph, acc and IoU stand for augmentation, morphological, segmentation accuracy and intersection over union, respectively.

References	Methods/Architectures	Dataset	Pre-Processing	Post-Processing	Results
Classical computer vision-based					
[14]	Region growing	PRIVATE	Local contrast & hole filling	-	83% acc
[15]	Active contours	LIDC-IDRI	Thresholding & morph operations	Markov random field	69% IoU
[16]	Level sets	LIDC-IDRI	Statistical intensity	Region condition	94% acc
[17]	Graph cuts	PRIVATE	Gaussian smoothing	-	98.74% dice
[18]	Adaptive thresholding	LIDC-IDRI	Histogram equalization & noise filtering	Morph operations	96% acc
[19]	GMM fuzzy C-means	LIDC-IDRI & GHGZMCPLA	Non-local mean filter & gaussian pyramid	Random walker	86% dice
[20]	Region-based fast marching	LIDC-IDRI	Convex hulls	Mean threshold	61–93% dice
Deep learning-based					
[10]	U-Net	LIDC-IDRI	Nodule ROI selection	-	74% dice
[7]	iW-Net	LIDC-IDRI	Nodule ROI selection	-	55% IoU
[8]	U-Det	LUNA16	Data aug.	-	82.82% dice
[11]	Nodulenet	LIDC-IDRI & LUNA16	Nodule ROI selection & Data aug.	-	71.85% IoU
[9]	DB-ResNet	LIDC-IDRI	Nodule ROI selection & Data aug.	Remove noisy voxel	82.74% dice
[13]	MRRN	TCIA & MSKCC & LIDC-IDRI	Nodule ROI selection & Data aug.	-	74% dice

2.1. Classical Computer Vision-Based Approaches

In the field of lung nodule analysis, many computer vision methods based on hand-crafted features have been used, such as region growing [14], active contours [15], level sets [16], graph cuts [17], adaptive thresholding [18], Gaussian mixture models (GMM) with fuzzy C-means [19], and region-based fast marching [20]. However, it is difficult to generalize a nodule segmentation model based on hand-crafted features that can be useful for CT images. All the aforementioned traditional approaches are semi-automated or depend on several image pre-processing and post-processing techniques. For instance, a contrast-based region growing method and fuzzy connectivity map of the object of interest (i.e., nodule) were used in [14] to segment various types of pulmonary nodules. This method did not perform adequately with irregular nodules due to merging different criteria in the region growing technique that needed a fine-tuning for parameters of the setting. Geometric active contours with a marker-controlled watershed as well as Markov random field (MRF) was used in [15] to segment the lung nodule. In turn, ref. [16] used a shape prior hypothesis along with level sets that iteratively minimized the energy needed to segment juxtapleural pulmonary nodules. However, these two methods depend on manually selected seeds in the nodule region, and thus the precision of the segmentation process depends on the proper selection of seeds. A graph cut algorithm with an expectation-maximization (EM) algorithm was proposed in [17] for lung segmentation on chest CT images. This algorithm yields acceptable segmentation results; however, it has a high computational cost because it focuses on training a GMM and searching on the corresponding graph using a heuristic searching algorithm. Ref. [18] used an adaptive thresholding technique along with a watershed transform to detect lung nodules. However, this approach mainly relies on different pre-processing and post-processing procedures. Ref. [19] combined GMM knowledge within the conventional fuzzy C-means method to improve the robustness of pulmonary nodule segmentation. The major disadvantage of fuzzy C-means algorithms is that they are sensitive to noise, outliers, and primary cluster selection. A region-based approach was introduced in [20] by using the fast marching method, which gives a precise segmentation of the nodule and can properly handle juxtapleural and juxtavascular nodules. The main disadvantage of region growing segmentation is the fact that the resulting histograms do not provide any spatial knowledge of the input images.

2.2. Deep Learning-Based Approaches

Recently, many researchers have developed various deep learning-based systems for lung nodule detection and segmentation. The authors of [21] introduced a lung CT image segmentation method using the U-net architecture proposed in [22], consisting of encoder and decoder networks. With the LIDC-IDRI dataset, they achieved a Dice score coefficient (DSC) of 0.9502. It is worth noting that the model presented in [21] and our method are based on U-Net; however, each one is designed for solving a different problem: ref. [21] fine-tuned U-Net to segment the whole lung from the CT images, while our method integrates two new blocks (CAWE and PAWE) with U-Net to improve the segmentation accuracy of nodules. Ref. [10] used a simple version of the U-Net model for lung nodule segmentation by utilizing only two convolutional layers in the encoder network and two deconvolutional layers in the decoder network, U-Net. The model used different receptive field sizes to enhance nodule feature extraction. Their model yields a DSC improvement of 2% compared to the original U-Net. Besides, ref. [9] presented a dual-branch residual network (DB-ResNet) that achieved results similar to [10]. The major differences between [9] and [10] are the use of convolutional blocks of ResNet [23] in the encoder networks and slightly modified pooling layers.

In turn, ref. [7] combined two U-Net models (named iW-Net) based on user interactions. Their architecture was designed by expecting nodules of only round shapes. The authors combined the weight map and the feature of the model output as a loss function. The iW-Net model gave a final competition performance metric score of 87.27% on nodule detection and a DSC score of 83.10% on nodule segmentation. In addition, ref. [13] presented a multiple resolution residual network (MRRN) that is a modification of the ResNet [23]. The modified MRRN network is used as the backbone of the U-Net model. Ref. [13] achieved a DSC score of 0.76. A slightly transformed version of U-Net called U-Det was presented in [8], where many hidden layers were used to filter the residual blocks located within the encoder and decoder. With the LUNA16 testing dataset, they achieved a DSC of 0.82 that was improved to 0.83 when U-Det applied the Mish activation function proposed in [24] for smoothing and non-monotonic activation.

Most aforementioned lung nodule segmentation methods use deep learning models that segment nodules from the whole-input CT images. However, this can degrade the segmentation precision. This is because the input images are usually resized before feeding to the deep learning model, which yields too many artifacts and badly affects the objects' boundaries and details. Consequently, in this work, we attempt to build a cascaded lung nodule detection and segmentation system that can outperform the accuracy of the state-of-the-art. Firstly, fine-tuning the important parameters of the state-of-the-art object detection (i.e., Faster-RCNN) is applied to be more appropriate for lung nodule detection and to have an automated system to localize the nodule in CT images. The output of the object detection model is to enable ROI that involves the nodule region. A segmentation model will then be fed with ROIs to segment the exact potential nodule region and properly determine its boundaries. Thus, improving the U-Net model is then achieved by integrating the position attention module (PAWE) and channel attention module (CAWE) to encode contextual information into spatial and channel features. These modules help our segmentation system to accurately distinguish nodules from non-nodules regions and also help in facilitating the model's training process, since they encourage the model to learn nodules' relevant features.

3. Proposed Methodology

The main components of the proposed framework are shown in Figure 1. As shown, as a pre-processing stage, we represented the 3D CT volumes as 2D CT slices. The Dicom CT slices are transformed into images of ".png" file format. A global thresholding technique is used for separating the lung region from the background in CT images. In order to detect the nodule ROIs from the input CT images, we fine-tuned the Faster R-CNN object detection model to adapt it for lung nodule detection. The detected nodule ROI is then fed to the segmentation model to precisely segment the nodule and its boundaries.

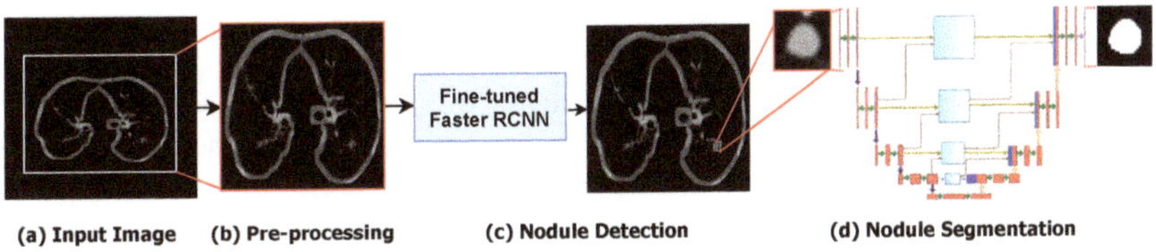

Figure 1. The step-by-step workflow of the proposed method. (**a**) The converted input image is extracted from the original CT slice. (**b**) The pre-processing step for selecting ROI of lung. (**c**) The detection of lung nodules using Optimized Faster R-CNN. (**d**) The segmentation of lung nodules using the proposed AWEU-Net model.

3.1. Pre-Processing

The raw CT scans' data are always available in the Dicom file format. However, to make the images more meaningful and useful for deep learning models, the pylidc [25] library is used for converting the Dicom images to a ".png" file format. Afterwards, a global Otsu binary thresholding technique and morphological dilation are applied on the CT images to separate the lung region from the background, as shown in Figure 1b. Experimentally, we found that the image size of 512 × 512 yields the best accuracy with lung nodule detection. Thus, we resized all images to that image size. Finally, we split the dataset into 70% for training, 10% for validation, and 20% for test. For training Faster-RCNN, we convert all ground-truth images to ms-coco format [26], where the dataset is formatted in JSON and is a collection of "info", "images", "annotations", and "categories".

3.2. Nodule Detection Model

Among all two-stage object detection models, the single-shot detector (SSD) is the fastest detection model, but it is not the most accurate one [27]. In our framework, we aimed at developing an accurate segmentation model; Faster RCNN is considered one of the most accurate detection models in inferring the locations of the target in the input image [28]. Thus, we preferred to choose the Faster RCNN model rather than SSD for localizing the nodules in CT images. In this stage, we attempted to fine-tune the important parameters of the Faster RCNN detection model. We focused on finding the best combination of the learning rate, step size, factor of dropped learning rate γ, and drop-out ratio to make Faster RCNN more appropriate for lung nodule detection.

As shown in Figure 2, the Faster-RCNN detection model is a two-stage detection network containing three main blocks: a backbone network, a region proposal network (RPN), and a box head. We used ResNet50 [23] as a backbone network to extract feature maps from the input image. The feature map is then fed into the RPN to perform boundary regression and classification analysis, and the output is a set of ROI candidates. The classification principle is based on whether a candidate ROI is either related to background or to the object (i.e., in our case, tumour nodules). The position and score of the candidate ROI are forwarded to the box head, where the final regression and classification of the object is performed. Finally, the bounding box of the target (nodule) with the classification score is returned from the detection model.

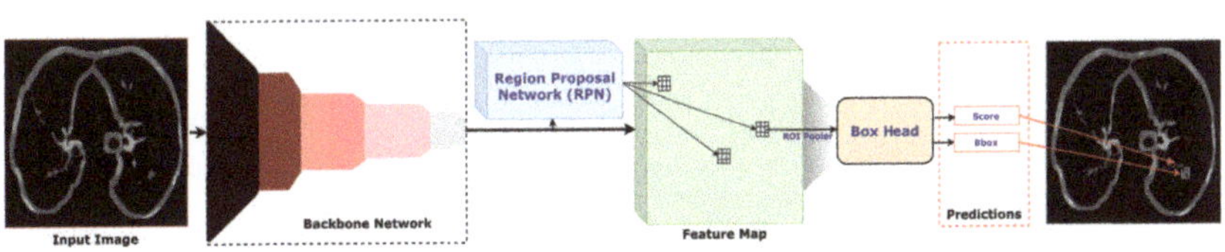

Figure 2. The detailed architecture of Optimized Faster R-CNN.

3.3. Nodule Segmentation Model

We cropped the ROI based on the bounding box provided by the nodule detection model in the first stage. We resized the ROI to 224 × 224 and fed it into the proposed nodule segmentation model. It is obvious that we scaled up the ROI of the nodule, which we did since we believe that is the best way to enhance spatial features, especially for small objects like nodules. By scaling up the input ROIs rather than downsampling them, the deep segmentation network can be better adapted to detect even tiny object boundaries.

In the second stage of our framework, we propose an attention-aware weight excitation U-Net, AWEU-Net, for our lung nodule segmentation, as shown in Figure 3. This network is based on the U-Net [22], which is a well-known deep learning model for medical image segmentation. The AWEU-Net model learns to segment the input ROIs by determining the boundaries of the nodule region to discriminate between nodules and non-nodule regions. The output of AWEU-Net is a binary image that contains ones for nodule regions and zeros for the others.

The proposed model integrates PAWE and CAWE blocks with U-Net in order to capture the correlation between both spatial and channel features, as well as to enhance the ability to discriminate between nodule and non-nodule feature representations. On the one hand, the feature map resulting from a convolution layer contains a set of channels; each can be a class-specific response comprising high-level features. Since some channels can be correlated, CAWE, which models inter-dependencies among channel maps, is able to highlight inter-dependent feature maps. On the other hand, PAWE helps to capture discriminant feature representations by encoding contextual information into local features extracted by convolution layers. The details about PAWE and CAWE will be discussed in Sections 3.4.1 and 3.4.2, respectively.

The AWEU-Net architecture is composed of two successive networks: an encoder and a decoder. The encoder consists of four convolution layers. Each encoder layer is composed of a convolution of 3 × 3 followed by a PAWE block and a ReLU as an activation function. Four down-sampling blocks with a max pooling of 2 × 2 followed by a stride of 2 are used after each encoder layer.

In turn, the decoder consists of four layers, which each also consist of a convolution of 3 × 3 followed by a PAWE block, a ReLU and a deconvolution of 2 × 2. In the original UNet, the outputs of the decoder layers (if exiting) were concatenated with the features extracted by the corresponding encoder layers to input to the next decoder layer. In AWEU-Net, we inserted the CAWE blocks between the corresponding layers of the encoder and decoder networks. Each CAWE block is fed with a feature map of the same size as the corresponding encoder layer. The CAWE blocks are fed by the concatenation of the features extracted by the corresponding encoder layer and the upsampling features extracted by the next encoder layers (e.g., *Up1* , *Up3* and *Up5* in Figure 3). The output of the CAWE block is fed as an input to the corresponding decoder layer. Additionally, the output feature map of the previous CAWE block is scaled up and also fed to the decoder layer as an input (e.g., *Up2*, *Up4*, and *Up6* in Figure). In this mechanism, we depend on the features extracted by PAWE and CAWE blocks to enhance the positional and channel low- and high-level features extracted by the encoder network and utilise them for the reconstruction means in the decoder network.

The final output layer of the model applies a convolution of 1 × 1 to map the final feature map of 64 channels to the number of targeted segmentation classes (i.e., in our case two classes related to the nodule and the background).

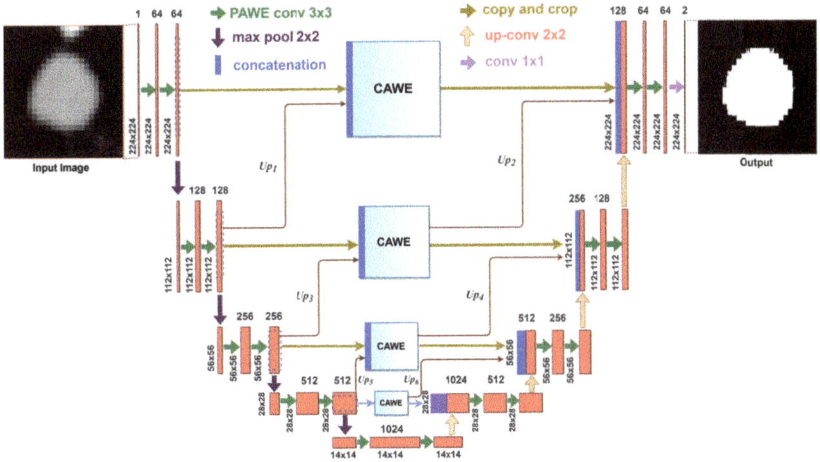

Figure 3. The architecture of the proposed AWEU-Net. The PAWE and CAWE block refers to the position attention-aware weight excitation and channel attention-aware weight excitation, respectively.

3.4. *Attention Mechanism*

For semantic segmentation, the scene can involve objects (e.g., cars) which are different in views, scales, and lighting. Thus, the features extracted by CNNs corresponding to the same object could have diversities, since CNN filters yield diverse local receptive fields. These diversities in the pixels corresponding to the same label/object cause intra-class inconsistency and affect the segmentation accuracy [29]. Regarding nodule segmentation, the nodules, in general, have different sizes that can yield features with intra-class inconsistency. Consequently, with our framework, we inserted global contextual attention models in both spatial and channel dimensions in the UNet network to explore relationships between features extracted in different layers of the encoder network before feeding to the decoder to reconstruct the segmented images.

In the next sub-sections, we will present the two attention modules PAWE and CAWE, which help our network to capture contextual information in spatial and channel dimensions, respectively.

3.4.1. Position Attention-Aware Weight Excitation (PAWE)

For accurate nodule semantic segmentation, deep learning models have to capture discriminant features of the nodules and background in a CT image. These features can be captured by aggregating the spatial context information from local features [30]. To model contextual relationships over local features, PAWE is able to enhance the local feature representation through encoding long-range contextual information. The process of PAWE can be elaborated as follows.

The PAWE block consists of two sub-blocks: the position attention block (PAB) and the weight excitation block (WEB). To demonstrate the proposed PAWE block, let the input feature be $Y \in \mathbb{R}^{C \times H \times W}$, where C, H and W are channel, height and width, respectively (see Figure 4). In the PAB block, Y is fed into 3 convolutions of 3×3 called A, B and C, respectively. The first 2 produced feature maps, $A^p, B^p \in \mathbb{R}^{C/8 \times H \times W}$ are provided by the first 2 convolutions A and B,where p superscript is referred to as "postion". A^p and B^p feature maps are then reshaped into $(H \times W) \times C/8$. A matrix multiplication is applied to the transposition of $\mathbf{A^p}$ and $\mathbf{B^p}$, producing a spatial attention map, $D^p \in \mathbb{R}^{(H \times W) \times (H \times W)}$, by using a softmax function:

$$d^p_{i,j} = \frac{\exp\left(Ai^p \cdot Bj^p\right)}{\sum_{i=1}^{H \times W} \exp\left(Ai^p \cdot Bj^p\right)}, \quad (1)$$

where $s_{i,j}$ indicates the ith position's associated position of jth. The softmax function D^p attempts to learn the relationship between two spatial positions in the input feature maps.

In addition, the output of the third convolutional layer $C^p \in \mathbb{R}^{C\times(H\times W)}$ is also reshaped to the same shape of the input feature map Y and then multiplied by a permuted order of the spatial attention map D^p of (1). The final output is reshaped to $\mathbb{R}^{C\times(H\times W)}$ to provide the final feature map of PAB block, F, as

$$F_{PAB,j} = \alpha_p \sum_{i=1}^{H\times W} s_{ij}^p Cj^p + Y_j, \tag{2}$$

where α_p is defined as 0 as explained in [29]. The resulting feature **F** at each feature position is a weighted sum of all the neighbours of the original features.

In the WEB, a sub-block for location-based weight excitation (LWE) proposed in [31] is used. The LWE provides fine-grained weight-wise attention during back propagation. The WEB shown in (Figure 4) can be defined as:

$$F_{WEB,j} = Re_2(FC_2(Re_1(FC_1(AP(W_{WEB,j}))))), \tag{3}$$

where $W_{WEB,j}$ is the weights across the jth output channel. The average pooling layer, AP, averages the values of each $H \times W$. Re_1 and Re_2 are two ReLU activation functions. FC_1 and FC_2 are two fully connected layers.

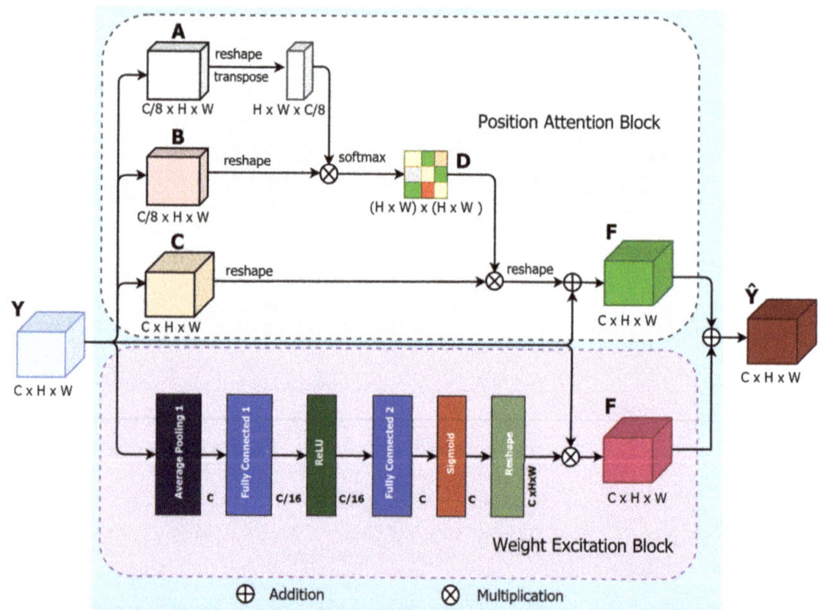

Figure 4. Illustration of the proposed PAWE block.

The output feature from WEB is reshaped and multiplied to the input feature map. Finally, an element-wise sum operation is performed between the feature maps from the PAB and WEB to produce the final PAWE features, as follows:

$$\hat{Y}_{PAWE,j} = F_{PAB,j} + F_{WEB,j}, \tag{4}$$

This process generates a global contextual description and aggregates the context according to a spatial weighted attention map by creating spatial-relevant weighted features, which provide common weight excitation and enhance the intra-class semantic coherence of the input features maps.

3.4.2. Channel Attention-Aware Weight Excitation (CAWE)

Each class-specific response (in our case, there are two classes of nodules and background) is related to each channel of local features extracted by the encoder layers. Different semantic class responses are correlated with each other [29]. Thus, to improve the feature representation of each class-specific response, the proposed CAWE block can properly highlight the interdependence between channels in feature maps and explicitly model inter-dependencies between channels. The process of CAWE shown in Figure 5 and can be illustrated as follows.

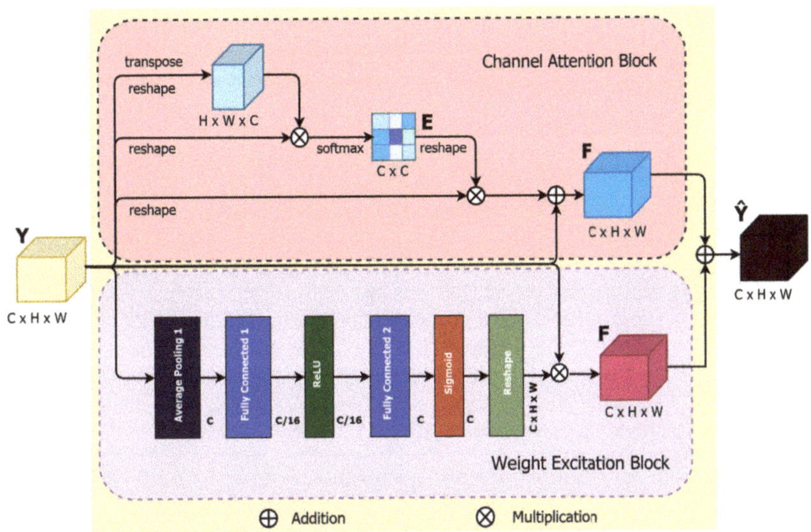

Figure 5. Illustration of the proposed CAWE block.

Like PAWE, the proposed CAWE block includes two sub-blocks, a channel attention block (CAB) and a weight excitation block (WEB). In the CAB block, the input $Y \in \mathbb{R}^{C\times H\times W}$ is reshaped in the initial two steps and permuted in the second part into $Y_1^c \in \mathbb{R}^{(H\times W)\times C}$ and $Y_2^c \in \mathbb{R}^{C\times(H\times W)}$, where the superscript c is defined for "channel". Afterwards, a matrix multiplication between Y_1^c and Y_2^c is performed. The channel attention map $E^c \in \mathbb{R}^{C\times C}$ can be defined as:

$$e_{i,j}^c = \frac{\exp\left(Y_{1,i}^c \cdot Y_{2,j}^c\right)}{\sum_{i=1}^{C} \exp\left(Y_{1,i}^c \cdot Y_{2,j}^c\right)}, \tag{5}$$

where the outcome of the ith channel on the jth is produced by $e_{i,j}^c$. A multiplication of the transposed version of the input feature maps, Y_3^c reshaped to $\mathbb{R}^{C\times(H\times W)}$, and the resulting E^c of (5) is performed. Consequently, the final channel attention map can be defined as:

$$F_{CAB,j} = \alpha_c \sum_{i=1}^{C} e_{ij}^c Y_{3,j}^c + Y_j, \tag{6}$$

where α_c quantifies the weight of the channel attention map of the input feature map Y. The final WEB sub-network feature map can similarly be obtained from (3).

Finally, an element-wise sum operation is performed between the CAB and WEB output features maps to produce the final CAWE features, as follows:

$$\hat{Y}_{CAWE,j} = F_{CAB,j} + F_{WEB,j}, \quad (7)$$

This process emphasizes generating channel-dependent feature maps using weighted excitation versions of the features of all channels and boosting the feature difference among the channels.

4. Experimental Results and Discussion

4.1. Datasets

In this work, we used two publicly available datasets:

- Lung Image Database Consortium image collection (LIDC-IDRI) [32] consists of 1018 CT scans performed on 1010 patients from 7 different organisations. Each CT scan has been analysed by four radiologists, who individually identified the nodule and manually segmented the region of all the nodules with a diameter larger than three millimetres. Each CT scan can include one or more nodule regions, so the total segmented masks are 5066. Looking closely at the dataset, many nodules are very small and do not satisfy the malignancy index. Therefore, we used a diameter threshold larger than 20 mm to excluded all tiny nodules from our dataset. Afterwards, we split our final dataset, which contains 2044 nodule masks in total, into train, validation and test sets of 70%, 10%, and 20% respectively;
- LUng Nodule Analysis 2016 (LUNA16) [33] is derived from the LIDC-IDRI dataset [32]. It contains 888 CT scans from the LIDC-IDRI dataset for the grand challenge with round annotation masks for all the segmented nodules. The LUNA16 challenge dataset contains 1186 nodule annotations. We obtained 2300 nodule masks from the annotation after pre-processing. We split the dataset into train, validation and test sets similar to the LIDC-IDRI dataset.

4.2. Model Implementation

We individually trained the nodule detection and segmentation models on the PyTorch framework [34]. To train the detection model, the stochastic gradient descent (SGD) [35] optimizer with a learning rate of 0.002 was used. The binary cross-entropy (BCE) and the $L1$ norm loss functions were used to train the detection model with a batch size of 4. On the other hand, the Adam [36] optimizer with a learning rate of 0.0002, as well as the BCE and the IoU loss functions, were also used to train the segmentation model with a batch size of 4. Note that data augmentation was applied during training for both detection and segmentation models to increase the size of the training dataset. We augmented the datasets by random rotation, flipping horizontally and vertically and applying the elastic transform. Finally, all the experiments were carried out on an NVIDIA GeForce GTX 1080 GPU with $8GB$ memory and running about 10–15 h to train 100 epochs for each model.

4.3. Evaluation Measures

Two different procedures were used on both datasets to evaluate the proposed detection and segmentation models. For pixel-level evaluation, the segmentation model provides a pixel-wise output of the class probabilities for every pixel in the input nodule ROIs. The output is converted into a binary segmentation map using a threshold value. Regarding pixel-level evaluation metrics, accuracy (ACC), sensitivity (SEN) and specificity (SPE) are calculated to evaluate the performance of the segmentation model. We also plot a receiver operating characteristic (ROC) curve for calculating area under the curve (AUC). For object-level evaluation, we used the segmentation output to calculate the Dice coefficient (DSC) and intersection over union (IoU) for assessing the ability of the algorithm to previously segment the boundaries of the nodule. Note that in our case, there is no "true negative" class, since there is no "object" corresponding to the absence of nodules. Besides, we also plot the precision-recall (PR) curve instead of the ROC to compare the ground truth number and find the correlation.

4.4. Nodule Detection

To detect the nodule in the input CT images, we used different state-of-the-art deep learning detector models, such as R-CNN [37], Fast R-CNN [38], original Faster R-CNN [39] and Optimized Faster R-CNN. The aforementioned detection models were trained and tested on the LIDC-IDRI and LUNA16 datasets. To train the above models, we used the data splits as discussed in Section 4.1. We used all default parameters for training the R-CNN [37], Fast-RCNN [38], and original Faster R-CNN [39] models based on their original papers. We fine-tuned the parameters of the original Faster R-CNN to find the best parameters to achieve the highest performance, and named it Optimized Faster R-CNN. The best combination for this model was a learning rate of 0.001, step size of 70,000, gamma of 0.1, and a dropout ratio of 0.5. The model was trained by the pre-trained ResNet50 model to extract the features with a batch size of 64. We finally compared the average precision (AP) of the detection as shown in Table 2 to select the best detection model among the tested models. The Optimized Faster R-CNN model yielded the best results, with the highest AP on both datasets. In turn, R-CNN, Fast R-CNN, original Faster R-CNN models did not properly detect all nodules in the input CT images. Therefore, we have selected the Optimized Faster R-CNN model to detect nodules in CT images. Some examples of lung nodule detection using Optimized Faster R-CNN are shown in Figure 6. As shown, the Optimized Faster R-CNN model is able to detect the nodule regions, even for small nodules.

Table 2. The average precision (AP) comparison of the four detection models (bold represents the best performance).

Datasets	Models	AP(%)
LIDC-IDRI	Optimized Faster R-CNN	**91.44**
	Original Faster R-CNN	85.45
	Fast R-CNN	79.41
	R-CNN	75.48
LUNA16	Optimized Faster R-CNN	**92.67**
	Original Faster R-CNN	89.31
	Fast R-CNN	82.32
	R-CNN	78.17

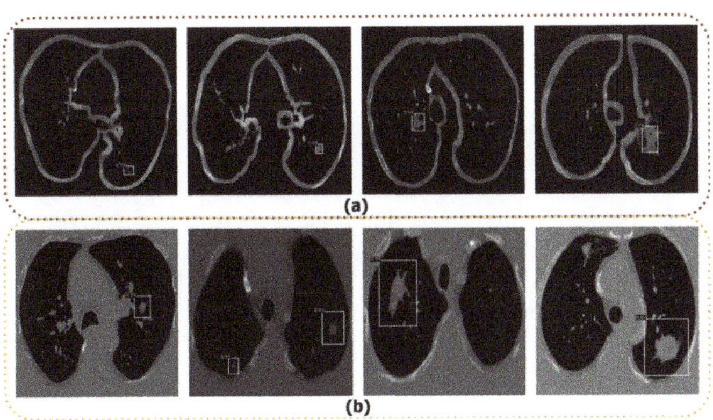

Figure 6. Examples of lung nodule detection using Optimized Faster R-CNN; (**a**) Detection results from the LIDC-IDRI dataset; (**b**) Detection results from the LUNA16 dataset.

4.5. Nodule Segmentation

The proposed lung nodule segmentation model was compared to the state-of-the-art approaches and evaluated in terms of quantitative and qualitative results. For the quantitative study, we used ACC, SEN, and SPE for pixel-level and DSC and IoU for object-level performance, respectively, as shown in Table 3. We compared the AWEU-Net to six different lung nodule segmentation models considering both datasets: PSPNet [40], MANet [41], PAN [42], FPN [43], DeeplabV3 [44], and U-Net [21,22]. As shown in Table 3, the integration of both PAWE and CAWE with the U-Net outperformed the segmentation results of the baseline model (U-Net). In general, AWEU-Net outperforms all tested models in terms of the ACC, SPE, DSC, and IoU metrics on the LUNA16 dataset. AWEU-Net yields ACC, SPE, DSC, and IoU scores of 91.32%, 93.46%, 89.79%, 82.32%, and 89.88%, respectively, which is 1.18%, 1.47%, 0.97%, 1.8%, and 0.93% points higher than the scores of the second-best method (i.e., U-Net). In turn, the DeeplabV3 achieved a SEN score of 93.01%, which is 1.32% points higher than AWEU-Net. However, the proposed segmentation model provides a comparable SEN score of 91.69%.

Table 3. Comparison between the proposed AWEU-Net and six other models on the LIDC-IDRI and LUNA16 test datasets (bold represents the best performance).

Datasets	Models	ACC	SEN	SPE	DSC	IoU
LUNA16	PSPNet	0.8718	0.8711	0.9012	0.8512	0.7513
	MANet	0.8874	0.8686	0.9285	0.8663	0.7743
	PAN	0.8604	0.8709	0.8873	0.8424	0.7354
	FPN	0.8846	0.9143	0.8905	0.8722	0.7806
	DeeplabV3	0.8918	**0.9301**	0.8910	0.8794	0.7916
	U-Net (baseline)	0.9014	0.9136	0.9199	0.8882	0.8054
	Proposed_AWEU-Net	**0.9132**	0.9169	**0.9346**	**0.8979**	**0.8234**
LIDC-IDRI	PSPNet	0.9309	0.8514	0.9620	0.8684	0.7783
	MANet	0.9327	0.8749	0.9557	0.8788	0.7905
	PAN	0.9268	0.8369	0.9603	0.8577	0.7653
	FPN	0.9393	0.8981	0.9562	0.8934	0.8127
	DeeplabV3	0.9429	0.9023	0.9602	0.8983	0.8191
	U-Net (baseline)	0.9436	0.8968	0.9635	0.8987	0.8200
	Proposed_AWEU-Net	**0.9466**	**0.9084**	**0.9641**	**0.9035**	**0.8321**

In addition, using the test set of the LUNA16 and LIDC-IDRI datasets, the box plots of DSC and IoU scores of the six models and AWEU-Net were drawn to demonstrate the segmentation ability of AWEU-Net as shown in Figure 7. On both datsets, the proposed AWEU-Net yields higher DSC and IoU mean scores and the lowest standard deviation with only two outliers; this is as compared to the other six segmentation models, which contain many outliers with lower mean and higher standard deviation scores.

Furthermore, to predict the probability of the binary segmented masks, the ROC and PR curves were constructed as shown in Figure 8. Using the LUNA16 test set, the proposed AWEU-Net model yields the highest AUC and PR of 97.10%, and 96.66%, respectively, among the seven segmentation models tested.

On the other hand, AWEU-Net outperforms all the tested models in terms of all evaluation metrics on the LIDC-IDRI dataset. The proposed model yields ACC, SEN, SPE, DSC, and IoU scores of 94.66%, 90.84%, 96.41%, 90.35%, and 83.21%, respectively. It has improved by 0.3%, 1.16%, 0.06%, 0.48%, and 1.21% in ACC, SEN, SPE, DSC, and IoU scores from the original U-Net. Again, the box plots of DSC and IoU scores of the LIDC-IDRI dataset to compare the models' performance is displayed in Figure 9. Likewise, the proposed AWEU-Net achieved the highest DSC and IoU mean scores and the smallest standard deviation with only one outlier. The proposed model achieved an AUC of the

ROC and PR on the LIDC-IDRI test dataset of 91.58%, and 82.02%, respectively, as shown in Figure 10.

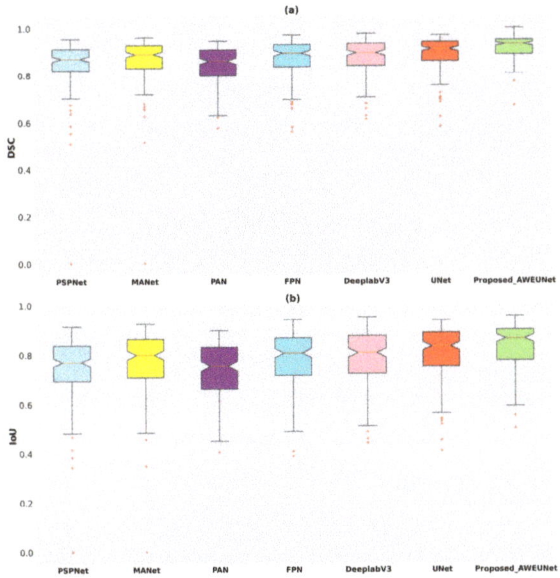

Figure 7. Boxplots of (**a**) Dice coefficient (DSC) and (**b**) intersection over union (IoU) scores for all test samples of the LUNA16 lung nodule segmentation dataset. Different boxes indicate the score ranges of several methods; the red line inside each box represents the median value, and all values outside the whiskers are considered outliers, which are marked with the (+) symbol.

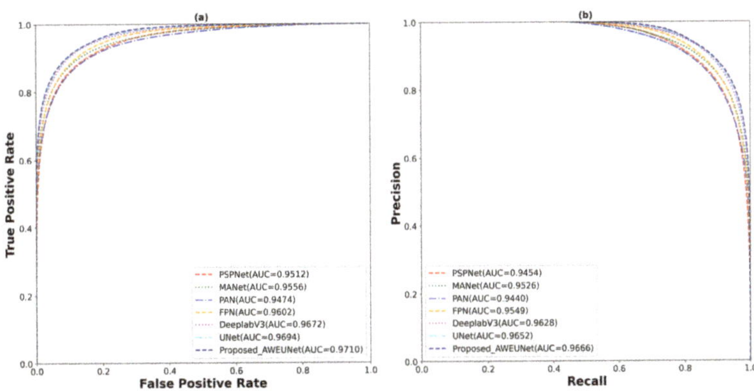

Figure 8. The (**a**) ROC and (**b**) PR curve for all test samples of the LUNA16 lung nodule segmentation dataset.

A qualitative comparison of the segmentation results of the AWEU-Net and the six segmentation models is shown in Figure 11. The segmentation results of the input nodule ROIs of CT images with a variety of difficult levels, including illumination variations and irregular shapes and boundaries of the nodule regions, were presented. As shown in Figure 11, four examples from the two datasets along with the ground truth and the predicted mask of the six tested models were compared to the proposed AWEU-Net model. AWEU-Net provides segmentation results very close to the ground truth with an average similarity of >86% (true positive (TP)). Our segmentation method also provides the lowest degrees of false negative (FN) and false positive (FP) results compared to the rest of the models. The AWEU-Net model yields more regular borders compared to PSPNet, MANet, FPN, since our model strives for higher accuracy on nodule region boundaries. The resulting segmentation of the six tested models may significantly differ from the ground truth in some cases, e.g., the second example of the LUNA16 dataset.

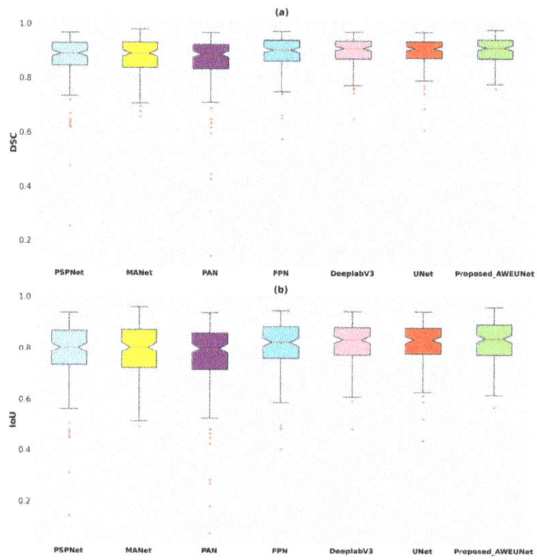

Figure 9. Boxplots of (**a**) Dice coefficient (DSC) and (**b**) intersection over union (IoU) scores for all test samples of the LIDC-IDRI lung nodule segmentation dataset. Different boxes indicate the score ranges of several methods; the red line inside each box represents the median value, and all values outside the whiskers are considered outliers, which are marked with the (+) symbol.

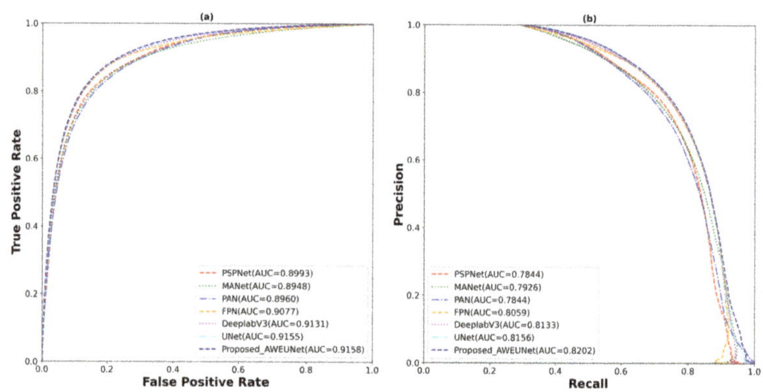

Figure 10. The (**a**) ROC and (**b**) PR curve for for all test samples of the LIDC-IDRI lung nodule segmentation dataset.

Finally, regarding the model efficiency, the total number of parameters, the sum of all the weights and biases on the proposed model, was around 34.5 million. Our model yielded a reduction of 40% compared to the baseline model, UNet (i.e., number of parameters of around 60 million). In order to assess the computational complexity of the model, we measured the number of resources that the proposed model used in training and inference by computing the multiply-accumulate operation (MACs) in billions of operations, (MACs(G)). The proposed model performed 65.3 billion MACs. Furthermore, our proposed model achieved an inference time of 10.8 ms (around 92.3 fps) on an NVIDIA GeForce GTX 1080 GPU. In summary, our framework can be executed on a single GPU, guaranteeing accurate nodule segmentation in real-time.

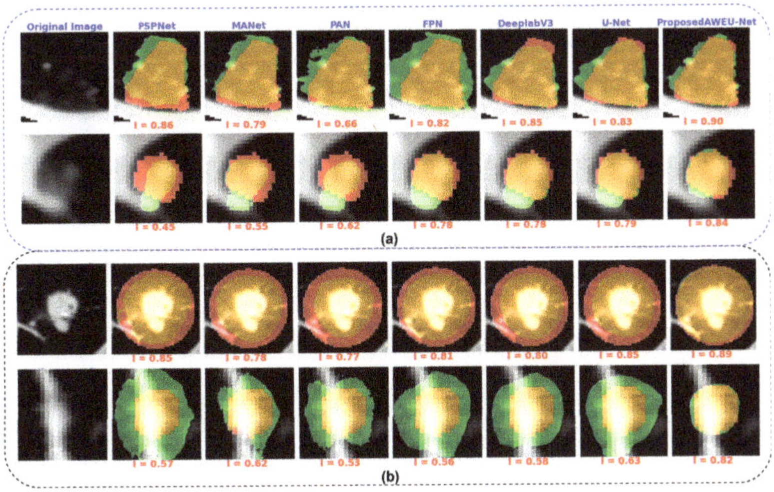

Figure 11. Examples of segmentation results by different state-of-the-art models. (**a**) Segmentation results on the LIDC-IDRI dataset and (**b**) segmentation results on the LUNA16 dataset. The colors of the segmentation visualization results are presented as follows: TP (orange), TN (background), FP (green), and FN (red).

5. Conclusions

This article proposed a reliable system for lung nodule detection and segmentation. The system contains two deep learning models. Firstly, the Optimized Faster R-CNN model [39] trained with lung CT scan images was used for detecting the nodule region in a CT image as an initial step. Secondly, a segmentation model, AWEU-Net, was proposed for segmenting the nodule boundaries of the detected nodule region. The proposed segmentation model, AWEU-Net, includes PAWE and CAWE blocks to improve the segmentation performance. Compared to the state-of-the-art models, the proposed AWEU-Net model yields the best segmentation accuracy with DSC and IoU scores of 89.79%, 90.35%, and 82.34%, 83.21% on the LUNA16 and LIDC-IDRI datasets, respectively. Although the proposed method provided promising nodule segmentation results, the number of parameters of the segmentation model is a bit high (around 34.5 million). Thus, it is not appropriate for computing devices with limited resources. Consequently, ongoing work will aim at developing a lightweight nodule segmentation model. In future work, we will develop a comprehensive nodule segmentation system, and it will be able to classify and grade nodule malignancy.

Author Contributions: Conceptualization, S.F.B. and M.M.K.S.; methodology, S.F.B. and M.M.K.S.; software, S.F.B. and M.M.K.S.; validation, S.F.B. and M.M.K.S.; formal analysis, S.F.B. and M.M.K.S.; investigation, S.F.B. and M.M.K.S.; resources, S.F.B. and M.M.K.S.; data curation, S.F.B. and M.M.K.S.; writing—original draft preparation, S.F.B. and M.M.K.S.; writing—review and editing, S.F.B., M.M.K.S., M.A.-N. and H.A.R.; visualization, S.F.B. and M.M.K.S.; supervision, M.A.-N., D.P. and H.A.R.; project administration, M.A.-N., D.P. and H.A.R.; funding acquisition, M.A.-N., D.P. and H.A.R.; All authors have read and agreed to the published version of the manuscript.

Funding: This research received no external funding.

Institutional Review Board Statement: The study was approved by URV.

Informed Consent Statement: Not applicable.

Data Availability Statement: The samples used are publicly available.

Acknowledgments: The Spanish Government partly supported this research through project PID2019-105789RB-I00.

Conflicts of Interest: All authors declare no conflict of interest.

Abbreviations

The following abbreviations are used in this manuscript:

CAD	Computer-aided Diagnosis
CADe	Computer-aided Detection
CASe	Computer-aided Segmentation
CT	Computed Tomography
AI	Artificial Intelligence
GMM	Gaussian Mixture Model
CNNs	Convolutional Neural Networks
R-CNN	Region-based Convolutional Neural Network
ROI	Region of Interest (ROI)
RPN	Region Proposal Network
FPN	Feature Pyramid Network
ReLU	Rectified Linear Unit
SGD	Stochastic Gradient Descent
BCE	Binary Cross-Entropy
Dice	Dice Coefficient
IoU	Intersection Over Union
ROC	Receiver Operating Characteristic
AUC	Area Under the Curve
ACC	Accuracy
SEN	Sensitivity
SPE	Specificity
GPU	Graphics Processing Unit
GB	Gigabytes
DL	Deep Learning

References

1. Cancer. Available online: https://www.who.int/news-room/fact-sheets/detail/cancer/ (accessed on 8 August 2021).
2. World Lung Cancer Day 2020 Fact Sheet. Available online: https://www.chestnet.org/newsroom/chest-news/2020/07/world-lung-cancer-day-2020-fact-sheet/ (accessed on 8 August 2021).
3. The National Lung Screening Trial Research Team. Reduced lung-cancer mortality with low-dose computed tomographic screening. *N. Engl. J. Med.* **2011**, *365*, 395–409. [CrossRef] [PubMed]
4. Yu, K.H.; Lee, T.L.M.; Yen, M.H.; Kou, S.; Rosen, B.; Chiang, J.H.; Kohane, I.S. Reproducible Machine Learning Methods for Lung Cancer Detection Using Computed Tomography Images: Algorithm Development and Validation. *J. Med. Internet Res.* **2020**, *22*, e16709. [CrossRef] [PubMed]
5. Seattle Cancer Care Alliance Proton Therapy Center. Available online: https://www.sccaprotontherapy.com/cancers-treated/lung-cancer-treatment (accessed on 8 August 2021).
6. Callister, M.; Baldwin, D.; Akram, A.; Barnard, S.; Cane, P.; Draffan, J.; Franks, K.; Gleeson, F.; Graham, R.; Malhotra, P.; et al. British Thoracic Society guidelines for the investigation and management of pulmonary nodules: accredited by NICE. *Thorax* **2015**, *70*, ii1–ii54. [CrossRef]
7. Aresta, G.; Jacobs, C.; Araújo, T.; Cunha, A.; Ramos, I.; van Ginneken, B.; Campilho, A. iW-Net: An automatic and minimalistic interactive lung nodule segmentation deep network. *Sci. Rep.* **2019**, *9*, 11591. [CrossRef] [PubMed]
8. Keetha, N.V.; Babu P, S.A.; Annavarapu, C.S.R. U-Det: A Modified U-Net architecture with bidirectional feature network for lung nodule segmentation. *arXiv* **2020**, arXiv:2003.09293.
9. Cao, H.; Liu, H.; Song, E.; Hung, C.C.; Ma, G.; Xu, X.; Jin, R.; Lu, J. Dual-branch residual network for lung nodule segmentation. *Appl. Soft Comput.* **2020**, *86*, 105934. [CrossRef]
10. Wu, B.; Zhou, Z.; Wang, J.; Wang, Y. Joint learning for pulmonary nodule segmentation, attributes and malignancy prediction. In Proceedings of the 2018 IEEE 15th International Symposium on Biomedical Imaging (ISBI 2018), Washington, DC, USA, 4–7 April 2018; pp. 1109–1113.

11. Tang, H.; Zhang, C.; Xie, X. Nodulenet: Decoupled false positive reduction for pulmonary nodule detection and segmentation. In Proceedings of the 22nd International Conference on Medical Image Computing and Computer-Assisted Intervention, Shenzhen, China, 13–17 October 2019; pp. 266–274.
12. Kumar Singh, V.; Abdel-Nasser, M.; Pandey, N.; Puig, D. Lunginfseg: Segmenting COVID-19 infected regions in lung ct images based on a receptive-field-aware deep learning framework. *Diagnostics* **2021**, *11*, 158. [CrossRef]
13. Jiang, J.; Hu, Y.-C.; Liu, C.J.; Halpenny, D.; Hellmann, M.D.; Deasy, J.O.; Mageras, G.; Veeraraghavan, H. Multiple resolution residually connected feature streams for automatic lung tumor segmentation from CT images. *IEEE Trans. Med. Imaging* **2018**, *38*, 134–144. [CrossRef]
14. Dehmeshki, J.; Amin, H.; Valdivieso, M.; Ye, X. Segmentation of pulmonary nodules in thoracic CT scans: A region growing approach. *IEEE Trans. Med. Imaging* **2008**, *27*, 467–480. [CrossRef]
15. Tan, Y.; Schwartz, L.H.; Zhao, B. Segmentation of lung lesions on CT scans using watershed, active contours, and Markov random field. *Med. Phys.* **2013**, *40*, 043502. [CrossRef]
16. Farag, A.A.; Abd El Munim, H.E.; Graham, J.H.; Farag, A.A. A novel approach for lung nodules segmentation in chest CT using level sets. *IEEE Trans. Image Process.* **2013**, *22*, 5202–5213. [CrossRef]
17. Dai, S.; Lu, K.; Dong, J.; Zhang, Y.; Chen, Y. A novel approach of lung segmentation on chest CT images using graph cuts. *Neurocomputing* **2015**, *168*, 799–807. [CrossRef]
18. Navya, K.; Pradeep, G. Lung Nodule Segmentation Using Adaptive Thresholding and Watershed Transform. In Proceedings of the 2018 3rd IEEE International Conference on Recent Trends in Electronics, Information & Communication Technology (RTEICT), Bangalore, India, 18–19 May 2018; pp. 630–633.
19. Li, X.; Li, B.; Liu, F.; Yin, H.; Zhou, F. Segmentation of pulmonary nodules using a GMM fuzzy C-means algorithm. *IEEE Access* **2020**, *8*, 37541–37556. [CrossRef]
20. Savic, M.; Ma, Y.; Ramponi, G.; Du, W.; Peng, Y. Lung nodule segmentation with a region-based fast marching method. *Sensors* **2021**, *21*, 1908. [CrossRef]
21. Skourt, B.A.; El Hassani, A.; Majda, A. Lung CT image segmentation using deep neural networks. *Procedia Comput. Sci.* **2018**, *127*, 109–113. [CrossRef]
22. Ronneberger, O.; Fischer, P.; Brox, T. U-net: Convolutional networks for biomedical image segmentation. In Proceedings of the 18th International Conference on Medical Image Computing and Computer-Assisted Intervention, Munich, Germany, 5–9 October 2015; pp. 234–241.
23. He, K.; Zhang, X.; Ren, S.; Sun, J. Deep residual learning for image recognition. In Proceedings of the IEEE Conference on Computer Vision and Pattern Recognition, Las Vegas, NV, USA, 27–30 June 2016; pp. 770–778.
24. Misra, D. Mish: A self regularized non-monotonic neural activation function. *arXiv* **2019**, arXiv:1908.08681.
25. Hancock, M.C.; Magnan, J.F. Lung nodule malignancy classification using only radiologist-quantified image features as inputs to statistical learning algorithms: Probing the Lung Image Database Consortium dataset with two statistical learning methods. *J. Med. Imaging* **2016**, *3*, 044504. [CrossRef]
26. Lin, T.Y.; Maire, M.; Belongie, S.; Hays, J.; Perona, P.; Ramanan, D.; Dollár, P.; Zitnick, C.L. Microsoft coco: Common objects in context. In Proceedings of the 13th European Conference on Computer Vision, Zurich, Switzerland, 6–12 September 2014; pp. 740–755.
27. Huang, J.; Rathod, V.; Sun, C.; Zhu, M.; Korattikara, A.; Fathi, A.; Fischer, I.; Wojna, Z.; Song, Y.; Guadarrama, S.; et al. Speed/accuracy trade-offs for modern convolutional object detectors. In Proceedings of the IEEE Conference on Computer Vision and Pattern Recognition, Honolulu, HI, USA, 21–26 July 2017; pp. 7310–7311.
28. Zeren, M.T.; Aytulun, S.K.; Kırelli, Y. Comparison of SSD and faster R-CNN algorithms to detect the airports with data set which obtained from unmanned aerial vehicles and satellite images. *Avrupa Bilim ve Teknoloji Dergisi* **2020**, 643–658. [CrossRef]
29. Fu, J.; Liu, J.; Tian, H.; Li, Y.; Bao, Y.; Fang, Z.; Lu, H. Dual attention network for scene segmentation. In Proceedings of the IEEE Conference on Computer Vision and Pattern Recognition, Long Beach, CA, USA, 15–20 June 2019; pp. 3146–3154.
30. Peng, C.; Zhang, X.; Yu, G.; Luo, G.; Sun, J. Large kernel matters–improve semantic segmentation by global convolutional network. In Proceedings of the IEEE Conference on Computer Vision and Pattern Recognition, Honolulu, HI, USA, 21–26 July 2017; pp. 4353–4361.
31. Quader, N.; Bhuiyan, M.M.I.; Lu, J.; Dai, P.; Li, W. Weight Excitation: Built-in Attention Mechanisms in Convolutional Neural Networks. In Proceedings of the 16th the European Conference on Computer Vision, Glasgow, UK, 23–28 August 2020; pp. 87–103.
32. Armato, S.G., III; McLennan, G.; Bidaut, L.; McNitt-Gray, M.F.; Meyer, C.R.; Reeves, A.P.; Zhao, B.; Aberle, D.R.; Henschke, C.I.; Hoffman, E.A.; et al. The lung image database consortium (LIDC) and image database resource initiative (IDRI): A completed reference database of lung nodules on CT scans. *Med. Phys.* **2011**, *38*, 915–931.
33. Setio, A.A.A.; Traverso, A.; De Bel, T.; Berens, M.S.; Van Den Bogaard, C.; Cerello, P.; Chen, H.; Dou, Q.; Fantacci, M.E.; Geurts, B.; et al. Validation, comparison, and combination of algorithms for automatic detection of pulmonary nodules in computed tomography images: The LUNA16 challenge. *Med. Image Anal.* **2017**, *42*, 1–13. [CrossRef]
34. Paszke, A.; Gross, S.; Massa, F.; Lerer, A.; Bradbury, J.; Chanan, G.; Killeen, T.; Lin, Z.; Gimelshein, N.; Antiga, L.; et al. Pytorch: An imperative style, high-performance deep learning library. *Adv. Neural Inf. Process. Syst.* **2019**, *32*, 8026–8037.
35. Gulcehre, C.; Sotelo, J.; Bengio, Y. A robust adaptive stochastic gradient method for deep learning. In Proceedings of the 2017 International Joint Conference on Neural Networks (IJCNN), Anchorage, AK, USA, 14–19 May 2017; pp. 125–132.

36. Kingma, D.P.; Ba, J. Adam: A method for stochastic optimization. *arXiv* **2014**, arXiv:1412.6980.
37. Girshick, R.; Donahue, J.; Darrell, T.; Malik, J. Rich feature hierarchies for accurate object detection and semantic segmentation. In Proceedings of the IEEE Conference on Computer Vision and Pattern Recognition, Columbus, OH, USA, 23–28 June 2014; pp. 580–587.
38. Girshick, R. Fast r-cnn. In Proceedings of the IEEE International Conference on Computer Vision, Santiago, Chile, 7–13 December 2015; pp. 1440–1448.
39. Ren, S.; He, K.; Girshick, R.; Sun, J. Faster R-CNN: Towards real-time object detection with region proposal networks. *IEEE Trans. Pattern Anal. Mach. Intell.* **2016**, *39*, 1137–1149. [CrossRef] [PubMed]
40. Zhao, H.; Shi, J.; Qi, X.; Wang, X.; Jia, J. Pyramid scene parsing network. In Proceedings of the IEEE Conference on Computer Vision and Pattern Recognition, Honolulu, HI, USA, 21–26 July 2017; pp. 2881–2890.
41. Fan, T.; Wang, G.; Li, Y.; Wang, H. Ma-net: A multi-scale attention network for liver and tumor segmentation. *IEEE Access* **2020**, *8*, 179656–179665. [CrossRef]
42. Li, H.; Xiong, P.; An, J.; Wang, L. Pyramid attention network for semantic segmentation. *arXiv* **2018**, arXiv:1805.10180.
43. Lin, T.Y.; Dollár, P.; Girshick, R.; He, K.; Hariharan, B.; Belongie, S. Feature pyramid networks for object detection. In Proceedings of the IEEE Conference on Computer Vision and Pattern Recognition, Honolulu, HI, USA, 21–26 July 2017; pp. 2117–2125.
44. Chen, L.C.; Papandreou, G.; Schroff, F.; Adam, H. Rethinking atrous convolution for semantic image segmentation. *arXiv* **2017**, arXiv:1706.05587.

applied sciences

MDPI

Article

An Approach for Pronunciation Classification of Classical Arabic Phonemes Using Deep Learning

Amna Asif [1,*], Hamid Mukhtar [2], Fatimah Alqadheeb [3], Hafiz Farooq Ahmad [3] and Abdulaziz Alhumam [3]

1 Information Systems Department, College of Computer Sciences and Information Technology (CCSIT), King Faisal University, P.O. Box 400, Al-Ahsa 31982, Saudi Arabia
2 Department of Computer Science, College of Computers and Information Technology, Taif University, P.O. Box 11099, Taif 21944, Saudi Arabia; h.mukhtar@tu.edu.sa
3 Computer Science Department, College of Computer Sciences and Information Technology (CCSIT), King Faisal University, P.O. Box 400, Al-Ahsa 31982, Saudi Arabia; 218002332@student.kfu.edu.sa (F.A.); hfahmad@kfu.edu.sa (H.F.A.); aahumam@kfu.edu.sa (A.A.)
* Correspondence: aarkhan@kfu.edu.sa

Abstract: A mispronunciation of Arabic short vowels can change the meaning of a complete sentence. For this reason, both the students and teachers of Classical Arabic (CA) are required extra practice for correcting students' pronunciation of Arabic short vowels. That makes the teaching and learning task cumbersome for both parties. An intelligent process of students' evaluation can make learning and teaching easier for both students and teachers. Given that online learning has become a norm these days, modern learning requires assessment by virtual teachers. In our case, the task is about recognizing the exact pronunciation of Arabic alphabets according to the standards. A major challenge in the recognition of precise pronunciation of Arabic alphabets is the correct identification of a large number of short vowels, which cannot be dealt with using traditional statistical audio processing techniques and machine learning models. Therefore, we developed a model that classifies Arabic short vowels using Deep Neural Networks (DNN). The model is constructed from scratch by: (i) collecting a new audio dataset, (ii) developing a neural network architecture, and (iii) optimizing and fine-tuning the developed model through several iterations to achieve high classification accuracy. Given a set of unseen audio samples of uttered short vowels, our proposed model has reached the testing accuracy of 95.77%. We can say that our results can be used by the experts and researchers for building better intelligent learning support systems in Arabic speech processing.

Keywords: deep learning; classical Arabic; short vowels; audio dataset; convolutional neural networks; optimization; regularization

Citation: Asif, A.; Mukhtar, H.; Alqadheeb, F.; Ahmad, H.F.; Alhumam, A. An Approach for Pronunciation Classification of Classical Arabic Phonemes Using Deep Learning. *Appl. Sci.* **2022**, *12*, 238. https://doi.org/10.3390/app12010238

Academic Editors: Aida Valls and Keun Ho Ryu

Received: 11 November 2021
Accepted: 21 December 2021
Published: 27 December 2021

1. Introduction

Speech processing technology has received considerable attention recently due to a variety of applications in the areas of automated speech recognition, information retrieval, and assisted communication. A lot of diverse research work has been done on speech processing for different human languages globally. As such, in recent years, deep learning has increasingly enabled autonomous speech processing including speech recognition and synthesis. However, the Arabic language has witnessed less research work in this domain due to its unique challenges.

Arabic is the fifth widely used language in the world and there are around 422 million speakers of Arabic as their first language on the globe [1]. In the broader term, Arabic language can be categorized into Classical Arabic (CA) and modern standard Arabic (MSA) dialects. MSA is a modified version of CA currently used in everyday communication in Arabic speaking countries. Classical Arabic is the language of the Holy Quran [2] and is still used largely in religious context and studies despite being more than 1400 years old.

Millions of Arabic and non-Arabic speakers around the world practice CA in their daily routine in the form of recitation of the Holy book or studying in a formal educational setting.

1.1. Arabic Phonemes and Their Pronunciation

A phoneme is the smallest unit of sound in human speech [3]. Phonemes include all the distinct units of sound spoken in a language. The Arabic language consists of 34 phonemes that can be broken down into 28 consonants, three short, and three long vowels [4] known by their common names of Fatha, Damma, and Khasra. Pronunciation is a general term that includes several distinct features present in human languages. The correct pronunciation is hard and challenging to measure because there is no universal definition for correctness in the context of human languages, but there has been some research in this domain [5,6].

In the Arabic language, the accurate pronunciation of phonemes is required in learning the language to void change in the meaning of the sentence. In many countries around the world, the CA is part of the schools' syllabus. Correct pronunciation of phonemes as per defined rules is an essential requirement to preserve the meaning of the words [7,8]. Mispronunciation results in two types of errors; firstly, it changes the meaning of the word completely. Secondly, it defies the rules of pronunciation, and both errors are forbidden in CA. Learning the correct pronunciation of Arabic alphabets is challenging, and it requires each learner to follow an individual teacher, who listens and corrects the mistakes separately for each learner [9]. These corrections belong to the pronunciation of all the 84 variations of phonemes. A major distinction of the CA from MSA is the emphasis on the correct use of vowels. When speaking CA, the speakers tend to produce all the sounds according to the existing rules for CA pronunciation. CA has precise and explicitly defined rules for correct pronunciation to conserve the accurate meaning of the words and provides a framework to facilitate both the natives and non-natives learning the language [10]. These rules are standardized, widely available, and recognized by the Arabic speaking world [11]. In CA, alphabets' articulation points and characteristics and massive practicing of vocals play a significant role in correct pronunciation. On the contrary, it has been studied previously that in the case of MSA, the different dialects such as Saudi, Egyptian, and Sudanese have different qualities of vowel pronunciation [12].

1.2. Motivation for Pronunciation Classification

In a typical CA learning institute, a single teacher may be responsible for listening to and correcting the mistakes of several dozen learners on a single day. This is quite a laborious job leading to fatigue and is prone to diminished error correction by the instructor over time. Moreover, each student has to wait for their turn to speak or read the learned parts. If this job can be automated by letting the students' work be evaluated through an automated mechanism, it will improve the efficiency of the learning process and the overall productivity of the students and instructors. Moreover, this will open new ways of distant and home-based learning with little reliance on the presence of an instructor. As learning can take place anytime and anywhere, it is required to develop automatic approaches that can detect pronunciation errors and give feedback to the learner.

It is worth mentioning that in the Arabic language, the correct pronunciation is vital and requires a lot of practice by the learner of the language. Currently, the students not only rely on in-class participation but also practice using digital technology. In this regard, we have evidence of the development of many applications supporting the students' learning process. There is a need for research in developing systems and applications for improving Arabic reading and speaking knowledge. In various parts of the world when there is an unavailability of real instructors, such applications play an important role in supporting the student's learning process. The improved model can easily be integrated with existing applications to increase the accuracy of Arabic short vowels classification in the learning process. The Arabic-speaking countries are a recent place of interest for many tourists around the world. Therefore, many people are interested to learn this language for their survival in this part of the world. Any applications developed using better speech recognition models will be beneficial for people who want to learn Arabic using the available online resources.

1.3. Objectives and Contributions

This research aims to develop a classification system for correct and incorrect pronunciation of the Arabic short vowels. It is a novel research idea focusing on the subtle pronunciation differences in Arabic speech processing. The outcomes of this research can be useful in developing advanced systems that can autonomously classify words and sentences to effectively facilitate CA learning with accurate pronunciation. As there are 28 alphabets in the Arabic language and each alphabet has three possible vowel states, they make a total of 84 unique phonemes. Thus, given audio that utters any of these 84 phonemes, our task is to accurately map the audio to the correct phoneme.

Although there are a few existing datasets related to Arabic pronunciation, there is no dataset that serves our purpose of containing the different possible vowel pronunciations. Thus, we created a new dataset of the recordings of the Arabic alphabets through an online audio recording system. After passing the data through various stages of preprocessing, we augmented the available audio data by generating synthetic audio to make a sufficiently bigger audio dataset. We trained a deep convolutional neural network (CNN) over this data for audio classification. The trained network can classify unseen audio data with a testing accuracy of 95.77% into one of the 84 classes. Our research is different from the previous works in terms of the dataset, features engineering, proposed architecture, and evaluation performance.

The major contributions of this work are:

1. Collection of an audio dataset for the Arabic alphabet focused on the three states of vowels for each alphabet.
2. Classification of Arabic short vowels by recognizing the correct short vowels from a recorded phoneme.
3. Constructing a general CNN architecture for phoneme classification. This allows replicating the architecture for similar tasks or a different number of classes.
4. Sharing our experience of model optimization and fine-tuning with the researchers and practitioners to aid their knowledge of building better models in the future.

The remainder of the article is organized as follows. Section 2 discusses the related work. Section 3 describes the materials and methods used in this research. Section 4 explains the development of the CNN model for Arabic short vowel classification and describes the techniques of data augmentation and fine-tuning applied to improve the model's performance. Section 5 reports the results of the classification. Section 6 contains discussions in view of the related work and implications of our work. Section 7 concludes this article.

2. Related Work

There are several research studies on using machine learning and deep learning for speech recognition and pronunciation detection of various types of Arabic datasets. Many of such techniques used audio features for further audio processing and classification. Several tools have been designed and developed to support Arabic learning and teaching. In the following, we report the previous research contributions in the above-mentioned areas.

2.1. Pronunciation Detection

In the direction of pronunciation detection (correct pronunciation and mispronunciation), the previous studies have investigated the machine learning techniques of SVM (support vector machine), KNN (k-nearest neighbors) and NN (neural network) [13], and DNN (deep neural network) techniques of CNN (convolutional neural networks) with transfer learning, AlexNet, BLSTM (bidirectional long short-term memory) [10]. The machine learning algorithms for pronunciation detection had achieved an accuracy of 74.37%

for KNN, 83.50% for SVM, and 90% for NN. Similarly, the DNN techniques reported that CNN, AlexNet, and BLSTM give 95.95%, 98.41%, and 88.32% accuracy for recognizing each alphabet, respectively. As well as detecting the quality of pronunciation of each alphabet using CNN, AlexNet, and BLSTM attained the accuracy of 97.88%, 99.14%, and 77.71%, respectively. In DNNs experiments, the dataset was limited to 29 classes of pronunciation, excluding the vowels. Therefore, it is required to fill this gap to support teaching and learning of CA basics. DNN algorithms and models such as CNN consist of multiple hidden layers capable of efficiently extracting important features from a large set of data. These features can be transformed from one layer to the next in a series of several layers with varying weights in the neural network until it results in a set of layers that can be used to initialize a deep learning algorithm for speech recognition [14]. A proposed method of a DNN based on articulatory models, with a multi-label learning scheme, shows promising results in speech error detection [15]. The researchers consider measuring all attributes responsible for generating the sound related to the movement of the tongue, lips, and other organs. This experiment observes 74% of accuracy for pronunciation error detection. We have found significant improvements in pronunciation detection accuracy in DNN techniques compared to machine learning algorithms from the results.

Besides machine learning and DNN techniques, pronunciation detection was performed using statistical methods. In [16], a system was proposed to detect how badly an Arabic word was pronounced using different scores of pronunciation measurement. The system used the GLL (Global Average Log Likelihood) score, the LLL (Local Average Log Likelihood) score, the RoS (Rate of Speech score), and the RoA (Rate of Articulation) score to assess the pronunciation quality of the learner quantitatively. The dataset consisted of three Arabic corpora, spoken by six young Algerian learners. The pronunciation of expert learners is used as a benchmark to assess the other five learners. Evaluation measurement was used to decide whether the score calculated by the system could detect mispronunciation; the researchers used the CA (Correct Acceptance), CR (Correct Rejection), FA (False Acceptance), and FR (False Rejection). The results showed that the system could detect mispronunciation, using the GLL score method, with 86.66% of correct rejection, and the GLL had the higher CA + CR (76.66%) and the lower FA + FR (23.32%).

In the direction of pronunciation detection of non-native speakers, mispronunciation of Arabic phonemes has been investigated for non-native speakers by analyzing the Arabic speech of Pakistani and Indian speakers from the KSU (King Saud University) database [17]. Research findings of this study highlight that non-native Arabic speakers often mispronounce five Arabic phonemes. The system trained with native and non-native speakers of Arabic phonemes and tested with only non-native speakers. A threshold was set to be compared with the calculated score of GOP (Goodness of Pronunciation) to decide whether the phoneme was pronounced correctly or not. Five experiments were conducted to set up the suitable parameters for the system using HMM (Hidden Markov model). The system used 16 mixtures with 19 HMM re-estimation. Moreover, it extracted 12 MFCC (Mel-frequency Cepstral Coefficients) from sound data. The result of the GOP showed a high accuracy from 87% to 100%, and the false rejection was zero to less than 10%. HMM for automatic speech recognition system has been proposed to help improve the pronunciation for Malaysian teachers of the Arabic language [18]. The aim is to develop a computer system for standard Arabic pronunciation learning by estimating the pronunciation score based on the HMM log-likelihood probability model. The system is designed to extract feature vectors from speech utterances using the MFCC technique; then, the Baum Welch Algorithm is applied to train the system and build the HMMs set. The pronunciation scoring system uses HMMs with the test speech features to perform classification of the speech utterances by applying the Viterbi Algorithm to calculate word pronunciation score. The dataset consisted of 200 words recorded by 20 native Arabic speakers and 10 non-native speakers. The accuracy performance of the proposed system was 89.69%.

2.2. Audio Features

In audio classification, features identification and extraction for audio processing is one of the most popular techniques. A system was developed in [19] to recognize the *Makhraj* (the areas of the mouth from which the Arabic alphabets are pronounced) pronunciation using MFCC features extraction techniques to build a database of features from the audio dataset. Then, the SVM classifier has been used to classify the Arabic alphabet's *Makhraj* pronunciation. The system is trained with the recorded audio of the Arabic alphabet *Makhraj* pronunciation. For new input data, the system extracted the features and matched them with the trained data. Then, it was classified and analyzed using the SVM method with RBF (Radial Basis Function) kernel. The audio data used in the research is a collection of 28 Arabic alphabets' audio and 12 features coefficients were extracted to distinguish between *Makhraj* pronunciation of Arabic alphabet. Different waveforms analysis is used to present the audio data, using audio visualization, FFT (Fast Fourier Transform), and Mel waveform. The result showed that using audio visualization, all letters had a similar representation. On the other hand, using FFT and Mel waveform, each Arabic *Makhraj* pronunciation showed different representations, which can be used to distinguish between different Arabic alphabets. Another research study proposed a CNN feature-based model to detect mispronunciation in Arabic words [20]. The proposed system extracted features from different layers of the AlexNet network. Researchers collected Quranic verses words that cover all Arabic alphabet letters 30 times by speakers of different ages. The participators of the collected dataset were native/non-native Arabic speakers. After removing the noise from the dataset, it converted to a 2D spectrogram and was used to input the CNN model. Then, discriminative features were extracted from fully connected layers 6, 7, and 8 of AlexNet containing high dimensions features. This method showed a significant result compared with the CNN model. It achieved 85% on the complete Arabic dataset. The MDD (Mispronunciation Detection and Diagnosis) task was performed using the RNN (Recurrent Neural Network) [21] and CTC (Connectionist Temporal Classification) model. The model consisted of five parts, the input layer, which accepts the framewise acoustic features. Then is the convolution, which contains a total of four CNN layers and two max-pool layers. The third part is a bi-directional RNN that captures the temporal acoustic features. The fourth part is MLP layers (Time Distributed Dense layers), which ends with a soft-max layer for the classification output. The last part is the CTC output layer that generate the predicted phoneme sequence. The experiment results showed that the proposed approach significantly outperformed previous approaches. Some researchers used LDA (Linear Discriminant Analysis) to classify the data into the correct class and draw a decision region between the given classes [22]. The first step in this system is preprocessing speech signals, which helps prepare the data for the next processing. The preprocessing includes end-point detection, pre-emphasis, and normalization. The second step is features extraction and uses MFCC techniques to extract different coefficients order 12, 20, and 35. The main contribution of this research is to test a different number of MFCC coefficients with varying percentages of training and testing in LDA. The best performance achieved is 92% for the Arabic phoneme (*Taa*) when using 35 MFCC coefficients and 80% of training data. A recent work uses APDM (Acoustic and Phonetic Decoding Model) for recognizing vowels for naturally uttered MSA-based Gas (Genetic Algorithms) [23]. They use MFCC and the LPC (Linear Prediction Coding) techniques to obtain speech parameters from the speech signal. GA based on Manhattan distance decision rule is applied on several Algerian male and female speakers' recordings and classify phonemes with accuracy of 98.02%. The studies that used the audio features data require extra preprocessing efforts to identify and extract the features from the audio dataset for further processing. Therefore, these methods are costly in terms of processing performance.

2.3. Tools in Arabic Learning

Many previous studies have developed tools for recitation assessment; in this regard, the HAFSS system [24] took user recitation of some Holy Quran phrases (in Arabic) as an input and then assessed the quality of the users' recitation. It provides feedback messages to help users know their pronunciation errors and improve their recitation. This system included a speech recognizer to detect errors in user's recitation. For each decision from the speech recognizer, there is a confidence score that is used to choose the suitable feedback. One of the main components in the system is the automatic generation of the pronunciation hypotheses model, which is used to generate pronunciation errors in user's recitation and detect the pronunciation patterns. The system has been tested in a school with two student groups, and the results showed improvement in students' recitation when using the HAFSS system. The performance of the students has been increased from 38% to 77% while getting lessons from a teacher and using the HAFSS system to practice recitation. In [25], a tool was developed that detects pronunciation errors in young Algerian students. The idea was to differentiate between young Algerian students who have difficulties in pronunciation from those who have standard pronunciation. Since the native language of Algerian is a delicate Arabic language, it is very difficult for Algerians to formulate the equivalent sound in standard Arabic. The researchers proposed a system based on a decision tree that provides a decision for the pronunciation of Arabic if it is correct. Moreover, the system provides feedback if the pronunciation contains articulations problems to enhance the pronunciation skills of the learners—the system trained with the acoustical model built on MFCC representation and HMM models. Three scores were used for the decision tree, first the GLL (global average log-likelihood). The second score was the TDS (Time Duration of the Speech), the total time to produce the sound, and the total number of phonemes of each pronounced word. All the scores are input to the decision tree to accept or reject the pronunciation sound. The system was trained on correct Arabic pronounced words and tested with eight young Algerian students who read 16 Arabic words to test the system. The result showed a 95.8% TPR (true positive rate) for good pronunciation and 88.4% for bad pronunciation. Their dataset was too specific and smaller in size. So, more data with further experiments would be needed to validate their approach. Arafa et al. [9] developed a system for teaching Arabic phonemes employing ASR (Automatic speech recognition) by detecting mispronunciation and giving feedback to the learner. In the experimental study, the authors recorded Arabic phonemes 10 times from 89 elementary school children, which resulted in 890 recordings for each Arabic phoneme. The previous studies supported the fact the automated tools are beneficial in support the CA learning process and improve the students' performance.

2.4. Audio Datasets in CA

From the previous research, we found that many studies have used a variety of datasets in developing CA audio processing systems. However, many of such studies are limited to Arabic alphabets and some basic words. Although, the challenging task in learning Arabic pronunciation is the correct use of Arabic short vowels. ASR is a way of automatically transcribing the speech into text [26]. One of the major challenges for ASR for Arabic is the predominance of non diacritized text material where diacritics [27] is the use of vowels (e.g., short vowels Fatha, Damma, and Khasra) for both acoustic and language modeling that can change the meaning of the sentence. However, the majority of the acoustic features for ASR are available without diacritized form [26]. It means that there is no vowel information, which results in the loss of variations in the pronunciation of the speech. In this scenario, it becomes difficult to train a reliable acoustic model without knowing short vowels. Authors report encouraging results for classifying Arabic phonemes, but they do not consider vowels in their study. The research on the Arabic speech for vowels focuses that the MSA [28] used the formants and consonant-vowels-consonant (CVC) utterances to identify vowel similarities and differences. The HMM technique was applied to classify vowels, and the resulting performance of phonetic features of vowels was analyzed. Al-Anzi and AbuZeina [28] highlighted that different researchers had considered different phonemes for the Arabic language; for instance, some take 34 phonemes (28 consonants and six vowels) while others take 112 phonemes by considering four diacritics for each alphabet. So, indeed, Arabic short vowels are crucial in CA; however, the research is limited in this area.

It has been emphasized that autonomous Arabic speech recognition poses challenges due to the vast lexical forms of vowels, which are semantically associated with a word in a sentence [2,29]. Thus, the correct classification of Arabic alphabets pronunciation with vowels is another important research challenge. None of the above research work has considered the correct pronunciation of the CA alphabet with vowels. Furthermore, the research is also limited in preprocessing performance due to features identification and extraction tasks. Therefore, it is required to improve the existing methods and algorithms for Arabic speech classification.

2.5. Summary of the Literature

The survey of the literature shows that the existing work in Arabic pronunciation can be categorized as mispronunciation detection [9,13], sometimes with a focus on non-native speakers [17,18], speech error detection [15], correct pronunciation detection [10], and detecting the similarities and differences between pronunciation of vowels and consonants for MSA [30]. As such, there have been a number of approaches from articulatory models [15] and transfer learning [10] to CNN [13] and acoustic models [31]. To the best of our knowledge, we could not identify any work that can detect the correct pronunciation of Arabic alphabets vowels in CA. One possible reason of not undertaking this work previously maybe because it makes a total of 84 cases to be distinguished from one another. This task is not only challenging in terms of data availability and retrieval but also in terms of model development. To carry out such a challenging task will remove many barriers in the correct utterance of CA words, which can benefit millions of people.

In this research work, we take up this challenging task and propose a deep convolutional neural network algorithm for the classification of Arabic alphabets with vowels. Our research is different from the previous works in terms of the dataset, features engineering, proposed architecture, and evaluation performance.

3. Materials and Methods

After exploring the existing literature, we have identified limited contributions in short Arabic vowels classification; because there was no existing dataset available, we began by collecting the Arabic short vowels audio data. We used a convenience sampling technique whereby the acquaintance, students, and faculty members were identified through personal contacts and social media. The data was collected in the form of audio clips from each participant.

3.1. Data Collection

One way to collect the audio dataset was to use an online audio recording tool like Phonic.ai (http://www.phonic.ai, accessed on 10 November 2021) website [32]. The website provides a service for collecting the participants' responses via sound recording. The website first collects user information through a survey followed by the instructions to record their audio. Then, demographic data such as age and gender are collected from the participants. The users also specify if they are native or non-native Arabic speakers. The survey is followed by a permission agreement to participate in this study; the participants must agree to continue and complete the survey. Then a sample of sound records for the Arabic alphabets with short vowels is played for the participants to listen and understand how their voice should be recorded. Finally, the recording page shows a video containing all the Arabic alphabets with a sequence of short vowels arranged in order and displaying them one by one with a gap of three to five seconds. The video lasts for two minutes and forty-nine seconds. Moreover, some participants shared their sound recordings through WhatsApp messenger application after reading and understanding the instructions.

The total number of the received audio was 85 individual recordings from 42 males to 43 females. There were 81 native Arabic speakers, and four non-native speakers. The received audio files were saved using a naming scheme consisting of the speaker identification code, their gender, native or non-native specification, the age range of the participant, the source of data collection, and the date of the recording. The dataset had 6229 records belonging to 84 classes. It contained an average of 74 examples per class. The data set was imbalanced in terms of the number of examples per class.

Table 1 describes the summary of statistics for the obtained recordings. The dataset is available in the form of audio clips on Kaggle (https://www.kaggle.com/amnaasif/arabic-short-vowels-audio-dataset, accessed on 10 November 2021) for researchers and interested stakeholders.

Table 1. Summary of data collection from participants.

Gender	Status	Number of Records	Age Distribution
Male	Native Non-native	40 2	7–50+ years old
Female	Native Non-native	41 2	7–40 years old

3.2. Data Preprocessing

After data collection, we applied the following preprocessing steps to the audio files.

Noise Reduction: The received audio recordings contained various noise and background sounds, as they were collected in different environments. The Audacity (https://www.audacityteam.org, accessed on 10 November 2021) software was used to identify the noise in our dataset and helped in reducing it.

Data Segmentation, Resampling, and Silence Truncation: Each recorded audio by a participant contained all the pronunciations, which were to be segmented into separate pronunciations for each vowel. This was done using audio segmentation based on the silence between each segment. We made all the segments in equal time duration of one-second padding the shorter one with silence. Since the received recordings were collected from different sources, the audio sampling rate varies from one sample to another; so, the audio recordings were resampled to a 16 kHz sample rate. Finally, we tested the segments of audio files to ensure intelligibility and clarity of pronunciation by listening to each segment and removing any sounds that were not clear or had the incorrect pronunciation of an alphabet.

Data Labeling: To label each recorded instance belonging to a class, each short vowel is coded with a unique number for further processing. Table 2 presents the short vowels and their class labels.

3.3. Spectrogram Conversion

Instead of using audio signals as input, we converted each input instance into its equivalent waveform and then to a spectrogram of 32×32 pixels. "Spectrograms are 2D images representing sequences of spectra with time along one axis, frequency along with the other, and brightness or color representing the strength of a frequency component at each time frame" [33]. The advantage of spectrograms over audio signals is that they retain more information than the hand-crafted features for audio analysis, they are of lower dimension than the raw audio and have found their role in neural networks [33].

Table 2. Representation of Arabic Short Vowels (ASV) with Class Label (CL) in IPA and native script along.

Arabic short vowel	أَ	إِ	أُ	بَ	بِ	بُ	تَ	تِ	تُ	ثَ	ثِ	ثُ	جَ	جِ
Class label	1	2	3	4	5	6	7	8	9	10	11	12	13	14
IPA symbol	aa	ai	au	Ba	bi	bu	Ta	Ti	Tu	θ_a	θ_i	θ_u	dʒa	dʒi
Arabic short vowel	جُ	حَ	حِ	حُ	خَ	خِ	خُ	دَ	دِ	دُ	ذَ	ذِ	ذُ	رَ
Class label	15	16	17	18	19	20	21	22	23	24	25	26	27	28
IPA symbol	dʒu	ħa	ħi	Ħu	xa	xi	Xu	Da	Di	du	ða	Đi	ðu	ra
y Arabic short vowel	رِ	رُ	زَ	زِ	زُ	سَ	سِ	سُ	شَ	شِ	شُ	صَ	صِ	صُ
Class label	29	30	31	32	33	34	35	36	37	38	39	40	41	42
IPA symbol	ri	ru	za	Zi	zu	sa	Si	Su	ʃa	ʃi	ʃu	sˤa	sˤi	sˤu
Arabic short vowel	ضَ	ضِ	ضُ	طَ	طِ	طُ	ظَ	ظِ	ظُ	عَ	عِ	عُ	غَ	غِ
Class label	43	44	45	46	47	48	49	50	51	52	53	54	55	56
IPA symbol	dˤa	dˤi	dˤu	tˤa	tˤi	tˤu	ðˤa	ðˤi	ðˤu	ʕa	ʕi	ʕu	ʁa	ʁi
Arabic short vowel	غُ	فَ	فِ	فُ	قَ	قِ	قُ	كَ	كِ	كُ	لَ	لِ	لُ	مَ
Class label	57	58	59	60	61	62	63	64	65	66	67	68	69	70
IPA symbol	ʁu	fa	fi	Fu	qa	Qi	Qu	Ka	Ki	ku	la	Li	lu	ma
Arabic short vowel	مِ	مُ	نَ	نِ	نُ	هـ	هـ	هـ	وَ	وِ	وُ	يَ	يِ	يُ
Class label	71	72	73	74	75	76	77	78	79	80	81	82	83	84
IPA symbol	mi	Mu	na	Ni	nu	Ha	Hi	Hu	Wa	wi	wu	Ja	ji	ju

4. CNN Model for Arabic Short Vowels Classification

Inspired by existing approaches to developing CNN architecture, we started with an initial architecture consisting of five convolutional and two pooling layers. With our constructed dataset and the first CNN model, we achieved a training accuracy of 84.27% and validation accuracy of 40.15%. Based on previous approaches [34,35], and the low validation accuracy, it was evident that the small dataset could not give high classification accuracy. Because obtaining more user data was not an option, for achieving good accuracy, we developed a two-phase approach as shown in Figure 1. In the first phase, our focus was on improving the validation results as much as the initial training results by generating more training and validation data. Data augmentation is a useful technique to improve the performance of model and expand limited datasets to take advantage of deep learning models. The validation error must continue to decrease with the training error to develop a useful deep learning model. However, data augmentation is not only a technique to improve accuracy by avoiding the overfitting. Many other alternative solutions are model fine-tuning and hyperparameter tuning to achieve higher accuracy by avoiding the overfitting [36]. Therefore, once our desired validation accuracy of 85% could be achieved using a bigger dataset, in the second phase our focus was to improve the deep learning model to achieve a test accuracy of 95% on the test data. As can be seen in the figure, model development is an iterative process in each of these phases. In phase 2, after each cycle of training, validation, and testing, the results are analyzed, and the model is modified by fine-tuning it. At the same time, the hyperparameters are optimized to achieve the desired accuracy above 95%.

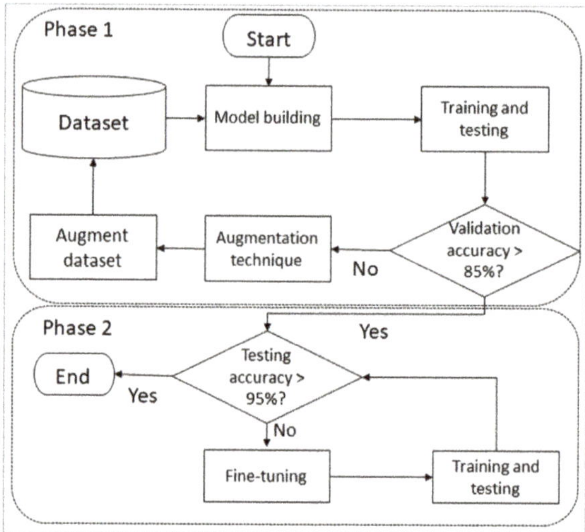

Figure 1. A two-phase approach of developing a CNN model for classifying Arabic short vowels.

4.1. Data Augmentation

Data augmentation is a technique for increasing the number of records in the dataset by slightly modifying the copies of original data for producing synthetic data using statistical methods and algorithms. Many data augmentation techniques exist in the literature [37], so we applied a few techniques one-by-one. Each time an augmentation technique was applied, it resulted in the gradual increase of the audio instances in the dataset; we also kept evaluating the validation accuracy of the base model, which was also improving. We used the augmentation techniques of noise injection, time-shifting, and changing the audio speed using time stretch, as shown in Figure 2. Other audio data augmentation techniques, such as pitch shift, were not ideal for our dataset because it required randomly changing the pitch, thus, distorting the pronunciation. The techniques are described as follows.

Noise injection: This technique introduces white noise in the audio dataset [38] as a ratio between the signal and noise. This method is appropriate for our model as we can assume that given environmental variations, user input is not noise-free in most circumstances. We applied random noise augmentation by adding two types of noise values X to audio files using the NumPy library in Python, with X = 0.005, and X = 0.0005. Figure 2b illustrates the noise injection results of an audio file, where noise is injected at the rate of X = 0.005.

Time-shifting: The method of time-shifting is applied to shift the audio forward or backward [38]. We used the roll method in Python's Numpy library to shift the start of an audio file S milliseconds. If D_a is audio data, and $D_a = [x_1, x_2, \ldots, x_n]$, by applying Silent $S_t = [S_1, S_2, \ldots, S_m]$, it becomes $D_{shifted} = [S_1, S_2, \ldots, S_m, x_1, x_2, \ldots, x_{n-m}]$, Where $S_t \leq D_s$. Figure 2c presents the result of the time-shifting function; the audio file is shifted forward for 2 milliseconds at the beginning of the graph by replacing it with silence.

Changing speed: This technique allows adjusting the speed of audio signal S by a certain rate R as $S = S \div R$. We used the values of R = [1.25, 1.4, 1.5, 1.6]. Figure 2d illustrates the result of changing the audio speed where R = 1.5. It is investigated in [39] that short vowels can be varied on shorter and longer duration.

After applying the data augmentation techniques, we ended up with a total of 49,829 audio files. With the augmented dataset, we achieved a validation accuracy of more than 85%.

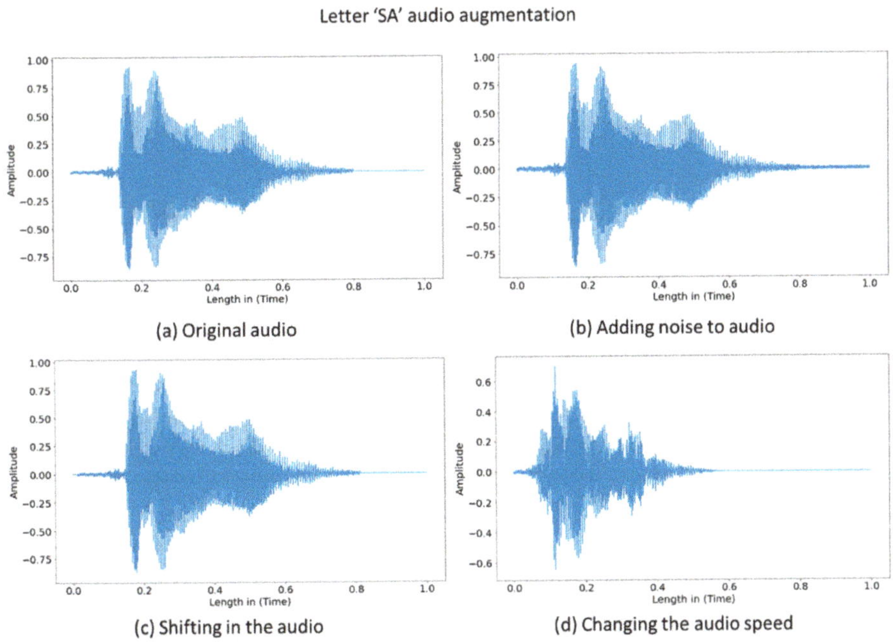

Figure 2. The results of different data augmentation techniques on an audio file: (**a**) the original audio data, (**b**) the audio signal after adding noise to original file, (**c**) the audio file after time-shifting, and (**d**) the audio file after increasing the speed.

4.2. Fine-Tuning the CNN Architecture

We started with a sequential CNN architecture with eight processing layers to develop our baseline model. The architecture is made up of convolutional layers such that each layer applies a set of convolution filters to an input followed by a non-linear activation function [40]. The convolutional layers are defined with a kernel of size 3 and the number of filters set to 32, 64, and 128 in different layers. A resizing layer is used to downsample each input to speed up the model training process. A normalization layer normalizes each pixel in the image using mean and standard deviation values. The other important layer of the CNN architecture is the pooling layer [41], and its function is to progressively reduce the input to decrease the number of parameters and increase the network performance. There are different types of pooling operations: max pooling, average pooling, global max pooling, and global average pooling used for downsampling of the input samples. The CNN can also have additional layers for optimization and improved performance. The batch normalization layer [42] allows normalizing the input of each layer, as the problem of internal covariate shift occurs due to constantly changing the distribution of activation, and each layer needs to learn to adapt to a new distribution. Similarly, the dropout layer [43] is used to reduce the model overfitting by randomly dropping out some percentage of the layer output. The flatten layer is used to convert the two-dimensional data into one-dimension for final classification by the dense layer.

Each layer has its associated activation function [44], whose task is to define how the weighted sum of input is transformed to the output from the nodes in a layer of the network. There are various activation functions, and the most popular ones are the ReLU (rectified linear unit), Sigmoid, and SoftMax. The ReLU function does not allow the activation of all the neurons simultaneously; when the output of the linear transformation is zero, the output neurons get deactivated. SoftMax function restricts the output values in the range 0 to 1, which are treated as probabilities of a particular class and usually used in the last layer of the neural network. Mathematically, the SoftMax function σ is applied to a vector of inputs, $\vec{z}$, where each component of z is converted into corresponding probability according to its weight, e^{z_i} is standard exponential function for input vector, k stands for the number of classes and e^{z_j} presents standard exponential function for output vector shown in Equation (1) as follows:

$$\sigma(\vec{z})_i = \frac{e^{z_i}}{\sum_{j=1}^{k} e^{z_j}} \quad (1)$$

The dataset was divided into a ratio of 80% for the training set and 10% for validation and testing sets each. The batch size was set to 32. With the CNN architecture having eight layers, we achieved a testing accuracy of 80%. An improvement strategy [35] was to update the baseline sequential model by adding more convolutional, max pooling, and dense layers and updating the network to 12 layers. To find the optimal CNN model, we ran our model by applying a random search for the best parameters, which allows finding the optimal values of filter, kernel, and learning rate to achieve maximum accuracy. We found that the accuracy can be increased by increasing the filter size. We tested many CNN model modifications in a trial-and-error [35,45] manner and kept increasing the testing and validation accuracy in small steps. That allowed us to improve the testing accuracy to 90.0%.

4.3. Hyperparameters Tuning

In deep learning neural networks, the function of optimizers [46] is to reduce the losses to achieve the most accurate results possible. We evaluated different optimizers: Adam, Nadam, RMSprop, and SGD, and found Adam [47] to achieve the best performance in the proposed network. Adam stands for adaptive moment estimation that adaptively estimates the first and second-order moments. It updates the network weights iteratively based on the training data. Specifically, Adam uses the update vector $\hat{v}_t$ and the past gradient $\hat{m}_t$ differently than the previous algorithms, shown in Equation (2):

$$\theta_{t+1} = \theta_t - \frac{\eta}{\sqrt{\hat{v}_t} + \epsilon}\hat{m}_t \quad (2)$$

Here θ_t represents the weights and bias parameters, η stands for the learning rate or step-size, $\hat{m}$ and $\hat{v}$ represent the first and second moment vectors and by parameterizing them by t we get their moving averages over time. Adam combines the advantages of AdaGrad and RMSProp algorithms [26] and compared to other optimizers, it requires less memory, is computationally efficient, and is suited for problems with large data or many parameters.

We utilized the learning rate scheduler to dynamically adjust the learning rate to achieve the desired accuracy. Initially, the learning rate was set to 0.001, and after 80 Epochs, the learning rate was reduced by 10% every Epoch. We continuously reduced the learning rate because it helped in achieving optimal weight update by gradually maintaining the training loss and avoiding its oscillating over training epochs [48]. We utilized the early stopping hyperparameter for stopping the model prematurely if the loss was not decreasing anymore. We also used the Tensorboard tool to get training details for analysis and further improvements to the model through the generated time-series graphs, histograms, and distributions [49]. We added 2 batch normalization and 1 dropout layer to this model. These steps are explained in Algorithm 1, where, labeled audio dataset is input to the model that output as training loss, accuracy, and predicted labels. The audio data was first converted to a labeled waveform dataset. Then the transformed to labeled spectrogram dataset. The spectrogram dataset was resized to 32 × 32 pixels and input to our optimized sequential DLNN model. The model was executed by passing the callback functions of learning rate schedular (lr_scheduler), early stopping (callback_Early_stopping), best model checkpoint (model_checkpoint_callback), and tensorboard (tensorboard_callback).

Together, hyperparameter tuning and model fine-tuning helped in achieving the testing accuracy of 95%.

Algorithm 1: Steps of classification of Arabic short vowels on a fully optimized and fine-tuned neural network

```
Input
        aDataset = Audio dataset
        labels = Labels of classes
        train_files, val_files, test_files = split(aDataset(80,10,10))
Output
        Accuracy = Model accuracy
        Loss = Model learning loss
        y_pred = Predicted labels
Algorithm
Begin
        waveform_ds = Map waveform and labels from aDataset
        spectrogram_ds = Map spectrogram and labels from waveform_ds

        Function preprocess_dataset(files)
        output_ds = Map waveform_ds from files_ds of files
        output_ds = Map spectrogram_ds of output_ds
        Return output_ds
        Endfunction
        train_ds = preprocess_dataset from train_files
                val_ds = preprocess_dataset from val_files
        test_ds = preprocess_dataset from test_files

        input_shape = Shape of spectrogram in spectrogram_ds
        norm_layer = Normalization in preprocessing

        model = Sequential(input_shape, Resizing(32, 32), norm_layer, layers)
        trainNetwork train_ds, val_ds, callbacks =
        [lr_scheduler,callback_Early_stopping, model_checkpoint_callback,
        tensorboard_callback]

        load weights of best_model
        model train accuracy = Evaluate(train_ds)
        model val accuracy = Evaluate(val_ds)
        model test_accuracy = Evaluate(test_ds)
        Loss_graph metrics['loss'], metrics['val_loss']
        Accuracy_graph metrics['Accuracy'], metrics['val_ Accuracy']
End
```

Figure 3 shows the resulting architecture after fine-tuning and applying hyperparameter tuning to the baseline architecture. As can be seen, hyperparameters have been added in the form of batch normalization and dropout layers in addition to the optimizer selection and learning weight. FC1 and FC2 represent the two fully connected layers for classification followed by the SoftMax activation function that assigns a probability to each of the output classes.

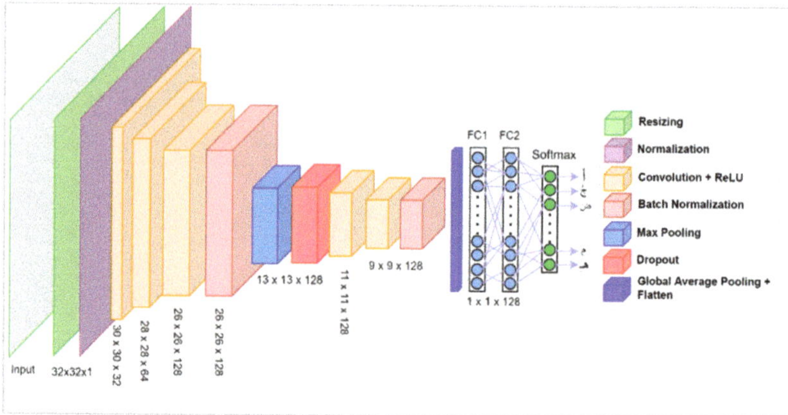

Figure 3. The architecture of the optimized CNN model.

4.4. Model Execution

The code is executed on a GPU-based desktop system with hardware configuration of Intel (R) Core (TM) i9 CPU @3.70, NVIDIA GeForce RTX 3070, and 64 GB RAM. The model is developed and run using the TensorFlow (https://www.tensorflow.org/, accessed on 10 November 2021) platform, mainly the Keras (https://keras.io/, accessed on 10 November 2021) API. The maximum number of epochs was set to 150. It took on average 91 s to complete one Epoch. We used the Jupyter notebook in conda (https://docs.conda.io/en/latest/, accessed on 10 November 2021) environment management system with miniconda installer for running the CNN model.

5. Results

In this study, we designed and executed four different experiments using combinations of two datasets and two model settings: (1) the original dataset trained on a sequential baseline CNN model, (2) the original dataset trained on an optimized and fine-tuned model, (3) the augmented dataset trained on the baseline CNN model, and (4) the augmented dataset trained on the optimized, fine-tuned model. Table 3 shows the comparative results (training, validation, and testing) of the four experiments. In the first experiment, the validation and testing accuracy gave very low results on the original dataset. The reason is that 6229 audio files are too few for 84 classes, which are on average 74 instances per class. Therefore, when performing validation and testing, the sample size becomes very small. However, after data augmentation, model optimization, and fine-tuning, the validation and testing accuracy increased to 95.87% and 95.77%, respectively. This implies that data augmentation and fine-tuning proved useful in model improvement.

Table 3. Model execution results on different parameters.

Exp. No.	Experiment Settings	Accuracy		
		Training	Validation	Testing
1	The original dataset on baseline CNN model (n = 6229)	84.27%	40.15%	36.0%
2	The original dataset on fine-tuned model (n = 6229)	99.74%	59.33%	56.91%
3	The augmented dataset using baseline CNN model (n = 49,829)	98.93%	90.63%	90.0%
4	Model optimization, hyperparameter tuning, and augmented dataset (n = 49,829)	99.9%	95.87%	95.77%

Figure 4a,b show the accuracy and loss of the optimized and fine-tuned CNN model. We can observe that the training and validation losses improve continuously. The optimized and fine-tuned CNN model has achieved the best training accuracy of 99.857% and loss of 0.0073, and the validation accuracy is 95.87% and loss of 0.2329.

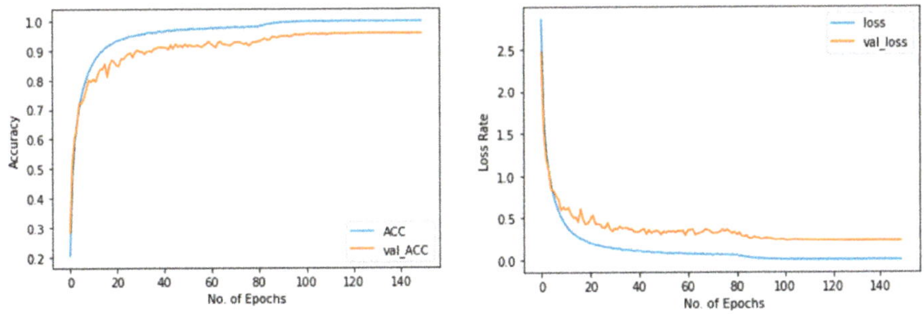

(**a**) Training and Validation Accuracy (**b**) Training and Validation Loss

Figure 4. (**a**) training and validation accuracy and (**b**) training and validation loss of the optimized CNN model on the augmented dataset.

Figure 5 shows the details of the misclassification by the classifier. Eleven of the 84 classes have 0% error rate. Only 20 classes have four or more mispredictions. The Figure shows that class 10 (phoneme (ثَ, θ_a)) and class 15 (phoneme (جُ, dʒu)) have been misidentified a maximum of eight times. From the analysis of the confusion matrix, we observe that the topmost misclassified classes are those not spoken in CA pronunciation by the native local people. For example, class 10 (phoneme (ثَ, θ_a)) is often pronounced as 7 (phoneme (تَ, ta)) in local Arabic dialect.

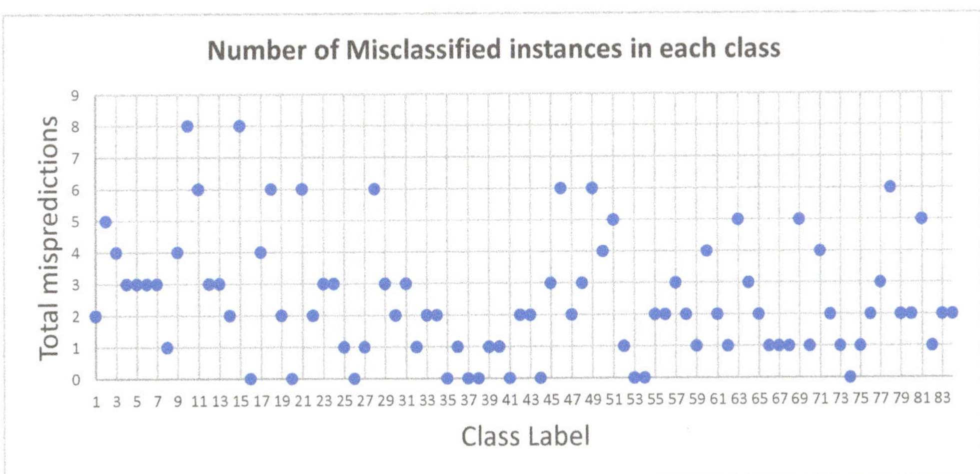

Figure 5. Scatter graph showing classes misclassified as other classes.

6. Discussion

This research addresses the recognition of short vowel phonemes in the Arabic language. The recognition task falls into the classification of classical Arabic words, a dialect of the Arabic language that is very old but is understood and applied to this day by millions of people around the world. Our work is one step ahead of the previous efforts in classifying a total of 84 Arabic short vowels. The authors in [10] achieved an accuracy of 97.88% using CNN for Arabic alphabets (28 classes) recognition without considering the vowels. We believe that the authors could achieve highly accurate results because without considering the vowel phonemes the chances of interclass misclassification are reduced significantly. Alotaibi and Hussain [30] have considered formants and consonant-vowels-consonant (CVC) utterances for identifying similarities and differences of vowels for MSA, but to the best of our knowledge, there is no work for CA vowels recognition despite their similarities with the MSA [12].

This research started with data collection followed by its augmentation through various techniques, which is lacking in the previous studies. In [9,10,20,50], the authors have contributed to the basic Arabic alphabets audio data collection, and they mostly performed manual feature extraction. In [51], the authors collected the audio dataset of Arabic words. However, Almisreb et al. [39] investigated Arabic short vowels' properties that helped us understand their characteristics and duration variations. We were inspired by this approach and applied data augmentation to our collected audio files. Our model supports the automatic classification of Arabic short vowels on 84 classes using deep learning neural networks instead of manually identifying the audio features as done in the studies [10,23,26]. The data augmentation helped us to achieve accuracy above 95%, as in the previous research the authors have used data augmentation on 28 classes of Arabic alphabet dataset, and improved DCNN model's accuracy from 65.89% to 95.95%, Alexnet's model accuracy from 78.03% to 98.41%, and BLSTM's model accuracy from 53.18% to 87.90%[10]. In the previous studies, the authors have identified different audio features for the classification of Arabic alphabets [10,18] into 28 classes, and Arabic words [2,7,25,29] using the CNN model. The studies in [13,17,19,20,26] identified appropriate features of Arabic audio for applying machine learning techniques [17,23,50] for classification purposes.

Mispronunciation detection of Arabic is a significant parameter in a Computer Assisted Language Learning (CALL) system [51]. This is mainly a problem for non-native speakers, and approaches like [52] try to detect confusing Arabic pronunciation of similar-sounding letters for non-native speakers. However, this problem even exists in the Arabic-speaking world due to the prevalence of different regional dialects in the various parts of the world.

Our proposed approach can be utilized for developing CNN models in a similar domain for learning support systems. This approach helps construct CNN models from scratch and improves them by applying various techniques of data augmentation, fine-tuning neural networks, and hyperparameters tuning. Similar methods are used in other domains [34] for CNN models improvements. Given that the model's performance improved with synthetic data, there are chances of achieving high accuracy if more real data can be retrieved. Given our experience in this research, we believe that it will be helpful for the researchers to save their time and make these processes simple by reducing the complexity of audio data preprocessing by bypassing features identification steps.

Due to a shortage of participants for audio collection we could only get maximum 85 audio records per class and total 6229 audio files. This dataset is smaller for the requirements of the CNN model training and validation process. However, the previous studies [10,20] on CA audios the authors utilized between 2k–4k audio files to investigate DLNN models. Thus, data augmentation helped us in obtaining sufficiently large dataset. Furthermore, in this paper, we have focused on data collection in single geographic region (95%) from Saudi Arabia, which has a native Arabic speaking population. It would be of interest to evaluate this approach on data from non-native speakers as well as natives from other Arab countries.

In the near future, we intended to improve our work by obtaining data from other Arabic-speaking groups to see the generalization of our approach in a wider context. We planned to expand our current work to participants from different nationalities as well as non-native speakers and age groups to explore their pronunciation similarity and differences from native speakers, areas of improvements to facilitate development of Arabic learning tools and applications. The analysis will be made on duration, variability, and overlapping attributes among CA learners. In addition, we also aim to quantify the standard duration of pronunciation of both short and long vowels in the classical Arabic language.

7. Conclusions

This article introduced a CNN architecture for the classification of Arabic short vowel alphabets. Using data augmentation techniques and hyperparameters tuning, we achieved a significant boost in our testing accuracy of 95.77% from a baseline model. Compared to previous approaches for Arabic alphabet classification, which classify only 28 basic alphabets, the current task was more challenging as it involved some similar sounding phonemes from 84 classes. The current work can be considered a significant leap in achieving highly accurate detection of mispronunciation of Arabic short vowels, which is considered an important step in learning classical Arabic. This contribution is beneficial for all interested stakeholders in CA to assist them in developing applications concerning Arabic pronunciation and learning recitation of the Holy Quran. Consequently, the CA learner will be benefitted for practicing Arabic short vowels using any tool based on our proposed model in unavailability of their real teacher. Furthermore, the comprehensive process of developing DLNN reported in this paper will help the developers and researchers to build learning tools by following the similar steps.

Author Contributions: Conceptualization, A.A. (Amna Asif), H.M. and H.F.A.; methodology, A.A. (Amna Asif) and H.M.; validation, A.A. (Amna Asif) and H.M.; formal analysis, A.A. (Amna Asif) and H.M.; investigation, A.A. (Amna Asif), H.M. and F.A.; resources, H.F.A.; data curation, F.A.; writing—original draft preparation, A.A. (Amna Asif), H.M., F.A. and H.F.A.; writing—review and editing, A.A. (Amna Asif), H.M.; visualization, F.A.; supervision, H.F.A.; project administration, A.A. (Abdulaziz Alhumam); funding acquisition, A.A. (Amna Asif) and A.A. (Abdulaziz Alhumam). All authors have read and agreed to the published version of the manuscript.

Funding: This work was supported through the Annual Funding track by the Deanship of Scientific Research, Vice Presidency for Graduate Studies and Scientific Research, King Faisal University, Saudi Arabia [Project No. AN000292].

Institutional Review Board Statement: Not applicable.

Informed Consent Statement: Informed consent was obtained from all subjects involved in the study.

Data Availability Statement: The collected dataset is available on (https://www.kaggle.com/amnaasif/arabic-short-vowels-audio-dataset, accessed on 10 November 2021).

Conflicts of Interest: The authors declare no conflict of interest.

References

1. Julian, G. What are the most spoken languages in the world. *Retrieved May* **2020**, *31*, 2020.
2. Ali, A.; Chowdhury, S.; Afify, M.; El-Hajj, W.; Hajj, H.; Abbas, M.; Hussein, A.; Ghneim, N.; Abushariah, M.; Alqudah, A. Connecting Arabs: Bridging the gap in dialectal speech recognition. *Commun. ACM* **2021**, *64*, 124–129. [CrossRef]
3. Twaddell, W.F. On defining the phoneme. *Language* **1935**, *11*, 5–62. [CrossRef]
4. Ibrahim, A.B.; Seddiq, Y.M.; Meftah, A.H.; Alghamdi, M.; Selouani, S.-A.; Qamhan, M.A.; Alotaibi, Y.A.; Alshebeili, S.A. Optimizing arabic speech distinctive phonetic features and phoneme recognition using genetic algorithm. *IEEE Access* **2020**, *8*, 200395–200411. [CrossRef]
5. Witt, S.M. Automatic error detection in pronunciation training: Where we are and where we need to go. In Proceedings of the International Symposium on Automatic Detection on Errors in Pronunciation Training, Stockholm, Sweden, 6–8 June 2012.
6. Huang, H.; Xu, H.; Hu, Y.; Zhou, G. A transfer learning approach to goodness of pronunciation based automatic mispronunciation detection. *J. Acoust. Soc. Am.* **2017**, *142*, 3165–3177. [CrossRef] [PubMed]
7. Al-Marri, M.; Raafat, H.; Abdallah, M.; Abdou, S.; Rashwan, M. Computer Aided Qur'an Pronunciation using DNN. *J. Intell. Fuzzy Syst.* **2018**, *34*, 3257–3271. [CrossRef]
8. Ibrahim, N.J.; Idris, M.Y.I.; Yusoff, M.Z.M.; Anuar, A. The problems, issues and future challenges of automatic speech recognition for quranic verse recitation: A review. *Al-Bayan J. Qur'an Hadith Stud.* **2015**, *13*, 168–196. [CrossRef]
9. Arafa, M.N.M.; Elbarougy, R.; Ewees, A.A.; Behery, G. A Dataset for Speech Recognition to Support Arabic Phoneme Pronunciation. *Int. J. Image Graph. Signal Process.* **2018**, *10*, 31–38. [CrossRef]

10. Ziafat, N.; Ahmad, H.F.; Fatima, I.; Zia, M.; Alhumam, A.; Rajpoot, K. Correct Pronunciation Detection of the Arabic Alphabet Using Deep Learning. *Appl. Sci.* **2021**, *11*, 2508. [CrossRef]
11. Czerepinski, K. *Tajweed Rules of the Qur'an: Part 1*; Dar Al Khair: Riyadh, Saudi Arabia, 2005.
12. Alghamdi, M.M. A spectrographic analysis of Arabic vowels: A cross-dialect study. *J. King Saud Univ.* **1998**, *10*, 3–24.
13. Nazir, F.; Majeed, M.N.; Ghazanfar, M.A.; Maqsood, M. Mispronunciation detection using deep convolutional neural network features and transfer learning-based model for Arabic phonemes. *IEEE Access* **2019**, *7*, 52589–52608. [CrossRef]
14. Nassif, A.B.; Shahin, I.; Attili, I.; Azzeh, M.; Shaalan, K. Speech recognition using deep neural networks: A systematic review. *IEEE Access* **2019**, *7*, 19143–19165. [CrossRef]
15. Duan, R.; Kawahara, T.; Dantsuji, M.; Nanjo, H. Efficient learning of articulatory models based on multi-label training and label correction for pronunciation learning. In Proceedings of the 2018 IEEE International Conference on Acoustics, Speech and Signal Processing (ICASSP), Calgary, AB, Canada, 15–20 April 2018; pp. 6239–6243.
16. Necibi, K.; Bahi, H. An arabic mispronunciation detection system by means of automatic speech recognition technology. In Proceedings of the 13th International Arab Conference on Information Technoloy Proceedings, Zarqa, Jordan, 10–13 December 2012; pp. 303–308.
17. Al Hindi, A.; Alsulaiman, M.; Muhammad, G.; Al-Kahtani, S. Automatic pronunciation error detection of nonnative Arabic Speech. In Proceedings of the 2014 IEEE/ACS 11th International Conference on Computer Systems and Applications (AICCSA), Doha, Qatar, 10–13 November 2014; pp. 190–197.
18. Khan, A.F.A.; Mourad, O.; Mannan, A.M.K.B.; Dahan, H.B.A.M.; Abushariah, M.A. Automatic Arabic pronunciation scoring for computer aided language learning. In Proceedings of the 2013 1st International Conference on Communications, Signal Processing, and their Applications (ICCSPA), Sharjah, United Arab Emirates, 12–14 February 2013; pp. 1–6.
19. Marlina, L.; Wardoyo, C.; Sanjaya, W.M.; Anggraeni, D.; Dewi, S.F.; Roziqin, A.; Maryanti, S. Makhraj recognition of Hijaiyah letter for children based on Mel-Frequency Cepstrum Coefficients (MFCC) and Support Vector Machines (SVM) method. In Proceedings of the 2018 International Conference on Information and Communications Technology (ICOIACT), Yogyakarta, Indonesia, 6–7 March 2018; pp. 935–940.
20. Akhtar, S.; Hussain, F.; Raja, F.R.; Ehatisham-ul-haq, M.; Baloch, N.K.; Ishmanov, F.; Zikria, Y.B. Improving mispronunciation detection of arabic words for non-native learners using deep convolutional neural network features. *Electronics* **2020**, *9*, 963. [CrossRef]
21. Leung, W.-K.; Liu, X.; Meng, H. CNN-RNN-CTC based end-to-end mispronunciation detection and diagnosis. In Proceedings of the ICASSP 2019–2019 IEEE International Conference on Acoustics, Speech and Signal Processing (ICASSP), Brighton, UK, 12–17 May 2019; pp. 8132–8136.
22. Zainon, N.Z.; Ahmad, Z.; Romli, M.; Yaacob, S. Speech quality based on Arabic pronunciation using MFCC and LDA: Investigating the emphatic consonants. In Proceedings of the 2012 IEEE International Conference on Control System, Computing and Engineering, Penang, Malaysia, 23–25 November 2012; pp. 398–403.
23. Aissiou, M. A genetic model for acoustic and phonetic decoding of standard Arabic vowels in continuous speech. *Int. J. Speech Technol.* **2020**, *23*, 425–434. [CrossRef]
24. Abdou, S.M.; Rashwan, M. A Computer Aided Pronunciation Learning system for teaching the holy quran Recitation rules. In Proceedings of the 2014 IEEE/ACS 11th International Conference on Computer Systems and Applications (AICCSA), Doha, Qatar, 10–13 November 2014; pp. 543–550.
25. Necibi, K.; Frihia, H.; Bahi, H. On the use of decision trees for arabic pronunciation assessment. In Proceedings of the International Conference on Intelligent Information Processing, Security and Advanced Communication, Batna, Algeria, 23–25 November 2015; pp. 1–6.
26. Abdelhamid, A.A.; Alsayadi, H.A.; Hegazy, I.; Fayed, Z.T. End-to-End Arabic Speech Recognition: A Review. In Proceedings of the 19th Conference of Language Engineering (ESOLEC'19), Alexandria, Egypt, 26–30 September 2020.
27. Fadel, A.; Tuffaha, I.; Al-Ayyoub, M. Arabic text diacritization using deep neural networks. In Proceedings of the 2019 2nd International Conference on computer Applications & Information Security (ICCAIS), Riyadh, Saudi Arabia, 1–3 May 2019; pp. 1–7.
28. Al-Anzi, F.S.; AbuZeina, D. Synopsis on Arabic speech recognition. *Ain Shams Eng. J.* **2021**, *13*, 9. [CrossRef]
29. Lamel, L.; Messaoudi, A.; Gauvain, J.-L. Automatic speech-to-text transcription in Arabic. *TALIP* **2009**, *8*, 1–18. [CrossRef]
30. Alotaibi, Y.A.; Hussain, A. Comparative analysis of Arabic vowels using formants and an automatic speech recognition system. *Int. J. Signal Process. Image Process. Pattern Recognit.* **2010**, *3*, 11–22.
31. Yu, D.; Li, J. Recent progresses in deep learning based acoustic models. *IEEE/CAA J. Autom. Sin.* **2017**, *4*, 396–409. [CrossRef]
32. Alqadheeb, F.; Asif, A.; Ahmad, H.F. Correct Pronunciation Detection for Classical Arabic Phonemes Using Deep Learning. In Proceedings of the 2021 International Conference of Women in Data Science at Taif University (WiDSTaif), Taif, Saudi Arabia, 30–31 March 2021; pp. 1–6.
33. Wyse, L. Audio Spectrogram Representations for Processing with Convolutional Neural Networks. In Proceedings of the First International Conference on Deep Learning and Music, Anchorage, AK, USA, 17–18 May 2017; pp. 37–41.
34. Mukhtar, H.; Qaisar, S.M.; Zaguia, A. Deep Convolutional Neural Network Regularization for Alcoholism Detection Using EEG Signals. *Sensors* **2021**, *21*, 5456. [CrossRef]

35. Tajbakhsh, N.; Shin, J.Y.; Gurudu, S.R.; Hurst, R.T.; Kendall, C.B.; Gotway, M.B.; Liang, J. Convolutional Neural Networks for Medical Image Analysis: Full Training or Fine Tuning? *IEEE Trans. Med. Imaging* **2016**, *35*, 1299–1312. [CrossRef]
36. Shorten, C.; M. Khoshgoftaar, T. A survey on Image Data Augmentation for Deep Learning. *J. Big Data* **2019**, *6*, 60. [CrossRef]
37. Wei, S.; Zou, S.; Liao, F. A Comparison on Data Augmentation Methods Based on Deep Learning for Audio Classification. *J. Phys. Conf. Ser.* **2020**, *1453*, 012085. [CrossRef]
38. Nanni, L.; Maguolo, G.; Paci, M. Data augmentation approaches for improving animal audio classification. *Ecol. Inform.* **2020**, *57*, 101084. [CrossRef]
39. Abd Almisreb, A.; Abidin, A.F.; Tahir, N.M. An acoustic investigation of Arabic vowels pronounced by Malay speakers. *J. King Saud Univ. -Comput. Inf. Sci.* **2016**, *28*, 148–156. [CrossRef]
40. Traore, B.B.; Kamsu-Foguem, B.; Tangara, F. Deep convolution neural network for image recognition. *Ecol. Inform.* **2018**, *48*, 257–268. [CrossRef]
41. Sun, M.; Song, Z.; Jiang, X.; Pan, J.; Pang, Y. Learning pooling for convolutional neural network. *Neurocomputing* **2017**, *224*, 96–104. [CrossRef]
42. Ioffe, S.; Szegedy, C. Batch normalization: Accelerating deep network training by reducing internal covariate shift. In Proceedings of the 32nd International Conference on Machine Learning, Lille, France, 7–9 July 2015; pp. 448–456.
43. Baldi, P.; Sadowski, P.J. Understanding dropout. *Adv. Neural Inf. Process. Syst.* **2013**, *26*, 2814–2822.
44. Sharma, S.; Sharma, S. Activation functions in neural networks. *Towards Data Sci.* **2017**, *6*, 310–316. [CrossRef]
45. Young, H.P. Learning by trial and error. *Games Econ. Behav.* **2009**, *65*, 626–643. [CrossRef]
46. Zhang, Z. Improved adam optimizer for deep neural networks. In Proceedings of the 2018 IEEE/ACM 26th International Symposium on Quality of Service (IWQoS), Banff, AB, Canada, 4–6 June 2018; pp. 1–2.
47. Kingma, D.P.; Ba, J. Adam: A method for stochastic optimization. *arXiv* **2014**, arXiv:1412.6980.
48. Brownlee, J. How to Configure the Learning Rate When Training Deep Learning Neural Networks. Available online: https://machinelearningmastery.com/learning-rate-for-deep-learning-neural-networks/ (accessed on 10 November 2021).
49. Google. TensorBoard: TensorFlow's Visualization Toolkit. Available online: https://www.tensorflow.org/tensorboard (accessed on 19 August 2021).
50. Lee, A.; Zhang, Y.; Glass, J. Mispronunciation detection via dynamic time warping on deep belief network-based posteriorgrams. In Proceedings of the 2013 IEEE International Conference on Acoustics, Speech and Signal Processing, Vancouver, BC, Canada, 26–31 May 2013; pp. 8227–8231.
51. Maqsood, M.; Habib, H.A.; Nawaz, T. An efficientmis pronunciation detection system using discriminative acoustic phonetic features for arabic consonants. *Int. Arab J. Inf. Technol.* **2019**, *16*, 242–250.
52. Maqsood, M.; Habib, H.; Anwar, S.; Ghazanfar, M.; Nawaz, T. A comparative study of classifier based mispronunciation detection system for confusing Arabic phoneme pairs. *Nucleus* **2017**, *54*, 114–120.

applied sciences

MDPI

Article

On Simulating the Propagation and Countermeasures of Hate Speech in Social Networks

Maite Lopez-Sanchez [1,*,†] and Arthur Müller [2]

1 Department of Mathematics and Computer Science, Universitat de Barcelona, 08007 Barcelona, Spain
2 Institut of Political science, University of the Bundeswehr Munich, 85579 Neubiberg, Germany; arthur.mueller@unibw.de
* Correspondence: maite_lopez@ub.edu; Tel.: +34-93-4037154
† Current address: Gran Via de les Corts Catalanes 585, 08007 Barcelona, Spain.

Abstract: Hate speech expresses prejudice and discrimination based on actual or perceived innate characteristics such as gender, race, religion, ethnicity, colour, national origin, disability or sexual orientation. Research has proven that the amount of hateful messages increases inevitably on online social media. Although hate propagators constitute a tiny minority—with less than 1% participants—they create an unproportionally high amount of hate motivated content. Thus, if not countered properly, hate speech can propagate through the whole society. In this paper we apply agent-based modelling to reproduce how the hate speech phenomenon spreads within social networks. We reuse insights from the research literature to construct and validate a baseline model for the propagation of hate speech. From this, three countermeasures are modelled and simulated to investigate their effectiveness in containing the spread of hatred: Education, deferring hateful content, and cyber activism. Our simulations suggest that: (1) Education consititutes a very successful countermeasure, but it is long term and still cannot eliminate hatred completely; (2) Deferring hateful content has a similar—although lower—positive effect than education, and it has the advantage of being a short-term countermeasure; (3) In our simulations, extreme cyber activism against hatred shows the poorest performance as a countermeasure, since it seems to increase the likelihood of resulting in highly polarised societies.

Keywords: hate speech; hate spread; countermeasures; social networks; opinion diffusion; education; deferring hate content; cyber activism

Citation: Lopez-Sanchez, M.; Müller, A. On Simulating the Propagation and Countermeasures of Hate Speech in Social Networks. *Appl. Sci.* **2021**, *11*, 12003. https://doi.org/10.3390/app112412003

Academic Editors: Aida Valls and Karina Gibert

Received: 17 November 2021
Accepted: 10 December 2021
Published: 16 December 2021

1. Introduction

In recent years, many concerns have arisen related to hate speech (which can take several forms and is known by different names such as derogatory language [1], bigotry [2], misogyny [3], bullying [4], or incivility [5]) and hate dissemination on the Internet (a.k.a. cyberhate). According to the United Nations, hate speech is defined as "the attack or usage of pejorative or discriminatory language with reference to a person or a group based on their religion, ethnicity, nationality, gender or other identity factor" (https://www.un.org/en/genocideprevention/documents/UN%20Strategy%20and%20Plan%20of%20Action%20on%20Hate%20Speech%2018%20June%20SYNOPSIS.pdf, accessed on 1 November 2021). These arisen concerns are well founded, as the usage of hateful language has become common on online social media. This is specially the case on community platforms, such as Gab, where the amount of hateful messages has steadily increased over the last years [6]. Gab.com is an American social networking service launched publicly in May 2017 that is known for its far-right userbase. It is critizised for using free speech as a shield for users and groups who have been banned from other social media. Other platforms, such as Twitter, also show a similar tendency in that hateful users have become more extreme [7]. In fact, some authors have noted that the spread of their messages seems to be inadvertently supported by the algorithms of the social networks [8].

To counter this problem researchers and politicians have proposed several measures with different temporal horizons. General long-term measures emphasise on education of democratic values and tolerance [9] as well as developing critical thinking [10]. They are proposed in order to introduce positive bias into the society, thus inhibiting the propagation of hate speech. Awareness campaigns by governmental organisations focus on mid-term effects and try to prevent forming negative prejudices against out-groups and minorities. Others propose expensive manual community management (i.e., moderation) or intrinsically motivated counter speech [11]. In contrast, the short-term measure of automatic message filtering (or blocking of hateful users) is criticised as, for some cases, it could be used against the freedom of expression human right. Moreover, this countermeasure bears hidden risks of having hateful users being just displaced to other platforms and not really eliminated [12]. In fact, despite the risks associated with banning users, there are some successful experiences that support it. For example, Reddit performed a massive banning of hateful groups in 2015 that did not lead to the displacement of haters to other subreddits/groups [13]. Additionally, the Twitter's user purge in 2017 neither lead to direct migration of haters to Gab—a more radical platform—since they were already there [6]. Exhaustive evaluations of these countermeasures, however, are still lacking and it is difficult to assess the effectiveness of these initiatives and, much less, to compare them among each other.

However, real social networks are too complex to experiment on, and therefore, this paper is devoted to propose an agent-based model [14] as a virtual experiment for the simulation and comparison of countermeasures against the spread of hatred (note that this paper is an extended version of [15]). Although multi-agent based simulations encompass several simplifications, they are definitely useful to conduct what-if analysis to assess the system's behaviour under various situations [16–18], which in our case correspond to different countermeasures. Specifically, we first build a hate speech propagation model based on current research insights on the behaviours of hateful users in social networks. Secondly, we simulate and compare three alternative countermeasures with different temporal effects: education, deferring hateful content and counter activism.

This paper is structured as follows. The next section introduces related work. Then, we bring forward some definitions in Section 3, which are used along the paper. Next, Section 4 defines the baseline propagation model so that Section 5 can then model the three alternative countermeasures. Subsequently, Section 6 presents the simulation results. Finally, the last section concludes the paper and discusses future work.

2. Related Work

This section is devoted to introduce the literature on hateful users and their behaviours, mathematical models of opinion spread research, and existing simulations related to the hatred.

2.1. Characterising Hateful Users in Social Networks

The literature has clearly identified that hateful users exhibit a very different behaviour than regular (normal) users. Their psychological profile describe them as being energetic, talkative, and excitement-seeking [19]. Nevertheless, in addition to these positive traits, they are also found to be narcissist, lack of empathy, and manipulative [20]. Moreover, haters show high activity on social media and follow more people than normal users. Despite the fact that hateful users gain 50% less back-followers for every spawned following relationship per day, they can receive much more followers over the lifetime of their accounts due to their high activity [21]. Surprisingly, although the amount of hateful persons is extremely limited and does not exceed 1% even on Gab, they are responsible for an non-proportionally high amount of content [6]. Moreover, their content can spread faster and diffuse in longer strains trough the network when forwarded by other users [22]. In fact, although hateful content seems to be less informative on Twitter, since this content contains less URLs and hashtags it is known to be more viral when enriched with images or videos [23]. Finally, hateful users turn out to be very densely connected and show more reciprocity (i.e., they follow back more often) than normal users [21,22].

In addition to characterising hateful content, researchers have also studied their effects on the content receivers. As expected, the impact varies depending on whether the receiver belongs to the targeted group (i.e., the victims of hatred content) or not (i.e., they just are listeners/followers receiving the content). Frequent and repetitive exposure to hate speech can lead to desensitization, decreaseing the listeners' harm perception. Moreover, it increases prejudices against the victims [24], attempting to construct and maintain a reality of domination of one group over another [25].

2.2. Models of Opinion Diffusion

When modelling social networks, researchers use a mathematical graph abstraction. Thus, a social network is built as a graph $G = (E, V)$ composed of a set of vertices V, which correspond to users, and a set of edges E representing how they relate (and communicate). Usually, individual opinions about a given topic are represented as numerical values in the interval $[0, 1]$. Both limits of the interval are associated with the extreme stances about the considered topic. In the case of hatred, 0 stands for a very non-hateful opinion whereas 1 stands for the most hateful opinion. Users influence one another by sending messages—through their connecting edges—that lead to a change on their opinion.

The literature has proposed different models for opinion change/diffusion. Here we just introduce a comprehensive (but not exhaustive) set of works. On the one hand, some models are based on the concept of consensus. Thus, Dimakis et al. [26] uses Average Consensus Gossiping (ACG) to force the entire network to converge to the average of all initial opinion values. Alternatively, the aim of DeGroot model [27] is to come to a consensus by using trust as a means to induce differences in the influence of users. DeGroot model was recently applied to the research on hateful behaviours on Twitter and Gab platforms to adjust the score for hate intensity of users [6,21]. On the other hand, bounded confidence models are proposed to follow the intuition that people usually do not accept opinions too far from their own, which is known as *confirmation bias*. For instance, Friedkin and Johnson [28] add some kind of stubbornness, distinguishing between an intrinsic initial opinion, which remains the same, and an expressed opinion, which changes over time. In contrast, Hegselmann and Krause (HK) [29] introduce confidence level—a threshold for opinion difference. Deffuant and Weisbuch (DW) [30] were the first to use asynchronous random opinion updates of two users considering the confidence level. Finally, Terizi et al. [31] conducted extensive simulations showing that Hegselmann–Krause and Deffuant–Weisbuch outperform other models in describing the spread of hateful content on Twitter.

2.3. Multi-Agent Simulations in the Context of Hatred and Polarisation

To the best of our knowledge, most multi-agent simulations devoted to the phenomenons of the hatred and polarisation use a two-dimensional grid as communication topology. Jager and Amblard [32] conducted a general simulation based on the Social Judgement Theory [33] to demonstrate consensus, bi-polarisation or the formation of multiple opinion groups as a result of the opinion forming dynamics. Stefanelli and Seidl [34] used the same theory to model opinion formation on a polarised political topic in Switzerland. The authors used empirical data to set up the simulation and validate their results. Bilewicz and Soral [1] proposed their own model of the spread of hatred which is dependent from the level of contempt, social norms and ability to identify hate speech. As apposed to this, Schieb and Preuß [35] employed a simplified version of the Elaboration Likelihood Model [36] on a message-blackboard and restricted the communication to a closed group of agents. In contrast to the models presented here, where the underlying psychological models use multiple influence factors to model opinion, works in previous Section 2.2 rely on a simple combination of one-dimensional opinion values. In general, none of mentioned models considered more complex topologies of social networks, neither they studied countermeasures against the spread of hatred in comparison to each other, which is the main contribution of this paper.

3. Terminology

This section is devoted to introduce the terminology required to properly describe our models.

Hate score is the central metric in our work and represents the users' attitude and opinion about some polarised topic, which is discussed within a social network. For instance, immigration laws or equal rights of men and women. It is a real number in $[0, 1]$, where both extremes correspond to a very non-hateful and hateful opinions, respectively. We use the hate score as a user opinion value in our diffusion models but also as a threshold to characterise users. The same concept was also employed by Mathew et al. [6] who showed that hate score distribution on Gab is positively biased towards a non-hateful stance. Similarly, we define a user as hateful when $\mathit{hate\ score} \geq 0.75$ and signal it as a red dot in the network graphical representation (see Figure 1). Otherwise, (i.e., when $\mathit{hate\ score} < 0.75$) we assume the user to be normal and represent it as a black dot. We take this threshold in accordance to—and for better comparability with—previous work. As stated before, the amount of haters is known to be a minority of ca. 1%. Therefore, we model the hate score using the Gamma distribution $\Gamma(\alpha, \lambda)$ as depicted in Figure 2, so that the area under curve for $x > 0.75$ is ca. 0.01. For those rare cases when the Gamma distribution naturally exceeds the value of 1 we artificially set users' hate score to the extreme stance of 1.

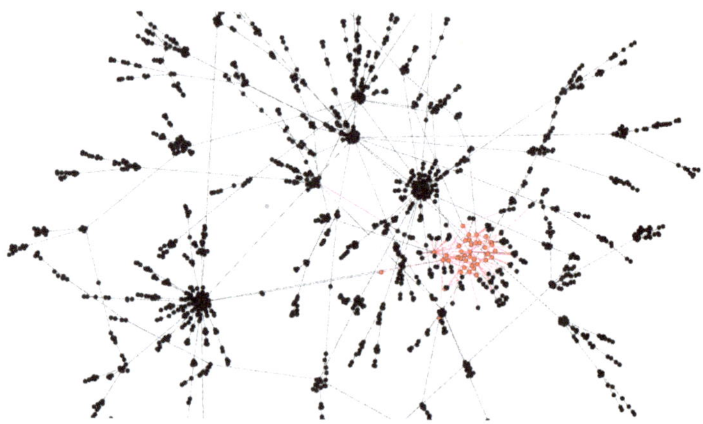

Figure 1. Hate core: Densely connected hateful users (red nodes) within the overall network mostly formed by normal users (black nodes).

Hate core is a network component consisting of densely connected hateful users. Figure 1 depicts how most hateful users (in red) are clustered together. Such components emerge from the high cohesiveness among hateful users as well as from their higher activity. It is also worth noticing that, although single users within a hate core do not exhibit the same influence as some famous mainstream users, as a compound, the hate core can achieve similar effects on the network and attract other users.

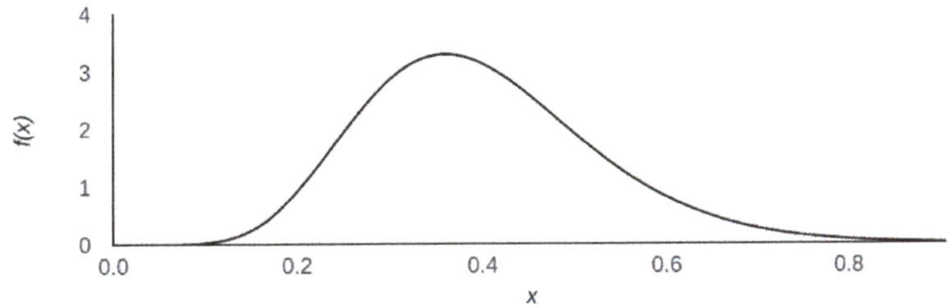

Figure 2. Gamma distribution $\Gamma(\alpha = 10, \lambda = 25)$ with the mean value $\mu = 0.4$ used to sample hate score values for new users who join the simulated social network.

Hate strains are extensions of hate cores. As illustrated by Figure 3, hate strains consist of connected hateful users, but exhibit less network density among them than the hate core that originated them. Most often, hate strains emerge from a hate core as the result of opinion diffusion under the negative influence of the hateful users in the core.

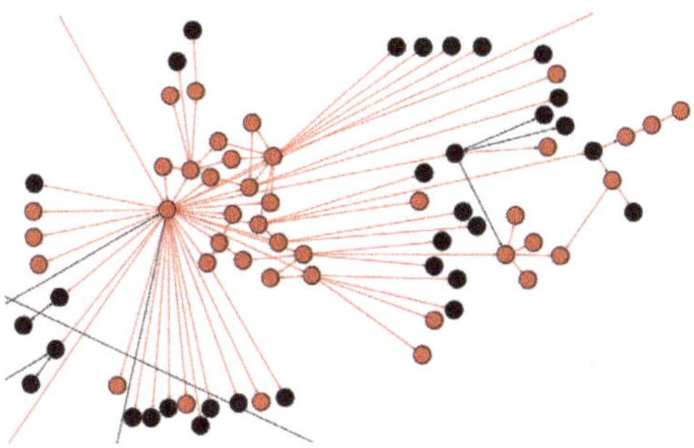

Figure 3. Hate strains (zoom): Hate core disseminating hatred in red strains to nodes with lesser network density.

Swap to a hateful society (by society we mean all the users in the social network). We identify such network transformation when hateful content floods the network and leads to a swap in the opinion of a significant number of users. In particular, we consider a society to be very hateful when the amount of hateful users exceeds 30% of all users in the social network. In fact, experiments have shown that after having trespassed this 30% limit, it becomes extremely difficult to return to a non-hateful society within the time scope of our simulation. Thus, although there may be some exceptions (as hate spread could still be stopped if strategical nodes with high influence within a hateful group were convinced to become non-hateful), we consider swap to a hateful society as the outcome of an irreversible process, which destabilises the society in a very severe way.

4. Our Multi-Agent Social Network: The Baseline Model

Agent-Based Modelling (ABM) [14] has been successfully used in social sciences to simulate and study social systems from the complex adaptive system perspective [37]. It is especially suited for those cases where it is difficult to predict the future behaviour of the whole system analytically, although insights of isolated behaviour of individuals are available. Considering this, we resort to agent-based modelling to simulate a social network where users distribute content. In this manner, each user—which formally corresponds to a vertex in the graph—is modelled as an agent, and agents interact by distributing content with their peer agents—edges in E. The type of distributed content depends on the user profile, which can be normal or hateful. In what follows, Section 4.1 details how such users are added and connected in the network. Then, Section 4.2 exploits the insights from the previous research briefly introduced in Section 2 to model the spread of hatred. Our aim is to ensure that our model is able to mimic those findings from previous work. We refer to the resulting model as the *baseline model* to stress the fact that subsequent sections enrich this baseline model with different countermeasures and study their effectiveness in containing the spread of hatred.

4.1. Network Construction

Initially, we create a very small predefined clique of two nodes. A clique is a maximal complete subgraph of a graph [38]. In this manner, the two distinct vertices in the clique are adjacent. Subsequently, we apply a network growth process iteratively, by connecting new nodes (i.e., agents/users) to an existing social network, which was grown in previous iteration steps. Following this process, nodes are created and connections are spawned to existing nodes according to some rule.

In particular, we reproduce the structure of a social network by applying the *preferential attachment* iterative method [39]. Figure 4 shows an example of the simulated network we build during this phase. Briefly, in each round, when joining the network, new users connect to existing users with a probability corresponding to the *node degree* (amount of followers). So that users with many followers are more likely to receive new followers. Since this preferential attachment method does not distinguish between different user profiles, we extend it for considering hateful users.

1. Firstly, we create, for every simulation round (tick), a new user node and assign it with a hate score that is sampled from the Gamma distribution $\Gamma(10, 25)$, the lavender (right-most) distribution depicted in Figure 5 (recall, from Section 3, that we do so to produce a proportion of about 1% of hateful users in the network). As a consequence of this hate score assignment, the new node becomes a normal user or a hater.
2. Secondly, for each tick, we also connect the newly created user node with some other users so to mimic their behaviour on Gab and Twitter as described in Section 2.1. Specifically, we proceed by defining several variables that help us tailor the network connections as follows:
   - When joining the network, a new hateful user creates twice new connections than a normal user. In our simulations, a hater sets $c_h = 2$ connections, whereas a normal user only establishes $c_n = 1$ connection (in the code, these limits are set with variables *n_following_conn_hater* and *n_following_conn_normal*, respectively). We adhere to Twitter's terminology and refer to the new created node as the *follower* and the node it connects to—i.e., the one being followed—as the *followee*.
   - A normal user connects to an existing user within the network according to the preferential attachment method, without considering its hate score. Conversely, a hateful user will prefer to attach to haters. In particular, a newly created hateful user opts in to connect to another hateful user with a probability $p_{h \to h} = 0.9$ (the arrow $\to$ in the notation indicates connection and, as for the code, this probability $p_{h \to h}$ appears as *p_hater_follows_hater*), and thus, it can still connect to a normal user with a probability of $p_{h \to n} = 0.1$. Preferential attachment is then used to choose the specific hater to connect to. As a response, the hateful followee spawn s a following connection with the same probability $p_{h \leftarrow h} = 0.9$ (the arrow $\leftarrow$ in the notation indicates following back and, as for the code, this probability $p_{h \leftarrow h}$ appears as *p_hater_back_follows_hater*).
   - Hateful users receive less followers from normal users per time interval. Hence, following back by normal users is modelled with a probability $p_{n \leftarrow n} = 0.8$ and $p_{h \leftarrow n} = 0.4$. Lastly, haters will be less likely to follow back normal users with $p_{n \leftarrow h} = 0.08$ (in the code, these probabilities appear as *p_normal_back_follows_normal*, *p_normal_back_follows_hater*, and *p_hater_back_follows_normal*, respectively).

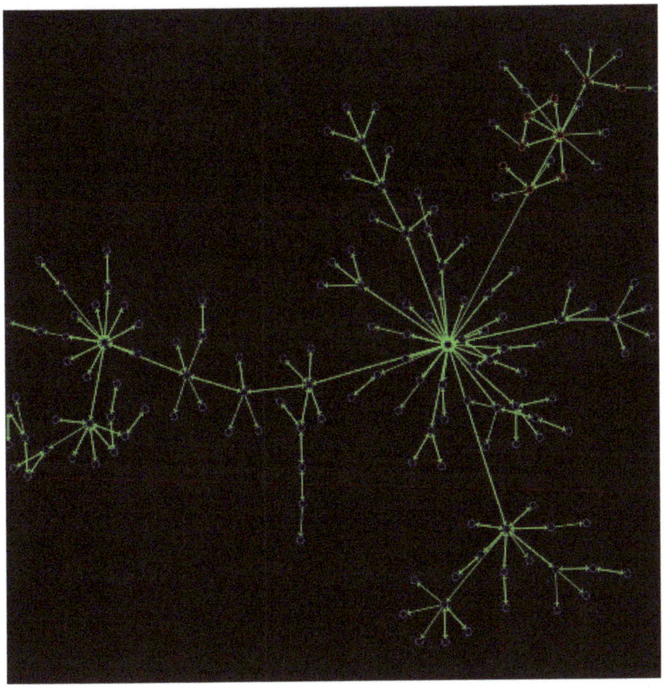

Figure 4. Representation of the network during the growth phase. Blue circles represent normal users, red circles correspond to hateful users. Arrows signal influence relations.

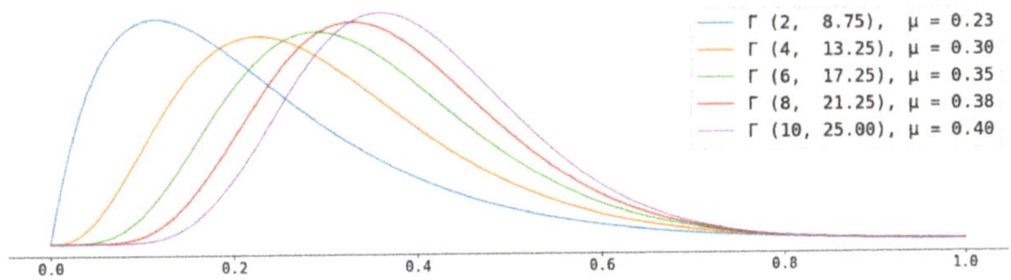

Figure 5. Alternative hate score probability distributions modelled with $\Gamma(\alpha, \lambda)$ distribution.

4.2. *Opinion Diffusion by Content*

The connections created in the network become the paths for opinion diffusion, since, if a user post some content, their followers will receive it and, as a result, may change their opinion. In the bounded confidence models described in Section 2.2, the influence of users is limited to its followers. However, in social networks such as Twitter, the content created by users can be reposted and, thus, arrive to and influence further audience.

Here we reuse the concept of confidence level, a threshold for opinion difference, defined in the Hegselmann–Krause (HK) model [29] and introduced in Section 2.2. Following the ideas of the heterogeneous HK model, we define the threshold τ_i for accepting the opinion of other users tailored for every user i depending on their hate score. The assumption is that: (i) extreme users tend to have rather fixed opinions, which cannot be changed by any influence; and (ii) alternatively, those with a middle hate score may be more open to accept opinions that differ from their own, and in fact could be still dragged to one of the extremes. Figure 6 depicts the function used for the threshold: extreme left and right sides of the possible hate score values exhibit fixed opinion (i.e., with a threshold of 0) whereas median users have a threshold of 0.49.

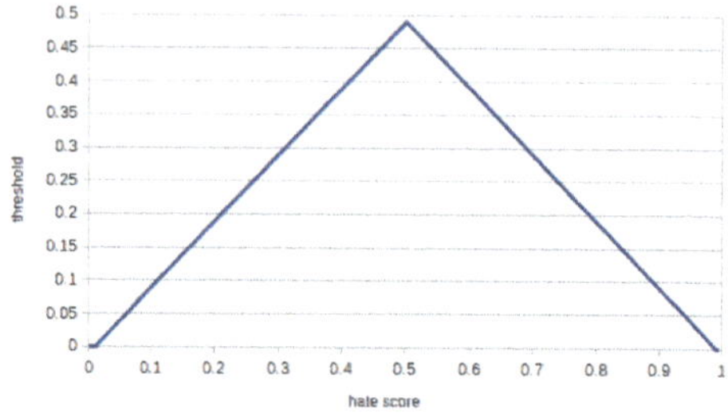

Figure 6. Threshold for accepting foreign opinion subject to the user's hate score.

Additionally, we borrow the formula of opinion adaption from the Deffuant–Weisbuch (DW) model [30] (see Section 2.2), but apply it to the author's opinion carried within a post. Thus, a post of user j will influence the opinion of user i by a factor $\mu = 0.05$ at the round k, if the difference of both opinions is below a confidence level τ_i:

$$x_{i,k} = x_{i,k-1} + \mu \cdot (x_{j,k-1} - x_{i,k-1}), \quad \text{iff} \quad |x_{i,k-1} - x_{j,k-1}| < \tau_i \tag{1}$$

where $x_{i,k}$ represents the opinion of user i at time k and the τ_i threshold is modelled as the previous triangular function on the users' opinion (hate score) in Figure 6.

As aforementioned, in our simulations of content diffusion, we consider that followers' opinions may change when the followees post new content but also when they repost. In particular characterise our diffusion model as follows:

- Hateful users are very active and post at every round (i.e., with a publication probability $p_{h\_pub} = 1$), whereas normal users only post with probability $p_{n\_pub} = 0.2$ (in the code, these probabilities appear as *p_publish_post_hater* and *p_publish_post_normal*, respectively).
- A post cannot be reposted twice by the same user. However, it can be reposted with some low probability even if the opinion does not correspond to reposter's own opinion. We align here with the retweet statistics provided by Ribeiro et al. [21] and set reposting probabilities between normal (n) and hateful (h) users to $r_{n \to n} = 0.15$, $r_{h \to h} = 0.45$, $r_{n \to h} = 0.05$ and $r_{h \to n} = 0.15$ (here, the arrow $\to$ indicates content flow and, in the code, these probabilities appear as *p_normal_reposts_normal*, *p_hater_reposts_hater*, *p_normal_reposts_hater*, and *p_hater_reposts_normal*, respectively). In this manner, a hater will repost a normal post with the lowest probability.
- In order to model different users' activity profiles, we limit the amount of reposts that a user can perform per round by setting variables to *max_reposts_by_normals* = 2 and *max_reposts_by_haters* = 6.

5. Modelling Countermeasures

This section describes how we enrich the baseline model in previous section with three alternative countermeasures aimed at containing the spread of hatred: *education, deferring hateful content*, and *counter activism*. The next subsections provide the necessary details about how these measures are modelled within our model.

5.1. Educational Bias

Education is considered the main long-term measure to counter the spread of hate speech. Indeed, advocates of free speech favour this measure over automatic filtering [40], which they strongly criticise. Education is aimed to develop critical thinking [10], enabling individuals to open discussions about any topic. Structured and argumented debates ought then lead to opinion forming and foster thinking [41]. Although education does not always result in tolerance (programmes fail to mitigate prejudice if they overlook factors such as values or ego defense [42]), from the perspective of the so called media literacy, education develops the skills to recognise hate speech itself [11]. Therefore, people need to be instructed in the usage and interpretation of modern digital media, and this is especially the case for youngsters.

Additionally, education can introduce an initial bias into the view of the population, e.g., by teaching democratic values and human rights [9]. However, rather than modelling how this positive bias is actually introduced (i.e., how education is implemented in the society), we can simply model its effect by skewing the hate score distribution used in the creation of the society. The mean value of the whole distribution should then move into the direction of non-hateful persons, hence decreasing the tendency towards the hatred. However, we assume that, despite the educational bias on the majority of the population, the group of very hateful persons will still be present in the population with the same proportion. So, the parameters of the Gamma probability distribution $\Gamma(\alpha, \lambda)$ are adjusted in such a way that the fraction of hateful persons stays invariably ca. 1% but their mean values μ are decreased. Figure 5 shows the baseline distribution $\Gamma(10, 25)$ and four further distributions with their corresponding mean values which vary from $\mu = 0.40$ down to $\mu = 0.23$. In this manner, we do not model how the positive bias is introduced into the population but take it for granted and simply apply the bias (i.e., the positively-skewed alternative distributions in Figure 5) to the population during the network construction phase (see Section 4.1).

5.2. Deferring Hateful Content

Hate motivated content seems to spread faster and farther through social networks than content generated by normal users. Thus, it can trespass community borders and reach out to wider audiences. The root of this behaviour seems to lie in the higher virality of the hate content, which is achieved, e.g., by usage of emotion triggering images [22]. Therefore, decelerating the publication of content or its visibility without filtering it out completely might already have a positive effect to deescalate conversations on polarised topics. This deceleration (or deferring) has the advantage of not infringing the freedom of speech as filtering and deleting of content would do.

Earlier theoretical work by Dharmapala and McAdam [43] proposed a utility-based model of hate speech influence that reveals that the perceived amount of hate speakers play a key role for motivating other users to join and engage in hate speech, making conversations more viral. It is also known that the lifetime of hateful conversations does not exceed a few days and has a culmination within the margins of one day [44]. This might be explained by people's fast emotional responses—triggered by some event—which settle down rapidly. Hence, we take the assumption that deferring hateful content (i.e., posts) might deescalate conversations on polarised topics due to decreased perceived amount of participants and stabilised emotional state of responders. As far as we know, such responses (i.e., reposting, replying or liking) give more weight to the content and lead to better promotion of it by internal algorithms of social networks [8,45]. In this work, the response to such content is interpreted as reposting and, thus, the countermeasure of deferring posts and deferring reposting should contribute to decrease the participation in "hot" topics as well as to make hateful content less viral. We model this countermeasure by decreasing the willingness of posting/reposting hateful deferred content: The longer it was deferred, the less the willingness to post/repost.

Deferring hateful content constitutes a short-term countermeasure that we apply during the opinion diffusion phase in our simulations (see Section 4.2). We implement this countermeasure by considering two variables:

- We employ a variable p_{defer} that stands for the probability of deferring a hateful post at each round (in the code, probability p_{defer} appears as *p_defer_hateful_post*). Any hateful post can be deferred again, if it is reposted in further rounds.
- In addition to parameters in Section 4.2, a cumulative factor $f\_repost\_deferred\_post = 0.5$ is used to decrease the probability of being reposted. This means that the proba-

bility of being reposted would diminish by a factor of 0.5 for posts deferred for one round, 0.25 for 2 rounds, 0.125 for 3 rounds and so on.

5.3. Counter Activism

Counterspeech is defined as a direct response to hateful or harmful speech which seeks to undermine it. Examples of counterspeech are the presentation of an alternative narrative, rebuking a person for inadequate expressions and convincing a person to change discourse [46].

Counterspeech can be organised by institutions or campaigns but can also be spontaneous, although counterspeech that is conducted by organised groups is associated with a more balanced discourse [7]. Overall, a large scale analysis of conversations related to current societal and political issues on German Twitter between 2013 and 2018 revealed a slight increase of counter tweets in recent years [7]. Additionally, investigations on the case of hate against the Roma Minority in Slovakia has shown that counterspeech can motivate other people to express their opinion against hate speech [47].

In addition to the expected positive effect of promoting anti-hate slogans and to spread positively influencing messages, counterspeech can have different outcomes [46]. On the one hand, extreme opinions of counter speakers tend to be rather ignored [35,48]. On the other hand, the conversation can escalate the hate and counter speakers may become victims themselves. This is especially the cases when utterances are perceived as a threat to one's own social identity [5]. Thus, moderate counterspeech may be more effective. Indeed, experiences from the activist group Red Levadura (https://redlevadura.net, accessed on 1 November 2021) reveal that emotional expression and empathy are the key factors for success of counterspeech [49]. In fact, counter speakers/activists have been identified as agreeable, altruistic, modest and sympathetic individuals [19].

Together, counter activists constitute a counter movement that act as the pole of '*the good*', which could subsume different counteractivities such as organised counterspeech or public awareness campaigns. In our model, counter activists spread positively influencing messages whose hate score are in the lowest values, so within the interval $[0, 0.25)$ from the default Gamma distribution $\Gamma(10, 25)$ in this work—the lavender (right-most) distribution depicted in Figure 5. As counter activists react against existing hate spreading groups, we implement them as a mid-term measure that starts during the opinion diffusion phase (see Section 4.2), where activists are sampled from the group of non-hateful persons with a probability $p_{convince}$ (in the code, probability $p_{convince}$ appears as *p_convincing_to_become_activist*), instead of from persons who are just joining the network. By default, their opinion is not fixed (however, activists' opinions can be fixed in the code by setting the *stubborn_activists?* variable to `true`) and can change due to opinion diffusion. When it exceeds *hate score* ≥ 0.25 they change to normal activity, but keep previously created connections. Furthermore:

- On becoming activist (denoted as a), a person spawns additional following connections c_a to the group of all activists (in the code, this c_a limit is set with the variable *n_following_conn_activist_additional*), which are answered with the probability $p_{a \leftarrow a} = 0.9$ (in the code, probability $p_{a \leftarrow a}$ appears as *p_activist_back_follows_activist*).
- Activists are as active as hateful users and, thus, they publish posts with the probability $p_{a\_pub} = 1$ at every round (in the code, probability p_{a_pub} appears as *p_publish_post_activist*).
- The maximal amount of reposts that an activist can perform per round is set to $m_a = 6$ so that they promote non-hateful content frequently (in the code, m_a appears as *max_reposts_by_activists*). However, they never repost any content of haters (and vice versa) and the reposting probabilities are $r_{a \to h} = r_{h \to a} = 0$, $r_{a \to a} = 0.45$ and $r_{a \to n} = r_{n \to a} = 0.15$ (in the code, these probabilities appear as *p_activist_reposts_activist* and *p_normal_reposts_activist*, respectively).

6. Simulation Results

We used NetLogo multi-agent modeling environment (http://ccl.northwestern.edu/netlogo/, accessed on 1 November 2021) to implement our model. The resulting simulator is publicly available (see Data Availability Statement) together with the tests we carried out to analyse the simulation results and the obtained data (associated tests and resulting simulation data are also publicly available).

As introduced by Section 4, the simulation is conducted in two phases. First, the network construction phase (see Section 4.1) is run until $t_1 \in [0, 500, 1000, 2000, 5000]$ rounds (ticks), which creates a network with about as many user nodes (more precisely, since we start with an initial network of two nodes and we add one network node per round, then the network finishes this phase having a size of $t_1 + 2$ nodes). Then, our opinion dynamics phase from Section 4.2 starts so that the network is grown for further 1000 rounds. Figure 7 depicts a screenshot of opinion diffusion in our simulation where both hateful and non-hateful posts are being distributed. Both posts and reposts are represented as solid circles, so our visualization (the graph layout algorithm is based on the Fruchterman–Reingold layout algorithm [50]) does not distinguish the publishing of original content from its subsequent reposting. Moreover, the direction of links indicates influence: we signal that a followee user A influences its follower B by a directed arrow (i.e., a link in the model) that goes from A to B. Thus, the amount of out-going links (i.e., the out-degree) of a user node corresponds to its amount of followers.

Each simulation is conducted 100 times for building the following average metrics:

- Fractions of normal and hateful persons, which correspond, respectively, to the proportion of the normal and hateful persons over the whole network population.
- Fractions of normal and hateful posts: the proportion of posts authored by hateful and normal users over the whole amount of posts that exist at the current round. Notice that the amount of posts per round can be much higher than the amount of existing persons, because of the possibility to repost multiple posts from the own neighbours.
- Mean and standard deviation of hate score distribution within the society.
- Ratio of network densities of hateful over normal users, which shows how much the group of hateful users is more cohesive than the group of normal users. Network density for each group is computed as the division of *actual_connections/potential_connections*, where $potential\_connections = n(n-1)/2$ and n is the network population. The overall ratio is then computed as the division of both network densities: *densities_ratio = network_density_haters/network_density_normals*.
- Reciprocity of following within normal or hateful users. When two users follow each other (i.e., a followee back-follows its follower) we count these two links as one reciprocal connection. Then, we compute the number of reciprocal connections divided by all connections within a specific group—be it the haters or the group of normal users.
- Mean followers and mean followees, e.g., mean followers corresponds to the average amount of out-going influence connections over a group of (hateful or normal) persons.
- Mean ratio follower/followee: the average of *out-going/in-coming* influence connections. This metric shows the connectivity profile in terms of following relations and is of interest because haters are known to have less followers than the following connections their create.
- Mean path length of reposts through the network: the average over all post path lengths through the network. It is computed considering that each post generates multiple paths if reposted by different followers.
- Fraction of swaps to a hateful society: proportion of runs which end with a swap (i.e., having more than 30% of hateful users, see Section 3). Those are not taken into account for none of the above metrics due to the instability they introduce. Instead, they are tracked separately through this specific metric.

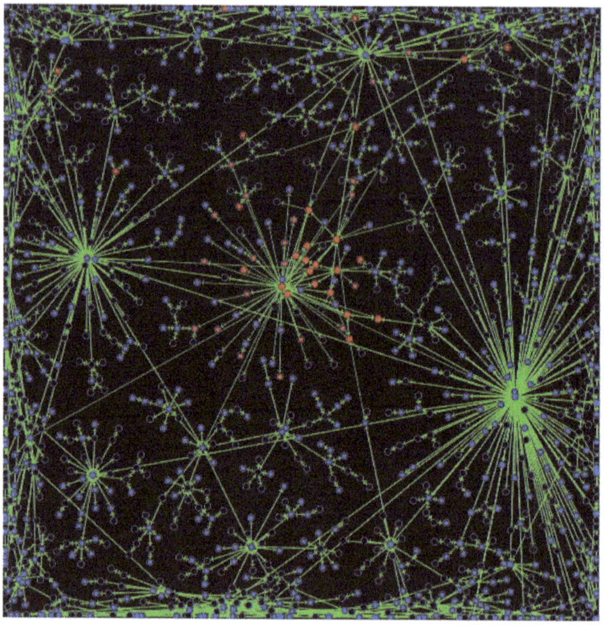

Figure 7. Representation of the network during the opinion diffusion by content. Blue and red dots represent content originally authored by normal and hateful users, respectively. In the center we can clearly see a densely connected hateful network component, which is infecting an influential normal user with hateful content. Infected influential users serves then as a proxy to spread content.

6.1. Validating the Baseline Model

Validation of the baseline model is an important step for this work, since it normalises the simulation with real statistics on hateful users. Regarding the first phase of network construction, multiple metrics could be satisfactorily reproduced in accordance to the state-of-the-art. However, runs resulted in extremely high network density ratios of hateful users over normal users which turned out to be ca. 11 times more than reported by [22]. Additionally, the amount of followers as well as the ratio between followers and followees of haters were too high compared to normal users. Subsequently, during the second phase of opinion diffusion, these metrics decreased until being very close to the reported values for higher network sizes. Table 1 details a subset of the resulting average values for simulations with varying number of ticks. As signaled in red, only the reciprocity among hateful users were too low compared to normal users. Although this might be repaired by introducing additional rewiring rules for users who switch from normal to hateful state, we advocate for the simplicity of the model and leave this for future work.

Simulation results also show an interesting fact in comparison between the phases of network growth without and with opinion diffusion. The switch from one phase to another demarcates a structural change in the sub-network of hateful users. It allows hate cores to disseminate hatred in strains to normal users with lesser network densities, showing that true hate cores might be even more densely connected than reported by statistics about real social networks. Overall, from these results, we can argue that our simulation represents hateful behaviours in a reasonable way.

Table 1. Simulation metrics for network growth with opinion diffusion by content. Diffusion dynamics are carried out for further 1000 ticks after starting. H and N stand for hateful and normal users, respectively. Red numbers in reciprocity rows signal values whose relation to each other is different than reported in the literature, whereas the remaining black numbers correspond to those similar to the reported values.

Metric	Opinion Diffusion Starts after t Ticks				
	0	500	1000	2000	5000
Fraction of H users	0.024	0.027	0.039	0.048	0.062
Fraction of posts by H	0.210	0.256	0.320	0.361	0.429
Ratio network density H/N	79.130	79.605	58.116	63.550	26.061
Reciprocity between N	0.888	0.886	0.888	0.887	0.889
Reciprocity between H	0.725	0.751	0.736	0.758	0.761
Mean followers of N	1.783	1.774	1.772	1.770	1.765
Mean followers of H	2.312	2.626	2.382	2.450	2.263
Mean followees of N	1.788	1.784	1.784	1.784	1.780
Mean followees of H	2.311	2.383	2.155	2.144	2.025
Mean follower/followee of N	0.884	0.880	0.883	0.878	0.878
Mean follower/followee of H	0.763	0.826	0.880	0.815	0.784
Mean path length N posts	0.699	0.693	0.706	0.704	0.705
Mean path length H posts	1.738	2.148	2.274	2.357	2.627

6.2. Countermeasures Simulation Results

Once we have been able to successfully replicate the behaviour of hateful users in a social network, we can now proceed to evaluate the influence of the three countermeasures we have implemented.

6.2.1. Educational Bias

As previously described in Section 5.1, simulations of the usage of education as a countermeasure are based on our definition of hateful user in Section 3. Specifically, we set a user to be hateful when its hate score is larger than a given threshold. As hate score is assigned considering a Gamma distribution, we then induce stronger positive bias in the users' hate score by decreasing the α parameter in the distribution (see Figure 5).

Our simulations show that the effect of this countermeasure is two-fold. On the one hand, considering the fraction of hateful persons, we can observe, from the top of Figure 8, that if we take as reference the hate scores set when $\alpha = 10$ in the Gamma distribution, then, all subsequent (lower) α values imply a substantial decrease in the amount of hateful persons, even if they are not completely removed from the society. Indeed, even with the strongest educational bias, which sets $\alpha = 2$ and reduces the fraction of hateful persons below 0.02 for all the considered network sizes, the amount of haters does not fall below of 1%. Similarly, as shown in the middle of Figure 8, the amount of hateful posts decreases down to below 0.27 for all network sizes when $\alpha = 2$. Additional tests show that the risk of swaps to a hateful society drops from 25% below 5% for the values of $\alpha = 6, 4, 2$ for all network sizes. These tests also show that the mean hate score have similar values for all network sizes and steadily decrease from ca. 0.4 for the original Gamma distribution (i.e., with $\alpha = 10$), down to a ca. 0.1 for $\alpha = 2$. This can be explained by the structure of the network that results from using preferential attachment, where some nodes have unproportionally higher influence. Hence, applying a skewed distribution upon it can skew the final distribution even more after opinion diffusion.

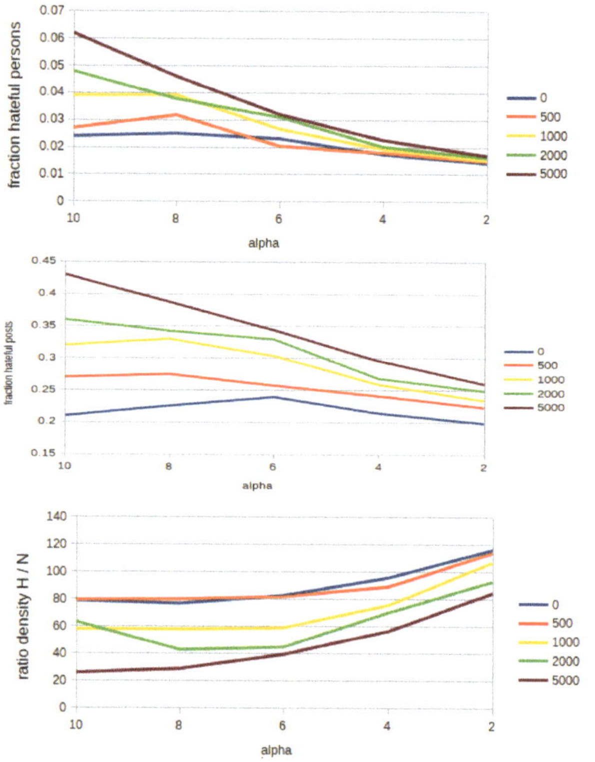

Figure 8. Effects of the education as countermeasure for different network sizes (0, 500, 1000, 2000, 5000) and depending on the α parameter of the Gamma distributions as shown in Figure 5: (**Top**) fraction of hateful users; (**Middle**) fraction of hateful posts; (**Bottom**) ratio of network densities of hateful to normal users.

On the other hand, the density among hateful users increases as depicted on the bottom of Figure 8. The same happens for the reciprocity and mean follower-followee ratio. This increase is due to the fact that education impedes the emergence of hate strains, so that hateful persons stay among like-minded within highly densely connected hate cores. Additionally, the mean path length of hateful posts increases linearly. This is a consequence of the fact that, although hate posts have much less room to unfold by reposting within hate strains (see Figure 3), hateful posts can still make very long paths by circulating posts between persons within a hate core (see Figure 1). Overall, and despite this increase of the density, the effect of the education countermeasure can be summarised as being very successful.

6.2.2. Deferring Hateful Content

As Section 5.2 details, the simulations aimed at studying the effect of deferring hateful content are conducted by varying the deferring probability p_{defer} and considering that a value of 0.7 deems realistic if we take into account the state-of-the-art accuracy in recognition of hate speech [51]. The obtained results show that, compared to the education, this countermeasure is less successful in decreasing the fraction of hateful persons as can be seen on the top of Figure 9, where a value of p_{defer} as high as 0.9 still results in a fraction of hateful users that varies from the ca. 4% for the largest network size down to a fraction of ca. 2% for the smallest one. Surprisingly, we can even observe some kind of reluctance and increase for $p_{defer} < 0.5$. As the middle of Figure 9 depicts, something similar happens with the fraction of hateful posts for the largest network size (i.e., 5000), although the main tendency for all the network sizes and probabilities is to decrease values.

As for the mean hate score, it follows a similar pattern, where highest mean hate score is ca. 0.425 for a $p_{defer} = 0.3$ in the 5000 network, and it goes down to ca. 0.39 for $p_{defer} = 0.9$ in all the simulated networks. However, this countermeasure has an obvious effect in decreasing the mean path length of hateful content from an initial range of ca. [1.75, 2.6] for a $p_{defer} = 0$ (i.e., without deferring) down to a range of ca. [0.25, 1.1] for a $p_{defer} = 0.9$ in all network sizes. More outstanding is its property in protection against swaps to a hateful society as shown on the bottom of Figure 8, as it goes below 0.025 for all network sizes when $p_{defer} = 0.9$.

Overall, we can observe that deferring results are similar to the ones from education, but they have the advantage of having a short-term effect. We argue this is very relevant because deferring content is aligned with the freedom of speech value, which is not the case of other short-term countermeasures such as automatic filtering.

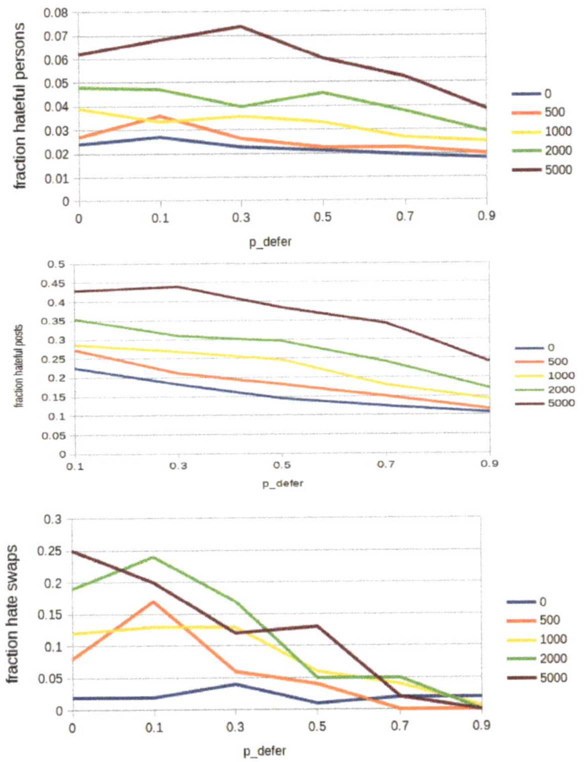

Figure 9. Effects of deferring hateful content as countermeasure for different network sizes (0, 500, 1000, 2000, 5000) and depending on p_{defer}, the probability of deferring hateful posts: (**Top**) fraction of hateful users; (**Middle**) fraction of hateful posts; (**Bottom**) fraction of swaps to a hateful society.

6.2.3. Counter Activism

In the case of counter activists (see Section 5.3), we used five different simulation setups with the aim of increasing the strength of the counter movement. Table 2 details how these setups where parameterised. Specifically, the strength of the counter movement strictly increases from setups #1, #2, #4, and #5 due to subsequent increases of the convincing probability $p_{convince}$, the number of activists' connections c_a, the inclusion of stubbornness, and the selection of activists by their influence, respectively. Setup #3 is in fact a variation of setup #2, as it has higher convincing probability but less connections.

Surprisingly, none of those simulations lead to a clear decrease of hateful users as depicted on the top of Figure 10, where values fluctuate within the [0.025, 0.08] interval for the different network sizes. Exceptionally, a decrease could be only recorded for settings with bigger networks over 5000 users in setups 2–4. The same happens to the fraction of swaps to a hateful society, whose values fluctuate in the [0.03, 0.41] interval for different setups and network sizes.

Even so, a drop of the mean hate score was recorded—especially for the settings with stubborn activists—as seen on the middle of Figure 10, the mean hate score drops from ca. 0.4 to values within the interval [0.21, 0.32]. Thus, activists seem to create higher polarisation within the society by dragging some persons into the positive direction without affecting hateful persons. This depletes representatives of the median opinion, so that people with higher hate scores are rather attracted by very hateful users.

Additionally, the bottom of Figure 10 plots an increase in the mean path length of activists' posts from values below 0.9 up to values of ca. 3.4. This can be attributed to the strengthening of the counter movement and replicates the tendency of activists to mimic the behaviour of hateful users. Overall, our simulations seem to suggest that activism needs to be carried out in a very sensible way. Otherwise, it may not lead to the desired results.

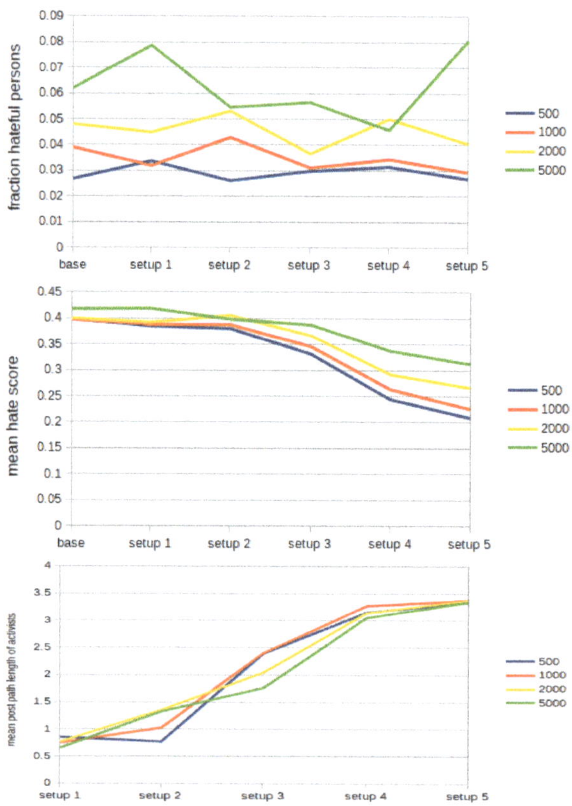

Figure 10. Effects of counter activists as countermeasure for different network sizes (0, 500, 1000, 2000, 5000) and for the five different setups from Table 2: (**Top**) fraction of hateful users; (**Middle**) mean hate score; (**Bottom**) mean path length of activists' posts.

Table 2. Experiment settings for activists' countermeasure.

	Experiment Setup				
	# 1	# 2	# 3	# 4	# 5
Convincing probability $p_{convince}$	0.01	0.01	0.04	0.01	0.01
Additional connections to other activists c_a	1	2	1	2	2
Fixed opinion (stubbornness)	false	false	false	true	true
Select activists by their influence	false	false	false	false	true

7. Conclusions and Future Work

This paper proposes a multi-agent model of the spread of hatred within social networks. We base our modelling on insights from previous research and take these as reference to successfully validate the resulting model, the so-called baseline model. Then, we enrich it by adding three countermeasures—education, deferring hateful content and conter activism—and conduct a series of experiments to assess their effectiveness in containing the spread of hatred. As a result, we conclude that: (i) Education proves to be very successful long-term countermeasure, although it still cannot eliminate hatred completely; (ii) Deferring hateful content shows a similar (lower) positive effect, but it also has the advantage of being a short-term countermeasure; (iii) Cyber counteractivism needs to be carefully articulated, as it can increase the society polarisation. Additionally, our simulations seem to indicate that hate cores—which are responsible for hate spread—in real-world social networks might be even more densely connected than reported by current statistics.

As future work, we plan to further refine our model to dive deeper in the study of counteractivism and the other implemented countermeasures. In addition, we find it particularly interesting to model the effects on the content receivers by differentiating passive listeners from the victims of the hateful content. Moreover, we also plan to incorporate additional countermeasures such as awareness campaigns or tight community management.

Author Contributions: Conceptualization, M.L.-S.; methodology, M.L.-S.; software, A.M.; validation, A.M. and M.L.-S.; formal analysis, A.M.; investigation, A.M.; resources, A.M.; writing—original draft preparation, M.L.-S. and A.M.; writing—review and editing, M.L.-S.; visualization, A.M.; supervision, M.L.-S.; project administration, M.L.-S.; funding acquisition, M.L.-S. All authors have read and agreed to the published version of the manuscript.

Funding: This research was funded by projects: 2017 SGR code 341 from the Agència de Gestió d'Ajuts Universitaris i de Recerca (AGAUR) from the Generalitat de Catalunya; MISMIS (Desinformación y agresividad en Social Media: Analizando el lenguaje, code PGC2018-096212-B-C33) from the Spanish Ministerio de Ciencia, Innovación y Universidades; Crowd4SDG (Citizen Science for Monitoring Climate Impacts and Achieving Climate Resilience, code H2020-872944) from the European Union; CI-SUSTAIN (Advanced Computational Intelligence Techniques for Reaching, code PID2019-104156GB-I00) from the Spanish Ministerio de Ciencia e Innovación; COREDEM (The Influence of Complex Reward Computation and Working Memory Load onto Decision-Making: A combined Theoretical, Human and Non-human primate approach code SGA2H2020-785907 within the Human Brain Project) from the European Union; and nanoMOOC (Nou format audiovisual amb funcionalitats tecnològiques avançades per a l'aprenentatge code COMRDI18-1-0010-02) from Agència de Suport a l'Empresa Catalana.

Institutional Review Board Statement: Not applicable.

Informed Consent Statement: Not applicable.

Data Availability Statement: Simulations in the paper were conducted with a NetLogo model developed by the authors that is publicly available from http://www.maia.ub.es/~maite/Students.html or directly at http://www.maia.ub.es/~maite/thesis/network_growth.nlogo. The data reported in the paper was generated during the study by using that simulation. Data and associated tests are also publicly available at https://github.com/agrizzli/simulation_countering_hate_speech, all accessed on 1 November 2021.

Conflicts of Interest: The authors declare no conflict of interest. The funders had no role in the design of the study; in the collection, analyses, or interpretation of data; in the writing of the manuscript, or in the decision to publish the results.

Abbreviations

The following abbreviations are used in this manuscript:

ABM	Agent-Based Modelling
ACG	Average Consensus Gossiping
a.k.a	also kwnon as
ca.	circa (about)
e.g.	exempli gratia (for example)
et al.	et alia (and others)
DW	Deffuant–Weisbuch
HK	Hegselmann–Krause
i.e.	id est (this is)

References

1. Bilewicz, M.; Soral, W. Hate speech epidemic. The dynamic effects of derogatory language on intergroup relations and political radicalization. *Political Psychol.* **2020**, *41*, 3–33. [CrossRef]
2. Cohen-Almagor, R. Fighting hate and bigotry on the Internet. *Policy Internet* **2011**, *3*, 1–26. [CrossRef]
3. Jane, E.A. 'Back to the kitchen, cunt': Speaking the unspeakable about online misogyny. *Continuum* **2014**, *28*, 558–570. [CrossRef]
4. Li, Q. Cyberbullying in schools: A research of gender differences. *Sch. Psychol. Int.* **2006**, *27*, 157–170. [CrossRef]
5. Kümpel, A.S.; Rieger, D. *Wandel der Sprach-und Debattenkultur in sozialen Online-Medien: Ein Literaturüberblick zu Ursachen und Wirkungen von inziviler Kommunikation*; Konrad-Adenauer-Stiftung e. V.: Berlin, Germany, 2019.
6. Mathew, B.; Illendula, A.; Saha, P.; Sarkar, S.; Goyal, P.; Mukherjee, A. Hate begets hate: A temporal study of hate speech. *Proc. ACM Hum.-Comput. Interact.* **2020**, *4*, 1–24. [CrossRef]
7. Garland, J.; Ghazi-Zahedi, K.; Young, J.G.; Hébert-Dufresne, L.; Galesic, M. Countering hate on social media: Large scale classification of hate and counter speech. *arXiv* **2020**, arXiv:2006.01974.
8. O'Callaghan, D.; Greene, D.; Conway, M.; Carthy, J.; Cunningham, P. Down the (white) rabbit hole: The extreme right and online recommender systems. *Soc. Sci. Comput. Rev.* **2015**, *33*, 459–478. [CrossRef]
9. Keen, E.; Georgescu, M. Bookmarks: Manual for Combating Hate Speech through Human Rights Education. Council of Europe. 2016. Available online: https://www.coe.int/en/web/no-hate-campaign/compendium/-/asset_publisher/PyHuON7WYezs/content/-bookmarks-a-manual-for-combating-hate-speech-online-through-human-rights-education-?inheritRedirect=false (accessed on 8 November 2012).
10. Isasi, A.C.; Juanatey, A.G. Hate Speech in Social Media: A State-of-the-Art Review. Available online: https://ajuntament.barcelona.cat/bcnvsodi/wp-content/uploads/2017/01/Informe_discurso-del-odio_ENG.pdf (accessed on 8 November 2012).
11. Gagliardone, I.; Gal, D.; Alves, T.; Martinez, G. *Countering Online Hate Speech*; Unesco Publishing: Paris, France, 2015.
12. Johnson, N.; Leahy, R.; Restrepo, N.J.; Velasquez, N.; Zheng, M.; Manrique, P.; Devkota, P.; Wuchty, S. Hidden resilience and adaptive dynamics of the global online hate ecology. *Nature* **2019**, *573*, 261–265. [CrossRef]
13. Chandrasekharan, E.; Pavalanathan, U.; Srinivasan, A.; Glynn, A.; Eisenstein, J.; Gilbert, E. You cannot stay here: The efficacy of reddit's 2015 ban examined through hate speech. *Proc. ACM Hum.-Comput. Interact.* **2017**, *1*, 1–22. [CrossRef]
14. Van Dam, K.H.; Nikolic, I.; Lukszo, Z. *Agent-Based Modelling of Socio-Technical Systems*; Springer Science & Business Media: Berlin/Heidelberg, Germany, 2012; Volume 9.
15. Müller, A.; Lopez-Sanchez, M. Countering Negative Effects of Hate Speech in a Multi-Agent Society. *Front. Artif. Intell. Appl. Artif. Intell. Res. Dev.* **2021**, *339*, 103–112.
16. Sulis, E.; Terna, P. An Agent-based Decision Support for a Vaccination Campaign. *J. Med. Syst.* **2021**, *45*, 1–7. [CrossRef] [PubMed]
17. Le Page, C.; Bazile, D.; Becu, N.; Bommel, P.; Bousquet, F.; Etienne, M.; Mathevet, R.; Souchere, V.; Trébuil, G.; Weber, J. Agent-based modelling and simulation applied to environmental management. In *Simulating Social Complexity*; Springer: Berlin/Heidelberg, Germany, 2013; pp. 499–540.
18. Ahrweiler, P.; Schilperoord, M.; Pyka, A.; Gilbert, N. Modelling research policy: Ex-ante evaluation of complex policy instruments. *J. Artif. Soc. Soc. Simul.* **2015**, *18*, 5. [CrossRef]
19. Mathew, B.; Kumar, N.; Goyal, P.; Mukherjee, A. Interaction dynamics between hate and counter users on Twitter. In Proceedings of the 7th ACM IKDD CoDS and 25th COMAD, Hyderabad, India, 5–7 January 2020; pp. 116–124.
20. Frischlich, L.; Schatto-Eckrodt, T.; Boberg, S.; Wintterlin, F. Roots of incivility: How personality, media use, and online experiences shape uncivil participation. *Media Commun.* **2021**, *9*, 195–208. [CrossRef]
21. Ribeiro, M.; Calais, P.; Santos, Y.; Almeida, V.; Meira, W., Jr. Characterizing and detecting hateful users on twitter. In Proceedings of the International AAAI Conference on Web and Social Media, Palo Alto, CA, USA, 25–28 June 2018; Volume 12.

22. Mathew, B.; Dutt, R.; Goyal, P.; Mukherjee, A. Spread of hate speech in online social media. In Proceedings of the 10th ACM Conference on Web Science, Boston, MA, USA, 30 June–3 July 2019; pp. 173–182.
23. Ling, C.; AbuHilal, I.; Blackburn, J.; De Cristofaro, E.; Zannettou, S.; Stringhini, G. Dissecting the Meme Magic: Understanding Indicators of Virality in Image Memes. *arXiv* **2021**, arXiv:2101.06535.
24. Soral, W.; Bilewicz, M.; Winiewski, M. Exposure to hate speech increases prejudice through desensitization. *Aggress. Behav.* **2018**, *44*, 136–146. [CrossRef]
25. Calvert, C. Hate speech and its harms: A communication theory perspective. *J. Commun.* **1997**, *47*, 4–19. [CrossRef]
26. Dimakis, A.G.; Kar, S.; Moura, J.M.; Rabbat, M.G.; Scaglione, A. Gossip algorithms for distributed signal processing. *Proc. IEEE* **2010**, *98*, 1847–1864. [CrossRef]
27. DeGroot, M.H. Reaching a consensus. *J. Am. Stat. Assoc.* **1974**, *69*, 118–121. [CrossRef]
28. Friedkin, N.E.; Johnsen, E.C. Social influence and opinions. *J. Math. Sociol.* **1990**, *15*, 193–206. [CrossRef]
29. Hegselmann, R.; Krause, U. Opinion dynamics and bounded confidence models, analysis, and simulation. *J. Artif. Soc. Soc. Simul.* **2002**, *5*, 1–33.
30. Weisbuch, G. Bounded confidence and social networks. *Eur. Phys. J. B* **2004**, *38*, 339–343. [CrossRef]
31. Terizi, C.; Chatzakou, D.; Pitoura, E.; Tsaparas, P.; Kourtellis, N. Angry Birds Flock Together: Aggression Propagation on Social Media. *arXiv* **2020**, arXiv:2002.10131.
32. Jager, W.; Amblard, F. Uniformity, bipolarization and pluriformity captured as generic stylized behavior with an agent-based simulation model of attitude change. *Comput. Math. Organ. Theory* **2005**, *10*, 295–303. [CrossRef]
33. Sherif, M.; Hovland, C.I. *Social Judgment: Assimilation and Contrast Effects in Communication and Attitude Change*; Yale University Press: London, UK, 1961.
34. Stefanelli, A.; Seidl, R. Opinions on contested energy infrastructures: An empirically based simulation approach. *J. Environ. Psychol.* **2017**, *52*, 204–217. [CrossRef]
35. Schieb, C.; Preuss, M. Considering the Elaboration Likelihood Model for simulating hate and counter speech on Facebook. *SCM Stud. Commun. Media* **2018**, *7*, 580–606. [CrossRef]
36. Petty, R.E.; Cacioppo, J.T. The elaboration likelihood model of persuasion. In *Communication and Persuasion*; Springer: Berlin/Heidelberg, Germany, 1986; pp. 1–24.
37. Janssen, M.A. Agent-based modelling. *Model. Ecol. Econ.* **2005**, *155*, 172–181.
38. Moon, J.W.; Moser, L. On cliques in graphs. *Isr. J. Math.* **1965**, *3*, 23–28. [CrossRef]
39. Barabási, A.L.; Albert, R. Emergence of scaling in random networks. *Science* **1999**, *286*, 509–512. [CrossRef] [PubMed]
40. Llansó, E.J. No amount of "AI" in content moderation will solve filtering's prior-restraint problem. *Big Data Soc.* **2020**, *7*, 2053951720920686. [CrossRef]
41. Howard, J.W. Free speech and hate speech. *Annu. Rev. Political Sci.* **2019**, *22*, 93–109. [CrossRef]
42. Leets, L. Experiencing hate speech: Perceptions and responses to anti-semitism and antigay speech. *J. Soc. Issues* **2002**, *58*, 341–361. [CrossRef]
43. Dharmapala, D.; McAdams, R.H. Words that kill? An economic model of the influence of speech on behavior (with particular reference to hate speech). *J. Leg. Stud.* **2005**, *34*, 93–136. [CrossRef]
44. Liu, P.; Guberman, J.; Hemphill, L.; Culotta, A. Forecasting the presence and intensity of hostility on Instagram using linguistic and social features. In Proceedings of the International AAAI Conference on Web and Social Media, Palo Alto, CA, USA, 25–28 June 2018; Volume 12.
45. Hrdina, M. Identity, activism and hatred: Hate speech against migrants on Facebook in the Czech Republic in 2015. *Nase Spol.* **2016**, *1*. [CrossRef]
46. Wright, L.; Ruths, D.; Dillon, K.P.; Saleem, H.M.; Benesch, S. Vectors for counterspeech on twitter. In Proceedings of the First Workshop on Abusive Language Online, Vancouver, BC, Canada, 4 August 2017; pp. 57–62.
47. Miškolci, J.; Kováčová, L.; Rigová, E. Countering hate speech on Facebook: The case of the Roma minority in Slovakia. *Soc. Sci. Comput. Rev.* **2020**, *38*, 128–146. [CrossRef]
48. Schieb, C.; Preuss, M. Governing hate speech by means of counterspeech on Facebook. In Proceedings of the 66th Ica Annual Conference, Fukuoka, Japan, 9–13 June 2016; pp. 1–23.
49. De Franco, M. #DecidimFest 2019: Strategies and Alliances to Curb Hate and Fear in a Polarized World. 2020. Available online: https://meta.decidim.org/conferences/decidimfest2020/f/1390/meetings/1453 (accessed on 19 November 2020).
50. Fruchterman, T.M.; Reingold, E.M. Graph drawing by force-directed placement. *Softw. Pract. Exp.* **1991**, *21*, 1129–1164. [CrossRef]
51. Aluru, S.S.; Mathew, B.; Saha, P.; Mukherjee, A. Deep Learning Models for Multilingual Hate Speech Detection. *arXiv* **2020**, arXiv:2004.06465.

applied sciences

MDPI

Article

Probabilistic Models for Competence Assessment in Education

Alejandra López de Aberasturi Gómez *, Jordi Sabater-Mir and Carles Sierra *

Artificial Intelligence Research Institute (IIIA-CSIC), 08193 Barcelona, Spain; jsabater@iiia.csic.es
* Correspondence: alejandra@iiia.csic.es (A.L.d.A.G.); sierra@iiia.csic.es (C.S.)

Abstract: Probabilistic models of competence assessment join the benefits of automation with human judgment. We start this paper by replicating two preexisting probabilistic models of peer assessment (PG_1-bias and PAAS). Despite the use that both make of probability theory, the approach of these models is radically different. While PG_1-bias is purely Bayesian, PAAS models the evaluation process in a classroom as a multiagent system, where each actor relies on the judgment of others as long as their opinions coincide. To reconcile the benefits of Bayesian inference with the concept of trust posed in PAAS, we propose a third peer evaluation model that considers the correlations between any pair of peers who have evaluated someone in common: PG-bivariate. The rest of the paper is devoted to a comparison with synthetic data from these three models. We show that PG_1-bias produces predictions with lower root mean squared error (RMSE) than PG-bivariate. However, both models display similar behaviors when assessing how to choose the next assignment to be graded by a peer, with an "$RMSE$ decreasing policy" reporting better results than a random policy. Fair comparisons among the three models show that PG_1-bias makes the lowest error in situations of scarce ground truths. Nevertheless, once nearly 20% of the teacher's assessments are introduced, PAAS sometimes exceeds the quality of PG_1-bias' predictions by following an entropy minimization heuristic. PG-bivariate, our new proposal to reconcile PAAS' trust-based approach with PG_1-bias' theoretical background, obtains a similar percentage of error values to those of the original models. Future work includes applying the models to real experimental data and exploring new heuristics to determine which teacher's grade should be obtained next to minimize the overall error.

Keywords: peer assessment; multiagent system; probabilistic model; comparative analysis; Bayesian network

Citation: López de Aberasturi Gómez, A.; Sabater-Mir, J.; Sierra, C. Probabilistic Models for Competence Assessment in Education. *Appl. Sci.* **2022**, *12*, 2368. https://doi.org/10.3390/app12052368

Academic Editors: Aida Valls and Agostino Forestiero

Received: 15 December 2021
Accepted: 19 February 2022
Published: 24 February 2022

1. Introduction

Automated assessment and feedback of open-response assignments remain a challenge in computer science despite recent milestones in natural language processing. Competence assessment is a sensitive topic (current "AI Act" under discussion at the European Parliament pinpoints student assessment based on AI as high risk) with a clear impact on the certification of students as competent professionals and on their educational career progress.

Despite the efforts on opening the black box of neural networks, current neural models are rarely equipped with logical narratives of the decision chains that lead them to a final prediction or classification. Nevertheless, transparency and explainability are desirable requisites for automated assessment systems.

As a result, many researchers propose hybrid solutions combining the benefits of automation with human judgment. Specifically, they propose using peer assessments to help the teacher in the evaluation of students in large classrooms. Furthermore, numerous studies from the field of psychology point out that peer evaluation methods positively impact students' formative process, leading toself-reflection [1–3].

Among the hybrid solutions in the bibliography, there are many that make use of Bayesian models to infer the grade of an assignment given a list of peer assessments ([4–7]). On the other hand, Gutierrez et al. [8] successfully built a network of trust among peers and the teacher that determines the relative importance that each peer's opinion has in the computation of the assignment's grade. This second approach, called PAAS, is very attractive as it exploits the naturalness with which a class of students evaluating each other can be modeled through multiagent theory.

The main objective of this article was to conceive a model that benefits from the Bayesian tradition in peer-assessment models while taking advantage of the concept of trust proposed by Gutierrez et al. [8]. As a result, we present *PG*-bivariate, a Bayesian model that translates the notion of trust among reviewers to a Bayesian approach thanks to its use of correlations among graders as the main feature to be learned by the model. Once *PG*-bivariate was implemented, we set out to compare it with one of the aforementioned Bayesian models and with PAAS in different experimental contexts. Given its theoretical robustness and the fact that it had been tested on an overwhelming number of more than 63,000 peers, the Bayesian model we chose to reimplement was PG_1-bias [4].

The three models (*PG*-bivariate, PG_1-bias and PAAS) were then compared in the context of competency assessment. All of them use a probabilistic approach to estimate a probability distribution for each automatic grade. Their inputs are always peer assessments and a given percentage of the teacher's grades (ground truths). Despite their commonalities, the relying idea varies for each model: whilst PG_1-bias [4] and *PG*-bivariate are Bayesian network models, PAAS [8] applies multiagent system theory. More specifically, it is a competency assessment model: Given a community of agents and a human leader whose assessing criterion is to be mimicked, PAAS builds a matrix representing each agent's trust in the rest.

The contributions of this work include (1) a Python reimplementation of PAAS, (2) a Python reimplementation of PG_1-bias, (3) a new model that integrates PAAS' use of trust measures with the Bayesian approach of *PG*-bivariate, and (4) a comparative analysis among the three models using simulated data and homogeneous units of measurement of the performance of the models.

This paper is structured as follows: In Section 1.2, we briefly review the concept of Bayesian network. After describing the materials and methods in Section 2, we present our models in Section 3. We experimentally evaluate the three models and compare them in Section 4. The discussion is presented in Section 4. Finally, we conclude in Section 5.

1.1. Related Work

The statistical models we present in this paper are part of a tradition of algorithms focused on score prediction. For instance, Bachrach et al. [6] makes use of a Bayesian model to grade tests. This model has two parts: a part that estimates the probability that a participant p will know the correct answer to a question q, represented by a variable c_{pq}; and a second part that models the probability of each potential answer r_{pq} of participant p to question q as a variable that depends on the correct answer to question q, y_q, and also on c_{pq}.

On the other hand, Sterbini and Temperini [5] exploited peer-evaluation among students to support the teacher when grading open-answers. To do so, they represented each student as a triplet of discrete variables: knowledge, judgment, and correctness, which are interrelated. Each student is asked to choose the best of three peer answers, and her choice is influenced by the triplet representing the student. The correctness of the three peer answers presented to her also makes an impact on that choice.

De Alfaro and Shavlovsky [9] created a web-based tool for collaborative grading and evaluation of homework assignments. Similarly to our setup, students played in this platform both the role of graders and gradees. In addition, to encourage quality peer reviews, they made the final grade received by the students dependent on (i) the consensus grade computed for the student's submission, (ii) an accuracy grade that measures the precision of the student in grading submissions, and (iii) a helpfulness grade that measures how helpful the reviews written by the student were. Such precision was computed by comparing the grades assigned by the student with the grades given to the same submissions by other students.

Perhaps more related to Piech et al. [4]'s work, Ashley and Goldin [10] devised a hierarchical Bayesian model to mine peer assessment data. However, although they do account for graders' biases similarly to the models in this work, they mainly use the resulting posterior distribution over the conjoint parameter space to answer queries regarding the suitability of the employed rubric criteria.

1.2. Bayesian Networks

Conditional probability distributions allow decoupling of joint probability distributions into sequences of conditional probability distributions over lower-dimensional spaces and marginal probability distributions. This decomposition eases the reasoning about the model and allows to better incorporate domain expertise. The methodology that constructs joint probability distributions by applying the chain rule to conditional and marginal probability distributions is known as generative modeling (Figure 1).

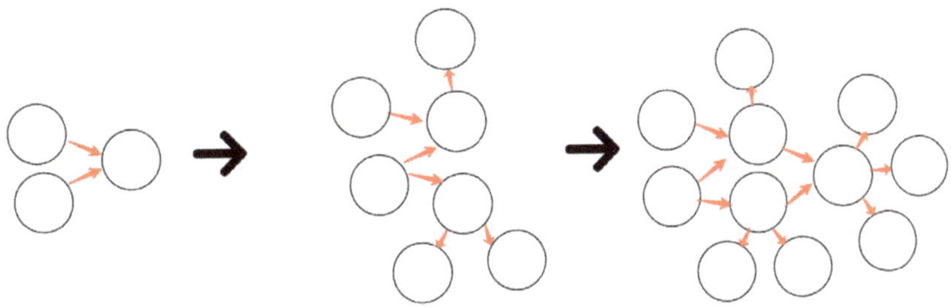

Figure 1. Iterative addition of variables to a generative model. This methodology constructs a joint probability distribution from intermediate conditional probability distributions.

Likewise, it is also possible to start with a coarse model represented by a joint probability distribution and then increase the complexity of the model by adding new variables as parents of that layer (Figure 2).

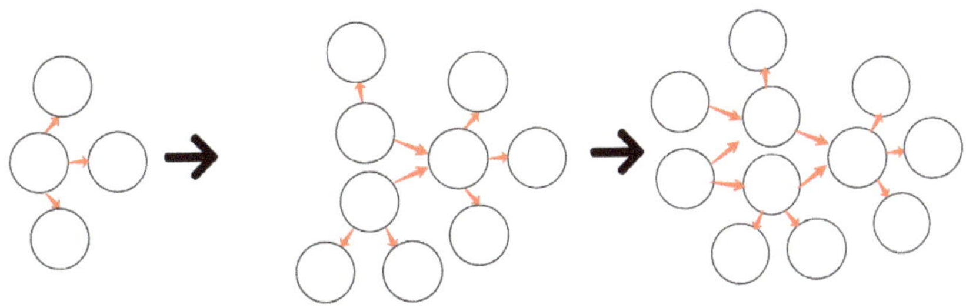

Figure 2. Reverse construction of a joint probability distribution. In this case, we begin with a marginal distribution over some few variables and then increase the complexity by conditioning this distribution over a new set of (ancestor) nodes.

A joint distribution that factorizes into marginal and conditional distributions can be represented by a probabilistic graphical model (PGM). In particular, this section will introduce the fundamentals of Bayesian networks, a kind of PGM that will be used during the description of two from three of the probabilistic models implemented for this research.

1.3. Representation

A Bayesian network (BN) is a directed acyclic graph $G = (V, E)$ and a set of parameters $\Theta = \{\theta_1, \theta_2, ..., \theta_n\}$ labeling the probability distributions associated with the nodes in the graph. This graph represents a decoupled joint probability distribution, such that

1. The vertices represent the random variables that we model.
2. For each vertex X_i, there is a conditional probability distribution $\pi(X_i|\mathbf{pa}_i)$.

The chain rule for Bayesian networks states that a Bayesian network $M = (G, \Theta)$ can be expressed as the product of marginal and conditional probability distributions associated to its nodes:

$$\pi_M(\overrightarrow{X}) = \prod_{i=1}^{n} \pi(X_i|\mathbf{pa}_i; \Theta_i) \tag{1}$$

where $\overrightarrow{X} = \{X_1, X_2, ..., X_n\}$ and $\mathbf{pa}_i$ stands for *parent* nodes and denotes the set of variables with a direct edge towards X_i. The symbol Θ_i represents the parameters of the probability distribution associated with that same vertex. If a node has no parents, then the probability distribution associated with it is marginal. As a side note, when depicting a Bayesian network, observed nodes will be shaded.

In addition, some representations will introduce the plate notation, a method of representing variables that repeat in a graphical model. Instead of drawing each repeated variable individually, a plate (rectangle) is used to group variables into a subgraph that repeat together, and a quantity is drawn on the plate to represent the number of times the subgraph repeats in the plate. It is assumed that the subgraph is duplicated that many times, the variables in the subgraph are indexed by the repetition number, and any links that cross a plate boundary are replicated once for each subgraph repetition.

Before deepening into the dependencies between variables in a Bayesian network, let us introduce some important concepts that will be used in the presentation of the models implemented in this research, namely parents, children, ancestors, ancestral ordering, and Markov blanket.

- Given a node X in a Bayesian network, its parent nodes are the set of nodes with a direct edge towards X.
- Given a node X in a Bayesian network, its children nodes are the set of nodes with an incoming edge from X.
- Given a node X in a Bayesian network, its ancestors are given by the set of all variables from which we can reach X through a directed, arbitrarily long path.
- Given the set of all the variables modeled in a Bayesian network, $X = \{X_1, X_2, ..., X_N\}$, an ancestral ordering of the variables is followed when traversing the network; if every time we reach a variable X, we have already visited its ancestors.
- Given a node X in a Bayesian network, its Markov blanket is given by its parents, its children, and the parents of its children.

1.4. Flow of Probabilistic Influence

The structure of a Bayesian network contains information regarding how variables in the model interact with each other. In other words, it is possible to determine whether the injection of information about a variable X updates our knowledge about another variable Y in the graph.

Considering the case of only two variables X and Y, if X is a parent of Y, then any update in the probability distribution associated with X will be reflected in changes in the probability distribution associated with Y. Likewise, knowing Y will update our information about X.

If we now think about three variables X, Y, W, three main situations can be distinguished:

1. If W is an intermediate node and all the edges go in the same direction (Figure 3), then an update in X will be reflected in Y if and only if W is not an observed variable, and vice versa: an update in Y will be reflected in X if and only if W is not an observed variable.
2. The same applies if W is a parent of two children X and Y (Figure 4). Again, there will be a flow of probabilistic influence from X to Y if and only if W is not observed.

3. Finally, if X and Y are parents of W (v-structure, Figure 5), then the situation reverses, and there is a flow of probabilistic influence from X to Y if and only if W is observed.

Figure 3. Case 1.

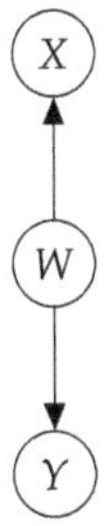

Figure 4. Case 2.

Figure 5. Case 3.

In short, it can be stated that in Bayesian networks, influence flow is stopped by observed nodes and nonobserved v-structures. A v-structure is observed if W or any of its descendants is observed.

Formally, we say that there is a flow of probabilistic influence from X to Y if there is an active trail from X to Y, where an active trail is defined as follows:

- Let $\mathcal{G}$ be a DAG.
- Let $X_1 \rightleftharpoons ... \rightleftharpoons X_2$ be a trail in $\mathcal{G}$.
- A trail is active given a set of observed variables W if
    1. Whenever there is a v-structure $X_{i-1} \rightarrow X_i \leftarrow X_{i+1}$, X_i or one of its descendants is in W.
    2. no other node along the trail is in W.

Let X, Y, and W be three disjoint sets of variables in $\mathcal{G}$. W d-separates X from Y in $\mathcal{G}$ if $XY|W$ holds in $\mathcal{G}$. $XY|W$ holds in $\mathcal{G}$ if there is no active trail between any variable in X and any variable in Y given W.

Given a variable X_j and its Markov blanket, $\mathbf{Mb}_j$, for any set of variables $\mathbf{X}_c \subseteq (\mathbf{V}\backslash\mathbf{Mb}_j)$:

$$\mathbf{X}_c X_j | \mathbf{Mb}_j \tag{2}$$

Hence, the Markov blanket of a variable in a BN d-separates that variable from the rest of the network.

2. Materials and Methods

All the analyses in this comparative study have been carried out with synthetic data. More specifically, for the generation of the data points injected into the Bayesian network models, ancestral sampling was used. According to this method, once a prior distribution $\pi_{\mathcal{S}}(\theta)$ and an observational model $\pi_{\mathcal{S}}(\tilde{y}|\theta)$ are specified, then we can iteratively generate an ensemble of reasonable model configurations and observations by sampling first from the prior

$$\tilde{\theta} \sim \pi_{\mathcal{S}}(\theta)$$

and then observations from the corresponding data generating process,

$$\tilde{y} \sim \pi_{\mathcal{S}}(\tilde{y}|\tilde{\theta})$$

Each simulated data point $\tilde{y}$ adds an independent sample to the synthetic database. This database is then used as input for the model to fit. Please refer to Section 3 for further information about the parameters of the Bayesian network models presented in this paper. With regards to the non-Bayesian model (PAAS), its equivalence with the original Java implementation in Gutierrez et al. [8] was tested in Section 3.2.2. As for the data fed to it, we replicated the synthetic experiment presented in Gutierrez et al. [8] for a class of 50 students.

3. Results

3.1. Probabilistic Models of Peer Assessment

In the last decade, the boom of massive online courses (MOOCs) has propitiated that some platforms that provide online educational resources direct efforts towards implementing technologies that somehow automate the assessment process. The goal is to help teachers evaluate usually large numbers of assignments in this teaching modality.

The variability of answers to open questions and their challenges to natural language processing (NLP) techniques make automatic assessment a limited tool to deal with this task. The more unique or creative an assignment is, the less appropriate it is to rely on purely computer-based assessment methodologies [11]. Many authors have proposed peer assessment techniques as a promising alternative to speed up evaluation in online courses. Moreover, according to these voices, this methodology may provide other potential benefits such as helping students see the task from an assessor's perspective and boosting self-reflection, as well as providing valuable feedback [3,12].

The three models of peer assessment that were studied in this paper are presented in this section. All of them are probabilistic and focus on an educational context and they all address a common question: How can we generate an automatic quality assessment of a submission not yet evaluated by the teacher in charge?

3.1.1. Personalized Automated Assessments

Provided a community (a class) and a leader or special member from that community (the teacher), the aim of Gutierrez et al. [8] is to predict as accurately as possible the personalized assessments that a teacher in a class would make. The key idea behind this model is to unequally weigh the opinion of the members in the community (the students) so that the closer a student's marking style is to the teacher's style, the more relevant that student's opinion will be to predict the teacher's opinion.

The closeness in marking styles will be represented as a trust value. That is, the teacher's trust in a student depends on the similarity between the teacher's (past) assessments and the student's (past) assessments of the same assignments. In what follows, a one-to-one mapping between assessed students (*a.k.a.* gradees) and assignments is presumed. In other words, for all the experiments in this work, every student completed one and only one assignment (*a.k.a.* exam).

In the case there were no commonly assessed exams by the teacher and a given student, the authors propose a key concept: indirect trust. In short, an indirect trust measure is computed as the reputation of the student within the community, where this reputation is biased towards the teacher's perspective.

Model

Let ε represent a person who needs to assess a set $\mathcal{I}$ of objects and let $\mathcal{P}$ be a set of peers able to assess objects in $\mathcal{I}$. In the context of our analysis, this would be a teacher that wants to assess a number of assignments ($\mathcal{I}$) completed by students that are evaluated by other students ($\mathcal{P}$). Assessments made by peers $v \in \{\varepsilon\} \cup \mathcal{P}$ on an object $u \in \mathcal{I}$, noted z_u^v, are elements from an ordered evaluation space $\mathcal{E}$.

An automated assessment of ε's opinion on an object u, noted e_u^ε, is a probability distribution $\mathbb{P} = \{x_1 \mapsto \alpha_1, x_2 \mapsto \alpha_2, ..., x_n \mapsto \alpha_n\}$, where $x_i \in \mathcal{E}$ and $\alpha_i \in [0,1]$, with $\sum_i \alpha_i = 1$. A value α_i represents the probability that ε's true assessment of u is x_i. Hence, the more peaked the grading distribution on an object is, the more confident we will be that the automated assessment closely approaches ε's. Inversely, the flat, equiprobable distribution will be the one denoting ignorance of ε's true assessment.

Given a history of past peer assessments over u, $\mathcal{O}^u$, the ultimate goal of PAAS is to compute

$$\mathbb{P}(X_u^\varepsilon = x|\mathcal{O}^u)$$

That is, we aim to compute the probability distribution representing ε's evaluation on every object given the assessments of peers on that object.

To that end, every individual evaluation z_u^v from $\mathcal{O}^u$ is taken into account, and the probability $\mathbb{P}(X_u^\varepsilon = x|z_u^v)$ is computed as a function of the trust (expected similarity between previous assessments) that ε has on v.

In our context, this trust is computed using the assignments graded in common by the teacher, i, and a student, j. In the case that there are no assignments in common, an indirect trust measure is obtained based on the notion of transitive trust: the trust that i has on student j can be computed from the trust i has on student k and the trust that student k has on student j. As there are many possible paths connecting i to j, appropriate aggregation functions have to be provided.

Direct Trust

When two agents i and j have commonly assessed one or more objects, a direct trust relationship between them $\mathbb{T}_{i,j}$ can be computed. This direct trust is modeled as a probability distribution on the difference between the evaluations performed by i and j. This way, we keep information about whether j, for instance, underevaluates with respect to i, or overevaluates, or shows any other possible pattern. Information about their evaluation dissimilarities is summarized in that probability distribution.

The evaluation difference between two assessments performed by i and j is defined as

$$diff(i,j) = z_u^i - z_u^j$$

If $diff(i,j) > 0$, it follows from the above definition that agent i over rates u with respect to j. The opposite occurs if $diff(i,j) < 0$.

From the previous definitions, it is clear that a situation of complete agreement between two agents is represented by

$$\mathbb{T}_{i,j} = \{0 \mapsto 1\}$$

In what follows, we will denote that situation as $\mathbb{O}$.

Indirect Trust

If the leader has no objects assessed in common with a student j, a relation of indirect trust is computed. Since the model deals with probability distributions, it becomes necessary to define aggregation operators for probability distributions. These operators compute an indirect trust distribution from two trust distributions by combining differences in an additive way. The basic idea is that if the difference of opinions between k and j is x, and the difference of opinions between k and the teacher is y, then the difference of opinions between j and the teacher is $x + y$. Translating the above to probabilities and assuming independence of opinions, a combined distance distribution operator $\otimes$ can be defined as follows.

Definition 1. *Given trust distributions $\mathbb{P}$ and $\mathbb{Q}$ over the numeric interval $[-b, b]$ we define their combined distance distribution, noted $\mathbb{R} = \mathbb{P} \otimes \mathbb{Q}$, as:*

$$r(X = x) = \begin{cases} \sum\limits_{x_1+x_2=x} p(X = x_1) * q(X = x_2) & \text{if } x \in (-b, b) \\ \sum\limits_{x_1+x_2 \leq -b} p(X = x_1) * q(X = x_2) & \text{if } x \leq -b \\ \sum\limits_{x_1+x_2 \geq b} p(X = x_1) * q(X = x_2) & \text{if } x \geq b \end{cases} \tag{3}$$

This aggregation combines the distributions along a path between two peers.

From that definition, it follows that $\otimes$ is commutative, and its neutral element is the probability distribution representing the complete agreement between two peers ($\mathbb{O}$). In case there are several possible paths from the teacher to a student j, an aggregation operator $\bigoplus$ is defined to combine the aggregations computed along the different paths:

Definition 2. *Given probability distributions $\mathbb{P}$ and $\mathbb{Q}$ over the numeric interval $[a, b]$, we define $\mathbb{P} \oplus \mathbb{Q}$, as*

$$\mathbb{P} \oplus \mathbb{Q} = \arg \min_{\mathbb{T} \in \{\mathbb{P}, \mathbb{Q}\}} (EMD(\mathbb{T}, \mathbb{O})) \tag{4}$$

with EMD standing for Earth mover's distance (a.k.a Wasserstein distance). (Informally, if the distributions are interpreted as two different ways of piling up a certain amount of dirt over the region D, the EMD is the minimum cost of turning one pile into the other, where the cost is assumed to be the amount of dirt moved times the distance by which it is moved [13]).

This aggregation is optimistically performed; it assumes that the combined distance that is closer to $\mathbb{O}$ (i.e., the one that brings the student's and teacher's opinions closer) is the true one. This operator is both commutative and associative. Thus, the order in which distributions (i.e., paths) are combined is irrelevant.

Incremental Updates

Every time an object is assessed by a peer, PAAS launches an overall update of the trust values and hence of the student marks. This update can be simplified as follows:

1. Initially, the default direct trust distribution $T_{i,j}$ between any two peers i and j is the one describing ignorance (i.e., the flat equiprobable distribution $\mathbb{F}$). When j evaluates an object α that was already assessed by i, $T_{i,j}$ is updated as follows:
2. Let $\mathbb{P}(X_u = x)$ for $x = \mathit{diff}(i, j)$ be the probability distribution of the assessment difference between i and j. The new assessment must be reflected in a change in the probability distribution. In particular, $\mathbb{P}(X_u = x)$ is increased a fraction of the probability of X not being equal to x:

$$\mathbb{P}(X_u = x) = \mathbb{P}(X_u = x) + \gamma \cdot (1 - \mathbb{P}(X_u = x)) \tag{5}$$

For instance, if the probability of x is 0.6 and γ is 0.1, then the new probability of x becomes $0.6 + 0.1 \times (1 - 0.6) = 0.64$. As in the example, the value of γ must be closer to 0 than to 1, for considerable changes can only be the result of information learned from the accumulation of many assessments.

3. The resulting $T_{i,j}$ is then normalized by computing the distribution that respects the new computed value and has a minimal relative entropy with the previous probability distributions:

$$\begin{aligned} \mathbb{T}_{i,j}(X) = \operatorname{argmin}_{\mathbb{P}'} \mathbb{P}'(X) \textstyle\sum_{x'} p(X^{\alpha}x') \log \frac{p(X^{\alpha}x')}{p'(X^{\alpha}x')} \\ \text{such that} \quad \{p(X^{\alpha}x) = p'(X^{\alpha}x)\} \end{aligned} \tag{6}$$

where $p(X^{\alpha}x')$ is a probability value in the original distribution, $p'(X^{\alpha}x')$ is a probability value in the potential new distribution $\mathbb{P}'$, and $\{p(X^{\alpha}x) = p'(X^{\alpha}x)\}$ specifies the constraint that needs to be satisfied by the resulting distribution.
These direct trust distributions between peers are stored in a matrix $\mathcal{C}$.

4. To encode the decrease in the integrity of information with time *(information decays with time, I may be sure today about your high competence playing chess, but maybe in five years time I will be no longer sure if our interactions stop. You might have lost your abilities during that period)*, the direct trust distributions in $\mathcal{C}$ are decayed towards a decay limit distribution after a certain grace period. In our case, the limit distribution is the flat equiprobable $\mathbb{F}$. When a new evaluation updates a direct trust distribution $T_{i,j}$, $T_{i,j}$ is first decayed before it is modified.

5. The indirect trust distributions between ϵ and each peer are stored in a distributions vector $\overrightarrow{t_{\varepsilon}}$.
Initially, $\overrightarrow{t_{\varepsilon}}$ contains the probability distributions describing ignorance $\mathbb{F}$. When matrix $\mathcal{C}$ is updated, $\overrightarrow{t_{\varepsilon}}$ is also updated as a product of its former version times matrix $\mathcal{C}$:

$$t_{\epsilon,j}^{k+1} = \bigoplus_{0<i\leq n} \mathbb{T}_{i,j} \otimes \mathbb{T}_{\epsilon,i}^{k} \tag{7}$$

6. If a direct trust distribution $T_{\varepsilon,j}$ exists between ε and j, the indirect trust distribution $t_{\varepsilon,j}$ is overwritten with $T_{\varepsilon,j}$ after the update of the indirect trust distributions.

Please notice that the decay in the direct distributions $T_{i,j}$ affects the indirect distributions $t_{\varepsilon,j}$ as well. This is because the indirect trust distributions are computed as aggregations of combinations of direct trusts.

3.1.2. Tuned Models of Peer Assessment in MOOCs

Looking for an answer to the same question posed by PAAS, the authors from *Tuned Models of Peer Assessment in MOOCs* [4] propose a Bayesian network model to represent the variables affecting the grade that a peer gives to another. Bayesian inference differentiates itself from other forms of inference by extending probability theory to the parameter space.

In this case, the space over which the probability is allocated is $Z \times \Theta$, where Z is the observational space and Θ is the parameter space. Observations are denoted as $z_u^v \in Z$, where u refers to the gradee and v to the grader. On the other hand, Θ is the space parameter.

More specifically, PG_1-bias considers that z_u^v is influenced by the following:

1. The assignment's true score, $s_u \in \mathbb{R}$. In the case of the implementation presented here, this is the teacher's grade.
2. The grader's bias, $b_v \in \mathbb{R}$. This bias reflects a grader's tendency to either inflate or deflate their assessment by a certain number of percentage points. The lower these biases, the more accurate the grades will be.
3. The grader's reliability, $\tau \in \mathbb{R}$, reflecting how close on average a grader's peer assessments tend to land near the corresponding assignment's true score after having corrected for bias. In this context, reliability is a synonym for precision or inverse

variance of a normal distribution. Notice that the reliability of every grader is fixed to be the same value.

The posterior probability distribution is computed as the product of the prior over the parameters times the likelihood function evaluated on the observations (times a constant depending only on the observations):

$$\pi_{\mathcal{S}}(s_u, b_v, \tau | z_u^v) \propto \int_{\Theta} \pi_{\mathcal{S}}(s_u, b_v, \tau) \pi_{\mathcal{S}}(z_u^v | s_u, b_v, \tau) d\theta \tag{8}$$

The authors propose that the prior distribution $\pi_{\mathcal{S}}(s_u, b_v, \tau_v)$ can be decoupled into three marginal probability distributions:

$$\pi_{\mathcal{S}}(s_u, b_v, \tau) = \pi_{\mathcal{S}}(s_u) \pi_{\mathcal{S}}(b_v) \pi_{\mathcal{S}}(\tau) \tag{9}$$

That is, the parameters are mutually independent. These priors have the form, according to Piech et al. [4], of

$$\pi_{\mathcal{S}}(b_v) = \mathcal{N}(0, \frac{1}{\eta_0}) \tag{10}$$

$$\pi_{\mathcal{S}}(s_u) = \mathcal{N}(\mu_0, \frac{1}{\gamma_0}) \tag{11}$$

$$\pi_{\mathcal{S}}(\tau) = \mathcal{G}^{-1}(\alpha_0, \beta_0) \tag{12}$$

Whereas the likelihood function is given by

$$\pi_{\mathcal{S}}(z_u^v | s_u, b_v, \tau) = \mathcal{N}(s_u + b_v, \frac{1}{\tau}) \tag{13}$$

The resulting PGM is shown in Figure 6.

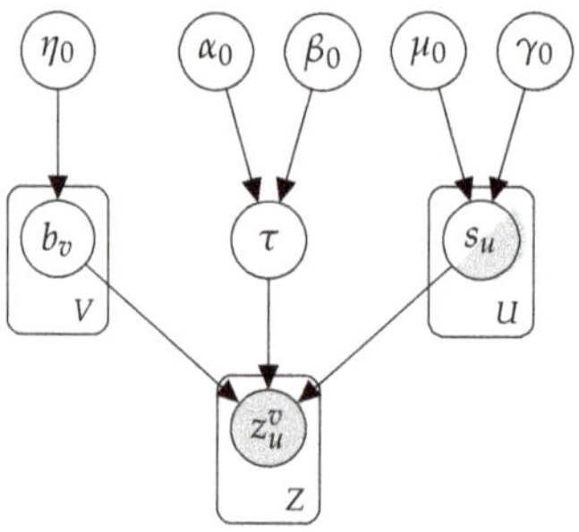

Figure 6. PG_1-bias [4].

Partially Known Parameters

As Figure 6 shows, once the evidence z_u^v is injected, there is an active trail towards s_v, which implies that we do not need an example of a teacher's grade to make inference about how they would grade any student in the class. This active trail does not mean, however, that we are not allowed to introduce some ground truths grades s_v in the Bayesian network. In that case, s_v is said to be a partially known parameter, and its injection should result in a reduction of the variance of the posterior distribution. Half-shaded nodes represent partially known parameters in *BN*s.

3.1.3. *PG*-Bivariate: A Bayesian Model of Grading Similarity

We contribute in this section with a novel Bayesian model of peer assessment, *PG*-bivariate. The approach adopted here tries to reconcile the benefits of Bayesian inference with the concept of trust posed in PAAS [8]. The key idea is to use a probability distribution to model the similarities as graders between any pair of peers in the system (teacher included). The chosen form for the probability density function modeling the similarities is a bivariate normal distribution. Figure 7 shows the graph representing the joint probability distribution of our model in a toy example with three students and a teacher, ϵ. The small, dark squares represent the relationship variables $\mathcal{R}$ between graders.

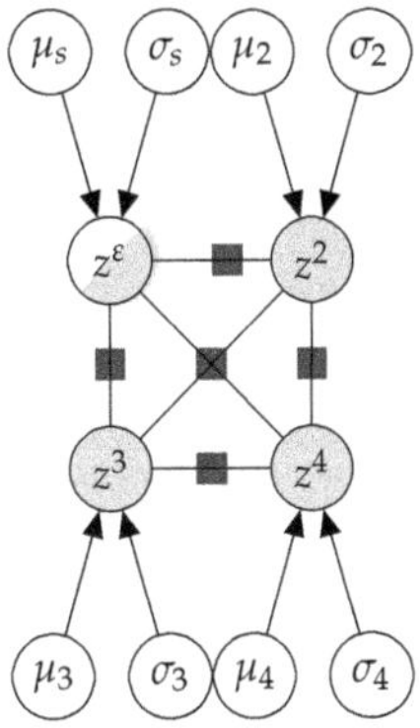

Figure 7. *PG*-bivariate.

More specifically, consider a class with S students and a teacher ε. Each student v_i assesses a number m_i of peer students. The allocation of assignments to graders is such that every student is assessed by the same number n of peers. Denoting the assessments by peer v as $\vec{z^v}$, we wish to compute the grades that the teacher would give to each student, $\vec{z^\epsilon}$. The model proposes that $\vec{z^v}$ is sampled from a normal probability density function with parameters μ_v and σ_v. We could add complexity to the model and encode causal reasoning over these parameters by populating the layers above μ_v and σ_v. We limit, however, to this basic level of characterization, without adding any more variables influencing location or scale. Please notice that in this case, z are vector variables containing the assessments of each peer to her assigned reviewees. In the case of z^ϵ, it is a partially known vector: making it a completely observed variable would imply that the teacher has assessed all the students in the class, which is not the case.

In line with the concept of direct trust coined by PAAS' authors, we define a bivariate vector formed by the set of common assessments $\{(z_i^{v_1}, z_i^{v_2}), (z_j^{v_1}, z_j^{v_2}), \ldots, (z_k^{v_1}, z_k^{v_2})\}$ between two peers v_1 and v_2, hereafter denoted as $\mathcal{R}_{v_1,v_2}$. This sequence is sampled from a bivariate normal distribution such that

- The location vector, $\vec{\mu}_{v_1 v_2} = (\mu_{v1}, \mu_{v2})$ is composed of each of the peers' location parameters separately.
- The covariance matrix $\Sigma_{v_1 v_2}$ contains the individual variances in its diagonal. The off-diagonal components codify the correlations between v_1 and v_2 when grading.

Putting it all together, the equations for *PG*-bivariate model are

$$z^{v_i} \sim \mathcal{N}(\mu_{v_i}, \sigma_{v_i}) \quad \forall v_i \tag{14}$$

$$\mathcal{R}_{v_i,v_j} \sim \mathcal{N}(\vec{\mu}_{v_i v_j}, \Sigma_{v_i v_j}) \quad \forall v_i, v_j \tag{15}$$

where $\vec{\mu}_{v_i v_j} = (\mu_{v_i}, \mu_{v_j})$, $\Sigma_{v_i v_j} = \begin{pmatrix} \sigma_{v_i}^2 & \sigma_{v_i v_j} \\ \sigma_{v_i v_j} & \sigma_{v_j}^2 \end{pmatrix}$, and (v_i, v_j) refers to a pair of graders having a set of evaluations in common.

We propose the following prior distributions for the hyperparameters:

$$\sigma_{v_i} \sim \text{Cauchy}(x_0, \gamma) \tag{16}$$

$$\mu_{vi} \sim \mathcal{N}(\bar{\preceq}, \bar{\supset}) \tag{17}$$

As for the modeling of the covariance matrix Σ, denoting by D a diagonal matrix whose elements are the square root of the elements in the diagonal of Σ, $D = \sqrt{Diag(\Sigma)}$, then the correlation matrix Ω is related to the covariance matrix Σ by

$$\Omega = D^{-1}\Sigma D^{-1} \tag{18}$$

Stan offers correlation matrix distributions, which have support on the Cholesky factors of correlation matrices. Cholesky's decomposition is unique: Given a correlation matrix Ω of dimension K, there is one, and only one, lower triangular matrix L such that $LL^T = \Omega$. We call such a matrix its Cholesky factor, and, according to Stan, even though models are usually conceptualized in terms of correlation matrices, it is better to operationalize them in terms of their Cholesky factors.

Hence, we used the prior

$$L \sim \text{LkjCorr}(\eta)$$

The observations in this system are vectors of peer grades $\vec{z^v}$, whereas the inference target is the vector of teacher grades, $\vec{z^t} = \vec{s}$. Please notice that in this case, no explicit mention is being made to the indirect trust. We limit our explicit modeling to the relationship between graders having a set of peers to assess in common. Nevertheless, from the graphical model, it can be seen that there exists a flow of probabilistic influence from any variable of type $\mathcal{R}$ to the rest of them. That is:, a change in the available information about the relationship between any pair of peers is reflected in an update of all the other variables $\mathcal{R}_{v_i v_j}$ for all pairs $\{v_i v_j\}$, including those containing the professor. In turn, this directs the flow of probabilistic influence towards the parameters of the distribution from which $\mathcal{R}_{v_i v_j}$ is sampled, namely $\vec{\mu}_{v_i v_j}$ and $\Sigma_{v_i v_j}$.

Partially Known Parameters

Similarly to PG_1-bias, PG-bivariate allows a flow of probabilistic influence from any peer's grades to the teacher's grading distribution without the need to have an example of how the teacher grades. However, it is possible to improve the inference results by introducing some of the teacher's grades in the system. Hence, the components in $\vec{s}$ will be considered partially known parameters as well.

3.2. Experiments

The following results refer to synthetic databases, though future work includes applying the models to real data. Similar to in previous sections, the term *ground truth* is used here. In this context, ground truth is a synthetic data point with a known value that serves as input to the model.

3.2.1. Experiments on Bayesian Networks

Posterior Predictive Sampling

Posterior predictive sampling (Equation (19)) is a useful method to compare predictions of new data based on past observations with the real data points. This comparison gives a sense of the estimations' quality and may help during the tuning of the hyperparameters.

$$\pi_S(y|\tilde{y}) = \int \pi_S(\theta|\tilde{y})\pi_S(y|\theta)\mathrm{d}\theta \tag{19}$$

We sampled from the posterior predictive distribution of the two tested BN models, namely PG_1-bias and PG-bivariate, comparing the histograms of predictions with the histogram of samples. Sampling from the posterior predictive distribution helped us calibrate the models' hyperparameters for the prior distribution. In general, weakly informative and very weakly informative priors were proposed. However, we introduced some domain knowledge by centering the distributions of grades around positive numbers and restricting the variance parameters to only positive values, to name a couple of examples.

Recalling the prior distribution equation for model PG_1-bias (Equation (10)), the chosen values in our implementation for $(\alpha_0, \beta_0, \gamma_0, \eta_0, \mu_0)$ were $\alpha_0 = 0.01$ (very weak prior), $\beta_0 = 0.01$ (very weak prior), $\gamma_0 = 10.0$ (weak prior), $\eta_0 = 1.0$, and $\mu_0 = 5.0$, whereas in the case of PG-bivariate, the chosen hyperparameters were $x_0 = 0.0$ (weak prior), $\gamma = 0.25$ (weak prior), $\bar{\mu} = 5.0$, $\bar{\sigma} = 10.0$ (weak prior), and $\eta = 1.0$ (weak prior).

Figure 8 represents a histogram of predicted peer grades z's from the posterior predictive distribution in model PG_1-bias for a class of 50 students (without ground truths). We observe the experimental histogram of observed peer grades in blue, $\tilde{z}$. The histogram of predictions (z) represents the distribution of data not part of the training set. This kind of prediction was repeatedly executed (20 times) to ensure that the results were consistent. Peer grades ranged from 0 to 10, following Equation (13).

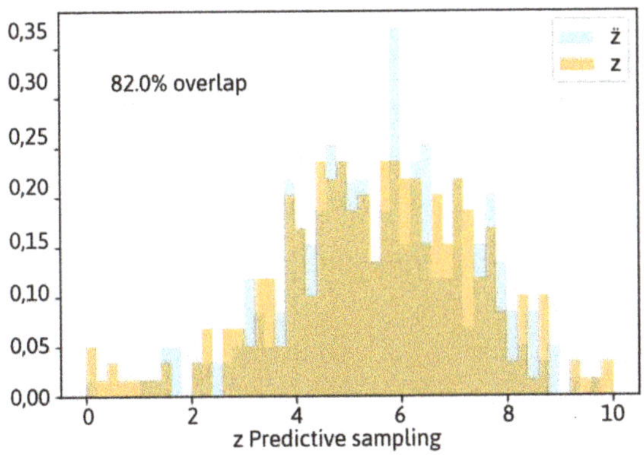

Figure 8. Posterior predictive sampling from PG_1-bias.

We calculated the percentage overlapping between the histogram of samples and the histogram of predictions using the vector of elementwise minima [14]. The overlapping range was between 0% (no overlap) and 100% (identical distributions). On average, we observed an overlapping of 82%. The range of values obtained for the overlapping in all the performed experiments was between 76% and 84%.

This value implies that the model captures the data generation system's dynamics, which results in samples z that follow a very similar distribution to that from which $\tilde{z}$ were sampled.

A similar figure to Figure 8 is obtained when we perform a posterior predictive sampling of the model presented in this paper, PG-bivariate. The overlapping, in this case, fell within the range [73–81%], with an average value of 77% in 20 different runs (see Figure 9).

Although in this case, the model does not offer a causal explanation of the parameters μ_v and σ_v from the normal distribution governing z^v (Equation (14)), the dynamics of the data generation system are captured. Intuitively, one would have expected a notably worse performance than PG_1-bias for several reasons. First of all, PG_1-bias models each use peer grade as a random variable that follows its normal distribution. On the other hand, the definition of the location parameter as the sum of the real grade and the graders' bias allows a direct ascension of the flow of probabilistic influence towards s_v. In the case of PG-bivariate, we adopt the perspective of a passive observer, representing the set of grades emitted by a referee as a random variable following a normal probability density function. Notice that in this case, no mention of who is evaluated by whom is being made. Hence, the location parameter μ_v cannot be decoupled as in the former case to further ease the probabilistic influence flow.

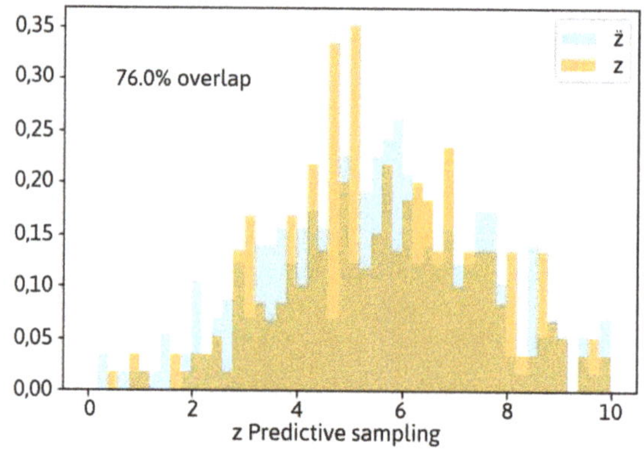

Figure 9. Posterior predictive sampling from *PG*-bivariate.

Studying the Error Evolution

To compare the performance of both models based on Bayesian networks, we studied the evolution of the models' root mean squared error ($RMSE$, Equation (20)) as a function of the number of observed true grades, keeping the number of students in a class, S, constant.

$$RMSE = \sqrt{\frac{1}{N}\sum_{n=1}^{N}(y_n - \tilde{y_n})^2} \tag{20}$$

where $\tilde{y}$ is the true value, y is the prediction, and N stands for the number of data points. Grades ranged from 0 to 10, both included. We examined the error evolution in both models as we introduced more teachers' grades as ground truths. We assessed two different criteria to determine which new grade to introduce in each step of the loop:

1. Random choice (baseline): The next observed ground truth is chosen randomly.
2. Total $RMSE$ decreasing policy: At each iteration, we picked and observed the true grade (i.e., the teacher's grade) of that student whose assessment was introducing the highest root mean squared error.

As a result, we obtain four curves for each model:

- The red line shows the evolution of the estimations' $RMSE$ as we introduce new ground truths following a random policy.
- The yellow, discontinuous line shows the evolution of the estimations' $RMSE$ without considering the known ground truths to correct for overly optimistic low error values. Additionally, in this case, a random policy for ground truth injection is followed.
- The blue line shows the evolution of the estimations' $RMSE$ as we introduce new ground truths following an $RMSE$ decreasing policy.
- The violet, discontinuous line shows the same information as the yellow line for the case of the $RMSE$ decreasing policy.

Figure 10 shows the resulting four lines in the case of PG_1-bias. Looking at the pair of continuous lines, we can see how $RMSE$ falls as the number of known ground truths increases in all cases. As expected, computing the $RMSE$ of all the grades (ground truths included) produces more optimistic errors. Moreover, it seems that the $RMSE$ of the unknown grades (discontinuous lines) decreases much more slowly than for the continuous counterpart, that is, adding new ground truths does not reflect quickly on better quality of the predictions. According to Figure 6, introducing a ground truth s_u has an impact on the bias b_v of those peers v who have made an assessment of u: z_u^v. There is also a flow of probabilistic influence towards the grader's reliability τ, which in the case of $PG1$-bias is common to all students. The informativeness of the prior determines the relative relevance that observed data and domain expertise have on the posterior distribution parameters.

Hence, we believe that the choice of an informative prior over b_v is preventing the model from showing an erratic behavior as new data is incorporated, but it also makes it more insensitive to the observations.

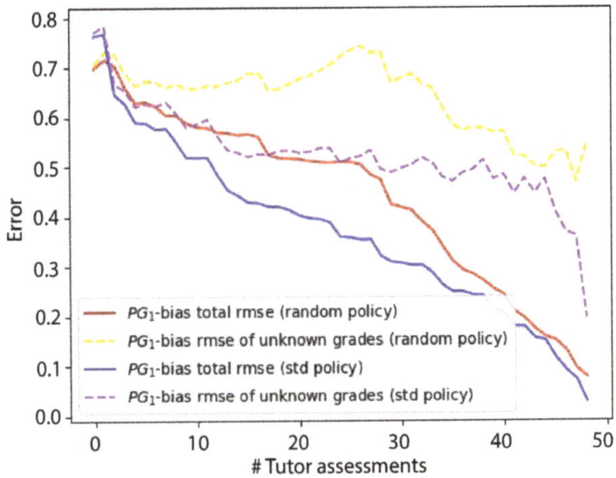

Figure 10. Reported $RMSE$ for model PG_1-bias.

In general, the random policy yielded worse results for the discontinuous and continuous pairs of lines (red line showing higher error than the blue line, and yellow line showing higher error than the violet line).

We can see in Figure 11 the corresponding curves for PG-bivariate. In this case, the value of the $RMSE$ was higher, which indicates that our predictions are, in general, worse than those of PG_1-bias. We see again that the random policy yields worse results than the $RMSE$-decreasing policy (red line above blue line, and yellow line above violet line). As in the previous case, the slopes of the continuous lines are sharper than those of the discontinuous lines, which implies that the probabilistic influence from the observed teacher grades towards the predictions does not flow easily. The discontinuous lines also make it evident that when the number of grades we are computing the $RMSE$ with is small (last portion of the graph), fluctuations of the $RMSE$ increase.

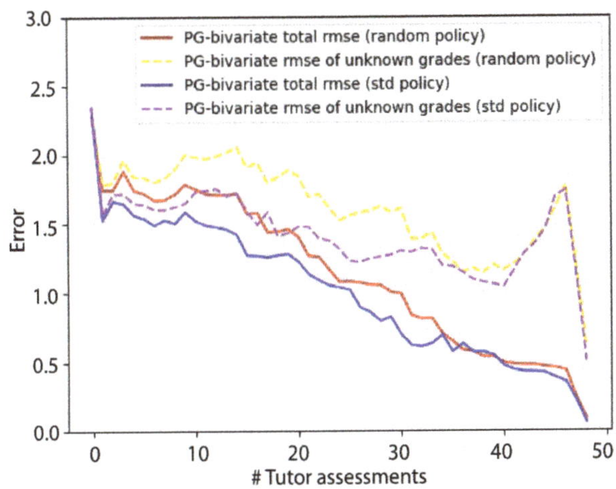

Figure 11. Reported $RMSE$ for model PG-bivariate.

3.2.2. Experiments on PAAS

To make sure that PAAS software was being correctly implemented in Python, the synthetic experiment reported in Gutierrez et al. [8] was reproduced. We worked here under the premise that obtaining similar figures to the originals with our implementation would be a guarantee that the model functioning was being correctly replicated. This experiment consisted of a simulation of a classroom of 200 students with 200 submitted assignments, where each assignment was evaluated by 5 students (1000 peer assessments performed). In order to show a critical case, the authors simulated that half of the assignments were evaluated accurately by half of the students (that is, those students provided the same mark as the teacher), and the other half of the assignments were evaluated poorly (that is, randomly) by the rest of the class. In the simulation, they followed two policies to pick the next ground truth to observe as well. Such policies were a random one and another seeking to reduce the entropy of the grades by selecting the assignment with the highest entropy in its probability distribution, where entropy is computed as in information theory:

$$\mathbb{H}(\mathbb{P}(s_u|\{z_u^v\}_{v\in S}) = \sum_{\tilde{s}} \mathbb{P}(s_u = \tilde{s}|\{z_u^v\}_{v\in S}) \cdot \ln\mathbb{P}(s_u = \tilde{s}|\{z_u^v\}_{v\in S}) \quad (21)$$

The reported error line by Gutierrez et al. [8] is shown in Figure 12 (up). Down, we represent the obtained error for our implementation of PAAS in Python and a class of 50 students. We can see that although the original implementation shows a lower error for the first iterations, the behavior of both the original and our Python implementation is similar: both errors decrease as new ground truths are introduced in the system, and the entropy heuristic decreases the error faster than the random one. Furthermore, in both implementations, a crossing of the lines occurs when the number of observed ground truths is between 20% and 30%. From that point on, the performance of the heuristic-driven model is better.

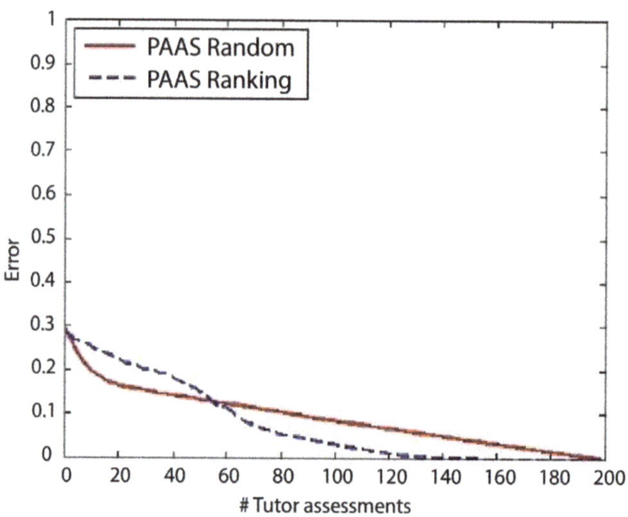

Figure 12. *Cont.*

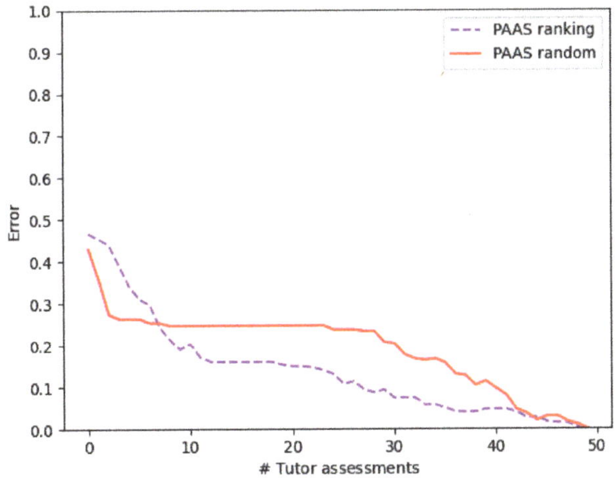

Figure 12. Percentage error using random (red line) and entropy decreasing (violet, discontinuous line) assessment order as a function of the number of observed teacher's grades for classes of 200 and 50 students using synthetic data. The **upper** figure represents the original implementation by Gutierrez et al. [8]. **Below**, our results are shown.

3.2.3. Comparison of the Three Models

In order to compare the results, it was necessary that all of them were in the same unit. We adapted the *BN* models and computed their percentage of error to make comparisons (Equation (22)). This is the error function that will be represented in the following figures:

$$\epsilon = \frac{\Sigma_{u \in \mathcal{S}} |\tilde{s}_u - s_u|}{S} \tag{22}$$

where S is the number of students, $\tilde{s}_u$ is the u's true grade, and s_u is the prediction. The experimental setups on which PAAS was tested considered quality assessments (that is, the graders were asked up to what degree, in their opinion, did an assignment meet a set of criteria). Although the model allowed both qualitative (e.g., $\{\text{bad}, \text{good}, \text{excellent}\}$) or quantitative assessments, in the latter case, only integer grades were considered. Hence, for the comparisons to be fair, the Bayesian network models were adapted to predict integer grades ranging from 0 to 3 (both included). We ran this comparison mimicking the experimental arrangement described in the previous Section 3.2.2 for a synthetic class of 50 students.

Figure 13 shows the models' comparison using the best policy of each of them. From all of them, PG_1-bias keeps showing the lowest error when the number of known ground truths is small. Once past that region, both PG_1-bias and PG-bivariate show similar shapes downwards, although PG_1-bias keeps being solidly and continuously superior. Looking at PAAS' results and despite its initial results being poor, we can see a sharp decrease in error once some evaluations were introduced as knowledge in the network, showing, from that point on, a similar, though fluctuating, performance to PG_1-bias.

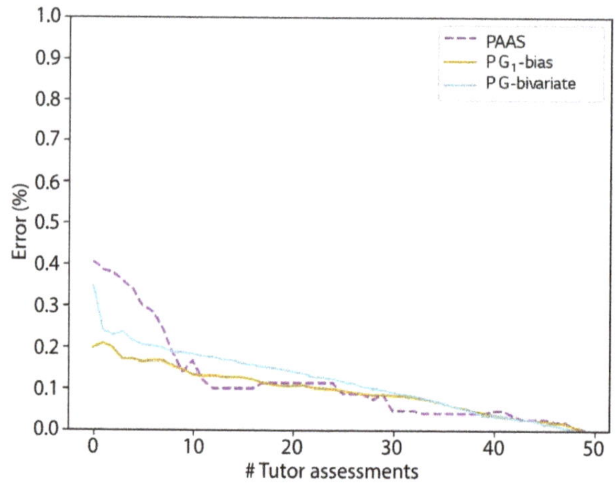

Figure 13. Percentage error as a function of the number of observed ground truth grades reported by the three studied models.

4. Discussion

The difficulties faced by neural models to make logical narratives of the decision chains that lead them to a final prediction pose a controversy on their applications in sensitive topics. One example is automatic grading and the ethical implications that it entails. These limitations have led the research to bet on hybrid solutions combining the benefits of automation with human judgment.

The first goal of this paper was to reproduce two models (PG_1-bias, by Piech et al. [4] and PAAS, by Gutierrez et al. [8]) of peer assessment. Both proposals estimate a probability distribution for each automatic grade using a probabilistic approach. These models' inputs are peer assessments and a percentage of the teacher's grades (ground truths). However, their theoretical foundations are different, and a common experimental setting was designed to make fair comparisons. All the code was written in Python, which implied the conversion of PAAS' original code from Java and the familiarization with Stan's Python interface (pystan) to reimplement PG_1-bias.

Our second goal was to implement a model that combined the first's powerful theoretical background with the second's multiagent-based ideas. As a result, PG-bivariate translates the notion of trust among reviewers to a Bayesian approach thanks to its use of correlations among graders as the main feature to be learned by the model. This allows for the obtention of automatic grades of the same quality as those of PG_1-bias without the need for complex causal reasoning over the distribution parameters representing each grader's grading style.

Despite the similarities between previous literature and the algorithms presented here, some important distinctions should be noted: first, in contrast to the Bayesian model by Bachrach et al. [6], the variables considered in PG_1-bias and PG-bivariate to predict the scores of the submitters are not only related to circumstances entailing the submitters, but also to the grades given by their peers and to the grading ability of such peers. Plus, instead of relying on highly abstract variables such as Sterbini and Temperini [5] do (e.g., judgment, knowledge), PG_1-bias and PG-bivariate work with more traceable variables such as true scores and grader's accuracy and reliability when correcting. Finally, and perhaps most importantly, all the algorithms in this work try to mimic the behavior that an expert would have had when faced with the task of grading a set of assignments. Hence, the trending opinion among peers is not what determines the quality of an assessment or what guides the prediction of an automatic score, in contrast to De Alfaro and Shavlovsky [9].

Regarding our experiments on Bayesian network models, the posterior predictive analysis showed that the distribution of new samples conditioned on the observed data was very similar to the distribution of samples. The posterior predictive distribution is influenced by the model configurations consistent with the relevant domain expertise and the observed data. This makes it ideal for informing predictions about future interactions with the latent system. Comparisons between the posterior predictive distribution and the observed data also measure how well our model approximates the latent system. Hence, we conclude that both models captured the data generating process.

When comparing these two models between them, specifically when analyzing the evolution of RMSE as a function of known ground truths (teacher's grades), we can see that PG_1-bias is slightly better. Not only does it report lower RMSE, but we observed that the computation times were shorter than in our model. According to the PyStan [15], most of the computation time during sampling with the NUTS algorithm is dedicated to calculating the Hamiltonian gradient. Looking at our computation times, *PG*-bivariate's phase space has a more problematic surface to the sampler than that one of PG_1-bias'. Hence, we conclude that although *PG*-bivariate uses a more simple definition of the distribution parameters (avoiding higher layers above μ_v and σ_v), in this case, a higher abstraction might be desirable to reduce the number of parameters to fit. Regarding the higher RMSE values reported by our model, our hypothesis is that better and more informed choices of the hyperparameters will yield lower errors in the future, further facilitating the flow of probabilistic influence between variables. However, it is worth noting that *PG*-bivariate showed a similar behavior to PG_1-bias in what policies are concerned: the random policy resulted in worse results than the one driven by the std error minimization policy. On the other hand, it is interesting to see that *PG*-bivariate offers remarkably good performance (the error of the predictions falls below 20% when only 20% of the ground truths are observed) despite its simplicity. The model can make predictions using the observed data and the correlations between graders, which somehow encode PAAS's direct trust. The flow of probabilistic influence between these *relationship* variables behaves in this case as PAAS' indirect trust. Given the success of PAAS in obtaining good automatic ratings, we find it interesting to continue exploring to what extent abstract variables can be dispensed within a Bayesian model that codifies the notion of trust between peers.

In Section 3.2.2, we showed that our Python implementation of PAAS, pyPAAS, reported a similar behavior to that by Gutierrez et al. [8]. To make fair comparisons among all models, we adapted the Bayesian inference models to compare with pyPAAS. More specifically, the synthetic data experiment was replicated for the three models. Under these conditions, PG_1-bias and *PG*-bivariate reported similarly low errors in situations of scarce ground truths. From 10% ground truths, the three models offer results of similar quality. It is worth mentioning that the calculation times for PAAS and our model were higher than those for PG_1-bias when the class size started to become relatively large.

Throughout this research, synthetic data was used. Future work will apply the models to real experimental settings. Exploring new heuristics to determine which ground truth to observe next is also an exciting research line to continue this work.

Finally, regarding the robustness of our method, the simultaneous concurrence of three circumstances guarantees that the model is shielded from potential malicious coalitions:

1. The small number (5) of peers assessing each assignment.
2. The fact that these graders are chosen randomly.
3. The fact that the process is entirely anonymous concerning the students (the graders do not know who are they assessing, and the gradees do not know the identity of their graders).

In future work, we plan to identify the scenarios where each of these models is preferable. Specifically, it would be interesting to confirm the hypothesis that the two models *PG*-bivariate and PAAS (that implicitly penalize bad raters) perform better than PG_1-bias under alliance and coalition dynamics among students. Applying the models to real experimental settings is also on the near horizon. Such settings may cover all those situations that can be modeled as consisting of multiple agents with different reputations issuing an opinion (e.g., peer review). Further improvements of *PG*-bivariate should find a point of compromise between the conceptual simplicity of the model and the number of parameters that it requires in exchange. Finally, exploring new heuristics to determine which ground truth to observe next is also an exciting research line to continue this work.

5. Conclusions

Our paper compares two state-of-the-art automatic evaluation methods ([4,8]) with a new model. This new model combines the Bayesian background of Piech et al. [4]'s with the use of a trust graph over the referees proposed by Gutierrez et al. [8]. As a result, we obtain a hierarchical Bayesian model that dispenses with the choice of abstract variables in favor of others that are easily interpretable. Similar to PAAS, this modeling through a trust graph explicitly shields the algorithm against bad graders.

There remain several issues to be addressed in future work. First, it is necessary to find a point of compromise between the conceptual simplicity of the model and the number of parameters that it requires in exchange for said simplicity. For example, in the case of *PG*-bivariate, dispensing with variables above the parameters μ_s and σ_s induces the necessity to calculate the parameters of all the bivariate distributions that describe correlations between students with some correction in common. This seems to cause a difficult phase space for the sampler because of the relatively long computation times. Second, we believe that further research that helps identify the scenarios where each of the compared models performs better is an interesting continuation line. For instance, it would be interesting to test whether the models that implicitly penalize bad raters perform better than PG_1-bias under alliance and coalition dynamics among students.

Similar to other studies of the area ([5,6,8]), we find that following a greedy policy that maximizes the entropy reduction induced by new observations yields better predictions of the grades. Exploring new heuristics to determine which ground truth to observe next is also an exciting research line to continue this work.

Author Contributions: Conceptualization, A.L.d.A.G., J.S.-M. and C.S.; methodology, A.L.d.A.G., J.S.-M. and C.S.; software, A.L.d.A.G.; validation, A.L.d.A.G., J.S.-M. and C.S.; formal analysis, A.L.d.A.G., J.S.-M. and C.S.; investigation, A.L.d.A.G.; resources, J.S.-M. and C.S.; writing—original draft preparation, A.L.d.A.G.; writing—review and editing, J.S.-M. and C.S.; visualization, J.S.-M.; supervision, J.S.-M. and C.S.; project administration, C.S. All authors have read and agreed to the published version of the manuscript.

Funding: This research has been funded by ACCIO through the projects NanoMOOCs (COMRDI18-1-0010—RIS3CAT MEDIA), and ADDIA (ACE014/20/000039—INNOTEC) and by the CSIC through the project MARA (Intramural 202050E132). It has been funded also by the European Union Horizon 2020 FET Proactive projects "WeNet" (grant agreement Number 823783), "TAILOR" (grant agreement 952215) and "AI4EU" (grant agreement Number 825619).

Institutional Review Board Statement: Not applicable.

Informed Consent Statement: Not applicable.

Data Availability Statement: The models presented in this study are openly available in GitHub at https://github.com/aloaberasturi/Probabilistic-Models-of-Competency-Assessment (accessed on 14 December 2021).

Conflicts of Interest: The authors declare no conflict of interest.

Abbreviations

The following abbreviations are used in this manuscript:

AI	Artificial intelligence
BN	Bayesian network
DAG	Directed acyclic graph
MOOC	Massive open online course
NLP	Natural language processing
PAAS	Personalized automated assessments
PGM	Probabilistic graphical model
RMSE	Root mean squared error

References

1. Schön, D.A. *The Design Studio: An Exploration of Its Traditions and Potentials*; International Specialized Book Service Incorporated: London, UK, 1985.
2. Tinapple, D.; Olson, L.; Sadauskas, J. CritViz: Web-based software supporting peer critique in large creative classrooms. *Bull. IEEE Tech. Comm. Learn. Technol.* **2013**, *15*, 29.
3. Kulkarni, C.; Wei, K.P.; Le, H.; Chia, D.; Papadopoulos, K.; Cheng, J.; Koller, D.; Klemmer, S.R. Peer and self assessment in massive online classes. *ACM Trans. Comput.-Hum. Interact. (TOCHI)* **2013**, *20*, 1–31. [CrossRef]
4. Piech, C.; Huang, J.; Chen, Z.; Do, C.; Ng, A.; Koller, D. Tuned models of peer assessment in MOOCs. *arXiv* **2013**, arXiv:1307.2579.
5. Sterbini, A.; Temperini, M. Correcting open-answer questionnaires through a Bayesian-network model of peer-based assessment. In Proceedings of the 2012 International Conference on Information Technology Based Higher Education and Training (ITHET), Istanbul, Turkey, 21–23 June 2012; pp. 1–6.
6. Bachrach, Y.; Graepel, T.; Minka, T.; Guiver, J. How to grade a test without knowing the answers—A Bayesian graphical model for adaptive crowdsourcing and aptitude testing. *arXiv* **2012**, arXiv:1206.6386.
7. Mi, F.; Yeung, D.Y. Probabilistic graphical models for boosting cardinal and ordinal peer grading in MOOCs. In Proceedings of the AAAI Conference on Artificial Intelligence, Austin, TX, USA, 25–30 January 2015; Volume 29.
8. Gutierrez, P.; Osman, N.; Roig, C.; Sierra, C. Personalised Automated Assessments. In Proceedings of the 2016 International Conference on Autonomous Agents & Multiagent Systems, Singapore, 9–13 May 2016; Jonker, C.M., Marsella, S., Thangarajah, J., Tuyls, K., Eds.; ACM: New York, NY, USA, 2016; pp. 1115–1123.
9. De Alfaro, L.; Shavlovsky, M. CrowdGrader: A tool for crowdsourcing the evaluation of homework assignments. In Proceedings of the 45th ACM Technical Symposium on Computer Science Education, Atlanta, GA, USA, 5–8 March 2014; pp. 415–420.
10. Ashley, K.; Goldin, I. Toward ai-enhanced computer-supported peer review in legal education. In *Legal Knowledge and Information Systems*; IOS Press: Amsterdam, The Netherlands, 2011; pp. 3–12.
11. Balfour, S.P. Assessing Writing in MOOCs: Automated Essay Scoring and Calibrated Peer Review^{TM}. *Res. Pract. Assess.* **2013**, *8*, 40–48.
12. Admiraal, W.; Huisman, B.; Pilli, O. Assessment in Massive Open Online Courses. *Electron. J. E-Learn.* **2015**, *13*, 207–216.
13. *The Earth Mover's Distance (EMD)*; The Stanford University: Stanford, CA, USA, 1999.
14. Swain, M.J.; Ballard, D.H. Color indexing. *Int. J. Comput. Vis.* **1991**, *7*, 11–32. [CrossRef]
15. Stan Development Team. PyStan: The Python Interface to Stan. 2021. Available online: http://mc-stan.org/2 (accessed on 14 December 2021).

applied sciences

MDPI

Article

Measuring Polarization in Online Debates

Teresa Alsinet *, Josep Argelich, Ramón Béjar and Santi Martínez

INSPIRES Research Center, University of Lleida, Jaume II, 69, 25001 Lleida, Spain; josep.argelich@udl.cat (J.A.); ramon.bejar@udl.cat (R.B.); santi.martinez@udl.cat (S.M.)

* Correspondence: teresa.alsinet@udl.cat

Abstract: Social networks can be a very successful tool to engage users to discuss relevant topics for society. However, there are also some dangers that are associated with them, such as the emergence of polarization in online discussions. Recently, there has been a growing interest to try to understand this phenomenon, as some consider that this can be harmful concerning the building of a healthy society in which citizens get used to polite discussions and even listening to opinions that may be different from theirs. In this work, we face the problem of defining a precise measure that can quantify in a meaningful way the level of polarization present in an online discussion. We focus on the Reddit social network, given that its primary focus is to foster discussions, in contrast to other social networks that have some other uses. Our measure is based on two different characteristics of an online discussion: the existence of a balanced bipartition of the users of the discussion, where one partition contains mainly users in agreement (regarding the topic of the discussion) and the other users in disagreement, and the degree of negativity of the sentiment of the interactions between these two groups of users. We discuss how different characteristics of the discussions affect the value of our polarization measure, and we finally perform an empirical evaluation over different sets of Reddit discussions about diverse classes of topics. Our results seem to indicate that our measure can capture differences in the polarization level of different discussions, which can be further understood when analyzing the values of the different factors used to define the measure.

Keywords: Reddit; user-based model; polarization; local search optimization

Citation: Alsinet, T.; Argelich, J.; Béjar, R.; Martínez, S. Measuring Polarization in Online Debates. *Appl. Sci.* **2021**, *11*, 11879. https://doi.org/10.3390/app112411879

Academic Editors: Aida Valls and Karina Gibert

Received: 17 November 2021
Accepted: 12 December 2021
Published: 14 December 2021

1. Introduction

Nowadays, there is growing controversy regarding the emergence of polarization in discussions on social networks, and the responsibility of companies in this problem. For example, Facebook researchers have studied the spread of divisive content on the platform [1,2]. One of their findings is that users with hyperactive engagement are far more partisan on average than normal users. Facebook has launched some initiatives to try to mitigate the factors that may be helping the spread of such content [3]. For example, they re-calibrated the prioritization of content in users' News Feeds to give more preference to posts of friends and family over news content, justified by the findings that suggest that people derive more "meaningful social interactions", or MSI, when they engage with people they know, although it is not clear what exactly comprises a meaningful social interaction for Facebook. Other actions include improving the detection of fake accounts and removing recommendations of pages that violate Facebook Community Standards or are rated false by their fact checkers.

However, the findings of Facebook suggest that completely fixing the polarization problem may be difficult. As they commented in an internal company presentation [4], Facebook algorithms exploit the human brain's attraction to divisiveness and tend to feed users with more and more divisive content, which seems to ensure that users will spend more time on the platform. Publishers and political parties reorient their posts toward outrage and sensationalism since this produces high levels of comments and reactions. Some political parties in Europe event told Facebook that the algorithm made them shift their policy positions. However, because engagement is of course fundamental for the profit of the company, social network companies should be cautious to not react disproportionately when trying to mitigate polarization.

In spite of all these issues, it is worth noticing that it is not clear how much responsibility we can give to social networks on these problems. For example, there is some work that indicates that, at least for polarization in people's views of political parties, the increased use of social networks and the internet may not be necessarily increasing polarization [5]. Putting aside whether specific social network platforms are more or less responsible for polarization, it is clear that what companies, or policymakers, can do is to define more transparent ways to monitor such possible non-desirable behaviors so that we can decide to act only in situations where there is some objective measure that polarization is taking place and to a certain level of severity.

Previous work has studied the presence of polarization in different concrete examples, trying to analyze the relevant characteristics in these cases. For example, the works [6,7] studied the emergence of so-called echo chambers, where users are found to interact mainly only with the users that agree with them and to have very few interactions with other groups. However, online discussions in social networks can also show polarization where there are answers with negative sentiment between different groups, which can be considered the most worrying situation. For example, in [8] they studied hyperpartisanship and polarization in Twitter during the 2018 Brazilian presidential election. Their findings showed that there was more interaction within each group (pro/anti Bolzonaro) but there was also an interaction between both groups. Actually, there are also cases where the interaction between groups can be more relevant than those within groups, like in the case studied in [9] where the analysis of the 2016 U.S. elections on Reddit showed a significant number of interactions between pro-Trump and pro-Hillary supporters. It is worth noting that the existence of polarization in debates could be partially explained as a consequence of the spiral of silence theory [10]. In this theory, the public opinion about emotionally and morally laden issues is considered to be driven toward one opinion that receives the strongest support from people and mass media. In contrast, people with contrary opinions suffer isolation pressure from the supporters of the majority opinion, which makes them hide their opinions (or not defend them strongly). Thus, the existence of polarization around a certain topic can be explained as the result of a spiral of silence, where two strongly defended, but opposed, opinions tend to attract the attention of all the people and the isolation pressure forces them to join one of the sides. So, we may consider that, in fact, two opposing spirals emerge as a result. However, it is clear that around certain topics, in certain societies, there may exist a clear majority opinion that can makes it difficult for the emergence of secondary publicly supported opinions. For example, in [11], it was studied the effect of governmental internet surveillance around the controversial topic of U.S. airstrikes against ISIS terrorists.

Our main focus in this work is to give a more clear and quantitative model for measuring polarization in an online debate such that this behavior can be monitored for generating a warning signal when necessary, that is, to detect communication patterns where users seem to interact positively only with a fixed group of users and negatively with the rest. Among the many online debating platforms that exist, in this work, we consider Reddit. Reddit (available at http://www.reddit.com/) is a social news aggregation, web content rating, and discussion website. Users submit content to the site, such as links, text posts, and images, which are then voted up or down by other members who, in turn, can comment on others' posts to continue the conversation. This online debating platform is widely used to create long and deep debates with comments and answers to comments, where, thanks to the almost unlimited text length of Reddit comments (40,000 characters), users can express their opinions more accurately, compared to other online debating platforms, such as Twitter which restricts the number of characters to 280.

As previous works seem to indicate that in a debate we can have interactions in two ways (within groups and between groups), an important aspect of our model is that it allows us to quantify the relevance of the kinds of interactions present. Our model is inspired by the model used in [12] to identify supporting or opposing opinions in online debates, based on finding a maximum cut in a graph that models the interactions between users. In our case, as we are interested in quantifying polarization, we define a model that is based on a weighted graph and with labeled edges, where node weights represent the side of the user in the debate and edge labels represent the overall sentiment between two users. Then, given a bipartition of this graph, in our model, the polarization degree of the bipartition is based on how homogeneous each partition is and how negative the interactions are between both partitions. Finally, our measure of debate polarization is based on the maximum polarization we have in all the possible bipartitions of the graph.

The structure of the rest of the paper is as follows. In Section 2, we present our user-based model to represent Reddit debates, the User Debate Graph, where a node represents a user's opinion as to its whole set of comments in the debate. In Section 3, we define a measure to quantify the polarization in a debate that is based on a value defined over bipartitions of the user debate graph. In Section 4, we introduce a greedy local search optimization algorithm for finding the polarization of a debate, based on searching a bipartition with the highest possible polarization value. Finally, in Section 5, we perform an empirical evaluation of the polarization value obtained with our algorithm with different Reddit debates.

2. User-Based Model for Reddit

In this section, we present a computational model to represent a Reddit debate that allows us to study the polarization of the debates based on the interactions between the participants. In Reddit, debates are generated from a main (root) comment that contains a link to some news and a set of comments, where each comment, except for the root comment, answers exactly one previous comment, usually by another user or author. These answers between comments lead to positive, negative or neutral interactions, and their analysis will allow us to classify users into three groups: users that are in agreement with the root comment of the debate, users that are in disagreement with it, and undecided users.

In order to analyze the agreement and disagreement between user comments in Reddit debates, recently in [13], we defined an analysis system that represents a Reddit debate as a two-sided debate graph. In this graph, comments are divided into two groups, the ones that agree with the root comment of the debate, and the ones that disagree with it. The edges of the graph represent disagreement between the comments of the two groups. So, in this work, we use the two-sided debate model as a starting point to build the user-based model to study the polarization of the debates based on the interactions between the participants.

Following [13], we first formalize the notions of a comment and Reddit debate for a root comment.

Definition 1 (Reddit comment and debate). *A Reddit comment is a pair $c = (m, u)$, where m is the text of the comment and u is the user's identifier of the comment.*

Let $c_1 = (m_1, u_1)$ and $c_2 = (m_2, u_2)$ be two comments. We say that c_1 answers c_2 if c_1 is a reply to the comment c_2.

Let $r = (m_r, u_r)$ be a comment such that m_r contains a link to some news. A Reddit debate on r is a non-empty set Γ of Reddit comments such that $r \in \Gamma$ and every comment $c \in \Gamma, c \neq r$, c answers a previous comment in Γ. We refer to r as the root comment of Γ.

Given the structure of a Reddit debate Γ on a root comment r, the next step is to extract the relationships between the comments in Γ. We represent Γ as a labeled tree, where each comment gives rise to a node, and edges denote answers between comments and are labeled with a value in the real interval $[-2, 2]$. The label for an edge (c_1, c_2) denotes the *sentiment* expressed in the text of the comment c_1 in response to the text of the comment c_2 so that the value -2 denotes a total disagreement and the value 2 a total agreement. We use the sentiment value 0 to denote both answers expressing a neutral position with respect to the opinion expressed in c_2, and answers expressing, at the same time, agreement with part of the opinion expressed in c_2, and disagreement with another part of c_2.

Definition 2 (Debate Tree). *Let* Γ *be a Reddit debate on a root comment* r*. A Debate Tree* (DebT) *for* Γ *is a tuple* $\mathcal{T} = \langle C, r, E, W \rangle$ *such that the following is true:*

- *For every comment in* Γ*, there is a node in* C*;*
- *Node* $r \in C$ *is the root node of* $\mathcal{T}$*;*
- *If* c_1 *answers* c_2*, then there is a directed edge* (c_1, c_2) *in* E*, and*
- W *is a labeling function of edges* $W : E \rightarrow [-2, 2]$*, where the value assigned to an edge denotes the sentiment of the answer, from highly negative* (-2) *to highly positive* (2)*.*

  Only the nodes and edges obtained by applying this process belong to C *and* E*, respectively.*

Given a Reddit debate, we make its corresponding DebT using the Python Reddit API Wrapper (PRAW, available at https://github.com/praw-dev/praw) to download its set of comments, and then we evaluate the sentiment for an edge (c_1, c_2) in the DebT using the sentiment analysis software of [14] using the text of the comment c_1, where the value assigned denotes the sentiment of the answer, from highly negative (-2) to highly positive (2).

In Figure 1, we show the DebT structure for a Reddit debate of the subreddit World News (r/worldnews) (https://www.reddit.com/r/worldnews/comments/f62i35 accessed on 4 October 2020). Each comment is represented as a node, and each answer between comments is represented as an edge. The graph has 42 nodes and 41 edges since each comment responds exactly to one comment, except for the root comment. These 41 answers between comments are classified as agreement edges (painted in green color) and disagreement edges (painted in red color), and there is no neutral answer. The darkness of the color is directly proportional to the sentiment of the answer concerning the maximum value. The root comment of the discussion is labeled with 0, and the other comments are labeled with consecutive identifiers according to their generation order. Notice that the graph is acyclic since a comment only answers a previous comment in the discussion. Additionally, most of the answers are disagreement answers, and they seem to generate the longest discussion paths in the DebT. However, this model does not allow to infer directly neither the set of the most dominant users' opinions between the different users in the debate, nor the position of every user concerning the root comment since the information of each user is represented by multiple nodes and edges in the DebT.

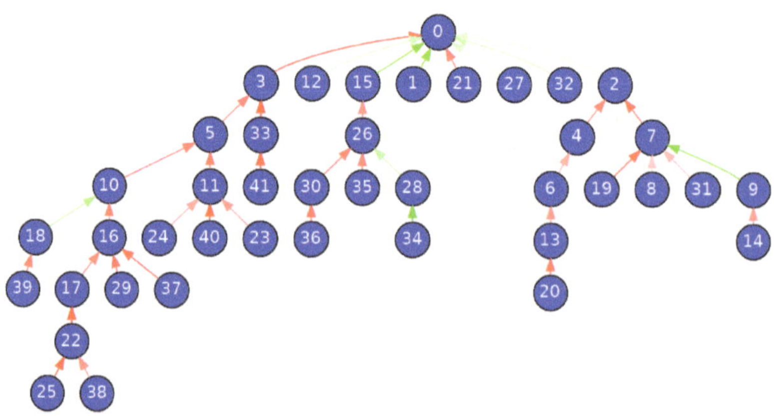

Figure 1. Debate tree for a Reddit debate of the subreddit World News.

Analyzing the relationships between the DebT nodes, we classify the comments into two groups: comments that support the root comment and comments that disagree with it. With this objective, we extend the DebT structure with a labeling function of nodes denoting the side of each comment in the debate. We label the comments that support the root comment with 1 and the rest of the comments with -1. We refer to this tree extension as a *two-sided debate tree.*

Definition 3 (Two-sided debate tree). *Let $\mathcal{T} = \langle C, r, E, W \rangle$ be a* DebT *for a Reddit debate Γ on a root comment r. A two-sided debate tree* (SDebT) *for $\mathcal{T}$ is a tuple $\mathcal{T}_S = \langle C, r, E, W, S \rangle$ such that S is a labeling function of nodes $S : C \rightarrow \{-1, 1\}$ defined as follows:*

- $S(r) = 1$;
- *For all node $c_1 \neq r$ in C, $S(c_1) = 1$ if for some node $c_2 \in C$, $(c_1, c_2) \in E$ and either $S(c_2) = 1$ and $W(c_1, c_2) > 0$, or $S(c_2) = -1$ and $W(c_1, c_2) \leq 0$; otherwise, $S(c_1) = -1$.*

To evaluate the side of a comment c_1 expressing a neutral position with respect the opinion expressed in c_2, i.e., with $W(c_1, c_2) = 0$, we consider a similar approach to the proposal of Murakami and Raymond [12]. In their proposal, they consider neutral opinions to be (slightly) negative. The motivation for this is the realization of Agrawal et al. [15] that, in newsgroups, people respond more frequently when they disagree. In Reddit, for example, when users agree with a comment, they can simply upvote without leaving any response. When users disagree, on the other hand, they can downvote, but it is more likely that they expose their reasons. Thus, from this perspective, a neutral response would be expressing more disagreement than agreement.

Figure 2 shows the SDebT structure we obtain for the Reddit debate of Figure 1. The nodes (comments) that support the root comment are colored in cyan, while the rest are colored in navy blue.

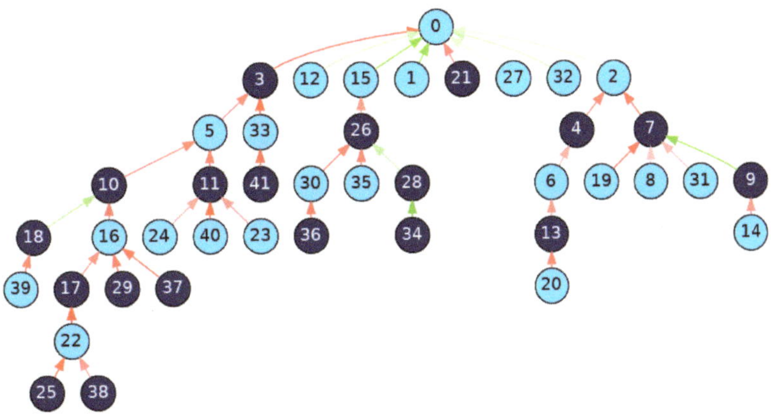

Figure 2. Two-sided debate tree for the Reddit debate of Figure 1.

At this point, we have classified the comments of a Reddit debate according to whether or not they offer support to the root comment and we have labeled the interactions between them according to the sentiment they express. So, to develop a quantitative model for measuring polarization in a Reddit debate, the next step is to introduce and investigate a suitable user-based model that allows us to represent the different interactions between the users to discover groups that either support or reject the root comment of the debate and how far these groups can be considered to be concerning each other.

To this end, once the comments are classified into two sides, in favor or against the root comment, we group comments by the user and we consider that the relationship between users' opinions of two users is defined from the agreement and disagreement relationships between the individual comments of these two users. When we represent a debate grouping comments by users, interactions between different users can give rise to circular agreement and disagreement relationships, and the agreement or disagreement of a user concerning the opinion of another user in the debate should be defined by aggregating the set of single interactions that have occurred between them during the debate.

Given a debate Γ on a root comment r with users' identifiers $U = \{u_1, \ldots, u_m\}$, we define the *opinion* of the user $u_i \in U$, denoted C_i, as the set of comments of u_i in Γ, except for the root comment r. We consider debates in which users are not self-referenced. That is, for all user $u \in U$ and each pair of comments $c_1 = (m_1, u, sc_1)$ and $c_2 = (m_2, u, sc_2)$, we assume c_1 does not respond to c_2, nor c_2 to c_1. Considering the particular case of root users (the users who post the root comment), we notice that when they do not participate in the debate, their opinion is empty, denoting that they have only posted the (root) news while staying passive throughout the debate. This is intentional since the root comment plays a special role in the debate, setting its topic. Thus, to be considered a "true participant" in the debate, the root user should contribute during the discussion. Notice that the Reddit platform itself distinguishes between root and non-root comments, as it provides two different global user metrics, Post Karma and Comment Karma, where the first one corresponds to the points achieved by posting interesting news (root comments) and the second one corresponds to the points achieved from non-root comments (debate generated on some root comment).

Next, we formalize the graph that we propose to represent user-based debates, called *User Debate Graph*, where the nodes are the users of the debate denoting their opinion concerning the root comment and the edges denote interactions between users mined from the prevailing sentiment among the aggregated comments of nodes. In addition, we define two weighting schemes: an *opinion weighting scheme* for nodes that associates every node of the graph with a side value representing the side of the user in the debate and an *interaction weighting scheme* for edges that associates every edge of the graph with a pair of values representing the overall sentiment of agreement or disagreement between users. For both schemes, we propose a *skeptical* approach based on stating that a user completely agrees or disagrees with a root comment or with another user if and only if one can be completely sure of it.

In a debate, a user can answer comments of different users, and thus, can agree or disagree with several users. This fact is represented in the User Debate Graph with a different edge for each user. However, if a user $u_i \in U$ answers several comments of a same user $u_j \in U$, the interaction between them is represented with a single edge in the User Debate Graph, and with a single sentiment value which is meant to be defined from the set of the sentiment of the answers of the user u_i to the user u_j, i.e., from the set of weights $\{W(c_1, c_2) \mid (c_1, c_2) \in E \text{ and } c_1 \in C_i \text{ and } c_2 \in C_j\}$. Moreover, a user can agree with part of a user's opinion and disagree with the rest. To reflect the possible ambivalence between the users' opinions, we also attach the edges of the User Debate Graph with a value that represents the ratio of interactions that agree with the users' opinions.

Retaining this information is crucial for the analysis of the debate since, depending on how we aggregate the sentiments, a single highly negative comment could outweigh several moderately positive ones (or vice versa). By storing both values (aggregated sentiment and ratio of positive answers), we can differentiate between consistent interactions with moderate opinions and interactions whose aggregated sentiment *seems* moderate but is a combination of inconsistent (positive and negative) responses.

Definition 4 (User Debate Graph). *Let Γ be a Reddit debate on a root comment r with users' identifiers $U = \{u_1, \ldots, u_m\}$ and let $\mathcal{T}_S = \langle C, r, E, W, S \rangle$ be a* SDebT *for Γ. A User Debate Graph* (UDebG) *for $\mathcal{T}_S$ is a tuple $\mathcal{G} = \langle \mathcal{C}, \mathcal{E}, \mathcal{S}, \mathcal{W} \rangle$, where:*

- *$\mathcal{C}$ is the set of nodes of $\mathcal{G}$ defined as the set of users' opinions $\{C_1, \ldots, C_m\}$; i.e., $\mathcal{C} = \{C_1, \ldots, C_m\}$ with $C_i = \{(m, u_i, sc) \in \Gamma \mid (m, u_i, sc) \neq r\}$, for all users $u_i \in U$.*
- *$\mathcal{E} \subseteq \mathcal{C} \times \mathcal{C}$ is the set of edges of $\mathcal{G}$ defined as the set of interactions between different users in the debate, i.e., there is an edge $(C_i, C_j) \in \mathcal{E}$, with $C_i, C_j \in \mathcal{C}$ and $i \neq j$, if and only if for some $(c_1, c_2) \in E$ we have that $c_1 \in C_i$ and $c_2 \in C_j$.*

- *$\mathcal{S}$ is an opinion weighting scheme for $\mathcal{C}$ that expresses the side of users in the debate based on the side of their comments. We define $\mathcal{S}$ as the mapping $\mathcal{S} : \mathcal{C} \to [-1,1]$ that assigns to every node $C_i \in \mathcal{C}$ the value*
$$\mathcal{S}(C_i) = \frac{\sum_{c \in C_i} S(c_i)}{|C_i|}$$
in the real interval $[-1,1]$ that expresses the side of the user u_i with respect to the root comment, from strictly disagreement (−1) to strictly agreement (1), going through undecided opinions (0).
- *$\mathcal{W}$ is an interaction weighting scheme for $\mathcal{E}$ that expresses both the ratio of positive interactions between the users' opinions and the overall sentiment between users by combining the individual sentiment values assigned to the responses between their comments.*
Let $\oplus$ be an aggregation operator of dimension n on the real interval $[-2,2]$, i.e., a bounded, monotonic, symmetric, and idempotent mapping $\oplus : [-2,2]_n \to [-2,2]$. We define $\mathcal{W}$ as the mapping $\mathcal{W} : \mathcal{E} \to ([0,1] \times [-2,2])$ that assigns to every edge $(C_i, C_j) \in \mathcal{E}$ the pair of values $(p, w) \in ([0,1] \times [-2,2])$ defined as follows:
$$p = \frac{|(c_1, c_2) \in E \cap (C_i \times C_j) \text{ with } W(c_1, c_2) > 0|}{|(c_1, c_2) \in E \cap (C_i \times C_j)|} \quad \textit{and}$$
$$w = \oplus_{\{(c_1,c_2) \in E \cap (C_i \times C_j)\}} W(c_1, c_2),$$
where p expresses the ratio of positive answers from the user u_i to the user u_j in the debate, and w expresses the overall sentiment of the user u_i regarding the comments of the user u_j, from highly negative (−2) to highly positive (2).
Only the nodes and edges obtained by applying this process belong to $\mathcal{C}$ and $\mathcal{E}$, respectively.

Notice that if a user only responds to the news of the debate (the root comment *r*), the user is mapped in the UDebG to a node in $\mathcal{C}$ with zero output degree denoting that the user starts no discussion with other users. In addition, users whose comments do not generate answers from other users are represented with nodes whose input degree is zero. Therefore, isolated nodes may appear in the UDebG that correspond to users who have neither generated nor participated in the debate, that is, users that have only answered to the news and whose opinions can be considered to be accepted by all users since they have not been discussed yet.

In the UDebG, each node denotes a user's opinion, and relationships between nodes are mined from the prevailing sentiment among the aggregated comments of those nodes. For instance, if we instantiate the aggregation operator $\oplus$ to the minimum function, we obtain a pessimistic interpretation of the degree of agreement and disagreement among users since it represents the fact that in a debate with multiple negative and positive responses from one user to another, a pessimistic analysis focuses on the most negative response and in the case of multiple positive responses, on the softest positive response. Similarly, the maximum operator corresponds to an optimistic interpretation and the mean operator to a more realistic interpretation based on an intermediate weight. Furthermore, the ratio of positive answers is either 1 or 0, only for strictly agreement or disagreement relationships, respectively, between users' opinions; thus, the ratio of positive answers is a value in $(0,1)$ only for ambivalent relationships.

Finally, since the edges between nodes in the UDebG can reflect ambivalent relationships between users' opinions, in our current user-based model, we do not consider indirect (transitive) relationships between users, and thus, the UDebG shows interactions between users only if there is enough evidence of such relationships through direct answers between authors' comments. Moreover, the UDebG for a given SDebT may contain cycles, and these cycles provide fundamental information about the different relationships that are established from the interactions between different users.

Figure 3 shows the UDebG structure we obtain for the two-sided debate tree of Figure 2. Each user's opinion is represented as a node, and each relationship between them is represented as an edge. The graph has 27 nodes and 31 edges; the overall sentiment for edges is evaluated using the mean operator, which leads to three agreement relationships, with the rest being disagreement. The edges are colored in green and red to denote agreement and disagreement relationships between users' opinions, respectively, and the darkness of the color is directly proportional to the ratio of positive and negative answers among analyzed users. The nodes are colored with different colors to denote the side of users' opinions in the debate based on the weighting scheme $\mathcal{S}$. The nodes colored in cyan correspond with users whose opinion supports the root comment, i.e., nodes $C_i \in \mathcal{C}$ with $\mathcal{S}(C_i) > 0$, the nodes colored in navy blue denote disagreement opinions, i.e., $\mathcal{S}(C_i) < 0$, and white nodes denote undecided opinions, i.e., $\mathcal{S}(C_i) = 0$. In this case, the darkness of the color reflects the proportion of user comments for and against the root comment.

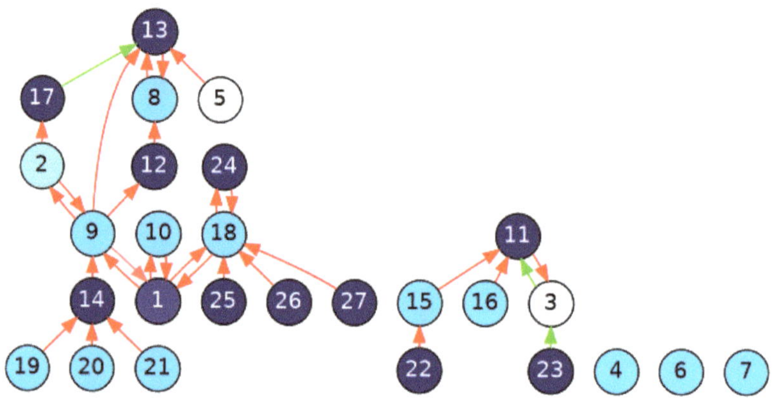

Figure 3. User Debate Graph for the two-sided debate tree of Figure 2.

3. Debate Polarization

Given a User Debate Graph $\mathcal{G} = \langle \mathcal{C}, \mathcal{E}, \mathcal{S}, \mathcal{W} \rangle$,, we propose a model to measure the level of polarization in the debate between its users. We identify two characteristics that a polarization measure should capture. First, a polarized debate should contain a bipartition of $\mathcal{C}$ into two sets (L, R) such that the set L contains mainly users in disagreement, the set R contains mainly users in agreement, and both sets should be similar in size. The second ingredient is the sentiment between users of L and R. A polarized discussion should contain most of the negative interactions between users of L and users of R, whereas the positive interactions, if any, should be mainly within the users of L and within the users of R.

We believe that both factors are relevant, as the existence of only the first characteristic can indicate simply a set of two *echo chambers*, where within each group, everybody agrees with the rest of the group but nobody pays attention to the other group. So, negative interactions between users of the two sides are the second ingredient for the right measurement of the polarization, such that the more negative the interactions between users of L and R, the higher the polarization we interpret between both sides. In our measure, the echo chambers case is considered as a case with middle polarization.

To capture these two characteristics with a single value, we define two different measures and their combination in a final one referred to as *the bipartite polarization level*. First, given a bipartition (L, R) of $\mathcal{C}$, we define its level of consistency and how balanced the sizes of L and R are as follows:

$$SC(L, R, \mathcal{G}) = LC(L, \mathcal{G}) \cdot RC(R, \mathcal{G}),$$

where $LC(L,\mathcal{G})$ evaluates the disagreement strength of the comments in L (regarding the root comment) and $RC(R,\mathcal{G})$ the agreement strength of the comments in R. They are defined as follows:

$$LC(L,\mathcal{G}) = \frac{\sum_{\substack{C_i \in L, \\ \mathcal{S}(C_i) \leq 0}} -\mathcal{S}(C_i)}{|\mathcal{C}|} \text{ and}$$

$$RC(R,\mathcal{G}) = \frac{\sum_{\substack{C_i \in R, \\ \mathcal{S}(C_i) > 0}} \mathcal{S}(C_i)}{|\mathcal{C}|}.$$

The value of $SC(L,R,\mathcal{G})$ lies in the interval $[0, 0.25]$. The minimum value (0) is achieved when either there are no users in L with $\mathcal{S}(C_i) < 0$ or no users in R with $\mathcal{S}(C_i) > 0$. The maximum value (0.25) is achieved when half of the users are users in disagreement with $\mathcal{S}(C_i) = -1$ and are found in L, and the other half are users in agreement with $\mathcal{S}(C_i) = 1$ and are found in R.

Secondly, the sentiment of the interactions between users of different sides is defined as follows:

$$SWeight(L,R,\mathcal{G}) = \frac{\sum_{\substack{e_i \in \mathcal{E} \cap \\ ((L \times R) \cup (R \times L))}} -c(p(e_i)) \cdot w(e_i)}{|\mathcal{E}|} + 2,$$

with $c(p(e_i)) = 2(p(e_i) - 0.5)^2 + 1/2$, and where $p(e_i)$ and $w(e_i)$ denote the values of p and w, respectively, in $\mathcal{W}(e_i) = (p, w)$. It is worth noticing that the value of $SWeight(L,R,\mathcal{G})$ lies in the interval $[0, 4]$:

- The minimum value (0) is achieved when all the interactions $e_i \in \mathcal{E}$ have $\mathcal{W}(e_i) = (1, 2)$ (pure positive interactions with maximum sentiment value), and all of them are found between L and R.
- The middle value (2) is achieved when the sum of negative values in the summatory of the numerator equals the sum of the positive ones. So, this represents a situation where positive and negative interactions between both partitions compensate for each other. As a special case, observe that if there are no interactions of any type between them (an echo chamber situation), then we also achieve this middle value.
- The maximum value (4) is achieved when all the interactions $e_i \in \mathcal{E}$ have $\mathcal{W}(e_i) = (0, -2)$ (pure negative interactions with minimum sentiment value), and all of them are found between L and R.

Finally, we combine the $SC(L,R,\mathcal{G})$ and the $SWeight(L,R,\mathcal{G})$ measures to define *the bipartite polarization level* of a given bipartition (L,R) as follows:

$$BipPol(L,R,\mathcal{G}) = SC(L,R,\mathcal{G}) \cdot SWeight(L,R,\mathcal{G}).$$

So, given the range of values for the previous measures, the value of $BipPol(L,R,\mathcal{G})$ lies in the interval $[0, 1]$:

- The minimum value (0) is achieved when either $SC(L,R,\mathcal{G}) = 0$ or $SWeight(L,R,\mathcal{G}) = 0$. If this happens with $SC(L,R,\mathcal{G}) = 0$ and $SWeight(L,R,\mathcal{G}) > 0$ then in L, we have only neutral and positive users, at least one user in R, and some interactions between L and R that are not extremely positive (such that $SWeight(L,R,\mathcal{G})$ can be positive). So, we consider that not having users in disagreement in L with $\mathcal{S}(C_i) < 0$ gives the lowest polarization level in the debate. We can also have $SC(L,R,\mathcal{G}) > 0$ and $SWeight(L,R,\mathcal{G}) = 0$ if both L and R contain nodes of both types (with $\mathcal{S}(C_i) < 0$ and with $\mathcal{S}(C_i) > 0$), and all the edges are found between L and R and have $\mathcal{W}(e_i) = (1, 2)$. This implies that the more edges we have in $\mathcal{E}$, the more different pairs of vertices (C_i, C_j) we will have with the same sign in $\mathcal{S}(C_i)$ and $\mathcal{S}(C_j)$, but in different sides of the partition. So, in this second case, a big quantity of edges implies an small value for $SC(L,R,\mathcal{G})$, when $SWeight(L,R,\mathcal{G}) = 0$.

- The middle value (1/2) can of course be achieved for many combinations of these two factors, but there are two canonical cases that reflect well the intended meaning of this middle case. They are the cases where one of the two factors has its maximum value and the other has its middle value. This can happen in two cases. First, this happens when $SC(L,R,\mathcal{G}) = 0.25$ and $SWeight(L,R,\mathcal{G}) = 2$. That is, when users in disagreement and users in agreement are perfectly balanced between L and R, as we discussed before when talking about the behavior of $SC(L,R,\mathcal{G})$, and either there are no interactions between L and R (echo chambers situation), or positive interactions between them are canceled by the negative ones (so we can say that, overall, the interactions are neutral between them). Second, this also happens when $SC(L,R,\mathcal{G}) = 0.25 \cdot 0.25$ and $SWeight(L,R,\mathcal{G}) = 4$. These values can be obtained when in L, the sum of negative $\mathcal{S}(C_i)$ values represents 0.25 of the total number of comments, and analogously for R with the positive values, (so the partitions are maximally heterogeneous), and all the interactions $e_i \in \mathcal{E}$ have $\mathcal{W}(e_i) = (0,-2)$ and are found only between L and R. That is, both partitions have a balanced representation of users in agreement and disagreement, and we have pure negative interactions between users in agreement of one partition and the users in disagreement of the other one.
- The maximum value (1) is achieved when both factors have its maximum value. That is, perfect balance between users in disagreement in L and users in agreement in R ($SC(L,R,\mathcal{G}) = 0.25$) and all the interactions have $\mathcal{W}(e_i) = (0,-2)$ and are found only between L and R. So, this extreme case represents a situation where there are two perfect homogeneous sides in the debate (agreement and disagreement sides) with sentiments that are maximally opposed (+1 and -1) and the only talking between them is to criticize the other side with maximum strength, so there are no positive answers between the two sides. This can be considered a very extreme situation because even in the most tense debates in a social network, one usually can find some positive answers, although these positive answers may be concentrated within each side, or not all the negative answers will have the extreme value (-2).

4. Finding the Most Polarized Partition

In this section, we present a greedy local search optimization algorithm for finding a bipartition (L,R) of $\mathcal{G} = \langle \mathcal{C},\mathcal{E},\mathcal{S},\mathcal{W}\rangle$ with a high value for $BipPol(L,R,\mathcal{G})$ that follows the same basic idea of common greedy algorithms for the maxcut problem, with the goal to find a partition with a polarization value that is close to the highest one. It seems very unlikely that we can have exact polynomial-time algorithms for our problem, as it can be shown that the maxcut problem can be reduced to a slight generalization of our optimization problem.

The algorithm follows these steps:

1. Initialize vertices in either L or R. The most obvious idea is to order vertices by polarity, and then assign the negative ones to L and the rest to R. Although this clearly gives the best possible value for $SC(L,R,\mathcal{G})$, other assignments could make improvements to the value of $SWeight(L,R,\mathcal{G})$. That is, it could be that the best partition (with respect to its value of $BipPol(L,R,\mathcal{G})$) is not a partition where all users' opinions of the same polarity are found on the same side of the partition because the value of $SWeight(L,R,\mathcal{G})$ can be significantly higher when some users' opinions with different polarity are placed on the same side. This class of users would be users that are, to some extent, inconsistent, in the sense that they seem to support one side of the debate, but at the same time, they seem to attack a significant number of users' opinions of the same side.
   To consider these kinds of partitions as possible solutions for our search algorithm, we propose two possible ways to initialize the partition (L,R):
   - g0: Place each user's opinion uniformly at random in either L or R. This way, we do not initially force a bias toward consistent partitions.

g1: Place each user's opinion C_i randomly in L with probability $P_L = \frac{1-\mathcal{S}(C_i)}{2}$ and in R with probability $1 - P_L$. In this second way, only users with a strict disagreement ($\mathcal{S}(C_i) = -1$) or a strict agreement ($\mathcal{S}(C_i) = 1$) will have a unique possible side on the initial partition.

2. Perform local improvement (regarding the value of *BipPol*) steps: at each iteration, find a good candidate from L (to be moved to R) and a good candidate from R (to be moved to L).

The pseudocode is shown in Algorithm 1.

Algorithm 1: Finding a local optimal solution for a bipartition of $\mathcal{G}= \langle \mathcal{C}, \mathcal{E}, \mathcal{S}, \mathcal{W} \rangle$ with high bipartite polarization.

Input: $\mathcal{G}= \langle \mathcal{C}, \mathcal{E}, \mathcal{S}, \mathcal{W} \rangle$
Output: a bipartition (L, R) of $\mathcal{G}$ with high bipartite polarization
$(L, R) := getInitBPart(\mathcal{G})$ `// Get Initial bipartition`
$improving := true$
$steps := 0$
while $(improving \ \&\& \ steps \leq maxsteps)$ **do**
  $improving := false$
  if $\exists v \in L, BipPol(L \setminus v, R \cup v, \mathcal{G}) > BipPol(L, R, \mathcal{G})$ **then**
    $(L, R) := (L \setminus v, R \cup v)$ `// improved bipartition`
    $improving := true$
  if $\exists v' \in R, BipPol(L \cup v', R \setminus v', \mathcal{G}) > BipPol(L, R, \mathcal{G})$ **then**
    $(L, R) := (L \cup v', R \setminus v')$ `// improved bipartition`
    $improving := true$
  $steps := steps + 1$
return (L, R)

Figure 4 shows the result of applying Algorithm 1 using the UDebG of Figure 3 as input. Nodes in red color are the nodes in the L set, and nodes in blue color are the nodes in the R set. The algorithm used to compute the initial partition is g1, and after this initialization, the algorithm performs a few steps depending on the randomness of g1. These steps are to reallocate nodes 1, 2, 3 and 5 to the sets that increase more the $BipPol(L, R, \mathcal{G})$ value, in case they are not already there. In this example, the reallocation of other nodes does not increase this value. We can observe that the only nodes within the same set that disagree (red arrows) between them are nodes 2 and 9 in the R set (blue nodes), and 3 and 11 in the L set (red nodes). All the other disagreement relations are between nodes of different sets, and all the agreement relations (green arrows) are between nodes of the same set. The final value of $BipPol(L, R, \mathcal{G})$ is 0.5161, being that the other values $LC = 0.4296$, $RC = 0.4419$, and $Sweight = 2.718$, as we will see in the next section, are common values for debates in the subreddit World News.

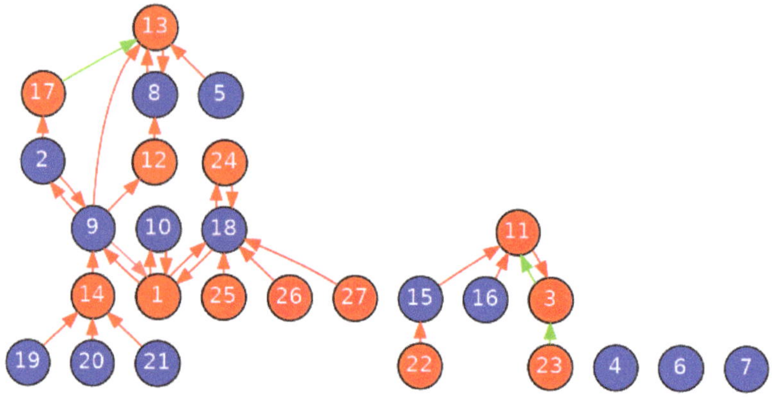

Figure 4. Bipartite graph showing sets L (red nodes) and R (blue nodes) after running Algorithm 1 using the UDebG of Figure 3 as input.

5. Empirical Evaluation

In this section, we present an empirical evaluation to detect if our optimization algorithm can find distinctive polarization in debates obtained from six representative subreddits with different topics: Halloween, travel, movies, Bitcoin, politics, and worldnews. Intuitively, we expect to have debates with low polarization in the Halloween and travel topics, those with higher polarization in the movies and Bitcoin topics, and those with the highest polarization in the politics and worldnews topics. From each subreddit, we downloaded debates from the last month, excluding the most recent debates that probably are still active, with comments per debate ranging from 50 to 500, and with a limit of 100 debates per subreddit.

Table 1 shows the average results for the polarization value (*BipPol*) obtained with our algorithm for each of the six sets of debates using the algorithm with the two initialization methods, so we have two rows per subreddit.

Table 1. Results with different initialization algorithms with debates from several subreddits. The numbers shown are average values over the debates of each subreddit.

Subreddit	#Debates	$\lvert\mathcal{C}\rvert$	$\lvert\mathcal{E}\rvert$	Alg.	*BipPol*	*LC*	*RC*	*SWeight*
Halloween	74	78.16	30.09	g0	0.4053	0.3681	0.5603	2.1594
				g1	0.4087	0.3685	0.5654	2.1643
travel	49	75.10	60.57	g0	0.4246	0.4835	0.4005	2.3808
				g1	0.4245	0.4836	0.4004	2.3808
movies	100	146.51	123.23	g0	0.4785	0.4826	0.4189	2.3930
				g1	0.4783	0.4825	0.4189	2.3926
Bitcoin	100	135.01	95.82	g0	0.4879	0.5332	0.3862	2.4360
				g1	0.4883	0.5333	0.3863	2.4367
politics	100	205.93	160.72	g0	0.5104	0.5713	0.3511	2.5742
				g1	0.5105	0.5712	0.3511	2.5752
worldnews	100	148.40	163.36	g0	0.5141	0.4950	0.4076	2.5732
				g1	0.5142	0.4949	0.4076	2.5732

In any case, the maximum number of iterations is fixed to the number of users in the debate graph. To analyze the reasons behind the *BipPol* values obtained, we also show the corresponding average values for the factors *LC*, *RC* (that make the *SC* value), and *SWeight*. The table also shows the total number of debates in each set, and the average values for the number of users and number of edges in their corresponding user debate graphs. The average results for the different subreddits show some differences in the *BipPol* values such that we can divide the six subreddits into three groups: subreddits with the lowest *BipPol*, the middle one, and the highest one.

In the first group (Halloween and travel), we observe average *BipPol* values around 0.4. However, looking into the value of the different factors, we can explain why there is an slightly higher *BipPol* value for the travel subreddit. For the Halloween subreddit, we observe that the agreement users group is more relevant (*RC* value around 0.56 in contrast to a *LC* value around 0.36) and a *SWeight* value around 2.15 (close to the neutral value for the global sentiment between users in agreement and disagreement). By contrast, for the travel subreddit, the *LC* value is higher than the *RC* value, but they are more balanced (0.48 versus 0.40), and the *SWeight* value is around 2.38. So now, negative interactions between both groups are more significant. These differences make sense if one looks into the nature of the discussions in both these subreddits. In the Halloween case, discussions are mainly friendly, discussing ideas about costumes and Halloween decorations. However, in the travel case, we find some controversies in many discussions about certain touristic destinations. Regarding the differences between using either the g0 or g1 initialization method in the algorithm, we observe no significant differences in the final value obtained; the only difference we observe is that when using the uniform random method (g0), the number of iterations needed by the algorithm can be about ten times higher, but in all the debates we have solved, the algorithm has never reached its maximum number of allowed iterations. So, even if the method (g0) makes the algorithm make more iterations, it seems that the quality of the solution obtained is similar to the one with g1.

In the second group (movies and Bitcoin), we observe average *BipPol* values around 0.47. The increase with respect to the first group can be explained due to slightly more balanced values for *LC* and *RC* (close to the 0.5 value), although in the more hot topic of Bitcoin, the disagreement group value (*LC*) is more significant, and also due to increased values for *SWeight*, but again, this is more significant in the Bitcoin sudreddit. It is worth noticing that a higher value for *SWeight* seems to be correlated with a higher size of the debate (in the values of $|\mathcal{C}|$ and $|\mathcal{E}|$). This is consistent with what we commented earlier about the tendency to respond more frequently when people disagree.

In the last group (politics and worldnews), we observe the highest *BipPol* average values, around 0.51. Here the values of *LC* and *RC* show a balance similar to the one of the second group (politics is similar to Bitcoin, and worldnews is similar to movies). However, the difference lies in the increase in the *SWeight* values, that again seems to be correlated with an increase in the size of the debates. The existence of the highest values for the *SWeight* in this last group of subreddits is somehow natural, as many of the topics discussed are highly controversial.

To further investigate the characteristics of the debates of these different subreddits, we also looked into the results obtained for the debates of each group with the minimum, median, and maximum *BipPol* values within the group. Table 2 shows these results, with three rows per subreddit, showing the results for the minimum, median, and maximum cases.

Looking at the results for the minimum, we observe that clearly the lowest *BipPol* values are obtained in the Halloween and travel subreddits (0.069 and 0.101, respectively), in contrast to the minimum *BipPol* value for the other subreddits, which is at least 0.29. The median values show differences similar to the ones we observed for the average value in Table 1, from the median value of 0.41 for Halloween to the median value of 0.51 for worldnews, so the differences in the minimum value are more significant. The maximum values are at least 0.56 (the maximum for Halloween), but the other maximum values are similar, around 0.6. Curiously, the highest maximum value is observed in a debate from the travel subreddit (0.68). So, these results indicate that one can find polarized debates on many topics, but when looking at the average values of the previous table, we can conclude that the average polarization tends to be higher for certain topics.

Table 2. Results for the min/median/max values of *BipPol* of the same debates of Table 1.

Subreddit	DebateID	$\vert\mathcal{C}\vert$	$\vert\mathcal{E}\vert$	*BipPol*	*LC*	*RC*	*SWeight*
Halloween	qdg24n	27	27	0.0692	0.0370	0.9246	2.0219
	qg6q98	84	39	0.4156	0.5765	0.3294	2.1890
	q71wxm	41	4	0.5626	0.3780	0.5935	2.5074
travel	n1uxcn	16	25	0.1019	0.6771	0.0625	2.4072
	muovcd	34	15	0.4379	0.4706	0.3725	2.4979
	mu25ze	27	16	0.6857	0.5556	0.4296	2.8730
movies	q7g2fh	135	156	0.3400	0.4547	0.3728	2.0054
	q1vw6l	125	111	0.4809	0.4680	0.4320	2.3786
	q8975v	87	86	0.5939	0.5594	0.4103	2.5874
Bitcoin	mmx2og	32	23	0.2953	0.7585	0.1875	2.0761
	mwglkl	110	62	0.4912	0.4818	0.4498	2.2668
	mtux4x	43	12	0.6107	0.5349	0.4651	2.4546
politics	qhm2di	139	153	0.4318	0.5454	0.3120	2.5374
	qcrq1v	216	201	0.5082	0.4961	0.3997	2.5630
	qerwa6	217	141	0.5927	0.5668	0.3944	2.6513
worldnews	morqje	78	91	0.3754	0.4983	0.2949	2.5553
	mq580o	195	257	0.5175	0.4521	0.4434	2.5816
	morpw5	59	68	0.6393	0.5036	0.4669	2.7186

6. Conclusions and Future Work

In this work, we introduce a quantitative model for measuring polarization in an online debate such that this behavior can be monitored for generating a warning signal when the debate polarization reaches some threshold value.

Among the many online debating platforms that exist, in this work, we consider Reddit; we model a Reddit debate as a weighted graph and with labeled edges, where the weights of nodes represent the side of the users' opinions in the debate and the labels of edges represent the overall sentiments between users' opinions. Given a bipartition of this graph, we quantify the degree of polarization of the bipartition by measuring how homogeneous each partition is and how negative the interactions are between both partitions. Then, our quantitative model is based on the maximum polarization we have in all the possible bipartitions of the graph, and it is computed with a greedy local search optimization algorithm.

Finally, we develop an empirical evaluation of the debate polarization value obtained with our algorithm with different Reddit debates. The results indicate that our quantitative model captures differences in the polarization level of different discussions, which can be further understood when analyzing the values of the different factors used to define the measure.

As future work, we plan to study the suitability of other schemes for the aggregation of individual interactions between two users, when computing the sentiment weighting scheme, as well as different features to capture the degree of polarization of users' opinions in online debates. For example, to quantify the weight that corresponds to each node of the graph, it can be interesting to consider parameters available on the Reddit platform, such as the user's Karma and the comment score, since active users that post interesting comments can be considered more relevant than users who are not very active or whose comments are valued or rated with negative scores. Moreover, we would also like to consider the identification of other communication patterns that may not be considered as polite, such as, for example, when users are *ratioed* on an online discussion. That is, users whose comments are ignored, but their replies receive much more attention. For a discussion about getting ratioed on Twitter see https://www.popbuzz.com/internet/social-media/ratioed-meaning-twitter/.

Author Contributions: Conceptualization, T.A., J.A., R.B. and S.M.; methodology, T.A., J.A., R.B. and S.M.; software, T.A., J.A., R.B. and S.M.; validation, T.A., J.A., R.B. and S.M.; formal analysis, T.A., J.A., R.B. and S.M.; investigation, T.A., J.A., R.B. and S.M.; data curation, T.A., J.A., R.B. and S.M.; writing—original draft preparation, T.A., J.A., R.B. and S.M.; writing—review and editing, T.A., J.A., R.B. and S.M.; visualization, T.A., J.A., R.B. and S.M.; supervision, T.A., J.A., R.B. and S.M.; project administration, T.A., J.A., R.B. and S.M.; funding acquisition, T.A., J.A., R.B. and S.M. All authors have contributed equally to the work. All authors have read and agreed to the published version of the manuscript.

Funding: This research was funded by Spanish Project PID2019-111544GB-C22 (MINECO/FEDER), by the European Union's Horizon 2020 Research and Innovation Program under Grant Agreements 723596, 768824, 764025 and 814945, and by 2017 SGR 1537.

Institutional Review Board Statement: We choose to exclude this statement since the study do not involve humans and animals.

Informed Consent Statement: We choose to exclude this statement since the study do not involve humans.

Data Availability Statement: We choose to exclude this statement since the study did not report any data.

Acknowledgments: The authors would like to thank the anonymous reviewers for providing helpful comments.

Conflicts of Interest: The authors declare no conflict of interest.

References

1. Newton, C. How to think about polarization on Facebook. *The Verge*, 28 May 2020.
2. Horwitz, J.; Seetharaman, D. Facebook Executives Shut Down Efforts to Make the Site Less Divisive. *The Wall Street Journal*, 26 May 2020.
3. Rosen, G. Facebook: Investments to Fight Polarization. 2020. Available online: https://about.fb.com/news/2020/05/investments-to-fight-polarization/ (accessed on 27 May 2020).
4. Hagey, K.; Horwitz, J. Facebook Tried to Make Its Platform a Healthier Place. It Got Angrier Instead. *The Wall Street Journal*, 16 September 2021.
5. Boxell, L.; Gentzkow, M.; Shapiro, J.M. *Cross-Country Trends in Affective Polarization*; National Bureau of Economic Research: Cambridge, MA, USA, 2020.
6. Bessi, A.; Zollo, F.; Vicario, M.D.; Puliga, M.; Scala, A.; Caldarelli, G.; Uzzi, B.; Quattrociocchi, W. Users Polarization on Facebook and Youtube. *PLoS ONE* **2016**, *11*, e0159641. [CrossRef] [PubMed]
7. Vicario, M.D.; Vivaldo, G.; Bessi, A.; Zollo, F.; Scala, A.; Caldarelli, G.; Quattrociocchi, W. Echo chambers: Emotional contagion and group polarization on Facebook. *Sci. Rep.* **2016**, *6*, 37825. [CrossRef] [PubMed]
8. Recuero, R.; Soares, F.B.; Gruzd, A.A. Hyperpartisanship, Disinformation and Political Conversations on Twitter: The Brazilian Presidential Election of 2018. In Proceedings of the Fourteenth International AAAI Conference on Web and Social Media, ICWSM 2020, Atlanta, GA, USA, 8–11 June 2020; Choudhury, M.D., Chunara, R., Culotta, A., Welles, B.F., Eds.; AAAI Press: Menlo Park, CA, USA, 2020; pp. 569–578.
9. Morales, G.D.F.; Monti, C.; Starnini, M. No echo in the chambers of political interactions on Reddit. *Sci. Rep.* **2021**, *11*, 2818. [CrossRef] [PubMed]
10. Noelle-Neumann, E. The Spiral of Silence a Theory of Public Opinion. *J. Commun.* **2006**, *24*, 43–51. [CrossRef]
11. Stoycheff, E. Under Surveillance: Examining Facebook's Spiral of Silence Effects in the Wake of NSA Internet Monitoring. *J. Mass Commun. Q.* **2016**, *93*, 296–311. [CrossRef]
12. Murakami, A.; Raymond, R. Support or Oppose? Classifying Positions in Online Debates from Reply Activities and Opinion Expressions. In Proceedings of the COLING 2010, 23rd International Conference on Computational Linguistics, Beijing, China, 23–27 August 2010; Huang, C., Jurafsky, D., Eds.; Chinese Information Processing Society of China: Beijing, China, 2010; pp. 869–875.
13. Alsinet, T.; Argelich, J.; Béjar, R.; Martínez, S. Measuring user relevance in online debates through an argumentative model. *Pattern Recognit. Lett.* **2020**, *133*, 41–47. [CrossRef]
14. Manning, C.D.; Surdeanu, M.; Bauer, J.; Finkel, J.; Bethard, S.J.; McClosky, D. The Stanford CoreNLP Natural Language Processing Toolkit; In Proceedings of the Association for Computational Linguistics (ACL) System Demonstrations, Baltimore, MD, USA, 22–27 June 2014; pp. 55–60.
15. Agrawal, R.; Rajagopalan, S.; Srikant, R.; Xu, Y. Mining newsgroups using networks arising from social behavior. In Proceedings of the 12th International Conference on World Wide Web, Budapest, Hungary, 20–24 May 2003; pp. 529–535.

applied sciences

MDPI

Article

Comparison Uncertainty of Different Types of Membership Functions in T2FLS: Case of International Financial Market

Zuzana Janková [1,*] and Eva Rakovská [2]

1 Institute of Informatics, Faculty of Business and Management, Brno University of Technology, 61200 Brno, Czech Republic
2 Faculty of Economic Informatics, University of Economics in Bratislava, 85235 Bratislava, Slovakia; eva.rakovska@euba.sk
* Correspondence: Zuzana.Jankova@vutbr.cz

Abstract: This article deals with the determination and comparison of different types of functions of the type-2 interval of fuzzy logic, using a case study on the international financial market. The model is demonstrated on the time series of the leading stock index DJIA of the US market. Type-2 Fuzzy Logic membership features are able to include additional uncertainty resulting from unclear, uncertain or inaccurate financial data that are selected as inputs to the model. Data on the financial situation of companies are prone to inaccuracies or incomplete information, which is why the type-2 fuzzy logic application is most suitable for this type of financial analysis. This paper is primarily focused on comparing and evaluating the performance of different types of type-2 fuzzy membership functions with integrated additional uncertainty. For this purpose, several model situations differing in shape and level or degree of uncertainty of membership functions are constructed. The results of this research show that type-2 fuzzy sets with dual membership functions is a suitable expert system for highly chaotic and unstable international stock markets and achieves higher accuracy with the integration of a certain level of uncertainty compared to type-1 fuzzy logic.

Keywords: computational finance; fuzzy logic; membership function; Type-1 fuzzy sets; T1FLS; Type-2 fuzzy sets; T2FLS

Citation: Janková, Z.; Rakovská, E. Comparison Uncertainty of Different Types of Membership Functions in T2FLS: Case of International Financial Market. *Appl. Sci.* **2022**, *12*, 918. https://doi.org/10.3390/app12020918

Academic Editors: Aida Valls and Karina Gibert

Received: 3 December 2021
Accepted: 10 January 2022
Published: 17 January 2022

1. Introduction

Fuzzy set theory was first introduced by Lotfi Zadeh in the 1960s as a way to capture the uncertainty and ambiguity often overlooked in complex systems. It can be considered as a generalization of the theory of classical sets. As research into fuzzy logic continues, fuzzy sets (FS) are gradually being refined to better reflect the original idea of modeling linguistic variables, which builds on the pillars of inaccuracy, vagueness and ambiguity that arise in language variables. This resulted in three main streams of fuzzy set representation: (a) type-1 fuzzy set (T1FS); (b) general type-2 fuzzy sets (GT2FS); (c) interval type-2 fuzzy sets (IT2FS). However, the computations with type-2 fuzzy sets are complex [1]. To solve this, simplification of type-2 is conducted through a special kind of type-2 fuzzy set called interval type-2 fuzzy set. Interval type-2 fuzzy sets manage uncertainty in membership functions, but computations are less intense, keeping the capabilities to process uncertainties [2,3]. The first mentioned fuzzy sets are considered to be a very simplified form of representation of linguistic variables, which is able to integrate exclusively a certain degree of uncertainty according to [4,5], while type-2 interval fuzzy sets are able to integrate uncertainty in the form of intervals, thus not limiting the level of uncertainty. Although IT2FS are more computationally intensive and more complex to design, according to [6], these sets are better able to deal with the additional noise and the nonlinear and chaotic environment of input data from various disciplines.

The reason for the widespread use and ever-increasing popularity of fuzzy logic in many scientific fields lies in the fact that it is able to better understand human thinking and, above all, is able to process uncertain, ambiguous and incomplete data. The selection and setting of the correct membership functions and the determination of the relevant fuzzy rules on the basis of which the model provides outputs is absolutely essential for the correct and accurate functionality of the fuzzy model. While setting fuzzy rules can be solved using a team of experts or through neural networks, such as the ANFIS model, the construction of fuzzy membership functions and the adequate setting of their parameters is a demanding task and prone to errors [7]. Thanks to the possibility of the parameterization of standard and widely used fuzzy functions (for example triangular, trapezoidal or Gaussian) it is possible to directly and consistently compare the performance of these functions in type-1 and type-2 fuzzy models. As described by [6], in addition to these standard functions, less common fuzzy functions can be found; however, these can also be easily converted from type-1 to type-2 by parameterization. As [8] point out, the choice of the correct fuzzy function, including the appropriate setting of parameters, thus plays a fundamental and indispensable role in the design, implementation, performance and subsequent interpretation of the fuzzy model. The authors add that these parameters of fuzzy functions are possibilistic in nature and depend on the specific area or domain in which the model is applied. Expert research or other sophisticated methods that set the parameters are necessary for the correct selection and setting of fuzzy functions. Study in [9] emphasizes that the correct choice of fuzzy functions depends to a large and not negligible extent on the subjective perception of inaccurate or vague data. In other words, there are no criteria or consensus to determine the appropriate fuzzy functions for a particular area, although, as [10] points out, these functions play a central role in the high performance of the fuzzy model. Thus, fuzzy functions can be considered as building blocks of fuzzy set theory. However, fuzzy functions must satisfy the necessary condition that they must be in the range 0 to 1. If this condition is met, fuzzy functions can take various shapes and forms.

It can thus be concluded that the correct choice of fuzzy functions is a specific problem, and many researchers and experts have justifiably given the attention of this issue. The next part provides an overview of the literature dealing with finding a suitable fuzzy function. When using different MFs for a given problem, it has usually been found that Gaussian and triangular MFs have very good results, better than other types of MFs. Research in [11] compared the response of the system with different MFs and expressed the view that a triangular MF is better than other MFs. In fact, if one does not have priority for the MF shape, triangular or trapezoidal shapes are easy to implement and calculate quickly. However, if someone has certain priorities for their shapes (e.g., from histograms on sampled data), it may be interesting to create MFs with shapes derived from these apriori shapes after some smoothing, if necessary, according to [10]. In their research, paper [12] used the Gaussian membership function in the ANFIS model to test data on investment instruments. The fuzzy model is applied to the Tehran stock exchange index (TEPIX). The created model was able to predict the development of the stock index with high accuracy. A similar recommendation can be found in [13], which also recommends Gaussian membership functions. The authors further emphasize that this feature best reflects the nature of stock market data. [14] modified the conventional adaptive neuro-fuzzy inference model (ANFIS) using Gaussian functions of layer 4 membership instead of standard polynomial functions. The modified model achieved better computational performance for stock market prediction tasks. In contrast, in the study by [15] integrated technical indicators into the fuzzy system as input variables broken down into seven attributes, which they described using triangular fuzzy functions.

Fuzzy sets are used to express ambiguities in the member function of fuzzy sets. However, so far, as follows from the above, there is no consensus on the geometry of the membership functions. These member functions are usually constructed using numeric data or a range of classes. However, there is uncertainty about the form of membership, i.e., whether it is a function of membership in a triangle or a function of trapezoidal membership, or other. When creating any fuzzy model, strong emphasis should be placed on membership functions, which are neglected and insufficiently researched in most research papers, which can cause significant problems with the functionality of the model. The expression of membership functions depends on both the subject (how deep the researcher's experience is) and the context (where the problem is). For this reason, the present research focuses on the use of membership functions with varying degrees of uncertainty, which is gradually increasing. Furthermore, not only one type of membership function is used, as is usual in the vast majority of studies, but triangular, trapezoidal, Gaussian and bell membership functions are gradually created. Not to rely solely on one membership function is intentional, because in addition to the knowledge base, membership functions are essential for the proper functioning of the model. Incorrect description of input data through membership functions can provide incorrect outputs. Especially when applied to the stock market, incorrect description of fuzzy functions can cause significant differences in the profitability of the investment strategy and can cause generating incorrect investment signals, which can cause losses to potential investors.

The paper is organized as follows: Section 1 introduction the subject matter, including the identification of the urgency of the problem and the definition of the aim of the work; Section 2 provides the theoretical background and examines already published works; Section 3 is focused on the theory of fuzzy mathematical background of fuzzy system type-1 and type-2; Section 4 is devoted to the methodology and experiments on the data of the international stock market; Section 5 discusses obtained results and evaluation of the created model; Finally, Section 6 summarizes outputs of the paper, recognized limits and suggestions for the subsequent research.

2. Theoretical Background

Numerous papers and research focused on the prediction of financial time series of stock indices using type-1 fuzzy logic can be found in the literature. However, classical type-1 fuzzy sets are not able to fully capture the additional uncertainty that is a feature of stock markets. For this reason, in recent years, research has focused on the application of type-2 fuzzy logic in economic areas. The latest research in this area is described below. The aim was to select research studies, taking into account the membership functions used, but there are not many articles focused on this topic. The type of function is not specified, as the vast majority of researchers do not pay much attention to choosing the appropriate membership function, but the accuracy of fuzzy model outputs and researchers state the function they use in their fuzzy models.

Authors in [16] use several stock indices to predict market trends: Bombay Stock Exchange; CNX Nifty; and S&P 500. The authors chose the type-1 fuzzy model, which integrates ANFIS, to predict them. The authors used fuzzy sets in the Gaussian fuzzy model. However, the authors point out that the integration of neural networks through the ANFIS model significantly reduces the interpretability of the created system. Researches [17,18] apply triangular fuzzy functions to the moving average prediction and the obtained forecast is implemented on the share at Bombay Stock Exchange. Paper [19] develops a fuzzy trading system integrating technical indicators on the basis of which they predict the development of the stock index. [20] predict the stock price of the State Bank of India (SBI) and the Dow-Jones Industrial Average (DJIA) using a fuzzy model. As part of fuzzification, the authors use a triangular fuzzy function, which is very simple and widespread, as the authors note. Other paper [21] integrates technical indicators into the model in order to improve the portability of the Athens Stock Exchange stock price forecast. The output of the strategy are signals to buy or sell a specific share. Authors [22] use two US stock indices, the DOW30 and the NASDAQ100 modified ANFIS model to predict. According to their empirical results, the bell functions of membership outperform trapezoids in performance. The authors also draw attention to a fundamental finding: the numbers and types of fuzzy functions have a fundamental and substantial influence in the process of predicting for financial time series. Paper [23] proposes a type-2 fuzzy model for stock index forecasting: Taiwan Stock Exchange Capitalization Weighted Stock Index, the Dow Jones Industrial Average and the National Association of Securities Dealers Automated

Quotation. As an input variable, they use a time series delayed by one day in the fuzzy model. [24] predict the closing prices of TAIEX and NASDAQ stock indices. They also note that membership functions are an important aspect of prediction. The authors use triangular MF, because according to their knowledge and experience they provide better performance than Gaussian. Authors [25] use T2FLS to form a profit analysis decision support model. The authors apply type-2 fuzzy sets because type-1 is too noisy. [26] in their research suggest the creation of T2FLS to improve and refine the effectiveness of the TAIEX and COVID-19 stock index prediction model.

As is mentioned above, the stock market, and therefore the entire financial system, is characterized by a highly unstable environment that is constantly changing. This is also one of the reasons why it is not possible to choose a universal form of membership functions within decision-making models or stock market prediction (see Table 1). However, this is a crucial area, as incorrect description of input data through membership functions results in incorrect output. In this case, incorrect investment signals or the indication of false signals to buy or sell, which may ultimately lead to the loss of investors or the impairment of their funds. The above literature review of scientific and research papers shows the following:

1. there is no consensus on the choice of fuzzy model membership functions in the stock market. Different authors choose different membership functions, mostly resorting to the simplest ones without deeper examination and deeper links to the nature of the input data sources;
2. it has been shown that type-2 fuzzy logic, the respective dual functions of this type of fuzzy sets, are insufficiently explored in the stock market, as this is an area that has only been widely applied in recent years;
3. in the case of dual membership functions, the uncertainty or degree of uncertainty contained between these functions is insufficient or not fully examined to provide more accurate outputs of the respective more accurate predictions.

Table 1. Summary of literature background.

Author	Fuzzy Sets	Type	Problem
[16]	Gaussian	T1FS	stock index prediction
[23]	Triangular	T2FS	stock index prediction
[17]	Triangular	T1FS	stock index prediction
[18]	Triangular	T1FS	stock index prediction
[19]	Trapezoidal	T2FS	stock index prediction
[20]	Triangular	T1FS	stock index prediction
[22]	Bell	ANFIS	stock index prediction
[21]	Bell	T1FS	stock prediction
[24]	Triangular	T1FS; T2FS	stock index prediction
[25]	Triangular	T2FS	profit prediction
[26]	Triangular	T2FS	stock index prediction

For this reason, the authors of this paper consider this area to be very important, especially a detailed examination of membership functions in terms of not only their appropriate shape (T1FL and T2FL) resulting from the nature of the selected dataset, but also the appropriate degree of uncertainty between dual membership functions (T2FLS) resulting from the nature of the stock market.

3. Type-2 Fuzzy Logic

The fuzzy inference is very similar for type-1 and type-2 fuzzy logic. First, a fuzzification process is performed, in which the measured real data are transformed into language variables. Authors [27] state that these language variables are represented by attributes, while the number of these attributes is normally in the range from 3 to 7. The degrees of attributes are then graphically represented by a mathematical function (see Figure 1). Within the higher level of fuzzy logic, there are three possible variants of fuzzification:

(a) sharp set; (b) type-1 fuzzy set; (c) type-2 fuzzy set. The former is used if the input data is perfect, i.e., it does not contain noise. Otherwise, fuzzy sets are selected. It is necessary to note that data with stationary noise are modeled as type-1 fuzzy sets and data with non-stationary noise are modeled as type-2 fuzzy sets.

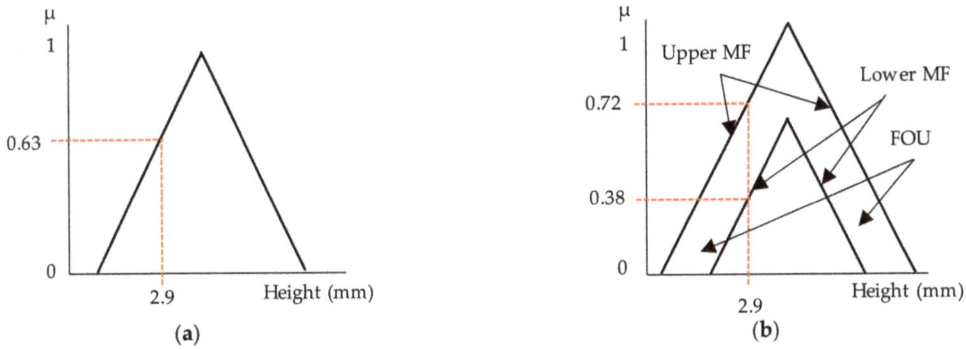

Figure 1. The difference between (**a**) T1FS and (**b**) T2FS.

The distribution function, which represents type-2 fuzzy set, can be according to [28,29] write as:

$$\widetilde{A} = \int_{x\epsilon X} \int_{u\epsilon J_x} \frac{\mu_{\widetilde{A}}(x,\mu)}{x,u} = \int_{x\epsilon X} \left[\int_{u\epsilon J_x} \mu_{\widetilde{A}}(x,\mu)/(x,u) \right] /x, \quad (1)$$

where x is the primary variable $J_x \epsilon [0,1]$ is the primary membership of x, u is the second variable, and $\int_{u\epsilon J_x} \mu_{\widetilde{A}}(x,\mu)/(x,u)$ is secondary possibility distribution at x.

Paper [30] generalized the fuzzy set interval and defined the term IT2FLS. The requirement for secondary possibility distribution is a condition of normality, which means that the X elements are fully distributed for x, which are defined as follows. IT2FLS is:

$$\widetilde{A} = \int_{x\epsilon X} \int_{u\epsilon J_x} \frac{1}{x,u} = \int_{x\epsilon X} \left[\int_{u\epsilon J_x} 1/(x,u) \right] /x, \quad (2)$$

where x is the primary variable, $J_x \epsilon [0,1]$ is the primary membership of x, u is the second variable, and $\int_{u\epsilon J_x} 1/(x,u)$ is secondary possibility distribution at x.

For interval type-2 fuzzy set $\widetilde{X}$ are upper possibility distribution defined as $\overline{\mu}(x)$ and lower possibility distribution defined as $\underline{\mu(x)}$ type-1 possibility distribution. According to

Mendel et al., 2006 the footprint uncertainty of $\widetilde{X}(FOU(\widetilde{X}))$ is defined as:

$$FOU\left(\widetilde{X}\right) = \bigcup_{x\in X} J_s = \left\{(x,y) : J_X = \left[\overline{\mu}(x), \underline{\mu}(x)\right]\right\}, \quad (3)$$

Interval type-2 fuzzy set $\widetilde{X} = \left[\overline{\mu}(x),\ \underline{\mu}(x)\right] = \left((a,b,d;h^U),\ (e,b,f;h^L)\right)$, where $\overline{\mu}(x)$

and $\underline{\mu}(x)$ are type-1 fuzzy sets of a,b,d,e,f are reference points of the IT2FLS, h^U indicates the reference point of a,b,d in the upper possibility function, h^L indicates the reference point of e,b,f in lower possibility function, $h^U \epsilon [0,1]$, and $h^L \epsilon [0,1]$. From a mathematical point of view, the lower and upper possibility distribution can be written as follows:

$$\overline{\mu}^U(x) = \begin{cases} \frac{\mu^U(x-a)}{b-a}, & a \le x \le b \\ \frac{\mu^U(d-x)}{d-b}, & b \le x \le d \\ 0, & otherwise \end{cases}, \quad (4)$$

$$\underline{\mu}^L(x) = \begin{cases} \frac{\mu^L(x-e)}{b-e}, & e \le x \le b \\ \frac{\mu^L(f-x)}{f-b}, & b \le x \le f \\ 0, & otherwise \end{cases}, \tag{5}$$

Subsequently, a comparative approach based on the possibility of uncertain mean and variation coefficient for IT2FLS is introduced, as described below.

As stated by [31], for all IT2FLS $\widetilde{X} = ((a,b,d;h^U), (e,b,f;h^L))$, whose possibility uncertainty mean value are as follows:

$$M\left(\widetilde{X}\right) = \frac{\widetilde{M(X^U) + M(\widetilde{X^L})}}{2}, \tag{6}$$

where possibility uncertainty means of the upper membership function $M(\widetilde{X^U})$ and lower membership function $M(\widetilde{X^L})$ are respectively written as:

$$M\left(\widetilde{X^U}\right) = 1/2 \int_0^{h^U} \left(\underline{X^U}(\alpha) + \overline{X^U}(\alpha) + 2b\right) f(\alpha) d\alpha, \tag{7}$$

$$M\left(\widetilde{X^L}\right) = 1/2 \int_0^{h^L} \left(\underline{X^L}(\alpha) + \overline{X^L}(\alpha) + 2b\right) f(\alpha) d\alpha, \tag{8}$$

$f(r)$ is an increasing function satisfying $f(0) = 0,\ f(1) = 1$ and $\int_0^{h^U} f(\alpha) d\alpha = 1/2$.

Authors [31] further describe that the variation coefficient is written for all IT2FLS as:

$$VC(\widetilde{X}) = \begin{cases} \frac{D(\widetilde{X})}{M(\widetilde{X})}, & if\ M\left(\widetilde{X}\right) \neq 0 \\ \frac{D(\widetilde{X})}{\epsilon}, & if\ M\left(\widetilde{X}\right) = 0 \end{cases}, \tag{9}$$

where ϵ is very small value to present the approximate $M\left(\widetilde{X}\right)$, $D\left(\widetilde{X}\right)$ is the variation values. The $D\left(\widetilde{X}\right)$ is mathematically written as:

$$\begin{gathered} D(X) = \sqrt{\widetilde{D(\widetilde{X^U})\, D\left(\widetilde{X^L}\right)}}, \\ D(\widetilde{X^U}) = 1/4 \int_0^{h^U} \left(\overline{X^U}(\alpha) - \underline{X^U}(\alpha)\right)^2 f(\alpha) d\alpha, \\ D(\widetilde{X^L}) = 1/4 \int_0^{h^L} \left(\overline{X^L}(\alpha) - \underline{X^L}(\alpha)\right)^2 f(\alpha) d\alpha, \end{gathered} \tag{10}$$

where $f(\alpha)$ is an increasing function satisfying $f(0) = 0,\ f(1) = 1$ and $\int_0^{h^U} f(\alpha) d\alpha = 1/2$.

According to [31], two IT2FLS $\widetilde{X}$ and $\widetilde{Y}$ are considered, whose benchmarks are defined as follows:

$$\begin{gathered} If\ M\left(\widetilde{X}\right) < M\left(\widetilde{Y}\right),\ then\ \widetilde{X} \prec \widetilde{Y}, \\ If\ M\left(\widetilde{X}\right) > M\left(\widetilde{Y}\right),\ then\ \widetilde{X} \succ \widetilde{Y}, \\ If\ M\left(\widetilde{X}\right) = M\left(\widetilde{Y}\right),\ then, \\ if\ VC(\widetilde{X}) < VC\left(\widetilde{Y}\right),\ then\ \widetilde{X} \prec \widetilde{Y}, \\ if\ VC(\widetilde{X}) > VC\left(\widetilde{Y}\right),\ then\ \widetilde{X} \succ \widetilde{Y}, \\ else\ \widetilde{X} \sim \widetilde{Y}, \end{gathered} \tag{11}$$

It expresses that > means "larger than", < means "less than" and $\sim$ means "same order".

3.1. Probability Information and Its Measure

Assume that x is a variable that expresses the value of X. The probability distribution is modeled as: $P : X \to [0,1]$. Here, for each $x \epsilon X$, $p(x)$ indicates the probability x, that the value is in X, and satisfies $\sum_{x_j \in X} P(x_j) = 1$.

Consequently, it can be stated that for each fuzzy subset A also has its probability, which indicates that x lies in A, that is $Prob(A) = \sum_{x_j \epsilon X} P(x_j)$. This is written as $Prob(X) = 1$ and $Prob(\varnothing) = 0$.

The calculation of the probability of a fuzzy set was proposed by [32]. Let B be a fuzzy subset of X such that $B(x_j)$ is membership grade of (x_j) in B. It is recommended, according to Wu and Mendel (2009), that:

$$Prob(B) = \sum_{j=1B}^{n} B(x_j)P(x_j), \tag{12}$$

Obviously, B is a crisp set if $B(x_j) = 1$ if $x_j \epsilon B$ and $B(x_j) = 0$ if $x_j \notin B$.

Let x be a prime variable from the $X = (x_1, x_2, \ldots, x_n)$ with the corresponding probabilities $P = (p_1, p_2, \ldots, p_n)$. Subsequently, the entropy of the distribution can be written as follows:

$$H(x) = -\sum_{i=1}^{n} P_i log_2 P_i,\ i = 1, 2, \ldots, n, \tag{13}$$

3.2. Combining Information about Possibilities and Probabilities

Considers a variable X taking values from $X = (x_1, x_2, \ldots, x_n)$, $(p_1, p_2, \ldots, p_n)$ provides probabilistic information, and $(\pi_1, \pi_2, \ldots, \pi_n)$ provides information on the distribution of options.

Author [32] originally proposed a probability for a fuzzy subset of F. Suppose that F is a fuzzy subset of X for $x_i \in X$, then the probability for fuzzy subset F is defined as:

$$P(F) = \sum_{i=1}^{n} F(x_i)p_i = \sum_{i=1}^{n} \pi_i p_i, \tag{14}$$

where π_i is the value of the uncertainty of the possibilities with respect to the fuzzy set theory, p_i is the probability value for the fuzzy event x_i.

A few years later, [33] generalized the probability theory and constructed a mathematical notation for calculating the probability of fuzzy event A, which is shown as:

$$P(A) = \begin{cases} \int \mu_A(x)f(x)dx, & if\ x\ is\ continuous \\ \sum_i \mu_A(x_i)f(x_i), & if\ x\ is\ discrete \end{cases}, \tag{15}$$

where $\mu_A(x)$ represents the probability distribution of X, $f(x_i)$ indicates the distribution function of the probability X.

Suppose that x is a variable that takes values in space X, the probability of uncertainty is modeled by $P = (p_1, p_2, \ldots, p_n)$, F is a fuzzy subset of X with represented possibility of distribution $\Pi = (\pi_1, \pi_2, \ldots, \pi_n)$. For each $x_i \epsilon X$ the conditional probability distribution $\widetilde{P}_i$ based on conditioning P with Π can be denoted as:

$$\widetilde{P}_i = P(x_i|F) = \frac{P((x_i) \cap F)}{P(F)} = \frac{p_i \pi_i}{\sum_{j=1}^{n} p_j \pi_j}, \tag{16}$$

where p_i indicates probability that x is taken of the V, and satisfying requirements $\sum_{i=1}^{n} p_i = 1$.

4. Data and Methodology

Stock market price change is a dynamic system that is constantly changing, and these sudden changes and movements of stock price movements doubles the complication of successful stock price prediction. In addition, stock markets are characterized by their highly non-linear nature that makes it difficult for investors to make quick decisions about the right investments. It is essential to develop an intelligent system for obtaining real-time price information, reducing one investor obsession and helping them maximize their profits.

However, financial time series show relatively complicated and non-stationary behavior, with the variable not having a linear trend or clear tendency to move to a fixed value according to [34]. Mainly for the reasons described above, researchers are increasingly focusing on the use of alternative techniques that can be used to predict the development of financial time series. These techniques certainly include fuzzy logic, which is able to include the uncertainty, nonlinearity and noise that occur in financial time series.

The Dow Jones Industrial Average (DJIA) is one of the best-known indicators of developments in the US stock market. The DJIA is named after its founder, Charles Dow, and its business partner, Edward Jones. The index was first calculated on 26 May 1896. At that time, it contained 12 exclusively industrial companies focused on areas such as the railway industry, cotton, tobacco, sugar cane, natural gas, etc. Today, the DJIA consists of 30 American companies, many of these are the largest and the most traded (blue-chip companies). The historical development of the DJIA index for the observed period is shown in Figure 2.

Figure 2. Historical development of the DJIA stock index.

Table 2 shows the basic statistics of selected stock index. The average price of the DJIA index for the selected period from January 2015 to January 2020 was 21,736.94. The price of the index rose to a maximum value of 28,645.26 and fell to a minimum value of 15,660.8. The standard deviation is a simple measure of the volatility of stock indices. For the DJIA index, the price fluctuated 3718.25. Skewness is a measure of asymmetry, or, more precisely, a lack of symmetry. Based on the values, it can be stated that the stock index has a positive asymmetric distribution of price probabilities. Furthermore, Kurtosis values show that it does not have a value close to 3, which is a theoretical value for the Gaussian probability distribution. This suggests that none of the probability distributions of this time series appear to be normally distributed. To confirm this assumption, the Jarque-Bera test was conducted with a zero hypothesis that the respective probability distribution was Gaussian (chi-square with 2 degrees of freedom). The reported values led to the rejection of the null hypothesis for all stock indices. This lack of normality is in line with the well-known "stylized facts" of market returns, as pointed out in previous studies [35].

Table 2. Summary statistics of DJIA index for the observed period.

Statistics	DJIA Index
Mean	21,736.94
Standard deviation	3718.25
Minimum	15,660.18
Maximum	28,645.26
Skewness	0.08
Kurtosis	−1.52
Jarque-Bera test	122.98 **

** indicates p-value lower than 0.01.

Daily stock index closing rates for the selected period January 2015 to January 2020 are used, a medium-term period of 5 years. It is also clear from the Figure 2 that this is a bullish trend or a rising trend without long-term bearish trends. The time period is deliberately ended in January 2020, before a sharp decline due to the COVID-19 pandemic situation. This is a sudden event that would significantly distort the output of the fuzzy model. Real data are processed through standard procedures applied in [36,37]. This step is purposeful, as the value of the stock index shows large fluctuations, which would not have to provide relevant outputs without pre-processing.

$$y_t = \frac{x_t - m}{M - m} \tag{17}$$

where x_t is the closing daily price of the stock index at time t, $M = max\{x_t\}$ and $m = min\{x_t\}$. Delayed data of the selected index are then selected as the input vector for the fuzzy model, up to a delay of three days, as described here:

$$(y_{t-3}, y_{t-2}, y_{t-1}, y_t) \tag{18}$$

The delay of up to three days proved to be the most accurate in an earlier survey by [38], which is why the application is also performed here.

The whole practical part is processed using MATLAB 2021a software and a special fuzzy logic toolbox was used to create the fuzzy model. The fuzzy model for stock market research consists of three input variables (value at time t-3, t-2 and t-1), one rule block and one output variable (value at time t).

5. Results and Evaluations

As previously described, fuzzy logic has a number of advantages, in particular being able to work with vague terms, chaotic and dynamic environments and nonlinear behaviors that are typical of everyday use in stock market analysis. Specifically, this paper applies a higher level of fuzzy logic known as type-2 fuzzy logic, whose milked fuzzy functions are able to include additional uncertainty.

The Mamdani fuzzy inference method has been widely applied in processing fuzzy rule-based systems. The Mamdani model is suitable if the modeled area is described using natural language. The input data is converted into fuzzy sets as part of the fuzzification process using mathematical functions. A total of nine models are created for each of the four shapes of the fuzzy sets: L—low, M—medium and H—high. Triangular, trapezoidal, Gaussian and bell member functions in the range [0; 1] were used to create fuzzy models and to examine the degrees of uncertainty in each of these MFs. A total of nine models were created for each MF, which differ in the magnitude of the uncertainty contained in the duplicate membership functions. Examined MF with varying levels of uncertainty are graphically illustrated in Appendix A.

When compiling fuzzy sets, the parameters lower membership function scaling factor, specified as a positive scalar less than or equal to 1, are used. Next, the lag value is set. This delay defines the point at which the lower membership function value starts increasing from zero based on the value of the upper membership function. For example, a lag value of 0.1 indicates that the lower membership function becomes positive when the upper membership function has a membership value of 0.1. For demonstration in this example, several values indefinitely are used to evaluate the accuracy of the created models. Gradually, a total of nine models are created, which differ in the degree of uncertainty contained in fuzzy sets describing the input variables. Model 1 contains membership functions with lower scale 1 and lower lag 0, essentially classical fuzzy sets of type 1. Model 2 consists of membership functions that already contain another degree of uncertainty, which was not mentioned for model 1. Specifically, model 2 contains fuzzy sets with lower scale 0.9 and lower lag 0.1. The following models contain more and more uncertainty, as can be seen from the figures corresponding to the membership functions. An overview of the plotting of the investigated types of IT2 FL MF is shown in Appendix A, where triangular, trapezoidal, Gaussian and bell MFs are plotted. The figure also shows how the distance between the individual functions gradually increases with the addition of uncertainty.

Thus, a total of nine models are constructed for each type of fuzzy membership function. The level of uncertainty with which the individual models differ is as follows:

- Model (1): T1FLS: LowerScale 1; LowerLag 0
- Model (2): T2FLS: LowerScale 0.95; LowerLag 0.05
- Model (3): T2FLS: LowerScale 0.90; LowerLag 0.10
- Model (4): T2FLS: LowerScale 0.85; LowerLag 0.15
- Model (5): T2FLS: LowerScale 0.80; LowerLag 0.20
- Model (6): T2FLS: LowerScale 0.75; LowerLag 0.25
- Model (7): T2FLS: LowerScale 0.70; LowerLag 0.30
- Model (8): T2FLS: LowerScale 0.65; LowerLag 0.35
- Model (9): T2FLS: LowerScale 0.60; LowerLag 0.40

The knowledge base represents rules in the form "IF-THEN" and expresses expert knowledge about the relationship between the delayed values of the DJIA stock index and the current value of this index. There are 27 defined fuzzy rules based on a team of experts. These rules are essential for the proper functioning of the fuzzy model and to describe the behavior of the fuzzy system. The antecedent part of fuzzy rules includes all the actual combinations of language values of the input variables. The result is the evaluation of all combinations, i.e., the assignment of linguistic values to the output variables.

The created fuzzy inference model is necessary to verify whether the presented outputs are sufficiently accurate and applicable in practice. For this purpose, several evaluation metrics are selected, which are used to verify and evaluate the error rate or accuracy of individual illustrative models of fuzzy logic. Specifically, the Root Mean Square Deviation (*RMSE*) indicator is selected, which compares the original data y_t and the data obtained from the output of the model $\dot{y}_t$. RMSE assigns a relatively higher weight to large prediction errors. This is especially useful when high error rates are undesirable. In the case of predicting the development of the stock index, this indicator can be considered the most appropriate. This indicator is also applied in many equally focused studies such as [16,17,23]. In addition to the above indicator, Mean Absolute Percentage Error (*MAPE*), Mean Absolute Error (*MAE*), Mean Squared Error (*MSE*) indicators are also applied and calculated. *MAPE* is suitable for comparing forecasts between different time series because it is expressed on a relative scale. The *MAE* indicates the size of the data set prediction error regardless of the direction of its development. It basically gives a linear error rate, which means that the differences have the same average weight. *MSE* is a suitable indicator to determine whether the examined data set does not show more outlying prediction values showing a high error rate. Unlike *MAE*, it places more weight on outliers due to the quadratic part of its calculation. However, this means that a single bad prediction can significantly increase the error rate. In particular, *RMSE* can be reported together with *MAE*, because the larger the size of the deviation between the two indicators, the greater the variance in the error rate of the examined data set. Based on the outputs of these indicators, it is then possible to evaluate which fuzzy model, or which type of fuzzy function and level of uncertainty, achieves the most accurate results on the stock market. In general, the lower the values of these indicators, the more accurate the model. The formulas of these evaluations are shown below:

$$RMSE = \sqrt{\frac{1}{n}\sum_{t=1}^{n}(y_t - \dot{y}_t)^2} \quad (19)$$

$$MAPE = \frac{1}{n}\sum_{t=1}^{n}\frac{|y_t - \dot{y}_t|}{y_t}x100 \quad (20)$$

$$MAE = \frac{1}{n}\sum_{t=1}^{n}|y_t - \dot{y}_t| \quad (21)$$

$$MSE = \frac{1}{n}\sum_{t=1}^{n}(y_t - \dot{y}_t)^2 \quad (22)$$

For subsequent comparison of models created on the basis of the above error and performance indicators, it is necessary to obtain predicted output values of individual models, based on which the performance of T2FLS models with membership functions with different levels of uncertainty is determined. These data are entered in Tables 3–6.

Table 3. Evaluation the level of uncertainty for triangular IT2 MF.

Model	RMSE	MSE	MAE	MAPE
Model (1)	0.1763	0.0311	0.1660	90.32
Model (2)	0.1753	0.0307	0.1659	89.57
Model (3)	0.1736	0.0301	0.1651	88.28
Model (4)	0.1705	0.0291	0.1629	86.05
Model (5)	0.1656	0.0274	0.1586	82.56
Model (6)	0.1585	0.0251	0.1523	78.72
Model (7)	0.1517	0.0230	0.1457	74.74
Model (8)	0.1512	0.0229	0.1456	75.19
Model (9)	0.1690	0.0286	0.1593	80.69

Table 4. Evaluation the level of uncertainty for trapezoidal IT2 MF.

Model	RMSE	MSE	MAE	MAPE
Model (1)	0.1250	0.0156	0.1213	63.75
Model (2)	0.1227	0.0151	0.1193	62.16
Model (3)	0.1200	0.0144	0.1167	60.25
Model (4)	0.1168	0.0137	0.1137	58.05
Model (5)	0.1132	0.0128	0.1101	55.50
Model (6)	0.1094	0.0120	0.1064	53.33
Model (7)	0.1052	0.0111	0.1025	51.12
Model (8)	0.1001	0.0100	0.0972	48.36
Model (9)	0.0985	0.0097	0.0957	48.05

Table 5. Evaluation the level of uncertainty for Gaussian IT2 MF.

Model	RMSE	MSE	MAE	MAPE
Model (1)	0.1918	0.0368	0.1812	99.66
Model (2)	0.1864	0.0347	0.1767	96.55
Model (3)	0.1847	0.0341	0.1756	95.45
Model (4)	0.1840	0.1840	0.1752	94.90
Model (5)	0.1839	0.0338	0.1755	94.74
Model (6)	0.1847	0.0341	0.1763	95.04
Model (7)	0.1864	0.0347	0.1780	95.90
Model (8)	0.1890	0.0357	0.1805	97.28
Model (9)	0.1924	0.0370	0.1837	99.15

Table 6. Evaluation the level of uncertainty for bell IT2 MF.

Model	RMSE	MSE	MAE	MAPE
Model (1)	0.1754	0.0308	0.1621	91.47
Model (2)	0.1675	0.0280	0.1566	87.23
Model (3)	0.1674	0.0280	0.1568	87.18
Model (4)	0.1693	0.0287	0.1586	88.18
Model (5)	0.1716	0.0295	0.1607	89.46
Model (6)	0.1741	0.0303	0.1629	90.83
Model (7)	0.1767	0.0312	0.1651	92.23
Model (8)	0.1795	0.0333	0.1701	95.25
Model (9)	0.1825	0.0333	0.1701	95.25

Table 3 shows the evaluation of accuracy and error rate for triangular MF IT2 FL. A total of nine models were examined, which differed in the degree of uncertainty involved in the membership functions. The RMSE indicator was used to compare the quality of models. A lower RMSE value indicates a better model. The table shows that model 8 had a value of 0.1512, which indicates a better model compared to another model form triangular IT2 MF. The value of the MAPE indicator is a dimensionless characteristic by which different models can be compared. The highest value was reached by model 1 with a MAPE indicator value of 90.32, while the lowest value was shown by model 7 with a MAPE indicator value of 74.74. Thus, in terms of this indicator, model 7 achieved a lower error rate and deviated less from reality in comparison with all other models with the same MF. The shortcoming in the MAE indicator was eliminated by first converting each error to a positive value using an absolute value, and the errors from all observations were then averaged as in the first cases. The MSE indicator converted negative error values to positive ones by amplifying them. The result is a large number and because large errors, i.e., large deviations of the forecasts from the actual situation, were given a large weight, while small errors were given much less weight. For both of these indicators, model 8 achieved the lowest error rate with a value of 0.1456.

The trapezoidal MF IT2 FL in Table 4 can be evaluated similarly. The same indicators as for triangular MFs were used for evaluation and comparison. The lowest RMSE is in the last model examined with a high level of uncertainty, namely with LowerScale 0.6 and LowerLag 0.4. This indicator showed a lower value (0.0985) than the triangular MF. This type of MF also achieved better results for other indicators, where MSE shows a value of 0.0097, MAE 0.0957 and MAPE 48.05. Even the worst model with a trapezoidal function (model 1) showed better results in all monitored metrics than the best model with a triangular function.

The penultimate form of the MF IT2 FL investigated is the Gaussian function. This form of function is very often used in stock market analysis see [38,39]. However, in this research, three-dimensional Gaussian functions showed the worst result. The best model in terms of Gaussian MFs was model 5 (LowerScale 0.8 LowerLag 0.2). It can be seen that, unlike the previous two types of MF, the Gaussian function showed the best results at the lowest level of uncertainty, while the other two types (triangular, trapezoidal) showed the best performance when including a large dose of uncertainty and uncertainty.

The last investigated shape of the MF IT2 FL is the bell function. Based on the comparison with the previous mathematical functions, the bell function achieved excellent results with a very low level of uncertainty LowerScale 0.95 LowerLag 0.05, resp. LowerScale 0.9 LowerLag 0.1. However, in terms of quality and error rate, even the bell MF could not be compared much with the trapezoidal function. This type of fuzzy set with triangular fuzzy sets showed very close results.

T2 FS are characterized by three-dimensional MF. The degree of membership for each element of a type-2 fuzzy set is a fuzzy set in [0,1]. The third dimension provides additional degrees of freedom to capture more information. Type-2 fuzzy sets are useful in circumstances where it is difficult to determine the exact membership function for fuzzy sets, which is useful for incorporating uncertainties. However, the use of type-2 fuzzy sets in practice has been limited due to the significant increase in computational complexity required to implement them. IT2 FS are characterized by secondary membership functions that have only values of 0 or 1. This limitation greatly simplifies the computational requirements associated with performing inferences with type-2 sets.

Based on the evaluation and comparison of individual forms of mathematical functions of fuzzy logic according to selected indicators, it can be stated that the trapezoidal form of MF is best suited to the analysis of the stock market. This is followed by a triangular and bell shape, and the worst results were achieved by the widely used Gaussian shape MF. Another finding resulting from the evaluation is the fact that while the trapezoidal MF shows the best performance, this excellent performance was only achieved with a large dose of uncertainty or uncertainty included in the three-dimensional functions that IT2 FL uses. It is similar with triangular MF. On the contrary, although the bell and Gaussian functions performed worse, they did not need as much uncertainty for their best performance.

6. Conclusions

The article shows which new and sophisticated methods should be used to achieve a high level of probability of successful stock market analysis with expected benefits. We present specific advanced methods for decision-making areas for successful investments. Specifically, IT2 FL was used with a focus on examining the appropriate membership function with different levels of uncertainty. According to the literature review, this area is insufficiently researched in the context of financial time series. Based on the results, it can be stated that when examining the stock index, the trapezoidal membership function showed the best results, but this was using the high level of uncertainty contained in three-dimensional MF. Triangular MFs also required similarly high levels of vagueness. On the other hand, the bell and Gaussian MF did not require such a high level of vagueness to be included in order to achieve their best performance at low uncertainty of the membership function, but the results were a bit worse than for a trapezoidal and triangular fuzzy set with a high level of uncertainty.

The benefits of the presented paper can be seen in three dimensions:

1. A detailed examination of the four basic forms of membership functions (triangular, trapezoidal, Gaussian and bell), which were chosen with regard to the nature of the input data originating from the stock market;
2. A detailed examination of the degree of uncertainty contained among the milked fuzzy membership functions in type-2 fuzzy logic. Several fuzzy models were created with varying degrees of uncertainty ranging from 0‰ to 40‰;
3. An application of type-2 fuzzy logic, which is neglected in the literature, lagging behind the more well-known and simpler type-1 fuzzy logic.

These outputs can be used in practical applications in the stock market and in education related to investments. The outputs of the expert system, specifically the fuzzy model, which provided the most accurate results or the lowest error rate, can be used to provide investment signals that can be placed directly on the stock market. The created model of support for investment decisions thus provides certain proposals for profitable investment strategies for investors, especially institutional investors, banks, investment companies and the like. This model is able to indicate the growth or decline of the DJIA stock index and thus send investment signals to buy or sell stock index shares.

However, it is also necessary to draw attention to the limits of research. This application of the model is made exclusively on the US stock market, which is characterized by high liquidity and efficiency. It would also be appropriate to verify the results on other US stock market stock indices, such as the S&P 500, or on other stock markets, whether equally developed or emerging. However, it can be assumed that a universal position regarding the shape of the fuzzy function is not possible, because the correct shape depends on the input characteristic data of the researched issue. It is necessary to pay close attention to this issue, as a well-chosen form of fuzzy function and degree of uncertainty can significantly affect the accuracy of outputs, which is extremely important, especially in stock markets, as even a slight improvement can cause investors significant profits or losses.

Author Contributions: Z.J.: conceptualization, methodology, software, validation, investigation, resources, data curation, writing—original draft preparation, visualization, project administration, funding acquisition; E.R.: formal analysis, writing—review and editing, supervision. All authors have read and agreed to the published version of the manuscript.

Funding: This paper was supported by project No. FP-S-20-6376 'Modeling and optimization of business processes in conditions of digital transformation' from the Internal Grant Agency at Brno University of Technology. The partial support of KEGA project No. 025EU-4/2021 of the Ministry of Education, Science, Research and Sport of the Slovak Republic is kindly appreciated.

Institutional Review Board Statement: Not applicable.

Informed Consent Statement: Not applicable.

Data Availability Statement: No new data were created or analyzed in this study. Data sharing is not applicable to this article.

Conflicts of Interest: The authors declare no conflict of interest.

Appendix A

Table A1. Examined MF with varying levels of uncertainty-.

	Triangular	Trapezoidal	Gaussian	Bell
1	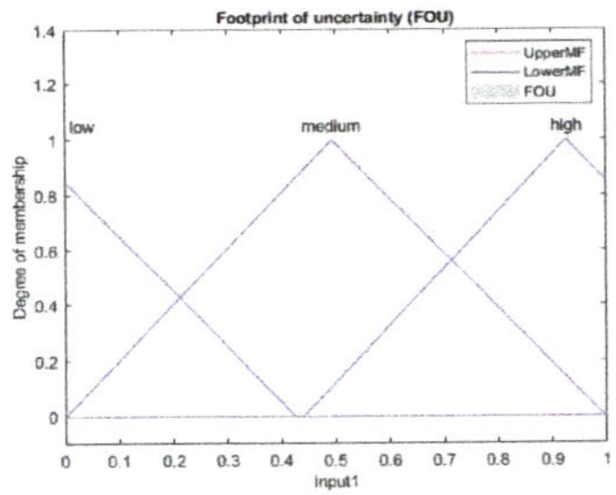	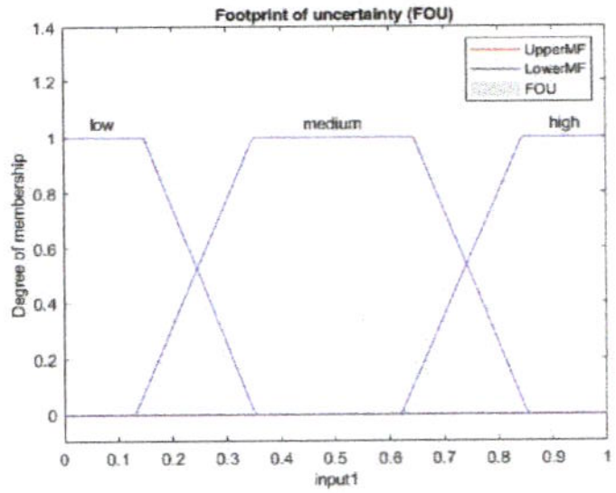	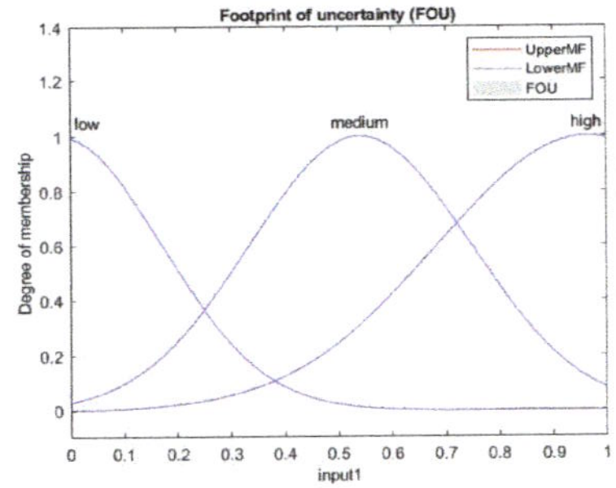	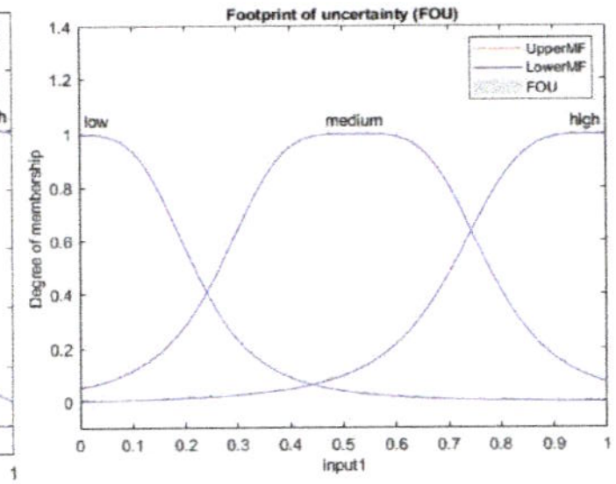
2	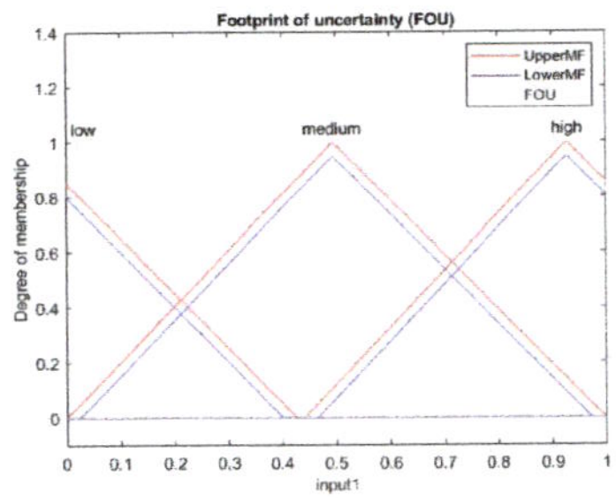	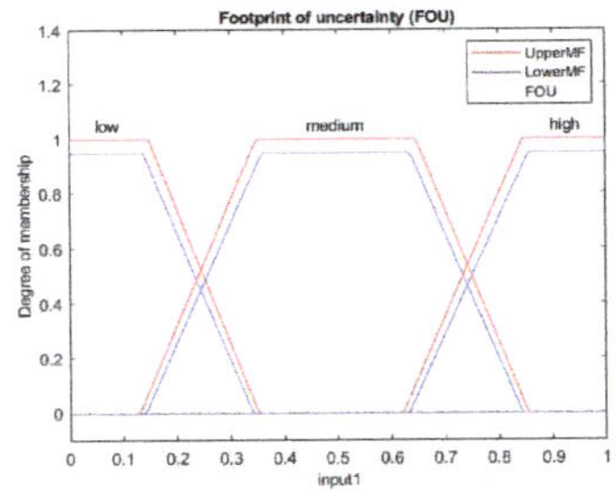	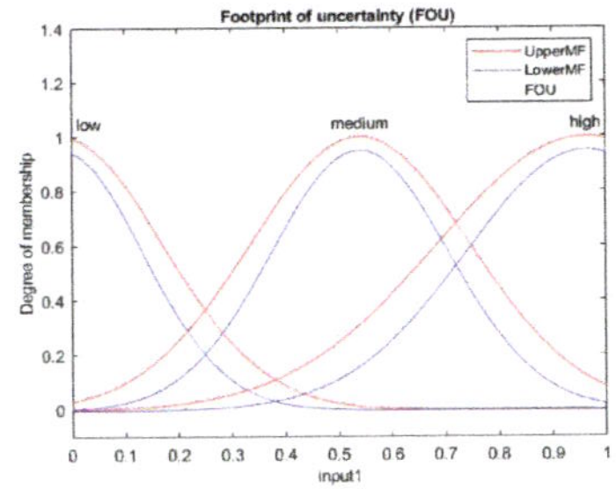	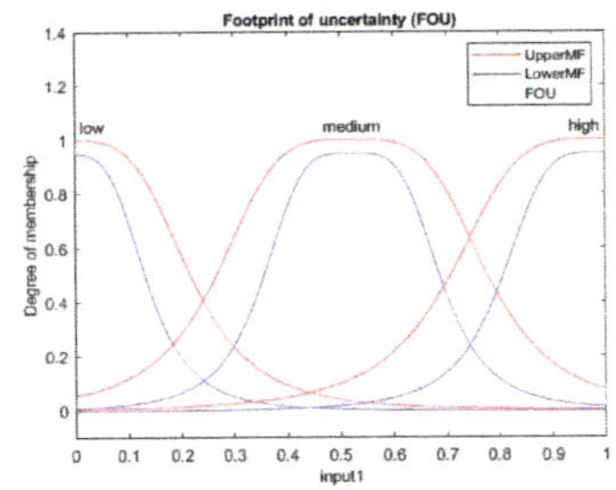

3

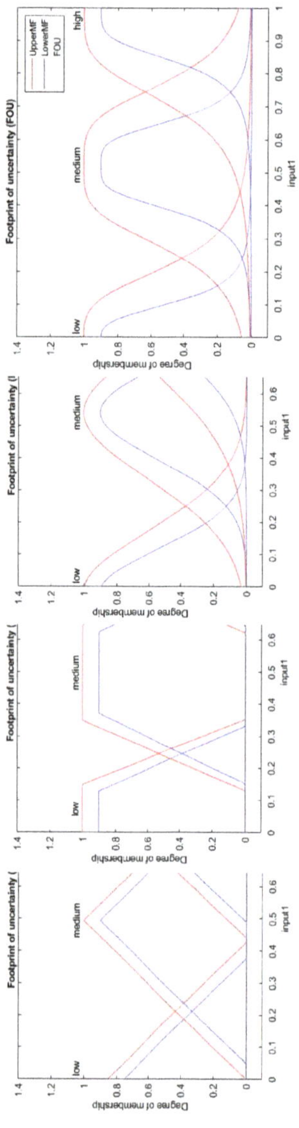
Footprint of uncertainty (FOU)
UpperMF
LowerMF
FOU
low
medium
high
Degree of membership
input1

4

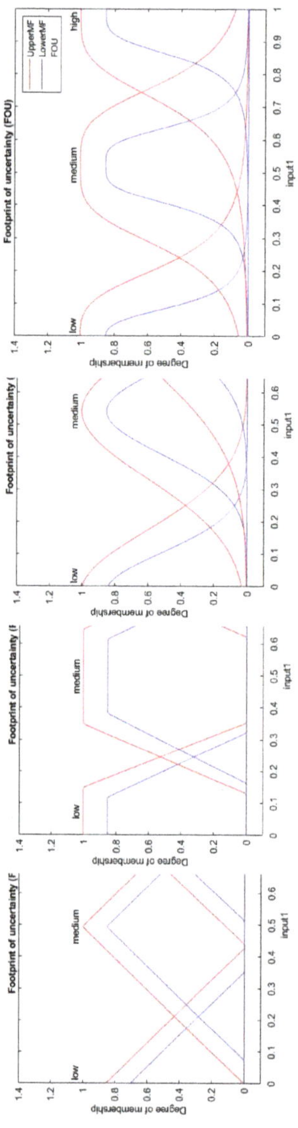
Footprint of uncertainty (FOU)
UpperMF
LowerMF
FOU
low
medium
high
Degree of membership
input1

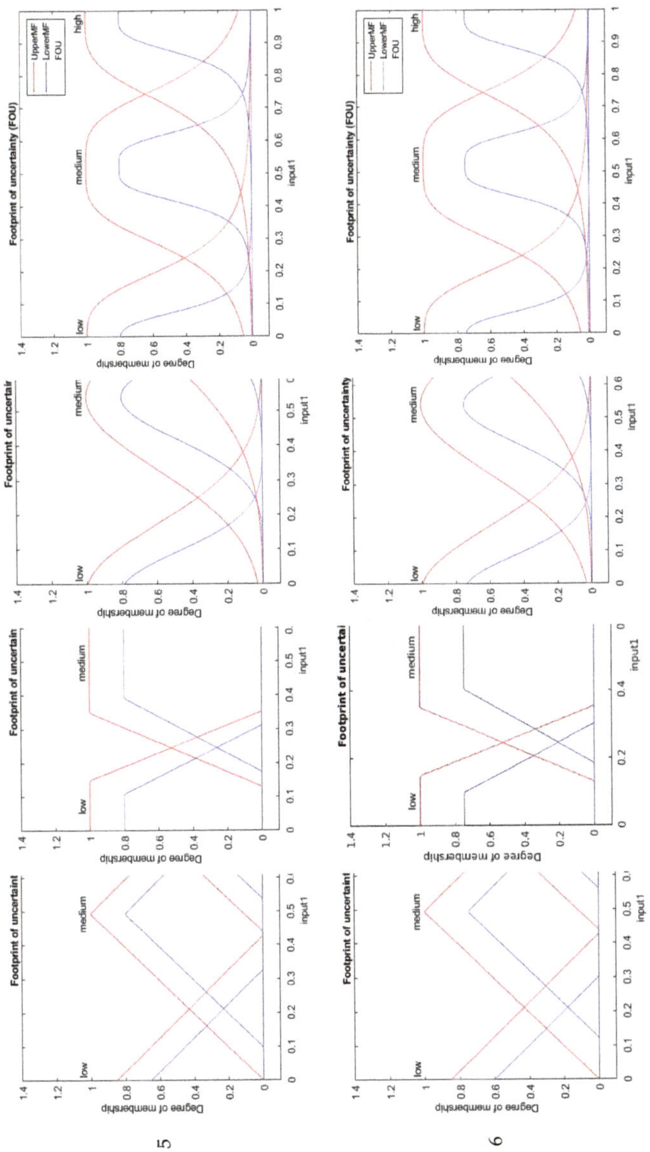

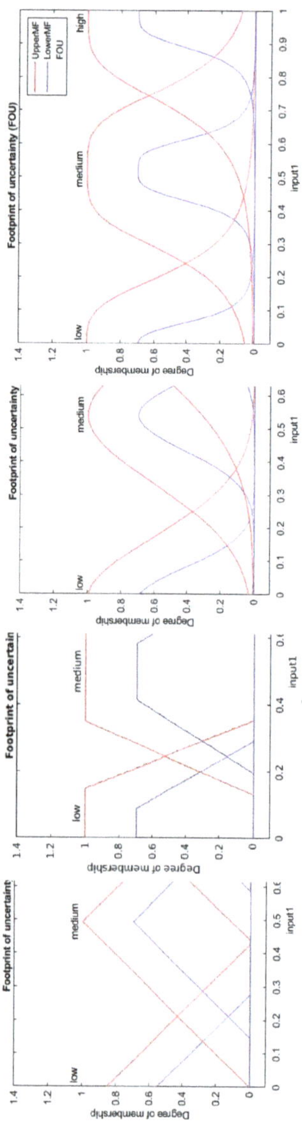

7

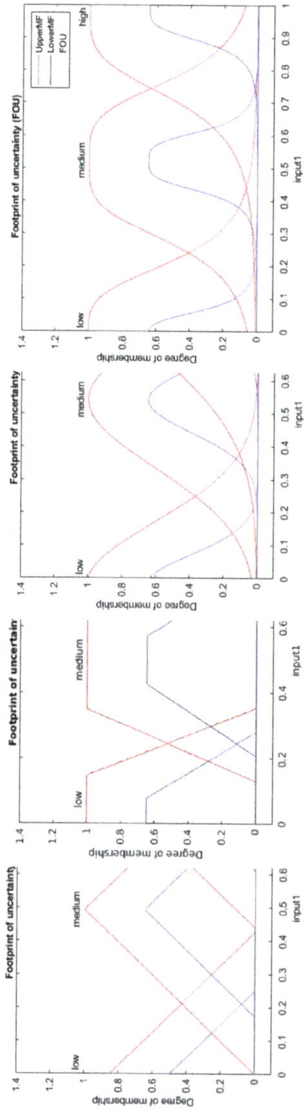

8

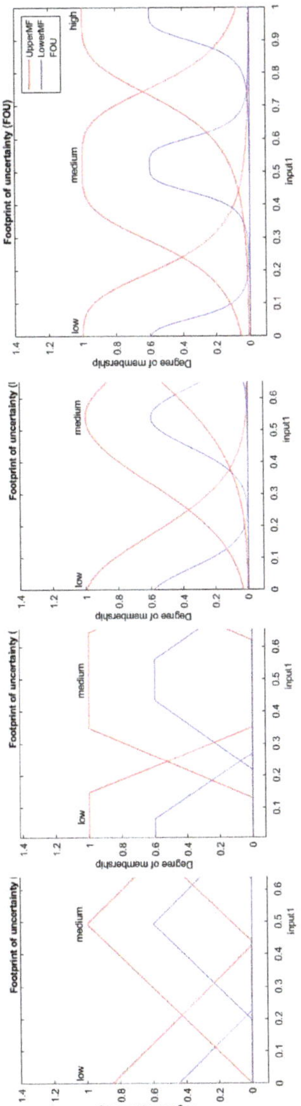
Footprint of uncertainty (FOU)
UpperMF
LowerMF
FOU
low
medium
high
Degree of membership
input1

References

1. Pincay, J.; Portmann, E.; Terán, L. Fuzzifying Geospatial Data to Identify Critical Traffic Areas. In Proceedings of the 19th World Congress of the International Fuzzy Systems Association, The 12th Conference of the European Society for Fuzzy Logic and Technology (IFSA-EUSFLAT 2021), Bratislava, Slovakia, 19–24 September 2021.
2. Li, R.; Jiang, C.; Zhu, F.; Chen, X. Traffic flow data forecasting based on interval type-2 fuzzy sets theory. *IEEE/CAA J. Autom. Sin.* **2016**, *3*, 141–148.
3. Mendel, J.M. Type-2 fuzzy sets and systems: An overview. *IEEE Comput. Intell. Mag.* **2007**, *2*, 20–29. [CrossRef]
4. Eren, M. Forecasting of the Fuzzy Univariate Time Series by the Optimal Lagged Regression Structure Determined Based on the Genetic Algorithm. *Econ. Comput. Econ. Cybern. Stud. Res.* **2018**, *52*, 201–215.
5. Puška, A.; Kozar, S.; Stevic, Ž.; Stovrag, J. A New Way of Applying Interval Fuzzy Logic in Group Decision Making For Supplier Selection. *Econ. Comput. Econ. Cybern. Stud. Res.* **2018**, *52*, 217–234.
6. Castro, J.R.; Sanchez, M.A.; Gonzalez, C.I.; Melin, P.; Castillo, O. A New Method for Parameterization of General Type-2 Fuzzy Sets. *Fuzzy Inf. Eng.* **2018**, *10*, 31–57. [CrossRef]
7. Yankova, T.; Ilieva, G.; Klisarova, S. The bezier curve as a fuzzy membership function shape. *Math. Appl. Ann. Acad. Rom. Sci.* **2018**, *10*, 245–265.
8. Wijayasekara, D.; Manic, M. Data Driven Fuzzy Membership Function Generation for Increased Understandability. In Proceedings of the IEEE International Conference on Fuzzy Systems, Beijing, China, 6–11 July 2014; pp. 133–140.
9. Kayacan, E.; Sarabakha, A.; Coupland, S.; John, R.; Khanesar, M.A. Type-2 fuzzy elliptic membership functions for modeling uncertainty. *Eng. Appl. Artif. Intell.* **2018**, *70*, 170–183. [CrossRef]
10. Sadollah, A. Introductory Chapter: Which Membership Function is Appropriate in Fuzzy System? In *Fuzzy Logic Based in Optimization Methods and Control Systems and Its Applications*; Sadollah, A., Ed.; IntechOpen: London, UK, 2018.
11. Zhao, J.; Bose, B.K. Evaluation of membership functions for fuzzy logic controlled induction motor drive. In Proceedings of the 28th Annual IEEE Conference of the Industrial Electronics Society, Sevilla, Spain, 5–8 November 2002.
12. Esfahanipour, A.; Aghamiri, W. Adapted Neuro-Fuzzy Inference System on indirect approach TSK fuzzy rule base for stock market analysis. *Expert Syst. Appl.* **2010**, *37*, 4742–4748. [CrossRef]
13. Talpur, N.; Salleh, M.N.M.; Hussain, K. An investigation of membership functions on performance of ANFIS for solving classification problems. *IOP Conf. Ser. Mater. Sci. Eng.* **2017**, *226*, 012103. [CrossRef]
14. Vlasenko, A.; Vynokurova, O.; Vlasenko, N.; Peleshko, M. A Hybrid Neuro-Fuzzy Model for Stock Market Time-Series Prediction. In *2018 IEEE Second International Conference on Data Stream Mining & Processing (DSMP)*; IEEE: New York, NY, USA, 2018; pp. 352–355.
15. Ghandar, A.; Michalewicz, Z.; Schmidt, M.; To, T.; Zurbrugg, R. Computational intelligence for evolving trading rules. *IEEE Trans. Evol. Comput.* **2009**, *13*, 71–86. [CrossRef]
16. Rajab, S.; Sharma, V. An interpretable neuro-fuzzy approach to stock price forecasting. *Soft Comput.* **2017**, *23*, 921–936. [CrossRef]
17. Gautam, S.S.; Abhishekh. A Novel Moving Average Forecasting Approach Using Fuzzy Time Series Data Set. *J. Control Autom. Electr. Syst.* **2019**, *30*, 532–544. [CrossRef]
18. Abhishekh, G.S.S.; Singh, S.R. A New Type 2 Fuzzy Time Series Forecasting Model Based on Three-Factors Fuzzy Logical Relationships. *Int. J. Uncertain. Fuzziness Knowl.-Based Syst.* **2019**, *27*, 251–276. [CrossRef]
19. Pinto, É.A.N.; Schnitman, L.; Reis, R.A. A Fuzzy Based Recommendation System for Stock Trading. In *Fuzzy Information Processing, Proceedings of the NAFIPS 2018: Communications in Computer and Information Science, West Lafayette, IN, USA, 7–9 June 2018*; Barreto, G., Coelho, R., Eds.; Springer: Cham, Switzerland, 2018; Volume 831.
20. Liu, Z.; Zhang, T. A second-order fuzzy time series model for stock price analysis. *J. Appl. Stat.* **2019**, *46*, 2514–2526. [CrossRef]
21. Chourmouziadis, K.; Chourmouziadou, D.K.; Chatzoglou, P.D. Embedding Four Medium-Term Technical Indicators to an Intelligent Stock Trading Fuzzy System for Predicting: A Portfolio Management Approach. *Comput. Econ.* **2021**, *57*, 1183–1216. [CrossRef]
22. Sharma, D.K.; Hota, H.S.; Rababaah, A.R. Forecasting US stock price using hybrid of wavelet transforms and adaptive neuro fuzzy inference system. *Int. J. Syst. Assur. Eng. Manag.* **2021**. [CrossRef]
23. Jiang, J.-A.; Syue, C.-H.; Wang, C.-H.; Wang, J.-C.; Shieh, J.-S. An Interval Type-2 Fuzzy Logic System for Stock Index Forecasting Based on Fuzzy Time Series and a Fuzzy Logical Relationship Map. *IEEE Access* **2018**, *6*, 69107–69119. [CrossRef]
24. Bhattacharya, D.; Konar, A. Self-adaptive type-1/type-2 hybrid fuzzy reasoning techniques for two-factored stock index time-series prediction. *Soft Comput.* **2018**, *22*, 6229–6246. [CrossRef]
25. Lathamaheswari, M.; Nagarajan, D.; Kavikumar, J.; Broumi, S. Triangular interval type-2 fuzzy soft set and its application. *Complex Intell. Syst.* **2020**, *6*, 531–544. [CrossRef]
26. Zare, A.; Shoeibi, A.; Shafaei, N.; Moridian, P.; Alizadehsani, R.; Halaji, M.; Khosravi, A. Accurate Prediction Using Triangular Type-2 Fuzzy Linear Regression. *arXiv* **2021**, arXiv:2109.05461.
27. Janková, Z.; Dostál, P. Type-2 Fuzzy Expert System Approach for Decision-Making of Financial Assets and Investing under Different Uncertainty. *Math. Probl. Eng.* **2021**, *2021*, 3839071. [CrossRef]
28. Sang, X.; Zhou, Y.; Yu, X. Uncertain possibility-probability information fusion method under interval type-2 fuzzy environment and its application in stock selection. *Inf. Sci.* **2019**, *504*, 546–560. [CrossRef]

29. Mendel, J.M.; John, R.I.; Liu, F. Interval type-2 fuzzy logic systems made simple. *IEEE Trans. Fuzzy Syst.* **2006**, *14*, 808–821. [CrossRef]
30. Mendel, J.M. *Uncertain Rule-Based Fuzzy Logic Systems*; Prentice Hall: Los Angeles, CA, USA, 2001.
31. Sang, X.; Liu, X. Possibility mean and variation coefficient based ranking methods for type-1 fuzzy numbers and interval type-2 fuzzy numbers. *J. Intell. Fuzzy Syst.* **2016**, *30*, 2155–2168. [CrossRef]
32. Zadeh, L.A. Probability measures of fuzzy events. *J. Math. Anal. Appl.* **1968**, *23*, 421–427. [CrossRef]
33. Zimmermann, H.J. *Possibility Theory, Probability Theory, and Fuzzy Set Theory*; Springer: Dordrecht, The Netherlands, 1996.
34. Chang, P.C.; Liu, C.H. A TSK type fuzzy rule based system for stock price prediction. *Expert Syst. Appl.* **2008**, *34*, 135–144. [CrossRef]
35. Polanco-Martínez, J.M.; Fernández-Macho, J.; Neumann, M.B.; Faria, S.H. A pre-crisis vs. crisis analysis of peripheral EU stock markets by means of wavelet transform and a nonlinear causality test. *Phys. A Stat. Mech. Appl.* **2018**, *490*, 1211–1227. [CrossRef]
36. Li, R.J.; Xiong, Z.B. Forecasting stock market with fuzzy neural networks. In Proceedings of the 2005 International Conference on Machine Learning and Cybernetics, Guangzhou, China, 18–21 August 2005; Volume 6, pp. 3475–3479.
37. Zhai, F.; Wen, Q.; Yang, Z.; Song, Y. Hybrid forecasting model research on stock data mining. In Proceedings of the 4th International Conference on New Trends in Information Science and Service Science, Gyeongju, Korea, 11–13 May 2010; pp. 630–633.
38. Janková, Z.; Dostál, P. Prediction of European Stock Indexes Using Neuro-fuzzy Technique. *Trendy Ekon. Manag.* **2020**, *14*, 45–57. [CrossRef]
39. Janková, Z.; Janková, Z.; Jana, D.K.; Dostál, P. Investment Decision Support Based on Interval Type-2 Fuzzy Expert System. *Inz. Ekon.-Eng. Econ.* **2021**, *32*, 118–129.

applied sciences

MDPI

Article

Towards Adaptive Gamification: A Method Using Dynamic Player Profile and a Case Study

Inmaculada Rodríguez [1,2,*,†], Anna Puig [2,3,*,†] and Àlex Rodríguez [2,†]

1 Departament de Matemàtiques i Informàtica, UBICS Research Institute, UB, 08028 Barcelona, Spain
2 Departament de Matemàtiques i Informàtica, Universitat de Barcelona, Avda Corts Catalanes, 585, 08007 Barcelona, Spain; alexrodriguez@ub.edu
3 Departament de Matemàtiques i Informàtica, IMUB Research Institute, UB, 08028 Barcelona, Spain
* Correspondence: inmarodriguez@ub.edu (I.R.); annapuig@ub.edu (À.P.)
† These authors contributed equally to this work.

Abstract: The design of gamified experiences following the one-fits-all approach uses the same game elements for all users participating in the experience. The alternative is adaptive gamification, which considers that users have different playing motivations. Some adaptive approaches use a (static) player profile gathered at the beginning of the experience; thus, the user experience fits this player profile uncovered through the use of a player type questionnaire. This paper presents a dynamic adaptive method which takes players' profiles as initial information and also considers how these profiles change over time based on users' interactions and opinions. Then, the users are provided with a personalized experience through the use of game elements that correspond to their dynamic playing profile. We describe a case study in the educational context, a course integrated on Nanomoocs, a massive open online course (MOOC) platform. We also present a preliminary evaluation of the approach by means of a simulator with bots that yields promising results when compared to baseline methods. The bots simulate different types of users, not so much to evaluate the effects of gamification (i.e., the completion rate), but to validate the convergence and validity of our method. The results show that our method achieves a low error considering both situations: when the user accurately (Err = 0.0070) and inaccurately (Err = 0.0243) answers the player type questionnaire.

Keywords: gamification; adaptive gamification; player types

Citation: Rodríguez, I.; Puig, A.; Rodríguez, À. Towards Adaptive Gamification: A Method Using Dynamic Player Profile and a Case Study. *Appl. Sci.* **2022**, *12*, 486. https://doi.org/10.3390/app12010486

Academic Editor: Aida Valls

Received: 15 November 2021
Accepted: 1 January 2022
Published: 4 January 2022

1. Introduction

The use of game mechanics in non-gaming contexts, which is known as gamification, has come into play to encourage and motivate user's behaviors. The design of gamification should pursue a clear objective. For example, a gamified fair [1] was designed to motivate users to rate the experience of visiting a fair. Another gamified application was created with the goal of connecting students online and was designed to engage them in sharing documents and insights about their classes [2]. The gamification of a massive open online course (MOOC) aimed to foster students' engagement and therefore increase the completion rate of courses [3].

The design of gamified experiences usually adopts the one-fits-all approach, which may fail as a result of considering that all users have the same profile. An alternative strategy is adaptive gamification, which considers that users have different motivations during their interaction in gamified systems. Adaptive gamification is based on player type classifications. A pioneering, representative work on adaptive gamification was proposed by Lavoue et al. [4]. They gathered information about the user profiles (i.e., player types) at the beginning of the experience and then suggested to the users game elements that fit those (static) profiles. However, this initial player profile may either be inaccurately input by the user or may evolve slightly over time [5] and therefore, sometimes the suggested game elements do not completely correspond to the real profile. Additionally, adaptive gamified systems usually take the most predominant player type to show the user game elements related to this predominant player type, with the drawback of ignoring other player types that may also characterize the user, so several player types should be considered, not only the predominant one [6].

In this paper, we extended our proposal of (dynamic) adaptive gamification presented in [7]. It used initial players' profiles, gathered from the HEXAD questionnaire [8], as well as information about users' interactions and while using the system. The goal of gamification was to foster users' engagement and thereby motivate the completion of online activities such as learning activities in a course or employees' progress report in a company. Concretely, we presented a case study focused on an STEM course integrated on Nanomoocs, a MOOCS platform. The case study first shows a real example of adaptive gamification in the educational field, and secondly, helps illustrate the elements that our method is based upon. Moreover, we depict the software architecture of the adaptive gamification system. Although evaluations of the effects of gamification are commonly carried out with users, this paper also presents an analysis of the proposed strategy using bots. The bots simulate different types of users, not so much to evaluate the effects of gamification (i.e., completion rate), but to validate the convergence and validity of our method. The results show that our method achieves a low error considering both situations: when the user accurately ($Err = 0.0070$) and inaccurately ($Err = 0.0243$) answers the player type questionnaire.

2. Previous Work

In this section, we first present research works concerning the data needed for the adaptation (player modeling), and then we focus on studies that describe the adaptation strategy: the context (educational systems, collaborative systems) and how the adaptation is carried out (static, dynamic).

2.1. Player Modeling

User-centered techniques have been proposed to correlate game elements and different user profiles [8,9]. Some of these techniques have focused on specific user characteristics such as motivation [10], personality traits [11,12], learning styles [13], player types [4], and types of interaction with different activities [14]. Others combined different characteristics, such as [15] which took into account learning styles and player types to determine the types of educational activities and game elements to include in a learning pathway. In contrast, ref. [16] suggested using the context, interactions, gender and player type to decide, through rules, the next game element to display. Other works instead focused on emotions to predict the individual's performance in gamified tasks, which is information that could potentially be used to adapt the game features [17].

Specifically, in current adaptive gamification approaches, the most commonly used taxonomies of player types are Bartle [18], BrainHex [19] and HEXAD [20]. These taxonomies allow us to easily identify player types from questionnaires as well as establish the correspondence between these player types and the game elements [9]. A recent study [21] proposed the so-called MoMo (motivational value model), a new prediction model that combines the four categorization models—Hexad, Bartle, Big Five and BrainHex. MoMo was validated along with Bartle and BrainHex using health applications. Results showed that it could predict players' preferences better than the individual models.

Questionnaires allow for the characterization of the player type of a user with a set of ratings. They determine the player type prior to the experience. For example, the HEXAD model distinguishes between six player types (achiever, player, philanthropist, disruptor, socializer and free spirit), and as result of the questionnaire, the user can obtain the following ratings: achiever 25%, player 18%, philanthropist 21%, disruptor 8%, socializer 10% and free spirit 5%. The decision of the final player type—and thus the most appropriate game element—usually relies on the predominant rating [4] or a combination of them [22]. As an alternative to questionnaires, a recent study [23] proposed to predict HEXAD player types through gameful applications. The authors created two applications. An interactive one resembling a questionnaire but with appealing look and gameful feedback, and another application in which users interacted with gamification elements by shooting snowballs. The former correctly predicted player types but the latter did not show sufficient variance to reliably predict HEXAD player types. However, users interacted with gamification elements that fitted their player type, which seemed promising and worthy of further study.

In this work, we relied on the HEXAD player model and used the questionnaire as proposed and validated by [24]. However, it should be noted that in all the studies analyzed, the no-player type of player was not taken into consideration, a fact that suggests that, in some cases, the use of gamification is more detrimental than beneficial [25]. Thus, we also added the no-player player type to the HEXAD taxonomy, and we also considered for each user the assessment of all their ratings included in their player type.

Certainly, as supported by different studies [26], the interpretation of the results of these questionnaires may not be very reliable. Even if the questionnaire is valid, the answers may be somewhat random or the results at the beginning may not persist during the experience depending on the moment or the mood of the user. In fact, in addition to questionnaires, some proposals in the literature also gathered user feedback on learning activities [14], or scores on different game elements during the experience. Thus, inspired by these works, we enriched the user model obtained from the initial questionnaire by means of user interactions and opinions during the course of the activity. Therefore, we will base the adaptation of the game mechanics to the "real" and "dynamic" user profile using two types of interactions: on the one hand, the interactions with the game elements; and on the other hand, the opinions that the user can give at certain moments regarding those elements. Thus, using both types of interactions, we will refine the player types during the experience, and consequently, the game elements will be activated.

2.2. Adaptation Strategy

Recent literature reviews have analyzed adaptive gamification in educational and collaborative systems [8,27]. In the educational context, different studies linked game elements to students' motivation and their player and learner type [11,13,28,29]. These studies laid the groundwork on which adaptation studies were constructed upon, which brought a variety of contributions from adaptation engine architectures [16,30] to evaluating gamification effectiveness [4,31,32]. The main needs of adaptive gamification systems emerging from the literature analysis in education are to enhance learners' models, to explore different adaptation methods, especially dynamic adaptation, and to assess the long-term impact of gamification in learner performance or motivation.

Furthermore, gameful applications for motivating ones; participation in collaborative systems spread between serious games [33,34] and gamified experiences [35]. There were different approaches to adaptation in gamified collaborative systems. These include difficulty adaptation, which that either be based on the player behavior (player's performance) [36,37] or based on the global behavior of a group of players [38], adaptive curriculum guidance [39,40], storytelling and content adaptation [41,42], adaptive presentation [43], and motivational interventions [32,44]. Our research work relates to the last two approaches as it presents to the user those game elements that fit their profile to motivate them to complete the course. Moreover, Ayastui et al. [27] formalized adaptation strategies and proposed a new taxonomy—gamification elements adaptation strategy (GEAS)—for the adaptation of gamification elements. The so-called full GEAS strategy refers to the adaptation that applies different gamification elements at different moments depending on the estimated user preferences [40,45]. Single GEAS adjusts some features of the gamification elements according to players' behavior [32,46]. Our research situates in full the GEAS element of the taxonomy as our adaptation provides different game elements depending on the MOOC's learners' profile and their behavior along the course.

Nevertheless, there is still room for improvement regarding dynamic adaptation, where we find few studies. Lavoue [4] proposed a matrix factorization model similar to those used in recommender systems. They used two matrices: one defining the player types of all users and the other representing how the game elements match the player types. They combined these two matrices to obtain game elements' scores for each user, and then they selected the element with the highest score. Another study [15] defined an off-line Q-Learning algorithm to generate an adaptive learning path for the user. The authors specified an S-Table and a Q-Table that represented the corresponding state at each taken action, and the Q-values of each action in each state, respectively. Both tables were the same for all the profiles. Nevertheless, the R-Table (the reward of each action in each state) was specific for each learning-player profile. All of them considered adapting game elements to the initial player's profile, keeping the player type static throughout the experience.

Our adaptive algorithm is based on a matrix factorization model similar to [4], which allows the recalculation of the player type, i.e., of all player ratings, during the experience in order to adapt the game element to the user model at any given moment. In this initial study, adaptation is based on activating one of the most appropriate game elements at any time depending on the recalculated player type.

3. Runtime Method for Adaptive Gamification

In this section, we present our proposal for adaptive gamification. First, we introduce the main definitions our method is based on: (i) the HEXAD player type model, including the non-player type; (ii) the selected game elements as a subset of those proposed by Tondello [24]; and (iii) we defined the matrices and vectors of ratings and interactions implied in our algorithm. Second, we describe how we match these player types with their corresponding game elements using an extended matrix factorization method [4].

3.1. Previous Definitions

3.1.1. Player Type

An essential concept for adaptive gamification is the player type model, which classifies what kind of game elements maximizes user motivation. As we mentioned above, we based ourselves on the HEXAD player typology, adding the non-player type [20]. Player types are defined as

$$PT = \{pt_1, pt_2, \ldots, pt_7\} \tag{1}$$

The player types pt_i are explained below:

1. Disruptor: motivated by the ability to modify the system;
2. Free spirit: motivated by the ability to freely explore the system;
3. Achiever: motivated by the ability to win challenges and unlock hidden content.;
4. Player: motivated by the game itself;
5. Philanthropist: motivated by the ability to share goods and help other users;
6. Socializer: motivated by social connections;
7. Non-player: users who do not like to play.

The player type of a particular user is represented as a vector of length 7, PR, where each component r_i represents its ratings for each player type, so the values vary between 0 and 1, $PR = (r_1, r_2, \cdots, r_7)$. For example, a user can be 20% disruptor, 10% free spirit, 30% achiever, 40% player, 40% socializer, 10% philanthropist and 0% no-player, which is encoded with the vector (0.2, 0.1, 0.3, 0.4, 0.4, 0.1, 0). Since the user's player type changes over time, we define $PR^{(t)}$ as the player ratings at the time t as

$$PR^{(t)} = (r_1^{(t)}, r_2{}^{(t)}, \cdots, r_7{}^{(t)}) \tag{2}$$

3.1.2. Game Element

Based on the correlation analysis of the HEXAD player types with 52 game design elements performed by Tondelo et al. [24], we selected a subset of 14 game elements (see Table 1) covering the whole spectrum of player types (see Figure 1). The 52 game design elements were grouped by player types using the correlation value of each player type's mean score and the corresponding design elements' mean score per user. In our study, considering the fact that all game elements are equally motivating for one player type, we selected at least two of those: one related to its nature itself (for example, the Easter Egg game element is specific to the free spirit players and not others), and another considered complementary or additional for another type of player, if it exists (for example, the social status is specific to the socializer but is also an additional game element for player types such as disruptor or player). It should be noted that in the analysis carried out by [24], they did not find game elements correlated to the philanthropist player type. Then, we opted to include two of the philanthropist game mechanics suggested by [20] (knowledge sharing and gifting). There are no game elements associated to non-player type. Moreover, the educational context of our application further helped us either select or discard the game elements for each player type. For example, a development tools mechanism can be implemented as a prize that allows disruptors to change the system, e.g., change the design of badges. Nevertheless the anarchic gameplay mechanism of disruptors is not adequate for a massive online open course because its implementation can go against the learning objectives.

Game elements are defined as

$$GE = (ge_1, ge_2, \ldots, ge_{14}) \tag{3}$$

The game elements, ge_i, are briefly explained below:

1. Development tool: allows the player user to create certain gamification mechanics such as badges, challenges and unlockables.
2. Challenge: the player must overcome a challenge, such as reaching a certain level and solving a problem in a certain time;
3. Easter egg: the mechanism consists of an image which, when pressed five consecutive times, allows access to a mini-game;
4. Unlockable: when a player overcomes a certain challenge, a hidden content is unlocked, which can be a message, a mini-game, etc.;
5. Badge: awarded to the player when they manage to complete a difficult task;
6. Level: shows the user's progress in completing a task, subdivided into levels;
7. Point: the player gains score, experience, virtual money, etc.;
8. Leaderboard: displays a ranking of scores;
9. Gift Opener: the player opens gifts they have received;
10. Lottery: game of chance (roulette) that allows players to increase their scores;
11. Social network: a small social network that allows players to create a profile, add friends and view their profiles;
12. Social status: collection of rankings of players based on their scores, especially those related to social interactions, such as the number of followers, visitors, etc.;
13. Share knowledge: the player sends help messages to everyone in a group;
14. Gift: the player sends gifts to other users.

It is worth mentioning that each game element does not target only one type of player but can motivate different types of players. Thus, the i-th game element, ge_i, has associated a vector of motivation indexes GM_i, where each component, m_j, is the percentage of motivation it can cause in one of the seven types of players, p_j. For example, the fifth game element "Badge" (see Figure 1) can motivate users with both achiever (pt_3) and player (pt_4) player types, then the $GM_5 = (0, 0, 0.5, 0.5, 0, 0, 0, 0)$:

$$GM_i = (m_1, m_2, \ldots, m_7) \;\; \forall i = 1 \ldots 14 \tag{4}$$

Therefore, we define the matrix, M as

$$M = (GM_i)_{1 \leq i \leq 14} = (m_{ij})_{1 \leq i \leq 14,\; 1 \leq j \leq 7} \tag{5}$$

where the rows of this matrix are indexed by the game elements, and the columns by the player types. Thus, $m_{i,j}$ represents the percentage of motivation (motivation index) that the i-th game element produces for the j-th type of player.

Finally, as can be appreciated on the right-hand side of Figure 1, the utility of using a game element, $U^{(t)}$, can be computed from the matrix M and the player type ratings, $PR^{(t)}$.

Table 1. Table of the distribution of game elements, ge_i, according to their primary player types and their additional player types.

Player Type	Game Elements	Additional Player Types [24]
Disruptor	Development Tools Creativity Tools (Challenges)	Free Spirit Player, Achiever, Free Spirit
Free Spirit	Unlockable Easter Egg	- Player
Achiever	Badge Level of Progression	Player Player
Player	Lottery Leaderboard Gift Opener (Prizes) Points	- - - -
Socializer	Social Network Social Status	Free Spirit -
Philanthropist	Share Knowledge Gifting	- -

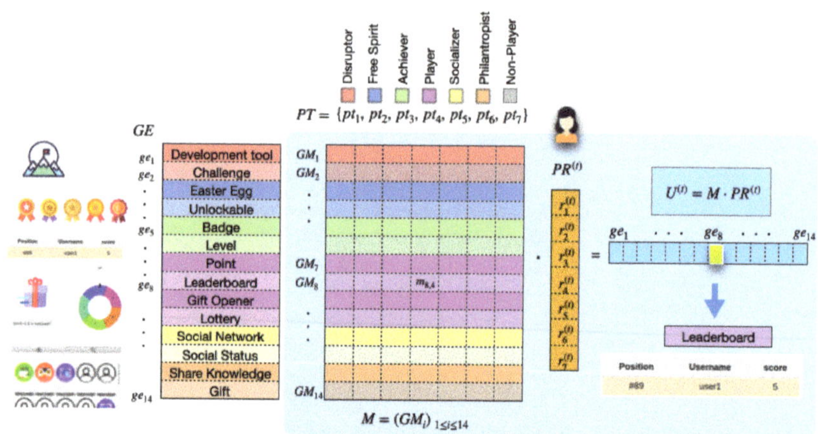

Figure 1. Fourteen selected game elements, ge_i, enumerated by $i = 1, \ldots, 14$. Each color represents one player type, pt_j, $(j = 1, \ldots, 7)$. The vector $PR^{(t)}$ defines the player ratings of a user and the matrix M stores all the motivation values that the i-th game element produces for the j-th player type. The game element with the highest utility is the item that will be presented to the user. As an example, the game element 8-th, the leaderboard, is highlighted to represent the next game element to be shown to the user.

3.1.3. Interaction Index

Our method uses the so-called interaction index to determine whether a game element motivates the user. We define it as the percentage of the user's interaction with each game element at time t. This is represented by a vector of length 14 (the number of game elements), $S^{(t)} = (s_1^{(t)}, s_2^{(t)}, \cdots, s_{14}^{(t)})$. The interaction index of the i-th game element, $s_i^{(t)}$, is defined by

$$s_i^{(t)} = 1 - e^{-\left(o_i^{(t)} \frac{n_i^{(t)} - n_i^{(t-1)}}{\tau_i^{(t)} - \tau_i^{(t-1)}}\right)} \tag{6}$$

where

$\tau_i^{(t)}$: the display time, i.e., the time interval for which the game element has been displayed until time t;

$n_i^{(t)}$: the number of interactions at time t;

$o_i^{(t)}$: the opinion, i.e., the user assessment of the game element. This is a value between 0 and 1. Opinions from 1 to 5 stars correspond to 0.2, 0.4, 0.6, 0.8 and 1, respectively.

Note that Equation (6) encodes the interaction index as a number between 0 and 1. If there are no interactions between time t and $t+1$, the interaction index is 0 (since $(n_i^{(t)} - n_i^{(t-1)})$ is 0). The interaction speed is $(\frac{n_i^{(t)} - n_i^{(t-1)}}{\tau_i^{(t)} - \tau_i^{(t-1)}})$. Then, the interaction index tends to 1 as the interaction speed increases.

The opinion $o_i^{(t)}$ modulates the interaction speed: $s_i^{(t)}$ tends to be faster than 1 when $o_i^{(t)}$ is near 1 than when $o_i^{(t)}$ is near 0.

3.1.4. Activities

Considering that any serious context to be gamified is composed of activities (e.g., exercises, videos, readings), game elements unfold as a result of users performing n activities:

$$A = \{a_1, a_2, \ldots, a_n\} \tag{7}$$

Then, users' experience in the system consists of a series of activities with intertwined gamification elements: a_1, a_2, ge_i, a_3, a_4, a_5, ge_j, ..., a_n, where $ge_i, ge_j, \ldots, \in GE$. The timing of ge_i appearing in this sequence of activities is actually defined by the gamification designer, who is the teacher in the case of MOOCs. Note that the teacher receives some recommendations regarding how to alternate activities and gamification elements in the course. For example, adding game elements after hard learning activities, such as reading a long text, is highly recommended, while in more entertaining activities, such as infographics, the subsequent use of game elements is not as essential.

3.2. Adaptive Method

Our method relies on the fact that the participant's inner player type (or real player type, RPT) may evolve slightly (i.e., is dynamic) during the experience. We therefore assume that this inner profile at the start of the experience, RPT_0, can be approximated by an initial player type, $PR^{(0)}$, using known and validated questionnaires [20]. Furthermore, we assume that the users' behavior—interactions with and opinions of game elements—reveal their real player type, RPT^t. Thus, we will take into account this behavior to approximate a different $PR^{(t)}$ during the experience. To recompute the player type in each iteration t, our algorithm takes into account the current $PR^{(t)}$ as well as the interactions that users made with the different game elements and how they rated them.

Once the type of initial player is defined, our method iteratively updates the player profile of the user at time t, $PR^{(t)}$, and thus calculates the utility of showing one or another game element to the user, $ge_i^{(t)}$. Each iteration consists of the three steps depicted in Figure 2 (considering the definitions introduced in Section 3.1). The steps are: (1) obtain $PR^{(t+1)}$; (2) score the utility of showing a game element to a user at a specific time $t+1$ (denoted by $U^{(t+1)}$); and (3) select which game element to activate based on the assigned scores.

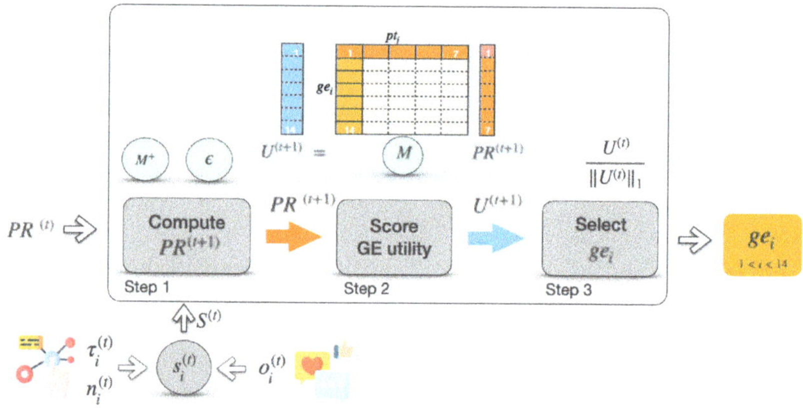

Figure 2. Steps to compute the utility of showing a ge_i associated to a user at time $t+1$. Blue circles indicate the constant data of our method. All other elements define dynamic values that change over time.

In the first step, we compute the new player type ratings, $PR^{(t+1)}$:

$$PR^{(t+1)} = (1-\epsilon)\ PR^{(t)} + \epsilon\ (M^{+} \cdot S^{(t)}) \tag{8}$$

where

$PR^{(t)}$: the player type of the user at time t;
$S^{(t)}$: the interaction indexes;
ϵ: to avoid extreme fluctuations between $PR^{(t)}$ and $PR^{(t+1)}$, where $0 < \epsilon < 1$—the value of this parameter should be tuned experimentally;
M^{+}: the Moore–Penrose pseudoinverse matrix of M, needed in order to interpret $S^{(t)}$ and $PR^{(t)}$ in the same space.

In the second step, once we calculated the new player profile, we compute the utilities as indicated in the top right-hand side of Figure 1, using the matrix M defined in Equation (5):

$$U^{(t+1)} = M \cdot PR^{(t+1)} \tag{9}$$

Finally, in the third step, we select the next game element to display considering the ith component of $\frac{U^{(t)}}{\|U^{(t)}\|_1}$ as the probability of choosing the ith game element using a weighted random choice ($\|\cdot\|_1$ is the ℓ_1-norm).

4. Case Study

4.1. Adaptive Gamification in Nanomoocs

Nanomoocs is an innovative massive open online course (MOOC) platform consisting of online courses designed as training pills. They are focused on well-identified user segments, with a systematized instructional design, high-quality audiovisual content, and incorporating technologies to improve learning such as peer review, personalization, gamification and emotions recognition. Concretely, our case study of adaptive gamification focuses on the course entitled "What can we do with the plastics in the sea?", intended for secondary school students aged between 14 and 15 years.

The course aims to develop the critical thinking and problem-solving skills of students regarding the topic of plastics in the sea. Students must follow a scientific method and reason about how to solve the problem of plastic discharges into the sea. They have to look for, contrast and select information sources and digital media in order to build new knowledge on the problem of plastics in the sea. They will analyze in a critical way how plastics affect the life of living beings, find out about and analyze different forms of extraction and transformation of plastic, and finally carry out a decision-making process in order to propose preventive and solution measures. The course consists of four formative units, each containing activities (A_i) such as videos, interactive info-graphics, readings and questionnaires (see Figure 3).

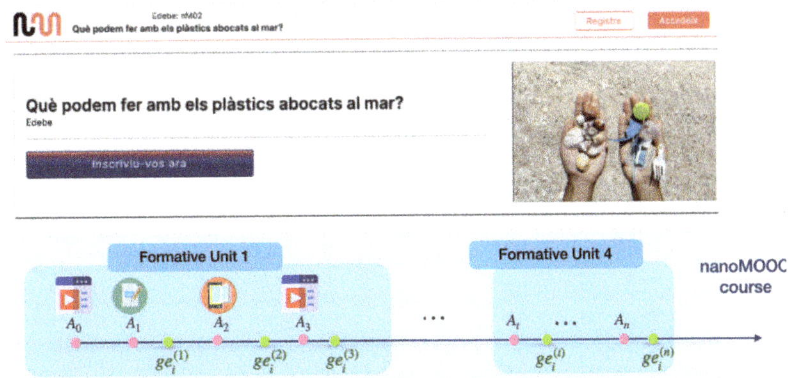

Figure 3. Nanomoocs about plastics in the sea.

Gamification elements (ge_i) appear when the participants complete a number of activities. For example, when students complete an interactive infographics and a posterior questionnaire and just after the completion of the questionnaire a gamification element (GE) unfolds to motivate and reward the student for the work performed. As the gamification system is adaptive, different students will be awarded with different gamification elements depending on their player type (see Figure 4).

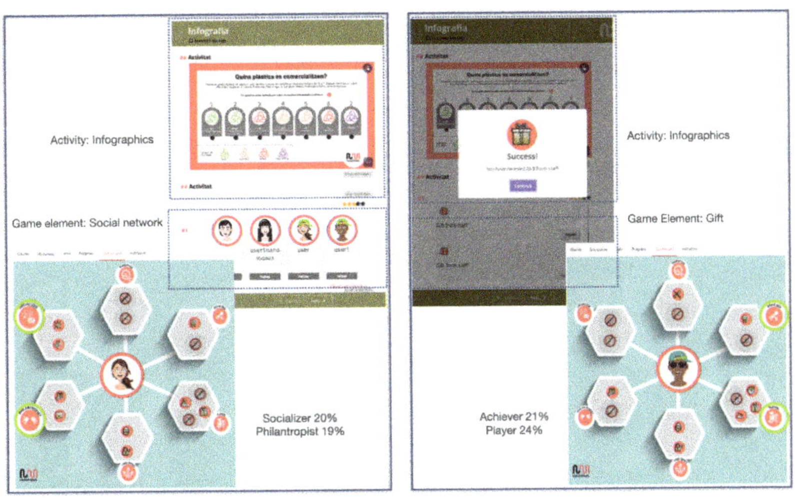

Figure 4. Two game scenarios exist for the same activity: on the left, the enabled game element for a socializer player; on the right, the activated game element for an achiever/player user.

The adaptive system needs an initial player profile of the participant which is gathered through a player type questionnaire (https://www.gamified.uk/UserTypeTest2016, accessed on 1 December 2020) [20] that the students answer at the beginning of the course. The result of the questionnaire is the initial information the system has about the participant, i.e., the way they like to play. The information gathered by this questionnaire has been referred to before in this paper as $PR^{(0)}$—player ratings at time 0. The gamified system then incorporates this profile in a dashboard that shows each user their player type (see Figure 4 and Table 2).

Table 2. The six game scenarios related to the activation of the fourteen game elements.

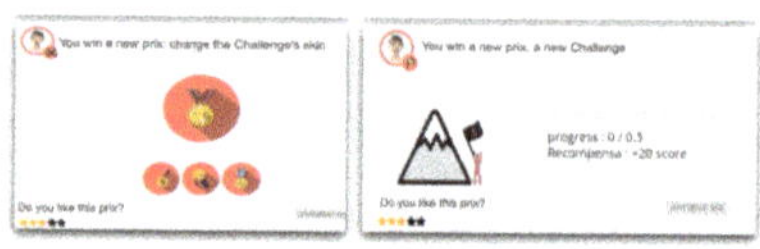 Disruptor	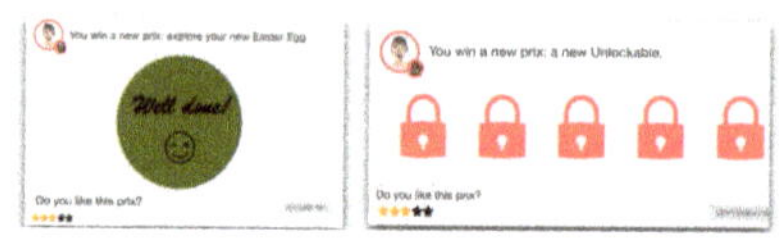 Free Spirit
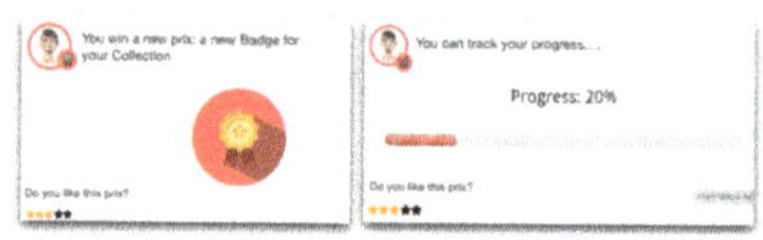 Achiever	 Philanthropist
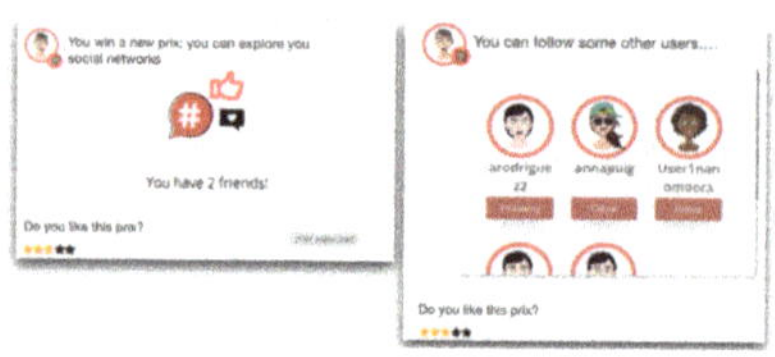 Socializer	 Player

Note that we face two different scenarios when asking participants to fill in the player type questionnaire. The first scenario is when the participant fills in the questionnaire conscientiously and thoroughly and the adaptive system then uses an initial player type data, PR^0, which is approximately equal to the real player type RPT_0, i.e., $RPT_0 \simeq PR^{(0)}$, as can be seen in the top of Figure 5. The second scenario occurs when the participant answers the questionnaire in an unreliable and untrustworthy way, i.e., the results of the questionnaire, $PR^{(0)}$, are far from the RPT_0, i.e., $RPT_0 \not\simeq PR^{(0)}$, as can be seen in the bottom of Figure 5.

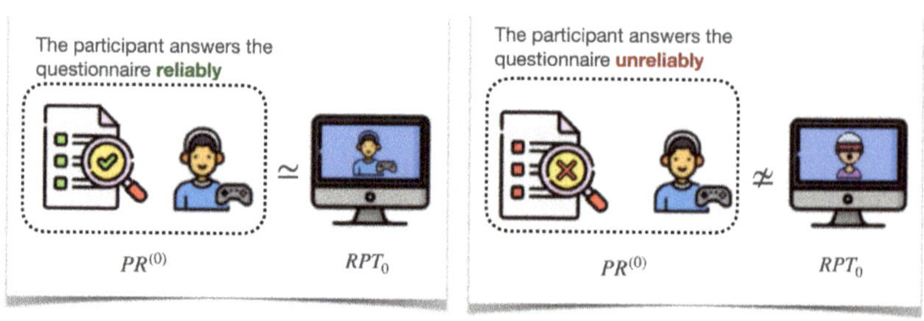

Figure 5. Player type questionnaire results as PR^0 (player type ratings at time 0) versus real player type RPT_0.

We also consider the case of participants that do not want to play (non-player type), and consequently do not answer the questionnaire. In this case, we have no initial information about the player type profile, but the system gives the user the opportunity to join in the gamified experience, when the user shows any interest in the random game mechanics that will appear during the experience. If there is no interest, the system will cancel the adaptive gamification for the non-player user.

4.2. Adaptive Gamification: Software Architecture

Our system is based on a service-oriented architecture in which the gamification system resides on an external server as it is shown in Figure 6. In our case, the learning management system edX (https://www.edx.org, accessed on 1 November 2021) hosts the Nanomoocs course and uses the gamification service through a restful API.

The edX platform uses so-called XBlocks as basic units to define activities within the course units. Thus, a specific gamification XBlock was created to encapsulate the calls to the external gamification API to ask for the next game element $ge_i^{(t)}$ to be displayed once the user completes the activity. Moreover, that XBlock is in charge of displaying the game element as well as monitoring the user's behavior with this game element. It captures the user's interactions $n_i^{(t)}$, the display time $\tau_i^{(t)}$ and the user's opinions $o_i^{(t)}$ that the user realizes in the game element. Every 15 s, all these interactions are sent to our adaptive method, running in an external gamification module (see the right part of Figure 6), to update the player profile, $PR^{(t)}$. This means that the iteration frequency of our method is approximately 15 s in an attempt to have enough time to gather information about changes in user behavior without overloading the system.

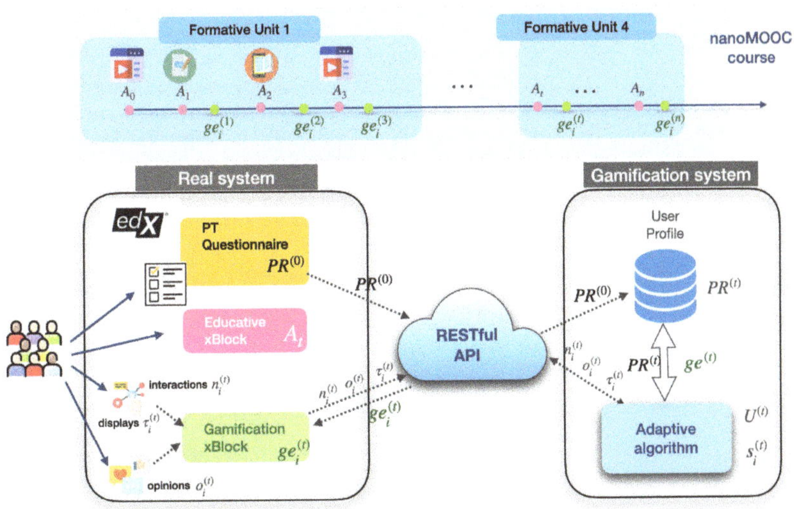

Figure 6. Software architecture to extend edX with adaptive gamification.

Therefore, when users access the course, they fill in the player type questionnaire (integrated within the course as another possible activity) and this first player rating, $PR^{(0)}$, is sent to the gamification server. From this point on, the user performs learning activities within the learning management system (LMS) and when some of them are finished, an event calls to the gamification service using the gamification XBlocks (see the =Nanomoocs timeline in Figure 3). At this point, edX obtains the next game element to be displayed, $ge_i^{(t)}$, according to the current player ratings, $PR^{(t)}$. This game element will be displayed in the edX course via an HTML file, which also contains JavaScript calls to the API to monitor the user's interactions, and the display time they spend on that game element. The users are also able to give their opinion about the active game element. Finally, the displayed game element will be enabled in the user's dashboard. Later on, the user will be able to consult all the game elements that they have already activated. Furthermore, the dashboard is an HTML file that performs calls to the gamification service to update user information at any time.

5. Adaptive Method Evaluation

5.1. Simulation System

We assume that an adaptive gamification strategy works if the game element proposed to the users fit their "real" player profile at any time t. Considering that we simulate the player using a bot, in the following, we note the "real" profile of the bot as RPT_0, and its player type rating at time t as $PR^{(t)}$. We also assume that the bot's real player type does not change over time, $RPT_t = RPT_0$, in such a way that we can measure the convergence of our algorithm to a particular player type.

The values of RPT_0 come from a dataset containing user types HEXAD test results (https://gamified.uk/UserTypeTest2016/user-type-test-results.php, accessed on 1 November 2021), where 42,782 tests were carried out, obtaining average type scores for all the modalities of players. We selected the eleven most representative modalities (see Table *Summary* in previous URL) and their corresponding average type scores (see also Table *Average Type Scores*). For instance, the Achiever modality appears in 12% of tests and its average scores are $RPT_0 = (0.12, 0.18, 0.20, 0.16, 0.16, 0.17, 0.0)$ meaning 12% disruptor, 18% free spirit, 20% achiever, 16% player, 16% socializer, 17% philanthropist, and 0% non-player.

Bearing in mind that a real user would either reliably (accurately) or unreliably (inaccurately) answer the HEXAD questionnaire at the beginning of the gamified experience, our bot simulates users' responses accurately or somewhat randomly. Therefore, we define $PR^{(0)}$ as being close to RPT_0 ($RPT_0 \simeq PR^{(0)}$) or far away from ($RPT_0 \not\simeq PR^{(0)}$). To do so, if the reliability is low, the bot rating $PR^{(0)}$ is the furthest non-null scores from RPT_0 in Table *Average Type Scores*. Otherwise, if the reliability is high, the bot takes its $PR^{(0)}$ as RPT_0. Note that when the bot simulates accurate responses to the questionnaire, it is desirable that the value of $PR^{(t)}$ remains close to RPT_0, while when it simulates inaccurate answers, it is convenient that $PR^{(t)}$ converges to RPT_0 when $t \to \infty$. Figure 7 shows the real profile of the bot and its initial player type rating, $PR^{(0)}$, computed far from or the same as RPT_0 in function of whether the questionnaire was reliably (accurately) or unreliably (inaccurately) "answered" respectively. Note that this figure is the counterpart of Figure 5 for the simulations using bots.

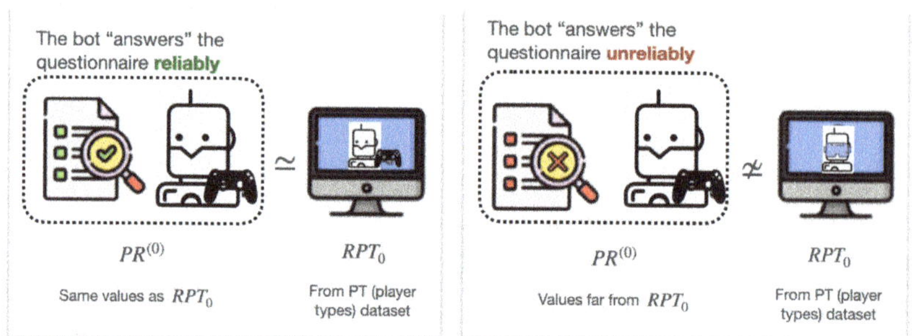

Figure 7. Bot's player type initial rating, PR^0, versus real bot's player type.

The bot simulates the interaction with the game elements (recall interaction indexes, $S^{(t)}$, and $\tau_i^{(t)}, n_i^{(t)}, o_i^{(t)}$ as input data of $s_i^{(t)}$ in the bottom left of Figure 2, and Equation 6).

To do so, we used two variables: the time between two consecutive interactions, $\mathrm{T}_i^{(t)}$, and the opinion, $\Theta_i^{(t)}$, defined as

$$\mathrm{T}_i^{(t)} = -9.5(GM_i \cdot RPT_0) + 10 \tag{10}$$

$$\Theta_i^{(t)} = \frac{1}{5}\,\mathtt{round}\big(4(GM_i \cdot RPT_0) + 1\big) \tag{11}$$

where i corresponds to the game element (ge_i) selected by the method. Thus:

- If the game element (ge_i) fits the real bot's profile (RPT_0), the bot interacts more frequently than otherwise. Therefore, $\mathrm{T}_i^{(t)}$ reflects this behavior by taking values from 0.5 (frequent interactions) to 10 (longer time between interactions);
- Regarding $n_i^{(t)}$, we consider that the bot interacts once every $\mathrm{T}_i^{(t)}$;
- The opinion can be calculated in a similar way using $\Theta_i^{(t)}$.

Figure 8 shows the main characteristics of the simulation system. Note that it is the counterpart of Figure 6, where edX and the real user are substituted by the bot generation system and their simulated behaviors, while the gamification system remains unchanged. That is, the bot's behavior and the adaptive algorithm work independently; therefore, the bot behaves exclusively in function of its player profile. The main differences in Figure 8 are: (1) the initialization of the $PR^{(0)}$ is performed by the most common cases of player types, which are stored in a database; (2) the real bot player type is defined from these initial player types (RPT_0); and (3) the inputs of our method, i.e., $PR^{(t)}$ and users' interactions—their game element display times and opinions—are based on the bot real player type (RPT).

Regarding the parameters of our method, we settled the ϵ value to 0.001. As mentioned above, we define this ϵ to compute the new player type ratings at each iteration. Setting ϵ to a low value, we achieve low fluctuations between two consecutive player types without losing the influence of the interactions and the opinions of the user (see Equation (8)). Moreover, to perform the simulations as close as possible to the real case with users, we generate interactions and the bot's opinions every 15 s to update the player type, although the execution time of each iteration is lower.

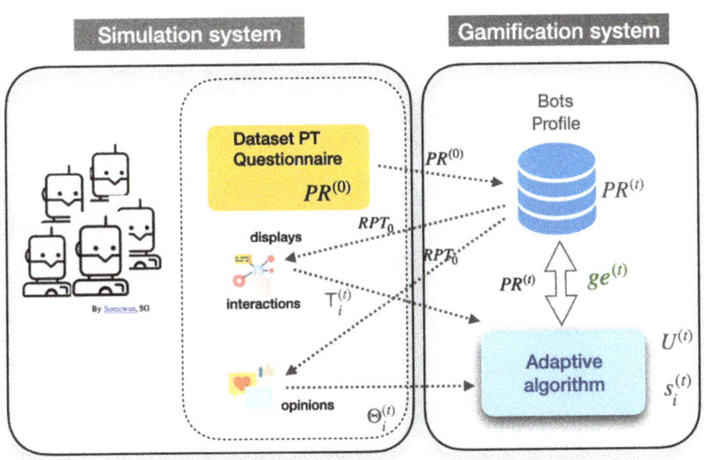

Figure 8. Evaluation infrastructure: on the left, the simulation system of bots; and on the right, the gamification system that takes bots' profiles instead of users' profiles.

5.2. Results

Once how the bot simulates a real user has been stated, we will consider three methods in function of whether the value of $PR^{(t)}$ is fixed or variable along the time (*A—constant player rating; B—random dynamic player rating; and C—our method: dynamic player rating*, assuming that the bot has previously answered the HEXAD questionnaire (fixing $PR^{(0)}$). Moreover, we define how the error is computed to compare the three methods. For each method, we analyze two cases (accurate answers, inaccurate answers).

1. **Method A**—Constant $PR^{(0)}$: a constant player rating, $PR^{(t+1)} = PR^{(0)}$, is assigned at any time. In this case, we use Equation (9) to calculate $U^{(t+1)}$ once and always use its maximal component.
   We calculate the error as follows. Since $PR^{(t+1)} = PR^{(0)}$, the distance is simply calculated as $Err = |RPT_0 - PR^{(0)}|$. Note that if the user has accurately answered the questionnaire, we have $Err = 0$;
2. **Method B**—Random dynamic $PR^{(t)}$: a dynamic player rating $PR^{(t+1)}$ is randomly chosen at each time t and then, the game element selected to be shown is also the maximal component of $U_i^{(t+1)}$. Note that this method is equivalent to picking a random game element.
   In this method, we compute the error as the average distance between RPT_0 and a random point $p \in \{(x_1, \ldots, x_7) \in [0,1]^7 : \sum_{i=1}^{7} x_i = 1\}$ using the average distance of random points in a unit hypercube (average distance of random points in a unit hypercube, https://martin-thoma.com/curse-of-dimensionality/, accessed on 1 November 2021).
3. **Method C**—Our method, dynamic $PR^{(t)}$: a dynamic player rating $PR^{(t+1)}$ is computed according the $s_i^{(t)}$ defined by Equation (6). The bot then simulates $\tau_i^{(t)}$, $n_i^{(t)}$, and $o_i^{(t)}$ using $\top_i^{(t)}$ and $\Theta_i^{(t)}$. The game element selected to be shown is a weighted random choice of $U^{(t+1)}$ (see step 3 in Figure 2).
   In this method, we calculate the error Err based on the distances between $PR^{(t)}$ and RPT_0 for all t from 1 to n_iter:

$$Err = \frac{1}{n\_iter} \sum_{t=1}^{n\_iter} |RPT_0 - PR^{(t)}| \tag{12}$$

Experiments were performed using the gamification software described in Figure 8 and run on a Windows 10, Intel Core i7 processor with 8GB RAM. Table 3 shows the mean errors (i.e., the distance between $PR^{(t)}$ and the real player profile RPT_0) and the standard deviation (SD) obtained by the three methods: (A) constant player ratings; (B) random dynamic player ratings; and (C) our method dynamic player ratings. In the simulations, the initial setup of a bot consists of its RPT_0 "real" player type, i.e., what define its playing behavior, and its approximated player type PR^0, i.e., the results of its player type questionnaire. We analyzed the three methods on two cases (accurate and inaccurate answers). Moreover, we define the worst and the best scenarios in all methods.

The *worst scenario* is defined by the bot that produces the biggest error in method C. Similarly, the *best scenario* is defined by the bot that produces the smallest error in method C. Then, Table 3 also depicts the errors obtained by those bots in methods A, B and C that started the simulation using the same RTP_0 and PR^0 as the bots in the best and the worst scenarios of C. In method *A*, as we take into account $PR^{(t)} = PR^{(0)} = RPT_0$, accurate answers are the ideal case ($Err = 0$). However, with inaccurate ones the mean error grows to 0.0311. Moreover, both cases of method *B* have similar errors ($Err = 0.08024$ and $Err = 0.08027$ in accurate and inaccurate answers, respectively), because there is a random selection of player type ratings and thus, they are almost independent of the reliability (accurate or inaccurate) of the answers. Finally, except in method *A* with accurate answers, method C behaves better than *A* and *B* ($Err = 0.0070$ and $Err = 0.0243$ in accurate and inaccurate answers, respectively) even in its worst scenario ($Err = 0.0333$). Note that the mean errors should be interpreted in the interval [0, 0.2857] since the maximum distance between two normalized points in a Unit Hypercube is 2/7 (approximately 0.2857). This indicates, for example, that the value of the error 0.0827 in case B-Random represents approximately 28%.

Table 3. Table of mean errors of methods A-Constant Player Type, B-Random Dynamic Player Type, and C-Our Method-Dynamic Player Type, when the questionnaire is accurately and inaccurately answered (the two cases in rows), together with the best and worst errors obtained with method C .

Cases/Methods	A—Constant Constant $PR^{(0)}$	B—Random Dynamic $PR^{(t)}$	C—Our Method Dynamic $PR^{(t)}$
Accurate Answers Case: $RPT_0 \simeq PR^{(0)}$			
Mean Error (SD)	0	0.08024 (0.00052)	0.0070 (0.00166)
Worst Scenario of C	0	0.0804	0.0105
Best Scenario of C	0	0.0797	0.0029
Inaccurate Answers Case: $RPT_0 \not\simeq PR^{(0)}$			
Mean Error (SD)	0.0311(0.00404)	0.08027 (0.00040)	0.0243 (0.00475)
Worst Scenario of C	0.0367	0.08012	0.0333
Best Scenario of C	0.0233	0.08018	0.0146

Regarding the results of the mean error in inaccurate answers, the difference between method A (Constant Player Type) and method C (Dynamic Player Type) is 0.0068. To analyze this difference, we conducted a statistical study. As the data are not normally distributed (Shapiro–Wilk p-value: 0.000005421, with an effect size large ES = 0.239), we used the Wilcoxon signed-rank test for paired samples. Results indicate that we can reject the null hypothesis ($mean_A = mean_C, p < 0.0001$, ES = 0.87) and then the difference is significant.

On the other hand, the player profiles of the worst (0.0333) and best (0.0146) cases of C in inaccurate answers were disruptor in both cases. Indeed, the disruptors, free spirit and philanthropists/achievers player types represent the most frequent results from the Marczewski questionnaire.

Figure 9 shows the values of a bot's player ratings $PR^{(t)}$ along 180 iterations for the best case of method C in inaccurate answers, as shown in Table 3. The real player type, RPT_0, is represented by dotted lines and colored stars (red—disruptor; green—free spirit; yellow—philanthropist) at the end of the X axis. As can be appreciated, the disruptor player type is the one that achieves the lowest error (distance between the red star and the end of the red line). The first values, 0.14 and 0.21, of the two initial vectors ($PR^{(0)}$ :: [0.14, 0.2, 0.17, 0.16, 0.15, 0.17, 0] RPT_0:: [0.21, 0.17, 0.16, 0.16, 0.14, 0.15, 0]), and the final value obtained, 0.21, in the vector $PR^{(180)}$ :: [0.21, 0.185, 0.156, 0.149, 0.136, 0.162, 0], are the vector of player ratings in the following order :: (disruptor, free spirit, achiever, player, socializer, philanthropist, non-player). Additionally, not only does the predominant player type converge to its real value, but other player types that are not predominant, such as the free spirit rating (initially at 0.20), also evolve by slightly decreasing their values with regard to the real one (0.17).

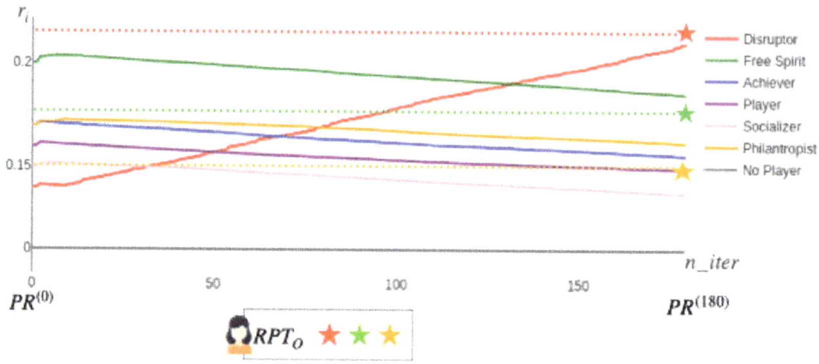

Figure 9. Evolution of the player rating in the best case in inaccurate answers (method C): philanthropist (yellow); free spirit (green); and disruptor (red).

6. Discussion

The results of our study conclude that the method *A*—where player type ratings are fixed along the experience—is the best when the users thoroughly and accurately answer the questionnaire, while our method, method C, works well in both cases (users either accurately or inaccurately respond to it). These results are aligned with other studies that considered player type ratings as fixed throughout the experience [4] (see method A of our experiments). Nevertheless, previous studies did not consider the dynamic player type method. Since it will not be possible to know whether (real) users accurately answer the player type questionnaire in real situations, our method is the most suitable for an adaptive gamified experience.

Our adaptive method needs information about the player type of the user at the beginning of the experience, so we used the HEXAD player types questionnaire proposed by [20]. It is worth noting that a recent study that describes a new method to predict player types using gameful applications [23]. As we mentioned in Section 3, we selected 14 game elements of the 52 presented in [24]. In the case of extending the proposal from 14 to 52 game elements, our adaptive method is still applicable, it would only be necessary to extend the GE and GM vectors of Equations (3) and (4) and the dimensions of the matrix M, Equation (5). Moreover, our approach is easily applicable to different player types taxonomies such as MoMo [21], a new promising model created as a combination of others, which will be considered in our future research.

However, our system has some limitations. Actually, it is necessary to test different values of updating the frequency of execution of player ratings (Step 1 of the method, as can be seen in Figure 2). Current simulations have used a frequency of one iteration every 15 s, ensuring that changes in the bots' behavior can be detected without overloading the system performance, but this value can be better tuned and even more so when considering real-life scenarios. Moreover, analyzing real users' experiences, we could tune some constants such as ϵ or the values of the matrix M to better adapt the users' profile and the gamified experience. Note that in the process of measuring the error *Err* in the simulated scenario, the average distance between $PR^{(t)}$ and RPT_0 is considered, and not the distance between $PR^{(t)}$ and RPT_t. However, it would be possible to calculate the error with respect to a dynamic RPT_t, in which case we would slightly randomly change some values of RPT_t, and therefore, from this value, the behavior of the bots (Θ, $\top$) would be dynamically calculated. Additionally, a simulation of a bigger change of the real player profile would be possible since it consists of carrying out our simulation n times with a different RPT_0 each time. Indeed, the error is currently calculated as the mean of the distances between $PR^{(t)}$ and RPT_0 of all the types of players (see Equation (12)), which may hide the particular behavior of individual player types (e.g., disruptor, free spirit).

The current evaluation performed using bots should be extended to real users. Thus, additional users' experience data, such as the course and activities completion rate and users' satisfaction, could be used to analyze the effectiveness of the method. The current case study, which is a short course composed of a few activities developed during a short period of time, is somehow limited in terms of the number of collected opinions and interactions. Our method should be applied to other larger gamified systems. Additionally, our method has a limited representation of user engagement behavior using just interactions, interaction speed and opinions. However, considering the more sophisticated variables in the definition of the interaction index, such as eye tracking, cursor tracking or emotion recognition could help us reveal more information about the real player type during the experience.

7. Conclusions

This research proposes a method to present game elements to users that fit their profile (player type). However, instead of taking a (static) picture of the profile at the beginning of the experience, we consider how it may change over the course of gamified activities. The method calculates the utility of showing a game element to the user based on their evolved player type (PT), their interactions with game elements, and the scores given by the users to those game elements. We developed a case study in the educative context, a course integrated on Nanomoocs, a massive open online course (MOOC) platform. We also present a preliminary simulation of the system using bots. A bot simulated three different methods: constant player type ratings, random dynamic player type ratings, and dynamic player type ratings, where the player type is recomputed using players' interactions and opinions at each step of the iterative method. The evaluation shows positive results as the comparative analysis of our proposal yields lower errors when converging towards the real player type for both cases with accurate and inaccurate answers in the player type questionnaire (mean $\pm$ SD with accurate answers–mean $\pm$ SD with inaccurate answers) (0.0070 ± 0.00166–0.0243 ± 0.00475) than the static player type (0 ± 0–0.0311 ± 0.00404) and the dynamic random player type (0.08024 ± 0.00052–0.08027 ± 0.00040). Note that the method of static player type with accurate answers gives 0 error because it is a static profile using accurate values of player types. In this paper, we evaluated the bots' simulations defining the error as a mean of distances between the current and real player type vectors. In future works, we plan to perform an in-depth analysis of the simulation error of each individual component of these vectors. Further work is planned to also include tests with users and incorporate new inputs to our method such as emotions and activity completion.

Author Contributions: The individual contributions for each author are: I.R. and A.P.: conceptualization, investigation, supervision, writing—review and editing, resources; À.R.: conceptualization, investigation, software, data curation, writing—original draft, writing—review and editing. All authors have read and agreed to the published version of the manuscript.

Funding: This research was funded by ACCIO project number COMRDI18-1-0010 and by MCIN/AEI/10.13039/501100011033 grants numbers PGC2018-096212-B-C33 (MISMIS project) and PID2019-104156GB-I00 (Ci-SUSTAIN project).

Acknowledgments: To Accio COMRDI18-1-0010, MISMIS-Language (PGC2018-096212-B-C33) and CI-SUSTAIN (Grant PID2019-104156GB-I00 funded by MCIN/AEI/10.13039/501100011033).

Conflicts of Interest: The authors declare no conflict of interest.

References

1. Rapp, A.; Alessandro, M.; Simeoni, R.; Console, L. Playing while Testing: How to Gamify a User Field Evaluation. In *Designing Gamification: Creating Gameful and Playful Experiences Held in Conjunction with SIGCHI Conference on Human Factors in Computing Systems (CHI'13)*; Gamification Research Network: Paris, France, 2013.
2. Cheng, R.; Vassileva, J. Adaptive Reward Mechanism for Sustainable Online Learning Community. In Proceedings of the Artificial Intelligence in Education, Amsterdam, The Netherlands, 18–22 July 2005; pp. 152–159.

3. Gené, O.; Núñez, M.; Blanco, A. Gamification in MOOC: Challenges, opportunities and proposals for advancing MOOC mode In Proceedings of the Second International Conference on Technological Ecosystems for Enhancing Multiculturality (TEEM'14 Salamanca, Spain, 1–3 October 2014; pp. 215–220. [CrossRef]
4. Lavoue, E.; Monterrat, B.; Desmarais, M.; George, S. Adaptive Gamification for Learning Environments. *IEEE Trans. Learn Technol.* **2019**, *12*, 16–28. [CrossRef]
5. Charles, D.; Black, M. Dynamic Player Modelling: A Framework for Player-centred Digital Games. In Proceedings of the 5th International Conference on Computer Games: Artificial Intelligence, Design and Education (CGAIDE'04), Microsoft Academi Campus, Reading, UK, 8–10 November 2004; pp. 29–35.
6. Hallifax, S.; Serna, A.; Marty, J.; Lavoué, G.; Lavoué, E. Factors to consider for tailored gamification. In Proceedings of the CHI PLAY 2019 Annual Symposium on Computer-Human Interaction in Play, Barcelona, Spain, 22–25 October 2019 pp. 559–572. [CrossRef]
7. Rodríguez, I.; Puig, A.; Rodríguez, A. We Are Not the Same Either Playing: A Proposal for Adaptive Gamification. In *Frontiers in Artificial Intelligence and Applications*; Villaret, M., Alsinet, T., Fernández, C., Valls, A., Eds.; Springer: Berlin/Heidelberg, Germany 2021; Volume 339, pp. 19–34. [CrossRef]
8. Hallifax, S.; Serna, A.; Marty, J.; Lavoué, E. Adaptive Gamification in Education: A Literature Review of Current Trends and Developments. *Lect. Notes Comput. Sci.* **2019**, *11722*, 294–307.
9. Klock, A.C.T.; Gasparini, I.; Pimenta, M.S.; Hamari, J. Tailored gamification: A review of literature. *Int. J. Hum.-Comput. Stud.* **2020**, *144*, 102495. [CrossRef]
10. Challco, G.C.; Moreira, D.A.; Mizoguchi, R.; Isotani, S. An ontology engineering approach to gamify collaborative learning scenarios. In *CYTED-RITOS International Workshop on Groupware*; Baloian, N., Burstein, F., Ogata, H., Santoro, F., Zurita, G., Eds. Springer: Berlin/Heidelberg, Germany, 2014; pp. 185–198.
11. Denden, M.; Tlili, A.; Essalmi, F.; Jemni, M. Does personality affect students' perceived preferences for game elements in gamified learning environments? In Proceedings of the IEEE 18th International Conference on Advanced Learning Technologies, ICALT 2018, Mumbai, India, 9–13 July 2018; pp. 111–115. [CrossRef]
12. Ferro, L.S.; Walz, S.P.; Greuter, S. Towards personalised, gamified systems. In Proceedings of the 9th Australasian Conference on Interactive Entertainment Matters of Life and Death—IE '13, Melbourne, Australia, 30 September–1 October 2013; ACM Press: New York, NY, USA, 2013; pp. 1–6. [CrossRef]
13. Borges, S.; Mizoguchi, R.; Durelli, V.H.S.; Bittencourt, I.; Isotani, S. A link between worlds: Towards a conceptual framework for bridging player and learner roles in gamified collaborative learning contexts. In *Advances in Social Computing and Digital Education, Croatia*; Koch, F., Koster, A., Primo, T., Guttmann, C., Eds.; Springer: Berlin/Heidelberg, Germany, 2016; pp. 19–34.
14. Knutas, A.; Ikonen, J.; Maggiorini, D.; Ripamonti, L.; Porras, J. Creating student interaction profiles for adaptive collaboration gamification design. *Int. J. Hum. Cap. Inf. Technol. Prof.* **2016**, *7*, 47–62. [CrossRef]
15. Chtouka, E.; Guezguez, W.; Amor, N.B. Reinforcement learning for new adaptive gamified LMS. In Proceedings of the International Conference on Digital Economy, Mumbai, India, 9–13 July 2018; Jallouli, R., Bach Tobji, M., Bélisle, D., Mellouli, S., Abdallah, F., Osman, I., Eds.; Springer: Berlin/Heidelberg, Germany, 2019; pp. 305–314.
16. Monterrat, B.; Lavoué, E.; George, S. Toward an Adaptive Gamification System for Learning Environments. In *Computer Supported Education*; Springer International Publishing: Cham, Switzerland, 2015; pp. 115–129.
17. Lopez, C.; Tucker, C. Toward Personalized Adaptive Gamification: A Machine Learning Model for Predicting Performance. *IEEE Trans. Games* **2020**, *12*, 155–168. [CrossRef]
18. Bartle, R.A. Hearts, Clubs, Diamonds, Spades: Players Who Suit Muds. *J. MUD Res.* **1996**, *1*, 19.
19. Nacke, L.E.; Bateman, C.; Mandryk, R.L. BrainHex: A neurobiological gamer typology survey. *Entertain. Comput.* **2014**, *5*, 55–62. [CrossRef]
20. Marczewski, A. Even Ninja Monkeys Like to Play: Gamification, Game Thinking and Motivational Design. In *Game Thinking & Motivational Design*; CreateSpace Independent Publishing Platform, Ed.; Blurb Inc.: London, UK, 2015; pp. 65–80.
21. Sienel, N.; Münster, P.; Zimmermann, G. Player-Type-based Personalization of Gamification in Fitness Apps. In Proceedings of the HEALTHINF, Vienna, Austria, 11–13 February 2021; pp. 361–368.
22. Mora, A.; Tondello, G.F.; Nacke, L.E.; Arnedo-Moreno, J. Effect of personalized gameful design on student engagement. In Proceedings of the IEEE Global Engineering Education Conference, EDUCON, Santa Cruz de Tenerife, Spain, 17–20 April 2018; pp. 1925–1933. [CrossRef]
23. Altmeyer, M.; Tondello, G.F.; Krüger, A.; Nacke, L.E. HexArcade: Predicting Hexad User Types by Using Gameful Applications. In Proceedings of the CHI PLAY 2020 Annual Symposium on Computer-Human Interaction in Play, Virtual Event, 2–4 November 2020; ACM: New York, NY, USA, 2020; pp. 219–230. [CrossRef]
24. Tondello, G.F.; Wehbe, R.; Diamond, L.; Busch, M.; Marczewski, A.; Nacke, L.E. The Gamification User Types Hexad Scale. In Proceedings of the 2016—CHI PLAY 2016 Annual Symposium on Computer-Human Interaction in Play, Austin, TX, USA, 16–19 October 2016; pp. 229–243.
25. Hallifax, S.; Lavoué, E.; Serna, A. To tailor or not to tailor gamification? An analysis of the impact of tailored game elements on learners' behaviours and motivation. *Lect. Notes Comput. Sci.* **2020**, *12163*, 216–227. [CrossRef]
26. Sabourin, J.; Lester, J. Affect and Engagement in Game-BasedLearning Environments. *IEEE Trans. Affect. Comput.* **2014**, *5*, 45–56. [CrossRef]

27. Dalponte Ayastuy, M.; Torres, D.; Fernández, A. Adaptive gamification in Collaborative systems, a systematic mapping study. *Comput. Sci. Rev.* **2021**, *39*, 100333. [CrossRef]
28. Škuta, P.R.; Kostolányová, K. Adaptive approach to the gamification in education. In Proceedings of the European Conference on Technology Enhanced Learning, Transforming Learning with Meaningful Technologies, Delft, The Netherlands, 16–19 September 2018; Springer International Publishing: Berlin/Heidelberg, Germany, 2018; p. 367.
29. Barata, G.; Gama, S.; Jorge, J.; Gonçalves, D. Gamification for smarter learning: Tales from the trenches. *Smart Learn. Environ.* **2015**, *2*, 1–23. [CrossRef]
30. Knutas, A.; Van Roy, R.; Hynninen, T.; Granato, M.; Kasurinen, J.; Ikonen, J. A process for designing algorithm-based personalized gamification. *Multimed. Tools Appl.* **2019**, *78*, 13593–13612. [CrossRef]
31. Hassan, M.A.; Habiba, U.; Majeed, F.; Shoaib, M. Adaptive gamification in e-learning based on students' learning styles. *Interact. Learn. Environ.* **2021**, *29*, 545–565. [CrossRef]
32. Jagušt, T.; Botički, I.; So, H.J. Examining competitive, collaborative and adaptive gamification in young learners' math learning. *Comput. Educ.* **2018**, *125*, 444–457. [CrossRef]
33. Kickmeier-Rust, M.D. *An Alien's Guide to Multi-Adaptive Educational Computer Games*; Informing Science: Santa Rosa, CA, USA, 2012.
34. Cooper, S.; Khatib, F.; Treuille, A.; Barbero, J.; Lee, J.; Beenen, M.; Leaver-Fay, A.; Baker, D.; Popović, Z. Predicting protein structures with a multiplayer online game. *Nature* 2010, *466*, 756–760. [CrossRef] [PubMed]
35. Eveleigh, A.; Jennett, C.; Lynn, S.; Cox, A.L. "I want to be a captain! I want to be a captain!" gamification in the old weather citizen science project. In Proceedings of the fIrst International Conference on Gameful Design, Research, and Applications, Stratford, ON, Canada, 2–4 October 2013; pp. 79–82.
36. Ismailović, D.; Pagano, D.; Brügge, B. Wemakewords-An adaptive and collaborative serious game for literacy acquisition. In Proceedings of the IADIS International Conference-Game and Entertainment, Rome, Italy, 20–26 July 2011.
37. Cantwell, D.; Broin, D.O.; Palmer, R.; Doyle, G. Motivating elderly people to exercise using a social collaborative exergame with adaptive difficulty. In Proceedings of the 6th European Conference on Games Based Learning, Cork, Ireland, 4–5 October 2012; pp. 4–5.
38. Llanos, J.; Carro, R.M. The squares: A multi-touch adaptive game for children integration. In Proceedings of the 2015 International Symposium on Computers in Education (SIIE), Setubal, Portugal, 25–27 November 2015; pp. 137–140.
39. Tranvouez, E.; Fournier, S. A MultiAgent architecture for collaborative serious game applied to crisis management training: Improving adaptability of non player characters. *EAI Endorsed Trans. Serious Games* **2014**, *1*, e7. [CrossRef]
40. Knutas, A.; van Roy, R.; Hynninen, T.; Granato, M.; Kasurinen, J.; Ikonen, J. Profile-Based Algorithm for Personalized Gamification in Computer-Supported Collaborative Learning Environments. In Proceedings of the 1st Games-Humans Interactions GHITALY@ CHItaly, Cagliari, Italy, 18 April 2017.
41. Yu, H.; Riedl, M.O. A sequential recommendation approach for interactive personalized story generation. In Proceedings of the AAMAS, Valencia, Spain, 4–8 June 2012; Volume 12, pp. 71–78.
42. Hastings, E.J.; Guha, R.K.; Stanley, K.O. Automatic content generation in the galactic arms race video game. *IEEE Trans. Comput. Intell. AI Games* **2009**, *1*, 245–263. [CrossRef]
43. Hassan, M.A.; Habiba, U.; Khalid, H.; Shoaib, M.; Arshad, S. An adaptive feedback system to improve student performance based on collaborative behavior. *IEEE Access* **2019**, *7*, 107171–107178. [CrossRef]
44. Feyisetan, O.; Simperl, E.; Van Kleek, M.; Shadbolt, N. Improving paid microtasks through gamification and adaptive furtherance incentives. In Proceedings of the 24th International Conference on World Wide Web, Florence, Italy, 18–22 May 2015; pp. 333–343.
45. Mora, A.; González, C.; Arnedo-Moreno, J.; Álvarez, A. Gamification of cognitive training: A crowdsourcing-inspired approach for older adults. In Proceedings of the XVII International Conference on Human Computer Interaction, Salamanca, Spain, 13–16 September 2016; pp. 1–8.
46. Emmerich, K.; Neuwald, K.; Othlinghaus, J.; Ziebarth, S.; Hoppe, H.U. Training conflict management in a collaborative virtual environment. In Proceedings of the International Conference on Collaboration and Technology, Raesfeld, Germany, 16–19 September 2012; Herskovic, V., Hoppe, H.U., Jansen, M., Ziegler, J., Eds.; Springer: Berlin/Heidelberg, Germany, 2012; pp. 17–32.

MDPI
St. Alban-Anlage 66
4052 Basel
Switzerland
Tel. +41 61 683 77 34
Fax +41 61 302 89 18
www.mdpi.com

Applied Sciences Editorial Office
E-mail: applsci@mdpi.com
www.mdpi.com/journal/applsci

www.ingramcontent.com/pod-product-compliance
Lightning Source LLC
Chambersburg PA
CBHW061615220926
51850CB00005B/517
9783036555317